AF352935

Gas Capture Processes

Gas Capture Processes

Special Issue Editors

Tohid N. Borhani
Zhien Zhang
Muftah H. El-Naas
Salman Masoudi Soltani
Yunfei Yan

MDPI • Basel • Beijing • Wuhan • Barcelona • Belgrade • Manchester • Tokyo • Cluj • Tianjin

Special Issue Editors

Tohid N. Borhani
Heriot-Watt University
UK

Zhien Zhang
The Ohio State University
USA

Muftah H. El-Naas
Qatar University
Qatar

Salman Masoudi Soltani
Brunel University London
UK

Yunfei Yan
Chongqing University
China

Editorial Office
MDPI
St. Alban-Anlage 66
4052 Basel, Switzerland

This is a reprint of articles from the Special Issue published online in the open access journal *Processes* (ISSN 2227-9717) (available at: https://www.mdpi.com/journal/processes/special_issues/gas_capture).

For citation purposes, cite each article independently as indicated on the article page online and as indicated below:

LastName, A.A.; LastName, B.B.; LastName, C.C. Article Title. *Journal Name* **Year**, *Article Number*, Page Range.

ISBN 978-3-03928-780-2 (Hbk)
ISBN 978-3-03928-781-9 (PDF)

Contents

About the Special Issue Editors

Tohid N. Borhani is currently an Assistant Professor at Heriot-Watt University teaching different subjects in chemical engineering. Before that, he was a Research Fellow at Cranfield University, Research Associate at The University of Sheffield, Research Associate at Imperial College London, and Postdoctoral Researcher at UTM. His main fields of study include separation process, process modelling and simulations, and chemometric and quantitative structure–property relationships (QSPRs). He has published several papers and book chapters in high level journals and publishers. He is an Associate Fellow of Higher Education Academy, U.K., and an Academic Member of the U.K. Carbon Capture and Storage Research Centre (UKCCSRC).

Zhien Zhang is currently a Research Fellow in the William G. Lowrie Department of Chemical and Biomolecular Engineering at Ohio State University, USA. His research interests include advanced processes and materials for gas capture, gas separation, and carbon capture, utilization and storage (CCUS). Dr. Zhang has published more than 80 peer-reviewed journal papers and 16 editorials in high impact journals such as Renewable & Sustainable Energy Reviews and Applied Energy. He is the Editorial Supervisor of the Journal of Natural Gas Science and Engineering and Editor of some international journals, e.g., Membranes, Environmental Chemistry Letters, etc.

Muftah El-Naas is the Director of the Gas Processing Center at Qatar University, where he also served as the QAFCO Industrial Chair Professor. His areas of expertise include CO_2 capture and sequestration, water treatment and purification, membrane separation, and plasma technology. Most of his recent research focuses on the development of new and environmental-friendly technologies for the oil and gas industry. Dr. El-Naas has authored more than 170 papers in international journals and conferences, in addition to several book chapters and patent applications. He has recently developed and patented a new process and a reactor system for the management of CO_2 emissions and desalination reject brine.

Salman Masoudi Soltani (Ph.D) is a lecturer in chemical engineering at Brunel University London, U.K., and a founding member of the Chemical Engineering Department at the university. His research has mainly centred on separation processes. Prior to this position, he worked as a postdoctoral research associate with the Clean Fossil & Bioenergy Research Group at Imperial College London, U.K., and as a postdoctoral Knowledge Transfer Partnership (KTP) research associate (industrial postdoc) at the University of Nottingham, UK. He is a Chartered Engineer (CEng/MIChemE), a Fellow of Higher Education Academy, U.K., and an Academic Member of the U.K. Carbon Capture and Storage Research Centre (UKCCSRC).

Yunfei Yan is a professor in the college of energy and power engineering at Chongqing University, and a professor in the Key Laboratory of Low Grade Energy Utilization Technology and System, Ministry of Education (Chongqing University) and Evaluation Center of Industrial Energy Conservation and Green Development, Ministry of Industry and Information Technology, China. Dr. Yan worked as a visiting scholar at Boston University, USA, from 2010 to 2011. His areas of expertise include catalytic combustion, micro energy and power system, heat and mass transfer, gas separation and carbon capture, multiphase flow and environmental protection, new energy, and renewable energy use and conversion. Dr. Yan has authored more than 60 papers in international journals and conferences, in addition to several patent applications.

Editorial

Gas Capture Processes

Zhien Zhang [1,*], **Tohid N. Borhani** [2], **Muftah H. El-Naas** [3], **Salman Masoudi Soltani** [4] **and Yunfei Yan** [5]

[1] William G. Lowrie Department of Chemical and Biomolecular Engineering, The Ohio State University, Columbus, OH 43210, USA
[2] School of Engineering and Physical Sciences, Heriot-Watt University, Edinburgh, Scotland EH14 4AS, UK; t.nborhani@hw.ac.uk
[3] Gas Processing Center, College of Engineering, Qatar University, Doha 2713, Qatar; muftah@qu.edu.qa
[4] Department of Chemical Engineering, Brunel University London, Uxbridge UB8 3PH, UK; salman.masoudisoltani@brunel.ac.uk
[5] Key Laboratory of Low-Grade Energy Utilization Technologies and Systems, Ministry of Education, Chongqing University, Chongqing 400044, China; yunfeiyan@cqu.edu.cn
* Correspondence: zhang.4528@osu.edu

Received: 31 December 2019; Accepted: 2 January 2020; Published: 4 January 2020

Abstract: The increasing trends in gas emissions have had direct adverse impacts on human health and ecological habitats in the world. A variety of technologies have been deployed to mitigate the release of such gases, including CO_2, CO, SO_2, H_2S, NO_x and H_2. This special issue on gas-capture processes collects 25 review and research papers on the applications of novel techniques, processes, and theories in gas capture and removal.

Keywords: global warming; gas emission; capture; CO_2

---

1. Introduction

The increasing trends in environmental gas emissions have had direct adverse impacts on human health and ecological habitats. Various technologies have been deployed to mitigate the release of such gases, including CO_2, CO, SO_2, H_2S, NO_x and H_2 [1–3]. Nevertheless, many of these technologies have demonstrated poor performance or are yet to become economically feasible for commercial deployment. Therefore, devising efficient capture processes and associated technologies has become significantly more urgent in the past few years.

In the current special issue of *Processes*, 25 review and research papers on the applications of novel techniques, processes, and theories in gas capture and removal are presented. This special issue is available online at: https://www.mdpi.com/journal/processes/special_issues/gas_capture. A concise summary of the presented works in this special issue is outlined hereafter.

2. Overview of Papers in This Special Issue

In the work of Mendoza et al. [4], iron ore was studied as a CO_2 absorbent. Carbonation was carried out via mechanochemical and high temperature–high pressure (HTHP) reactions. Kinetics of the carbonation reactions were studied for the two methods. In the mechanochemical process, the kinetics were analyzed as a function of the CO_2 pressure and the rotation speed of the planetary ball mill, while in the HTHP process, the kinetics were studied as a function of pressure and temperature. The highest CO_2-capture capacities achieved were 3.7341 mmol of CO_2/g of sorbent in ball-milling (30 bar of CO_2 pressure, 400 rpm, and 20 h) and 5.4392 mmol of CO_2/g of absorbent in HTHP (50 bar of CO_2 pressure, 100 °C, and 4 h). To overcome the kinetic limitations, water was introduced to all carbonation experiments. The calcination reactions were carried out in an argon atmosphere using thermogravimetric analysis (TGA) and differential scanning calorimetry (DSC) analysis. Siderite can be decomposed in the same temperature range (100 °C to 420 °C) as the samples produced by both methods. This range reached higher temperatures compared with pure iron oxides due to the decomposition temperature increase with decreasing purity. Calcination reactions yield magnetite and carbon. A comparison of recyclability (use of the same material in several cycles of carbonation–calcination), kinetics, spent energy, and the amounts of initial material needed to capture 1 ton of CO_2 revealed the advantages of the mechanochemical process compared with HTHP.

Osagie et al. [5] conducted steady-state simulation and exergy analysis of a 2-amino-2-methyl-1 -propanol (AMP)-based post-combustion capture (PCC) plant. Exergy analysis was able to identify the location, sources of thermodynamic inefficiencies, and magnitude in the thermal system. Furthermore, thermodynamic analysis of different configurations of the process helped to identify opportunities to reduce the steam requirements for each of the configurations. Exergy analysis performed for the AMP-based plant and the different configurations revealed that a rich split with intercooling configuration gave the highest exergy efficiency of 73.6%, while those of the intercooling and the reference AMP-based plant were 57.3% and 55.8%, respectively. Thus, exergy analysis of flowsheeting configurations could lead to significant improvements in plant performance and to cost reduction for amine-based CO_2-capture technologies.

Mahmud et al. [6] investigated the reaction kinetics of carbon dioxide with blends of N-methyldiethanolamine and L-arginine using the stopped-flow technique. The experiments were performed over a temperature range of 293 to 313 K, and at solution concentrations up to 1 mol/L of different amino acid/amine ratios. The overall reaction rate constant (k_{ov}) was found to increase with increasing temperature and amine concentration, as well as with increased proportion of L-arginine concentration in the mixture. The experimental data were fitted to the zwitterion and termolecular mechanisms using a nonlinear regression technique with average absolute deviations (AAD) of 7.6% and 8.0%, respectively. A comparative study of the promoting effect of L-arginine with that of the effect of glycine and diethanolamine (DEA) in N-methyldiethanolamine (MDEA) blends showed that the MDEA–arginine blend exhibited faster reaction rate with CO_2 with respect to the MDEA–DEA blend, while the case was reversed when compared to the MDEA–glycine blend.

In the study of Zhang et al. [7], four kinds of pore-forming material were screened and utilized to prepare sorbent pellets via the extrusion–spheronization process. In addition, the impacts of additional content of pore-forming material and their particle sizes were also investigated comprehensively. It was found that the addition of 5 wt. % polyethylene resulted in the highest CO_2-capture capacity (0.155 g CO_2/g sorbent in the 25th cycle) and a mechanical performance of 4.0 N after high-temperature calcination, results approximately 14% higher and 25% improved, respectively, compared to pure calcium hydrate pellets. Smaller particle sizes of the pore-forming material were observed to lead to a better performance in CO_2 sorption, while for mechanical performance, there was an optimal size for the pore-former used.

Gutierrez et al. [8] reported the conceptual design of an amine-based carbon dioxide (CO_2) separation process for enhanced oil recovery (EOR). A systematic approach was applied to predict the economic profitability of the system while reducing the environmental impacts. Firstly, they modeled the process with UniSim and determined the governing degrees of freedom (DoF) through a sensitivity analysis. They then proceeded with the formulation of the economic problem, where the employment of econometric models allowed them to predict the highest dynamic economic potential (DEP). In the second part of the study, the waste reduction (WAR) algorithm was applied to quantify the environmental risks of the studied process. This method was based on the minimization of the potential environmental indicator (PEI) by using the generalization of the waste reduction algorithm. Results showed that the CO_2 separation plant was promising in terms of economic revenue. However, the PEI value indicated that the higher the profitability, the larger the environmental risk. The optimal value of the DEP corresponded to 0.0274 kmol/h and 60 °C, with a plant capacity according to the molar flow rate of the produced acid gas. In addition, the highest environmental risk was observed at the upper bounds of the DoF.

Baena-Moreno et al. [9] presented a method for regeneration of a sodium hydroxide (NaOH) solution as a valuable byproduct from a biogas-upgrading unit through calcium carbonate ($CaCO_3$) precipitation, as an alternative to the elevated energy consumption required by the physical regeneration process. The purpose of this work was to study the main parameters that might affect NaOH regeneration using an aqueous sodium carbonate (Na_2CO_3) solution and calcium hydroxide ($Ca(OH)_2$) as a reactive agent for regeneration and carbonate slurry production, in order to outperform the regeneration efficiencies reported in earlier works. Moreover, Raman spectroscopy and scanning electron microscopy (SEM) were employed to characterize the solid obtained. The studied parameters were reaction time, reaction temperature, and molar ratio between $Ca(OH)_2$ and Na_2CO_3. In addition, the influence of small quantities of NaOH at the beginning of the precipitation process was studied. The results indicated that regeneration efficiencies between 53% and 97% could be obtained by varying the main parameters mentioned above, and both Raman spectroscopy and SEM images revealed the formation of a carbonate phase in the obtained solid. These results confirmed the technical feasibility of this biogas-upgrading process through $CaCO_3$ production.

Liu et al. [10] studied the effect of physical and mechanical activation on the physicochemical structure of coal-based activated carbons (ACs). In the stage of CO_2 activation, a rapid decrease of the defective structure and the growth of aromatic layers accompanied by the dehydrogenation of aromatic rings resulted in the ordered conversion of the microstructure and severe carbon losses on the surfaces of Char-PA, while the oxygen content of Char-PA, including C=O (39.6%), C–O (27.3%), O–C=O (18.4%), and chemisorbed O (or H_2O) (14.7%), increased to 4.03%. Char-PA presented a relatively low S_{BET} value (414.78 m^2/g) owing to the high value of non-V_{mic} (58.33%). In the subsequent mechanical activation from 12 to 48 h under N_2 and dry ice, the strong mechanical collision caused by ball-milling destroyed the closely arranged crystalline layers and caused the collapse of mesopores and macropores, resulting in disordered conversion of the microstructure and the formation of a defective structure; a sustained increase in the S_{BET} value from 715.89 to 1259.74 m^2/g was found with prolonged ball-milling time. There was a gradual increase in the oxygen content from 6.79% to 9.48% for Char-PA–CO_2-12/48 obtained by ball-milling under CO_2. Remarkably, the variations of physicochemical parameters of Char-PA-CO_2-12/48 were more obvious than those of Char-PA-N_2-12/48 under the same ball-milling time, which was related to the stronger solid–gas reactions caused by the mechanical collision under dry ice. Finally, the results of the SO_2 adsorption test of typical samples indicated that Char-PA–CO_2-48 with a desirable physicochemical structure can maintain 100% efficiency within 30 min and its SO_2 adsorption capacity could reach 138.5 mg/g at the end of the experiment. After the 10th cycle of thermal regeneration, Char-PA–CO_2-48 still had a strong adsorptive capacity (81.2 mg/g).

Liu et al. [11] studied the catalytic effect of NaCl (1 and 3 wt. %) in the presence of oxygen functional groups on the improvement of the physicochemical structure of coal-based activated carbons. A large quantity of Na can be retained in 1NaJXO and 3NaJXO with the presence of oxygen functional groups to promote further its catalytic characteristics during pyrolysis, resulting in disordered transformation of the carbon structure. In addition, the development of micropores was mainly affected by the distribution and movement of Na catalyst, whereas the growth of mesopores was mainly influenced by the evolution of oxygen functional groups. The active sites of 3NaJXO-800 were no longer preferentially consumed in the presence of Na catalyst during subsequent CO_2 activation to facilitate the sustained disordered conversion of the microstructure and the rapid development of the micropores, resulting in an obvious high S_{BET} value as activation proceeded. A high S_{BET}/burn-off ratio value (41.48 $m^2/g/\%$) of 3NaJXO-800 with a high value of S_{BET} (1995.35 m^2/g) at a low burn-off value (48.1%) was obtained, associated with a high efficiency of pore formation. Finally, the SO_2 adsorption efficiency of 3NaJXO-800-48.1 was maintained at 100% for 90 min. After 180 min, 3NaJXO-800-48.1 still presented a high adsorptive capacity (140.2 mg/g). It was observed that a large micropore volume in the case of hierarchical pore structure necessarily ensured optimal adsorption of SO_2.

Karamian et al. [12] investigated the effect of different nanofluids, such as water/Al_2O_3, water/Fe_2O_3, or water/SiO_2, on absorption rate. The results showed that the absorption of CO_2 and SO_2 in nanofluids significantly increased by up to 77% in comparison with the base fluid. It was also observed that the type of gas molecules and nanoparticles determined the mechanism of mass transfer enhancement by nanofluids. Additionally, the results indicated that the values of mass transfer coefficient of SO_2 in water/Al_2O_3, water/Fe_2O_3, and water/SiO_2 nanofluids were, respectively, 50%, 42%, and 71% higher than those of SO_2 in pure water ($k_{LSO2-water} = 1.45 \times 10^{-4}$ m/s). Moreover, the values for CO_2 in the above nanofluids were, respectively, 117%, 103%, and 88% higher than those of CO_2 in water alone ($k_{LCO2-water} = 1.03 \times 10^{-4}$ m/s). Finally, this study tried to offer a new, comprehensive correlation for mass transfer coefficient and absorption rate prediction.

Petrovic and Soltani [13] optimized a post-combustion carbon-capture unit using monoethanolamine (MEA), based on a Taguchi experimental design, to understand the impacts of the operational parameters on the energy consumption of the capture unit. An equilibrium-based approach was employed in Aspen Plus to simulate 90% capture of the CO_2 emitted from a 600 MW combined-cycle gas turbine power plant. The effects of inlet flue gas temperature, absorber column operating pressure, exhaust gas recycle ratio, and amine concentration on the energy demand were evaluated using signal-to-noise ratios and analysis of variance. The optimum parameters were found to be: flue gas temperature = 50 °C, absorber pressure = 1 bar, exhaust gas recirculation = 20%, and amine concentration = 35 wt. %, with a relative importance of amine concentration > absorber column pressure > exhaust gas recirculation > flue gas temperature. This configuration gave a total capture unit energy requirement of 5.05 GJ/t$_{CO2}$, with a reboiler energy requirement of 3.94 GJ/t$_{CO2}$. All the studied factors, except for the flue gas temperature, demonstrated a statistically significant association to the response (i.e., energy demand).

In the study of Ge et al. [14], low-cost activated carbons were prepared from waste polyurethane foam by physical activation with CO_2 and chemical activation with $Ca(OH)_2$, NaOH, or KOH. The activation conditions were optimized to produce microporous carbons with high CO_2 adsorption capacity and CO_2/N_2 selectivity. The sample prepared by physical activation showed CO_2/N_2 selectivity of up to 24, much higher than that of the sample prepared by chemical activation. This was mainly due to the narrower microporosity and the rich N content produced during the physical activation process. However, physical activation samples showed inferior textural properties compared to chemical activation samples, which led to a lower CO_2 uptake of 3.37 mmol/g at 273 K. Porous carbons obtained by chemical activation showed a high CO_2 uptake of 5.85 mmol/g at 273 K, comparable to the optimum activated carbon materials prepared from other wastes. This was mainly attributed to large volumes of ultra-micropores (<1 nm) up to 0.212 cm^3/g and a high surface area of 1360 m^2/g. Furthermore, in consideration of the presence of fewer contaminants, lower weight losses of physical activation samples, and the excellent recyclability of both physically and chemically activated samples, the waste polyurethane-foam-based carbon materials have potential application prospects in CO_2 capture.

Ghasem [15] studied the simultaneous absorption of CO_2 and NO_2 from a mixture of gases (5% CO_2, 300 ppm NO_2, balance N_2) by aqueous sodium hydroxide solution in a membrane contactor. For the first time, a mathematical model was established for the simultaneous removal of the two undesired gas solutes (CO_2, NO_2) from flue gas using a membrane contactor. The proposed model considers the reaction rate and radial and axial diffusion of both compounds. The model was verified and validated with experimental data and found to be in good agreement. The model was used to examine the effect of the flow rate of liquid, gas, and inlet solute molar fraction on the percent removal and molar flux of both impurity species. The results revealed that an increased liquid flow rate improved the percent removal of both compounds. A high inlet gas flow rate decreased the percent removal. It was possible to obtain the complete removal of both undesired compounds. The model was confirmed to be a dependable tool for the optimization of such process, and for similar systems.

Wang et al. [16] investigated the structures and electronic properties of monolayer arsenene doped with Al, B, S, and Si, based on first-principles calculations. The dopants exerted great influence on the properties of the arsenene monolayer. The electronic properties of the substrate were effectively tuned by substitutional doping. After doping, NO adsorption onto four kinds of substrate was investigated. The results demonstrated that NO exhibited a chemisorption character on Al-, B-, and Si-doped arsenene, and a physisorption character on S-doped arsenene with moderate adsorption energy. Due to the adsorption of NO, the band structures of the four systems had great changes; the adsorption reduced the energy gap of Al- and B-doped arsenene and opened the energy gap of S- and Si-doped arsenene. The large charge depletion between the NO molecule and the dopant demonstrated that there was a strong hybridization of orbitals at the surface of the doped substrate because of the formation of a covalent bond, except for S-doped arsenene. The charge depletion also indicated that Al-, B-, and Si-doped arsenene might be good candidate gas sensors to detect NO gas molecules, owing to their high sensitivity.

Shoukat et al. [17] studied various novel amine solutions both in aqueous and non-aqueous (monoethylene glycol (MEG)/triethylene glycol (TEG)) forms for hydrogen sulfide (H_2S) absorption. The study was conducted in a custom-built experimental setup at temperatures relevant to subsea operation conditions and atmospheric pressure. Liquid-phase-absorbed H_2S and amine concentrations were measured analytically to calculate H_2S loading (mol of H_2S/mol of amine). The maximum achieved H_2S loadings as the function of pKa, gas partial pressure, temperature, and amine concentration were presented. The effects of solvent type on absorbed H_2S were also discussed. Several new solvents showed higher H_2S loading compared to aqueous N-methyldiethanolamine (MDEA) solution, which is the current industrial benchmark compound for selective H_2S removal in natural gas sweetening processes.

Wang et al. [18] investigated the carbonaceous deposits on the surface of a coking chamber. Scanning electron microscopy (SEM), X-ray fluorescence spectrum (XRF), Fourier-transform infrared spectrometer (FTIR), Raman spectroscopy, X-ray diffraction spectrometry (XRD), and X-ray photoelectron spectroscopy (XPS) were applied to investigate the carbonaceous deposits. FTIR revealed the existence of carboxyl, hydroxyl, and carbonyl groups in the carbonaceous deposits. SEM showed that different carbonaceous deposit layers presented significant differences in morphology. XRF and XPS revealed that the carbonaceous deposits mainly contained C, O, and N elements, with smaller amounts of Al, Si, and Ca elements. It was found that the C content gradually increased in the formation of carbonaceous deposits, while the O and N elements gradually decreased. It was also found that the absorbed O_2 and H_2O took part in the oxidation process of the carbon skeleton to form =O and –O– structures. The oxidation and elimination reaction resulted in a change in the bonding state of the O element, and finally formed compact carbonaceous deposits on the surface of the coking chamber. Based on these analyses, the formation and evolution mechanisms of carbonaceous deposits were discussed.

Zhang et al. [19] proposed an oxygen recovery (OR) process for a supercritical water oxidation (SCWO) system based on the solubility difference between oxygen and CO_2 in high-pressure water. A two-stage gas–liquid separation process was established, using Aspen Plus software to obtain the optimized separation parameters. Accordingly, energy consumption and economic analyses were conducted for the SCWO process with and without OR. Electricity, depreciation, and oxygen costs were the major contributions to the cost of the SCWO system without OR, accounting for 46.18, 30.24, and 18.01 $/t, respectively. When OR was introduced, the total treatment cost decreased from 56.80 $/t to 46.17 $/t—a reduction of 18.82%. Operating costs were significantly reduced at higher values of the stoichiometric oxygen excess for the SCWO system with OR. Moreover, the treatment cost for the SCWO system with OR decreased with increasing feed concentration for increased reaction heat and oxygen recovery.

Szpalerski and Smoliński [20] presented a strategy for maximizing recovery of flare gases in industrial plants processing hydrocarbons. The functioning of a flare stack and the depressurization systems in a typical refinery plant was described, and the architecture of the depressurization systems and construction of the flares were shown in a simplified way. A proposal to recover the flare gases together with their output outside the industrial plant, in order to minimize impact on the environment (reduction of emissions) and to limit consumption of fossil fuels was presented. Contaminants that might be found in these depressurization systems were indicated. The proposal presented in the article assumed the injection of an excess stream of gases into an existing natural gas pipeline system. A method of monitoring was proposed, aiming to eliminate the introduction of undesirable harmful components into the systems.

Jin et al. [21] experimentally and theoretically studied the mechanism for gas transportation in emerging 2D-material-based membranes. They measured the gas permeances of hydrogen and nitrogen from their mixture through the supported MXene lamellar membrane. Knudsen diffusion and molecular sieving through straight and tortuous nanochannels were proposed to elucidate the gas transport mechanism. An average pore diameter of 5.05 Å in straight nanochannels was calculated by linear regression in the Knudsen diffusion model. The activation energy for H_2 transport in the molecular sieving model was calculated to be 20.54 kJ mol^{-1}. The model indicated that the gas permeance of hydrogen (with a smaller kinetic diameter) is contributed to by both Knudsen diffusion and the molecular sieving mechanism, but the permeance of larger molecular gases like nitrogen is from Knudsen diffusion. The effects of critical conditions such as temperature, the diffusion pore diameter of structural defects, and the thickness of the prepared MXene lamellar membrane on hydrogen and nitrogen permeance were also investigated to better understand the different contributions to hydrogen permeation of Knudsen diffusion and molecular sieving. At room temperature, the total hydrogen permeance was 18% due to Knudsen diffusion and 82% due to molecular sieving. The modeling results indicated that molecular sieving plays a dominant role in controlling gas selectivity.

Zhao et al. [22] studied the influence of gas molar fraction and activity in aqueous phase while predicting phase equilibrium conditions. In pure gas systems, such as CH_4, CO_2, N_2 and O_2, the gas molar fraction in the aqueous phase was proposed as one of phase equilibrium conditions, and a simplified correlation of the gas molar fraction was established. The gas molar fraction threshold maintaining three-phase equilibrium was obtained by phase equilibrium data regression. The UNIFAC model, the predictive Soave–Redlich–Kwong equation, and the Chen–Guo model were used to calculate aqueous phase activity and the fugacity of gas and hydrate phases, respectively. The calculations showed that the predicted phase equilibrium pressures were in good agreement with published phase equilibrium experiment data, and the percentages of absolute average deviation pressures were given. The water activity, gas molar fraction in the aqueous phase, and the fugacity coefficient in vapor phase were discussed.

Wang et al. [23] established a numerical calculation model for the rapid estimation of coal seam gas content based on the characteristic values of gas desorption at specific exposure times. Combined with technical verification, a new method which avoids the calculation of gas loss for the rapid estimation of gas content in the coal seam was investigated. Study results showed that the balanced adsorption gas pressure and coal gas desorption characteristic coefficient (K_t) satisfied the exponential equation, and the gas content and K_t were linear equations. The correlation coefficient of the fitting equation gradually decreased as the exposure time of the coal sample increased. Using the new method to measure and calculate the gas content of coal samples from two different working faces of the Lubanshan North mine (LBS), the deviation of the calculated coal sample gas content ranged from 0.32% to 8.84%, with an average of only 4.49%. Therefore, the new method meets the needs of field engineering technology.

In the work of Yang et al. [24], four samples with different coal ranks were collected and diffusion experiments were conducted under different pressures through the adsorption and desorption processes. Three widely used models, i.e., the unipore diffusion (UD) model, the bidisperse diffusion (BD) model, and the dispersive diffusion (DD) model, were adopted to compare their applicability and to calculate the diffusion coefficients. Results showed that for all coal ranks, the BD model and DD model could match the experimental results better than the UD model. Concerning the fast diffusion coefficient D_{ae} of the BD model, three samples displayed a decreasing trend with increasing gas pressure, while the other sample showed a V-type trend. The slow diffusion coefficient D_{ie} of the BD model increased with gas pressure for all samples, while the ratio β is an intrinsic characteristic of coal and remained constant. For the DD model, the characteristic rate parameter k_Φ did not change sharply and the stretching parameter α increased with gas pressure. Both D_{ae} and D_{ie} were in proportion to k_Φ, which reflected the diffusion rate of gas in the coal. The impacts of pore characteristic on gas diffusion were also analyzed. Although pore size distributions and specific surface areas were different between the four coal samples, correlations were not apparent between pore characteristic and diffusion coefficients.

Wang et al. [25] presented a mechanism analysis of air reactor (AR) coupling in a high-flux, coal-direct chemical looping combustion (CDCLC) system and provided a theoretical methodology for optimal system design with favorable operation stability and low gas leakages. First, they presented dipleg flow diagrams of the CDCLC system and concluded the feasible gas–solid flow states for solid circulation and gas leakage control. On this basis, semi-theoretical formulas of gas leakage were proposed to predict the optimal regions of the pressure gradients of the AR. Meanwhile, an empirical formula of critical sealing was also developed to identify the advent of circulation collapse, so as to ensure the operational stability of the whole system. Furthermore, the theoretical methodology was applied in condition design of a cold system. The resulting favorable gas–solid flow behaviors, together with the good control of gas leakages, demonstrated the feasibility of the theoretical methodology. Finally, the theoretical methodology was applied to carry out a capability assessment of a high-flux CDCLC system under a hot state in terms of the restraint of gas leakages and the stability of solid circulation.

Wang et al. [26] investigated the fundamental effects of air reactor (AR) coupling on oxygen carrier (OC) circulation and gas leakages with a cold-state experimental device of the proposed in situ gasification chemical looping combustion (iG-CLC) system. The system exhibited favorable pressure distribution characteristics and good adaptability of solid circulation flux, demonstrating the positive role of the direct coupling AR method in the stabilization and controllability of the whole system. The OC circulation and the gas leakages were mainly determined by the upper and lower pressure gradients of the AR. With an increase in the upper pressure gradient, the OC circulation flux increased initially and later decreased until the circulation collapsed. Additionally, the upper pressure gradient exerted a positive effect on the restraint of gas leakage from the FR to the AR, but a negative effect on the suppression of gas leakage from the AR to the FR. Moreover, gas leakage of the J-valve to the AR, which is directly related to the solid circulation stability, was exacerbated with an increase of the lower pressure gradient of the AR. In real iG-CLC applications, the pressure gradients should be adjusted flexibly and optimally to guarantee balanced OC circulation together with an ideal balance of all gas leakages.

Krištof et al. [27] presented preliminary results on the spray characteristics of a spiral nozzle used for gas absorption processes. First, they experimentally measured the pressure impact footprint of the spray generated. Effective spray angles were then evaluated from the photographs of the spray and via Archimedean spiral equation using the pressure impact footprint records. Using classical photography, areas of primary and secondary atomization were determined together with the droplet size distribution, which were further approximated using selected distribution functions. Radial and tangential spray velocities of droplets were assessed using laser Doppler anemometry. The results showed atypical behavior related to different types of nozzles. In the investigated measurement range, the droplet size distribution showed higher droplet diameters (about 1 mm) compared to those from, for example, air-assisted atomizers. The results were similar for the radial velocity, which was lower (a maximum velocity of about 8 m/s) compared to, for example, effervescent atomizers, which can produce droplets with a velocity of tens to hundreds of m/s. In contrast, the spray angle ranged from 58° and 111° for the inner small and large cones, respectively, to 152° for the upper cone and, in the measured range, was independent of the inlet pressure of liquid at the nozzle orifice.

Ibrahim et al. [28] reviewed the main research work carried out over the last few years on direct mineral-carbonation process utilizing steel-making waste, with emphasis on recent research achievements and potential for future research.

At the end of this editorial, the editors would like to express their sincere gratitude to the authors for their valuable contributions to this special issue and thank the editorial staff of *Processes* for their help and support during the review process.

References

1. Yan, J.; Zhang, Z. Carbon Capture, Utilization and Storage (CCUS). *Appl. Energy* **2019**, *235*, 1289–1299. [CrossRef]
2. Babu, P.; Linga, P.; Kumar, R.; Englezos, P. A Review of the Hydrate Based Gas Separation (HBGS) Process for Carbon Dioxide Pre-Combustion Capture. *Energy* **2015**, *85*, 261–279. [CrossRef]
3. Zhang, Z.; Li, Y.; Zhang, W.; Wang, J.; Soltanian, M.R.; Olabi, A.G. Effectiveness of Amino Acid Salt Solutions in Capturing CO_2: A Review. *Renew. Sustain. Energy Rev.* **2018**, *98*, 179–188. [CrossRef]
4. Mora Mendoza, E.Y.; Sarmiento Santos, A.; Vera López, E.; Drozd, V.; Durygin, A.; Chen, J.; Saxena, S.K. Siderite Formation by Mechanochemical and High Pressure–High Temperature Processes for CO_2 Capture Using Iron Ore as the Initial Sorbent. *Processes* **2019**, *7*, 735. [CrossRef]
5. Osagie, E.; Aliyu, A.M.; Nnabuife, S.G.; Omoregbe, O.; Etim, V. Exergy Analysis and Evaluation of the Different Flowsheeting Configurations for CO_2 Capture Plant Using 2-Amino-2-Methyl-1-Propanol (AMP). *Processes* **2019**, *7*, 391. [CrossRef]
6. Mahmud, N.; Benamor, A.; Nasser, M.; El-Naas, M.H.; Tontiwachwuthikul, P. Reaction Kinetics of Carbon Dioxide in Aqueous Blends of N-Methyldiethanolamine and L-Arginine Using the Stopped-Flow Technique. *Processes* **2019**, *7*, 81. [CrossRef]
7. Zhang, Z.; Pi, S.; He, D.; Qin, C.; Ran, J. Investigation of Pore-Formers to Modify Extrusion-Spheronized CaO-Based Pellets for CO_2 Capture. *Processes* **2019**, *7*, 62. [CrossRef]
8. Gutierrez, J.P.; Erdmann, E.; Manca, D. Optimal Design of a Carbon Dioxide Separation Process with Market Uncertainty and Waste Reduction. *Processes* **2019**, *7*, 342. [CrossRef]
9. Baena-Moreno, F.; Rodríguez-Galán, M.; Vega, F.; Reina, T.; Vilches, L.; Navarrete, B. Regeneration of Sodium Hydroxide from a Biogas Upgrading Unit through the Synthesis of Precipitated Calcium Carbonate: An Experimental Influence Study of Reaction Parameters. *Processes* **2018**, *6*, 205. [CrossRef]
10. Liu, D.; Hao, Z.; Zhao, X.; Su, R.; Feng, W.; Li, S.; Jia, B. Effect of Physical and Mechanical Activation on the Physicochemical Structure of Coal-Based Activated Carbons for SO_2 Adsorption. *Processes* **2019**, *7*, 707. [CrossRef]
11. Liu, D.; Su, R.; Hao, Z.; Zhao, X.; Jia, B.; Dong, L. Catalytic Effect of NaCl on the Improvement of the Physicochemical Structure of Coal-Based Activated Carbons for SO_2 Adsorption. *Processes* **2019**, *7*, 338. [CrossRef]
12. Karamian, S.; Mowla, D.; Esmaeilzadeh, F. The Effect of Various Nanofluids on Absorption Intensification of CO_2/SO_2 in a Single-Bubble Column. *Processes* **2019**, *7*, 393. [CrossRef]
13. Alexanda Petrovic, B.; Masoudi Soltani, S. Optimization of Post Combustion CO_2 Capture from a Combined-Cycle Gas Turbine Power Plant via Taguchi Design of Experiment. *Processes* **2019**, *7*, 364. [CrossRef]
14. Ge, C.; Lian, D.; Cui, S.; Gao, J.; Lu, J. Highly Selective CO_2 Capture on Waste Polyurethane Foam-Based Activated Carbon. *Processes* **2019**, *7*, 592. [CrossRef]
15. Ghasem, N. Modeling and Simulation of the Absorption of CO_2 and NO_2 from a Gas Mixture in a Membrane Contactor. *Processes* **2019**, *7*, 441. [CrossRef]
16. Wang, K.; Li, J.; Huang, Y.; Lian, M.; Chen, D. Adsorption of NO Gas Molecules on Monolayer Arsenene Doped with Al, B, S and Si: A First-Principles Study. *Processes* **2019**, *7*, 538. [CrossRef]
17. Shoukat, U.; Pinto, D.D.D.; Knuutila, H.K. Study of Various Aqueous and Non-Aqueous Amine Blends for Hydrogen Sulfide Removal from Natural Gas. *Processes* **2019**, *7*, 160. [CrossRef]
18. Wang, H.; Jin, B.; Wang, X.; Tang, G. Formation and Evolution Mechanism for Carbonaceous Deposits on the Surface of a Coking Chamber. *Processes* **2019**, *7*, 508. [CrossRef]
19. Zhang, F.; Chen, J.; Su, C.; Ma, C. Energy Consumption and Economic Analyses of a Supercritical Water Oxidation System with Oxygen Recovery. *Processes* **2018**, *6*, 224. [CrossRef]

20. Szpalerski, J.; Smoliński, A. Analysis of the Excess Hydrocarbon Gases Output from Refinery Plants. *Processes* **2019**, *7*, 253. [CrossRef]
21. Jin, Y.; Fan, Y.; Meng, X.; Zhang, W.; Meng, B.; Yang, N.; Liu, S. Theoretical and Experimental Insights into the Mechanism for Gas Separation through Nanochannels in 2D Laminar MXene Membranes. *Processes* **2019**, *7*, 751. [CrossRef]
22. Zhao, W.; Wu, H.; Wen, J.; Guo, X.; Zhang, Y.; Wang, R. Simulation Study on the Influence of Gas Mole Fraction and Aqueous Activity under Phase Equilibrium. *Processes* **2019**, *7*, 58. [CrossRef]
23. Wang, F.; Zhao, X.; Liang, Y.; Li, X.; Chen, Y. Calculation Model and Rapid Estimation Method for Coal Seam Gas Content. *Processes* **2018**, *6*, 223. [CrossRef]
24. Yang, X.; Wang, G.; Zhang, J.; Ren, T. The Influence of Sorption Pressure on Gas Diffusion in Coal Particles: An Experimental Study. *Processes* **2019**, *7*, 219. [CrossRef]
25. Wang, X.; Liu, X.; Jin, Z.; Zhu, J.; Jin, B. Theoretical Methodology of a High-Flux Coal-Direct Chemical Looping Combustion System. *Processes* **2018**, *6*, 251. [CrossRef]
26. Wang, X.; Liu, X.; Jin, B.; Wang, D. Hydrodynamic Study of AR Coupling Effects on Solid Circulation and Gas Leakages in a High-Flux In Situ Gasification Chemical Looping Combustion System. *Processes* **2018**, *6*, 196. [CrossRef]
27. Krištof, O.; Bulejko, P.; Svěrák, T. Experimental Study on Spray Breakup in Turbulent Atomization Using a Spiral Nozzle. *Processes* **2019**, *7*, 911. [CrossRef]
28. Ibrahim, M.; El-Naas, M.; Benamor, A.; Al-Sobhi, S.; Zhang, Z. Carbon Mineralization by Reaction with Steel-Making Waste: A Review. *Processes* **2019**, *7*, 115. [CrossRef]

Article

Siderite Formation by Mechanochemical and High Pressure–High Temperature Processes for CO$_2$ Capture Using Iron Ore as the Initial Sorbent

Eduin Yesid Mora Mendoza [1,2,*], **Armando Sarmiento Santos** [1], **Enrique Vera López** [1], **Vadym Drozd** [2], **Andriy Durygin** [2], **Jiuhua Chen** [2] **and Surendra K. Saxena** [2]

[1] Grupo de Superficies, Electroquímica y Corrosión, GSEC, Instituto para la Investigación e Innovación en Ciencia y Tecnología de Materiales, INCITEMA, Universidad Pedagógica y Tecnológica de Colombia UPTC, Tunja 150008, Colombia; asarmiento.santos@uptc.edu.co (A.S.S.); enrique.vera@uptc.edu.co (E.V.L.)

[2] Center for the Study of Matter at Extreme Conditions, Department of Mechanical and Materials Engineering, College of Engineering and Computing, Florida International University, Miami, FL 33199, USA; drozdv@fiu.edu (V.D.); durygina@fiu.edu (A.D.); chenj@fiu.edu (J.C.); saxenas@fiu.edu (S.K.S.)

* Correspondence: eduin.mora@uptc.edu.co

Received: 6 September 2019; Accepted: 5 October 2019; Published: 14 October 2019

Abstract: Iron ore was studied as a CO$_2$ absorbent. Carbonation was carried out by mechanochemical and high temperature–high pressure (HTHP) reactions. Kinetics of the carbonation reactions was studied for the two methods. In the mechanochemical process, it was analyzed as a function of the CO$_2$ pressure and the rotation speed of the planetary ball mill, while in the HTHP process, the kinetics was studied as a function of pressure and temperature. The highest CO$_2$ capture capacities achieved were 3.7341 mmol of CO$_2$/g of sorbent in ball milling (30 bar of CO$_2$ pressure, 400 rpm, 20 h) and 5.4392 mmol of CO$_2$/g of absorbent in HTHP (50 bar of CO$_2$ pressure, 100 °C and 4 h). To overcome the kinetics limitations, water was introduced to all carbonation experiments. The calcination reactions were studied in Argon atmosphere using thermogravimetric analysis (TGA) and differential scanning calorimetry (DSC) analysis. Siderite can be decomposed at the same temperature range (100 °C to 420 °C) for the samples produced by both methods. This range reaches higher temperatures compared with pure iron oxides due to decomposition temperature increase with decreasing purity. Calcination reactions yield magnetite and carbon. A comparison of recyclability (use of the same material in several cycles of carbonation–calcination), kinetics, spent energy, and the amounts of initial material needed to capture 1 ton of CO$_2$, revealed the advantages of the mechanochemical process compared with HTHP.

Keywords: CO$_2$ capture; iron ore; carbonation; calcination; recyclability; mechanochemical reactions; carbonation kinetics

---

1. Introduction

The planet temperature has been monitored since the 19th century, however only since the 1980s has the world realized that global warming was occurring, according to Goddard Institute for Space Studies (GISS) report [1]. The global temperature has increased in the last 50 years by about 1 °C. CO$_2$ atmospheric concentration rose to 401 ppm in 2015, which is an increase of 110 ppm since the start of the industrial age [1–3]. Even though some countries have defined policies using international deals, such as Rio de Janeiro (1992), Kyoto (1997), and Copenhagen (2009), the global warming problem remains. A premise in the Paris agreement (2016) is to invest more than 100 billion dollars a year to achieve a solution to the green-house effect of gasses emission. Indeed, these agreements have helped to develop and improve different strategies and technologies to reduce the amount of CO$_2$ emitted to the atmosphere, as it is the main cause of the green-house effect [3–5].

Renewable energy and energy-efficient use as preventive approaches to reduce CO_2 emissions are still far from suitable solutions. Several techniques can be used for CO_2 separation from flue gases and its subsequent sequestration. These are classified as either physical and chemical methods. The chemical absorption reactions use chemical absorbents, such as amines, and are the most mature technology. Although absorption by amine is the method mainly used in industry, the required energy for the absorbents' regeneration is considerably high [6–8]. Membranes are not considered as applicable solutions due to low mass transfer [9,10]. However, some recent studies have shown remarkable improvements in the CO_2/N_2 selectivity of the membranes [10]. Cryogenics and micro algals are laboratory-scale technologies [11].

As a result of an exothermic reaction, carbonates generated from metallic oxides, such as MgO, FeO, or CaO, and carbon dioxide, which can be taken from flue gases, are a suitable means of capturing CO_2. These carbonates can be heated, and after that, using an endothermic reaction, pure CO_2 is released. This pure gas can be used in valuable industrial applications, such as oil recovery oil in the food industry, etc., allowing the recovery of the metal oxides, which, in turn, can be used in a new cycle of carbonation–calcination until the active sites disappear [5,12]. Iron oxides have good thermodynamic properties to be a CO_2 sorbent. Kumar et al. [13] demonstrated that a mixture of magnetite and iron can be carbonated with 57% conversion efficiency by adding water as a catalyst. Graphite and iron are suitable reducing agents for carbonation. In addition, they are highly available in steel making industries. Recent research found a novel way to carbonate iron oxides and iron using a ball milling process, reaching almost pure siderite without the use of water [14].

In addition, wustite, hematite, magnetite, and even siderite, iron ore can contain different compounds, such as CaO, MnO, Al_2O_3, K_2O, S, SiO_2, P_2O_5 [15]. Siderite is used in industrial processes. Hence, its thermal decomposition has generated interest. Depending on the atmosphere, different calcination products are formed. Hematite (Fe_2O_3) is common in an oxidizing atmosphere. Magnetite is obtained in a CO_2 atmosphere, while magnetite and wustite (FeO) are found in an inert atmosphere or in vacuum [14,16,17].

In this work, the CO_2 capture capacity of iron ore from the El Uvo mine, Colombia, was studied. Thermodynamic simulations supported the results of carbonation–calcination reactions. The CO_2 capture capacity was studied for the HTHP process as a function of pressure, temperature, and reaction time, while in the ball milling process, it was studied as a function of pressure, revolution speed, and reaction time. The kinetics of mechanochemical reactions was studied, and finally, energy spend in both carbonation methods was compared.

2. Experimental

The iron ore samples were prepared from a mineral rock, which was crushed and ground using a mortar and pestle, and, finally, processed by ball milling in an inert atmosphere for two hours. The chemical composition of the ore was studied by X-ray fluorescence analysis (XRF), as shown in Table 1. The major component was iron, but there are some impurities, such as SiO_2, CaO, Al_2O_3, MgO, MnO, P_2O_5, Na_2O, K_2O, S, and Zn.

The reactor for carbonation in the HTHP process was a closed cylindrical vessel of length 31.75 mm and an internal diameter of 8.89 mm. The reactor was loaded with 0.25 g of iron ore and iron (Good Fellow, 99% purity, <60 μm) mixture. The molar ratio of iron ore/Fe was defined according to the stoichiometry of reactions (1) and (2).

$$Fe_2O_3(s) + Fe + 3CO_2(g) \rightarrow 3FeCO_3(s) \tag{1}$$

$$2FeOOH(s) + Fe + 3CO_2(g) \rightarrow 3FeCO_3(s) + H_2O \tag{2}$$

Zero-point one five milliliters of water was added to the vessel to improve the kinetics of the carbonation reaction. High purity CO_2 gas (99.99% purity, Airgas) was introduced into the system at desired pressures from 30 to 50 bar. Before the experiments, the reactor was flushed with CO_2 gas three times to ensure an inert atmosphere in the system.

Table 1. Chemical composition of iron ore in wt.% according to X-ray fluorescence (XRF) analysis.

Fe (Total)	46.98
SiO_2	9.58
CaO	4.38
Al_2O_3	5.43
MgO	0.43
MnO	0.23
P_2O_5	2.72
Na_2O	0.59
K_2O	0.04
S	0.89
Zn	0.08

Mechanochemical reactions between iron ore and carbon dioxide were performed at room temperature and elevated CO_2 pressure (10–30 bar). Planetary ball mill Retsch PM100 was operated at 200 to 400 revolutions per minute. The vessel for the ball milling was a stainless-steel jar of 50 mL volume capable of holding up to 100 bar gas pressure. High purity CO_2 gas (Airgas, 99.999%) was loaded into the reactor at different pressures together with 3.00 g of iron ore and 0.5 mL of water. The average temperature in the reactor during the ball milling process was 32 °C. The mechanochemical reaction was run for different periods, and each 1 h milling interval was followed by half an hour cooling interval to avoid the overheating of the sample. The powder to balls (stainless steel) weight ratio was 2:27. The reactor was flushed several times with CO_2 gas to ensure a pure CO_2 atmosphere inside the reactor. Thermogravimetric analysis (TGA) and differential scanning calorimetry (DSC) were conducted in a temperature range of 25 to 1000 °C using a TA Instruments SDT Q600 instrument. Experiments were performed in air and Ar atmospheres with a heating rate of 10 °C/min. Powder X-ray diffraction patterns were collected using Bruker GADDS/D8 diffractometer equipped with Apex Smart CCD Detector and molybdenum rotating anode. Collected 2D diffraction patterns were integrated using Fit2D software [18]. Quantitative phase analysis of the samples was performed using the Rietveld method and GSAS package [19,20]. The CO_2 sorption capacity was calculated using the results generated by the Rietveld refinement of XRD patterns. Raman spectroscopy characterization was used to identify carbon in siderite decomposition products. A continuous-wave (CW) argon ion (Ar+) laser (model 177G02, Spectra Physics) of 514.4 nm in wavelength was used as a source of monochromatic radiation. Backscattered Raman spectra were collected by a high-throughput holographic imaging spectrograph (model HoloSpec f/1.8i, Kaiser Optical Systems) with volume transmission gratings, a holographic notch filter, and thermoelectrically cooled charge-coupled device (CCD) detector (Andor Technology). The spectra were usually collected with 10 min exposure.

3. Results

3.1. Iron Ore Characterization

The iron ore sample was analyzed by XRD technique. As shown in Figure 1, Fe can be found in the form of hematite Fe_2O_3 (JCPDS card number #00-072-0469), goethite FeOOH (JCPDS card number#00-081-0462), and siderite $FeCO_3$ (JCPDS card number #00-029-0696).

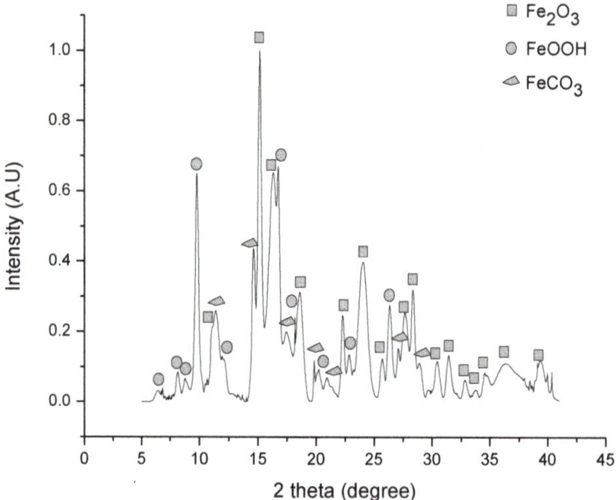

Figure 1. XRD pattern of iron ore sample.

Rietveld refinement shows that iron ore is composed of Fe_2O_3 (48.02%), $FeCO_3$ (21.15%), and FeOOH (30.83%). These results defined chemical reactions (1) and (2).

Reactions (1) and (2) use Fe iron as a reducing agent. Moreover, Sulfur (S) included in the iron ore in the form of sulfate or sulfide can act as a reducing agent. In García et al. [21], siderite is obtained from hematite, carbon dioxide, sulfur dioxide, and water at suitable conditions.

The regeneration reaction studied in this work is shown below:

$$6FeCO_3(s) \rightarrow 2Fe_3O_4(s) + C(s) + 5CO_2(g) \tag{3}$$

After regeneration, products can be further recycled back and used in a new carbonation reaction to complete the cycle.

3.2. Thermodynamics Simulation of Iron Ore Carbonation

FACTSAGE software and the databases therein, FACT- F*A*C*T 5.0, SGPS- SGTE, and SGSL were used to verify the thermodynamic feasibility of the carbonation process at equilibrium for the system iron ore–CO_2. According to XRD analysis and chemical composition of the ore, carbonation simulations were performed considering iron ore as a mixture of hematite and goethite. Carbonation in ball milling was simulated at a temperature of 32 °C (the average temperature measured in the reactor) and CO_2 pressures between 1 and 50 bar. As can be seen in Figure 2, siderite is stable at those conditions.

Figure 3 shows simulations at HTHP conditions for iron ore carbonation in the temperature range of 25 to 325 °C and CO_2 pressures of 10 and 50 bar. Figure 3a,b do not include Fe in the reaction, while Figure 3c,d include this metallic element. The presence of siderite is evident in all cases. The stability of siderite increased with pressure. This behavior was observed in both systems. It is clear that when Fe is included in the system, the decomposition temperature of siderite increased for the same CO_2 pressure. For example, the siderite decomposition temperature at 50 bar began at 150 and 225 °C for systems without and with Fe, respectively, revealing the advantage of including Fe due to higher temperatures favoring the thermodynamic conditions for carbonation.

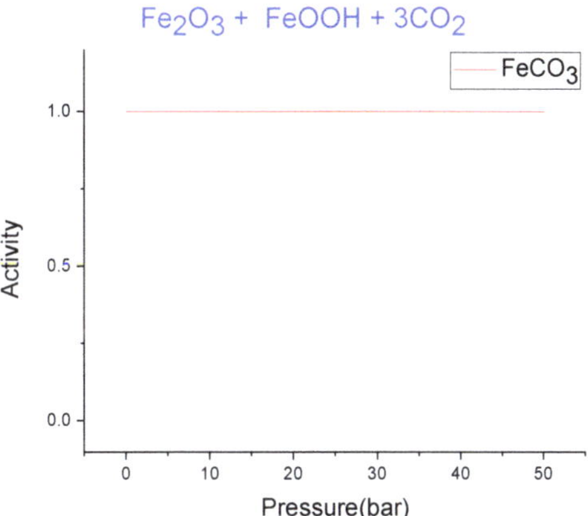

Figure 2. FACTSAGE simulation of siderite stability at 32 °C and various pressures for the system, iron ore–CO$_2$.

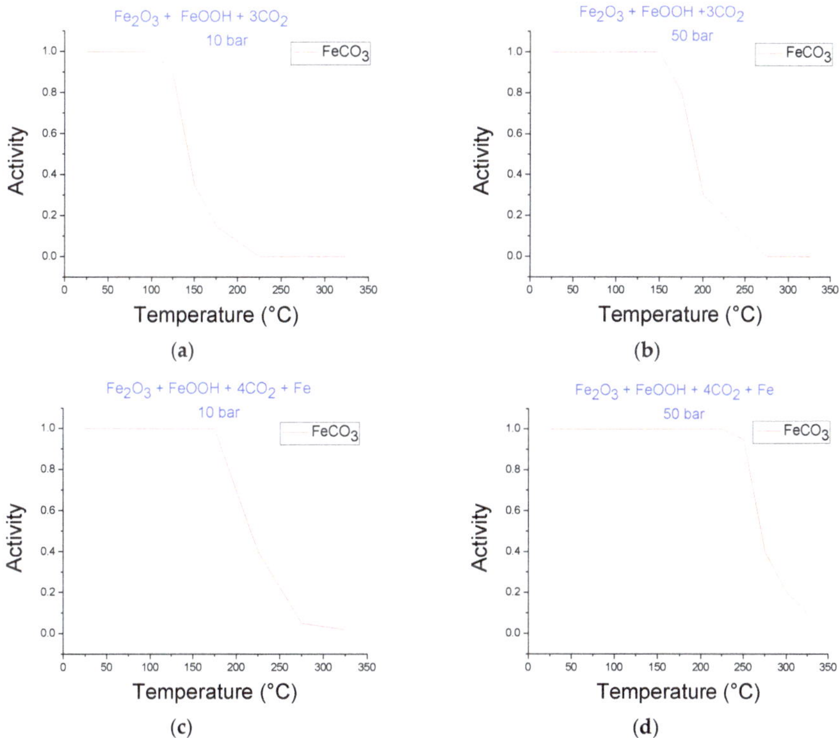

Figure 3. FACTSAGE simulations at various temperatures for the system iron ore–CO$_2$, (**a**) 10 bar, (**b**) 50 bar, and for the system iron ore–CO$_2$–Fe (**c**) 10 bar, (**d**) 50 bar.

3.3. Iron Ore Carbonation in Mechanochemical Process

Siderite yield in the carbonation reaction increased (JCPDS card number # 029-0696) as a result of carbonation by the ball milling method. Figure 4 shows the increases in iron carbonate formation at 30 bar, 400 rpm, 32 °C, and 20 h of time reaction.

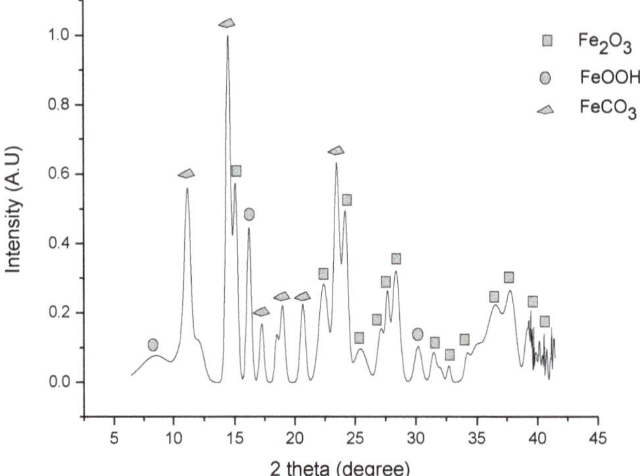

Figure 4. Carbonation of iron ore in ball milling at 30 bar, 400 rpm, 32 °C, and 20 h of reaction time.

Initially, the experiments were performed without water, but the siderite amount did not change. That can be related to kinetics limitations. It is clear that water acts as a catalyst in the carbonation process of metal oxides [13,22]. Figure 4 reveals that a considerable amount of siderite can be obtained by 20 h of milling. For those conditions, the CO_2 capture capacity of hematite and goethite is 0.1643 g CO_2/g sorbent or 3.7341 mmol CO_2/g sorbent calculated from the Rietveld refinement of the XRD pattern. This value translates to a 26.82% conversion rate. The calculation was performed, taking into account the initial amount of $FeCO_3$ contained in the ore and that the amount of initial absorbent is the sum of the weights of goetite and hematite.

Table 2 shows the calculations of CO_2 capture capacity at different conditions of pressure, revolution speed, and duration of the reaction. As can be seen in the table, the CO_2 capture at the same temperature by iron ore increased at higher pressures and longer reaction times.

Table 2. CO_2 capture capacity of iron ore at different conditions of pressure, revolution speed, and time reaction in the mechanochemical process.

Pressure (bar)	Revolution Speed (rev/min)	Time Reaction (h)	CO_2 Capture Capacity (mmol CO_2/g sorbent)
10	400	3	1.075
10	400	6	1.9523
20	400	3	1.7545
20	400	6	2.9204
20	200	3	0.4318
30	200	3	0.6864
30	200	6	0.8545

Additionally, values in Table 2 reveal that the ball milling process is affected by the revolution speed of the ball mill. At faster speeds, the siderite yield increased due to the transfer of higher kinetic energy, promoting the appearance of defects, which generate more active sites in goethite and hematite, facilitating the reaction with CO_2. According to [23] in the ball milling process, there is a critical revolution speed above which the balls will be pinned to the inner walls of the vial and do not fall to exert any impact force. In these experiments, the speeds of revolution were kept below the critical speed.

3.4. Iron Ore Carbonation in the HTHP Process

Using this method, neither using graphite as a reducing agent and water as a catalyst nor iron as a reducing agent and no water, the siderite yield was increased. Figure 5 confirms an increase of siderite yield (JCPDS card number #00-029-0696) as a result of carbonation by the HTHP method at 50 bar, 100 °C, and 4 h with the addition of metallic iron and water to iron ore. For these conditions, the CO_2 capture capacity was 0.2393 g CO_2/g sorbent or 5.4392 mmol CO_2/g sorbent calculated from the Rietveld refinement of the XRD pattern. This value translates to a 39.08% conversion rate.

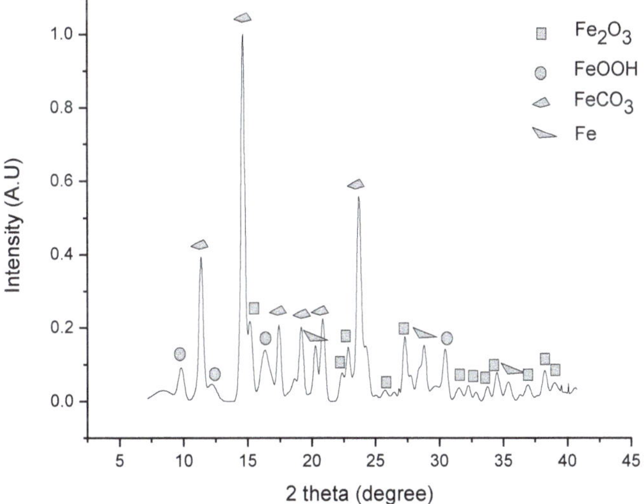

Figure 5. Carbonation of iron ore in the thermo pressure process at 50 bar, 100 °C, and 4 h of time reaction.

Table 3 presents the CO_2 capture capacity at different conditions of pressure, temperature, and reaction time. Longer reaction times increased the CO_2 capture for all cases. Siderite stability at higher temperatures decreased. For example, at 200 °C, 50 bar, and 4 h, a decrease of 53.2% in siderite formation was evidenced compared to 150 °C. These considerations can be confirmed with FACTSAGE simulations, which showed that at 50 bar, siderite began to decompose at temperatures around 200 °C. In this case, it can be supposed that the decomposition temperature is lower due to the presence of water. The CO_2 capture capacity of iron ore increased at higher pressures while keeping the temperature and reaction time constant, confirming the same effect observed in the mechanochemical process.

Table 3. CO_2 capture capacity of iron ore at different conditions of pressure, temperature, and reaction time in the high temperature–high pressure (HTHP) process.

Pressure (bar)	Temperature (°C)	Reaction Time (h)	CO$_2$ Capture Capacity (mmol CO$_2$/g Sorbent)
30	100	4	4.7927
30	200	1	1.6629
30	200	4	1.9945
50	100	1	2.9118
50	100	4	5.4392
50	150	4	5.7069
50	200	4	2.6713

3.5. Thermal Decomposition of Siderite Studied by Thermogravimetric Analysis

The siderite decomposition reaction was studied on two samples. The first one was obtained by the mechanochemical process at 30 bar of CO_2 pressure, 400 rpm, and 20 h of reaction time, and the second one was obtained at 50 bar of CO_2 pressure, 100 °C, and 4 h of reaction time by the HTHP process. The decomposition temperature of siderite was experimentally identified using thermogravimetric analysis. Figure 6a,b show the TG–DSC plots for siderite obtained in mechanochemical and HTHP processes, respectively, in an argon atmosphere.

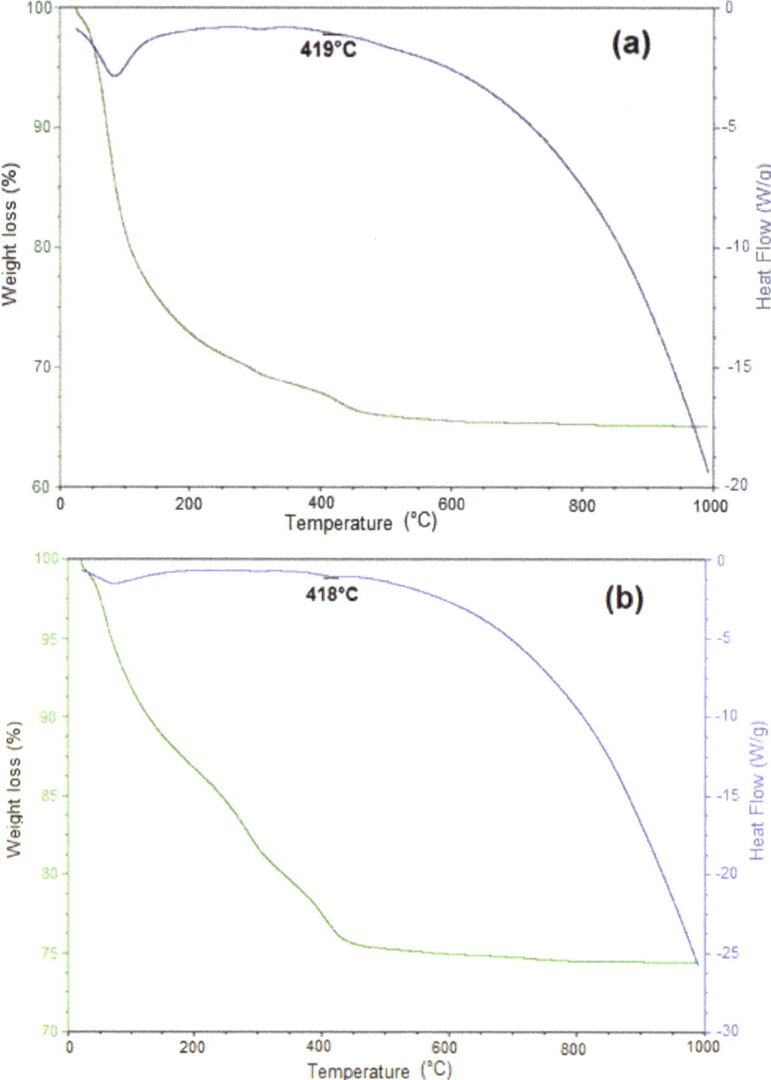

Figure 6. Thermogravimetric (TG)-heat flow plot of thermal decomposition of the product formed at 30 bar CO_2 pressure, 400 rpm, and 20 h (**a**), and at 50 bar, 100 °C, 4 h (**b**), heating rate 10 °C/min in an argon atmosphere.

The reaction mechanism for decomposition of siderite samples synthesized from pure iron oxides without water in an argon atmosphere was studied in [14]. Here, according to the thermal gravimetry (TG) plots, mass losses around 100 °C occurred for both plots. They were equivalent to 18 wt% and 6 wt%, respectively, and corresponded to the release of adsorbed water by iron ore. The release of CO_2 started from 100 °C. Some research works have reported that in the temperature interval between 250 and 375 °C, there are losses of weight corresponding to the dehydration of the goethite phase and iron hydroxides [24,25].

The thermal decomposition of siderite was significant at the temperature range of 100 to 420 °C. According to [14], this range reaches higher temperatures compared to the decomposition temperature of siderite synthesized from pure iron oxides, due to siderite decomposition temperature increasing with decreasing purity [15,26]. Patterson et al. [27] found that magnesium, manganese, or calcium increases the siderite decomposition temperature.

To identify the products after calcination, the siderite sample obtained by the mechanochemical reaction at 30 bar of CO_2 pressure, 400 rpm, and 20 h of reaction time was decomposed. The X-ray diffraction pattern in Figure 7 evidences the presence of magnetite (JCPDS # 001-1111), hematite (JCPDS # 00-001-1053), and graphite (JCPDS # 00-026-1079) after decomposition in a vacuum at 300 °C for 1 h. As can be seen, at this temperature, siderite was completely decomposed. The same products were identified for decomposed siderite, obtained by the HTHP process, and under the same decomposition conditions.

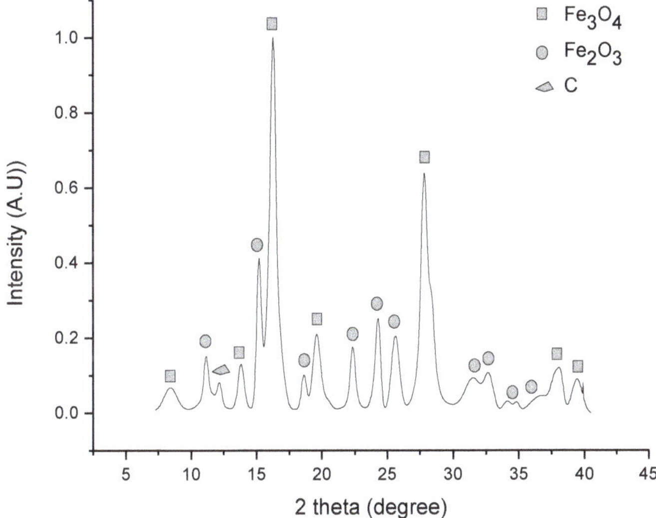

Figure 7. XRD pattern after siderite decomposition at 300 °C in a vacuum. Magnetite, hematite, and graphite were identified.

Raman spectroscopy is a suitable technique for graphite identification, due to its high sensitivity to highly symmetric covalent bonds with little or no natural dipole moment. The carbon–carbon bonds that make up these materials fit this criterion perfectly, and as a result, Raman spectroscopy is highly sensitive to these materials and able to provide a wealth of information about their structure. Every band in the Raman spectrum corresponds directly to a specific vibrational frequency of a bond within the molecule. The 1582 cm^{-1} band of graphite is known as the G band, and at 1370 cm^{-1}, a characteristic line appears, which is named D mode for a disorder-induced mode of graphite [28–30].

According to Figure 8, decomposed siderite at 300 °C in vacuum evidences the presence of graphite, by mean of peaks at 1582 cm^{-1} and 1370 cm^{-1}. Figure 8a,b show the Raman patterns of decomposed siderite produced by mechanochemical and HTHP processes, respectively. These results confirm that reaction (3) occurs during siderite decomposition obtained by both ball milling and HTHP.

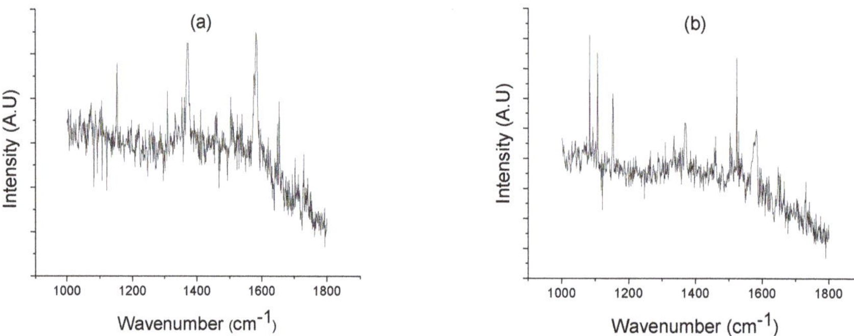

Figure 8. Raman pattern of decomposed siderite from processes, (**a**) mechanochemical (**b**), HTHP.

3.6. Carbonation–Calcination Cycles

Samples of decomposed siderite were studied in various cycles in CO_2 absorption/release reactions to confirm if the materials can be reused. Initially, using the two calcinated samples, there was no CO_2 capture, neither in mechanochemical nor in HTHP

For the mechanochemical method at 30 bar CO_2 pressure, 400 rpm, 20 h, adding water to the samples, siderite yield was accomplished. After recarbonation, samples were decomposed at 300 °C in a vacuum. Table 4 shows the CO_2 capture capacity of the transformed material for several cycles in the mechanochemical process.

Table 4. CO_2 capture capacity for several cycles of carbonation–calcination using the mechanochemical process.

N° Cycle, Carb-Calc	Added Extra Substances	CO_2 Capture Capacity (mmol CO_2/g Sorbent)
1	Water	3.7341
2	Water	4.1354
3	Water	6.2158
4	Water	6.9611

The addition of magnetite as a new chemical to the absorbent mixture and carbon as a new reducing agent improved the CO_2 capture capacity in subsequent cycles. Here, one additional carbonation reaction is:

$$2Fe_3O_4(s) + C(s) + 5CO_2(g) \rightarrow 6FeCO_3(s). \tag{4}$$

Hence, iron ore can be used for multiple cycles according to the combination of (1), (2), (3), and (4) reactions.

For HTHP, the recarbonation was studied at 50 bar, 100 °C, and 4 h. Graphite was used as a reducing agent in the first place due to its availability and cost. There was no siderite formation after adding extra-graphite to the mixture. It was necessary to add iron and water to achieve new carbonation. The iron addition is not propitious in terms of cost. After recarbonation, samples were decomposed at 300 °C in a vacuum. Table 5 shows the CO_2 capture capacity of material in the second and third cycles. As can be seen, it was necessary to include extra iron in both cycles. Here, only three cycles were studied because CO_2 capture capacity decreases dramatically during cycles.

Table 5. CO_2 capture capacity for several cycles of carbonation–calcination using the HTHP process.

N° Cycle, Carb-Calc	Added Extra Reactants	CO_2 Capture Capacity (mmol CO_2/g Sorbent)
1	Water, iron	5.4392
2	Water, iron	4.8687
3	Water, iron	2.7125

3.7. Discussion

It is clear that the addition of water to the mixtures facilitates CO_2 sorption and thus, affects the reactivity and capacity of the materials. The presence of moisture increases the mobility of alkaline ions and thus, accelerates the reactions [13,31–33]. Here, in the two methods of carbonation studied, reactions without water did not allow the siderite formation due to kinetics limitations. According to [13], water on the sorbent surface before and after calcinations facilitates the reaction with CO_2, which results in the formation of CO_3^{2-} and H^+ ions. Free Fe^{+2} ions can further react with CO_3^{2-} to form $FeCO_3$. The presence of water has a dual effect. It not only helps CO_2 uptake of sorbent but also affects the siderite stability [13,14].

With increasing siderite formation over time, its layer thickens, which inhibits the contact between Fe^{+2} and CO_3^{2-} harming the formation of new siderite. The mechanochemical process provides a way to remove the outer layer of $FeCO_3$; this layer is generally nonporous. This fluidization regime allows the carbonation reaction to remain more active [14,34,35]. This is likely the reason the carbonation process did not need an extra-reducing agent, such as iron, to obtain siderite in all of the cycles, which is an advantage compared with the HTHP process, which needed metallic iron. Initially, the presence of iron allowed high levels of CO_2 capture; however, with cycles, the actives sites vanished, and the CO_2 capture was practically negligible.

The advantage of the fluidization regime used in the mechanochemical process can be explained through kinetics. Figure 9 shows the siderite yield vs. time for two samples at 20 bar CO_2 pressure and 32 °C. The first one was treated at 400 rpm and the other one at 200 rpm. As was explained above, the carbonation depends strongly on the revolution speed for the same conditions of pressure and temperature, if the revolution speed is lower than the critical speed. The product yield at 3 h of reaction time and 400 rpm is about three times higher than the mass gained at 200 rpm speed.

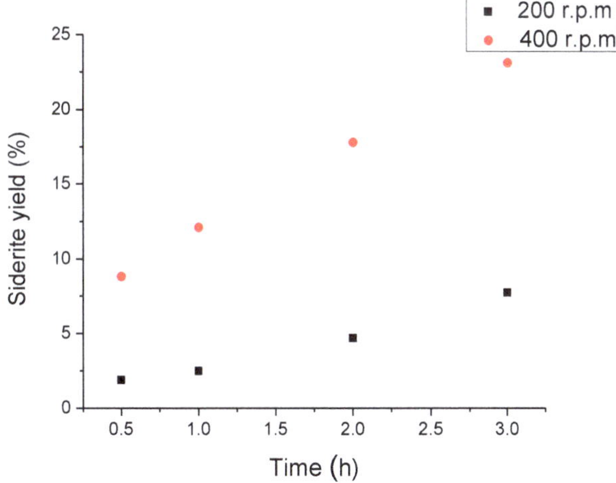

Figure 9. Siderite formation as a function of rotation speed at 20 bar CO_2 pressure and 32 °C in the ball milling process.

According to Alkaç and Atalay [36], using the mass fractional conversion x with respect to time, it is possible to calculate $f(x)$, the reaction model, which comprises the particular fractional conversion and related mechanism in terms of mathematical equations [15,36]. If $f(x)$ vs. t has high lineality, it indicates a suitable fitting for a given model and the slope gives the value of the rate constant, k, at a fixed temperature. Constant k is directly proportional to the reaction rate. For example, Figure 10 depicts $f(x)$ as a function of time, taking the Jander three dimensional diffusion model which presented high lineality. This model expresses $f(x)$ as

$$f(x) = \left[1 - (1 - x)^{\frac{1}{3}}\right]^2. \tag{5}$$

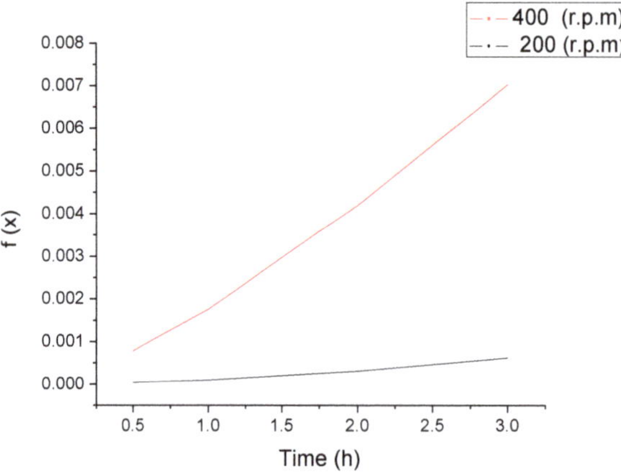

Figure 10. The Jander three-dimensional diffusion model evaluated for fractional mass conversion values (x) as a function of time at 20 bar and 32 °C in the ball milling process.

The slopes represent k according to the expression above. Hence, if the machine operates to 400 rpm, the rate constant increases 4.14 times compared to k at 200 rpm. Hence, if the kinetics energy transferred to the absorbent is bigger, carbonation conditions improve [37,38].

Another important consequence of the mechanochemical treatment is the improvement in calcination conditions. Some researches [39,40] have reported siderite, which has been treated in the ball milling process, as having lower decomposition temperatures.

According to carbonation reactions, it is possible to establish a projection of the amount of raw material that would be used in large-scale CO_2 emissions. Table 6 shows the needed material amount to capture 1 ton of CO_2 in carbonation reactions, assuming 100% conversion. The amount of formed siderite is 2.63 tons in both reactions.

Table 6. Amount of raw material needed to capture 1 ton of CO_2 in several carbonation reactions, assuming 100% of conversion yield.

Reaction	Material	Amount (ton)
$Fe_2O_3(s) + Fe + 3CO_2(g) \rightarrow 3FeCO_3(s) + H_2O$	Fe_2O_3 Fe	1.21 0.42
$2FeOOH(s) + Fe + 3CO_2(g) \rightarrow 3FeCO_3(s) + H_2O$	FeOOH Fe	1.35 0.42

In terms of steel production in a blast furnace, to produce one ton of steel, around 1.8 tons of CO_2 emissions are generated. Here, for carbonation by mechanochemical interaction, 3.43 tons of iron ore are needed operating at 30 bar, 400 rpm for 36 h for the almost total transformation of iron ore. Otherwise, using HTHP, 2.26 tons of iron ore and 0.75 tons of metallic iron are needed as initial materials. However, the capture is reduced to a maximum of 43.15% in HTHP due to kinetics limitations and surface area conditions. Since carbonation by the mechanochemical method has remarkable advantages, the amount of the recovered material is calculated from the calcination reaction (3). In this case, 4.37 tons of siderite will be transformed into 2.91 tons of magnetite, 0.073 tons of carbon, and 1.38 tons of CO_2. Magnetite and carbon will be used in the next cycle as sorbent and reducing agent, respectively, and pure CO_2 can be used in industrial applications.

Another important point to consider is the energy needed in each process. According to [41], the total spent energy in ball milling can be calculated as a function of the filling factor of the reactor, ball mass, ball diameter, number of balls, reaction time, rotation speeds of plate and reactor, the radius of plate and reactor, and sample mass. A sample processed at 30 bar, 400 rpm, and 3 h consumes 14.062 W-h per gram of absorbent having a CO_2 capture capacity of 2.7816 mmol CO_2/g sorbent. In HTHP, the energy can be calculated by multiplying the values of electrical current, electrical voltage, a factor of heat losses, and time reaction. The power factor was taken as one, due to the total impedance in the electrical circuit that warms the reactor being completely resistive. To compare the expenses of energy demands from the mechanochemical and HTHP processes, similar values of CO_2 capture capacities were taken. Hence, the spent energy to process a sample at 50 bar, 100 °C, and 1 h, which produced a CO_2 capture capacity of 2.9118 mmol CO_2/g sorbent was calculated. This process needed 41.58 W-h per gram. This energy is almost three times larger than the energy in the mechanochemical method. In addition, the CO_2 pressure is higher which is an extra penalty.

4. Conclusions

Iron ore was studied for CO_2 capture. The CO_2 capture capacity was evaluated for two methods, ball milling and HTHP. Water was always added to accomplish carbonation. Higher levels of capture were achieved at higher pressures and reaction times. Faster revolution speed allowed an increase in the siderite formation in ball milling. In HTHP, carbonation reactions were favored by temperatures between 100 and 150 °C, but at 200 °C, an inverse reaction was observed. The range of regeneration of iron oxides was identified between 100 °C and 420 °C in both methods, reaching higher temperatures than the siderite decomposition temperatures formed from synthetic iron oxides, due to decomposition temperature increases with decreasing purity. Magnetite and carbon were identified as decomposition products. It was necessary to add water to accomplish re-carbonation in the mechanochemical process while in the HTHP process, metallic iron and water were needed. Carbonation by the mechanochemical process was studied for four carbonation–calcination cycles presenting suitable conditions, while HTHP was only studied for three cycles due to its CO_2 capture capacity decreasing. The amounts of hematite, goethite, and metallic iron needed to capture 1 ton of CO_2 in carbonation reactions were calculated. A projection of the material needed to carbonate, as well as the material recovered in the calcination, for a real application was carried out. Kinetics and energy requirements confirmed the advantages of carbonation by mean of the mechanochemical process compared to the HTHP method.

Author Contributions: All authors contributed to the manuscript. E.Y.M.M. and V.D. proposed the initial idea; A.D. designed and managed the technological equipment; E.Y.M.M. and V.D. performed the simulations; A.D., E.Y.M.M. and V.D. performed the experimentation. V.D. and E.Y.M.M. wrote the manuscript; J.C., A.S.S., E.V.L. and S.K.S. improved the manuscript.

Funding: This research received no external funding.

Acknowledgments: The authors acknowledge the financial support by Departamento Administrativo de Ciencia, Tecnología e Innovación, Colciencias in Colombia. The work in part is supported by the NSF Geophysics Program (EAR 1723185).

References

1. Sanz-Pérez, E.; Murdock, C.; Didas, S.; Jones, C. Direct Capture of CO_2 from Ambient Air. *Chem. Rev.* **2016**, *116*, 11840–11876. [CrossRef] [PubMed]
2. Wiers, B. Capture of Carbon Dioxide from Air and Flue Gas in the Alkylamine-Appended Metal–Organic Framework mmen-Mg2(dobpdc). *J. Am. Chem. Soc.* **2012**, 7056–7065. [CrossRef]
3. Yu, C.H.; Huang, C.H.; Tan, C.S. A Review of CO_2 capture by absorption and adsorption. *Aerosol Air Qual. Res.* **2012**, *12*, 745–769. [CrossRef]
4. Kim, Y.; Worrell, E. International comparison of CO_2 emission trends in the iron and steel industry. *Energy Policy* **2002**, *30*, 827–838. [CrossRef]
5. Kumar, S.; Saxena, S. A comparative study of CO_2 sorption properties for different oxides. *Mater. Renew. Sustain. Energy* **2014**, *1*, 2. [CrossRef]
6. Han, K.; Ahn, C.; Lee, M.S. Performance of an ammonia-based CO_2 capture pilot facility in iron and steel industry. *Int. J. Greenh. Gas Control* **2014**, *27*, 239–246. [CrossRef]
7. Pannocchia, G.; Puccini, M.; Seggiani, M.; Vitolo, S. Experimental and modeling studies on high-temperature capture of CO_2 Using Lithium Zirconate Based Sorbents. *Ind. Eng. Chem. Res.* **2007**, *46*, 6696–6706. [CrossRef]
8. Kumar, A.; Ramaprabhu, S. Nano magnetite decorated multiwalled carbon nanotubes: A robust nanomaterial for enhanced carbon dioxide adsorption. *Energy Environ. Sci.* **2011**, *4*, 889–895.
9. Ho, M.; Leamon, M.; Allinson, D.; Wiley, D. Economics of CO_2 and mixed gas geosequestration of flue gas using gas separation membranes. *J. Membr. Sci.* **2006**, *45*, 2546–2552.
10. Zhao, L.; Rienache, E.; Menzer, R.; Blum, L.; Stolen, D. A parametric study of CO_2/N_2 gas separation membrane processes for post-combustion capture. *J. Membr. Sci.* **2008**, *325*, 284–294. [CrossRef]
11. Zaman, M.; Hyung Lee, J. Carbon capture from stationary power generation sources: A review of the current status of the technologies. *Korean J. Chem. Eng.* **2013**, *30*, 1497–1526. [CrossRef]
12. Salvador, C.; Lu, D.; Anthony, E.; Abanades, J. Enhancement of CaO for CO_2 capture in an FBC environment. *Chem. Eng.* **2003**, *96*, 187–195. [CrossRef]
13. Kumar, S.; Drozd, V.; Durygin, A.; Saxena, S. Capturing CO_2 Emissions in the iron industries using a magnetite–iron mixture. *Energy Technol.* **2016**, 1–5. [CrossRef]
14. Mora, E.; Sarmiento, A.; Lopez, E.; Drozd, V.; Durigyn, A.; Saxena, S.; Chen, J. Iron oxides as efficient sorbents for CO_2 capture. *J. Mater. Res. Technol.* **2019**, *8*, 2944–2956. [CrossRef]
15. Gotor, F.; Macias, M.; Ortega, A.; Criado, J. Comparative study of the kinetics of the thermal decomposition of synthetic and natural siderite samples. *Phys. Chem. Miner.* **2000**, *27*, 475–503. [CrossRef]
16. Dhupe, A.; Gokanrn, N. Studies in the Thermal Decomposition of Natural Siderites in the esence of Air. *Int. J. Miner. Process.* **1990**, *28*, 209–220. [CrossRef]
17. Fosbol, P.; Thompsen, K.; Stenby, E. Review and recommended thermodynamic properties of $FeCO_3$. *Corros. Eng. Sci. Technol.* **2013**, *45*, 115–135. [CrossRef]
18. Hammersley, A. *ESRF Internal Report, ESRF97HA02T, FIT2D: An Introduction and Overview*; European Synchrotron Radiation Facility: Grenoble, France, 1997.
19. Toby, B. EXPGUI, a graphical interface for GSAS. *J. Appl. Cryst.* **2001**, *34*, 210–221. [CrossRef]
20. Larson, A.; Von Dreele, R. *General Structure Analysis System (GSAS)*; Report LAUR 86-748; Los Alamos National Laboratory: Los Alamos, NM, USA, 2004.
21. Garcia, S.; Rosenbauer, R.; Palandri, J.; Maroto-Valer, M. Sequestration of non-pure carbon dioxide streams in iron oxyhydroxide-containing saline repositories. *Int. J. Greenh. Gas Control* **2012**, *7*, 89–97. [CrossRef]
22. Kumar, S.; Saxena, S.; Drozd, V.; Durygin, A. An experimental investigation of mesoporous MgO as a potential pre-combustion CO_2 sorbent. *Mater. Renew. Sustain. Energy* **2015**, *4*. [CrossRef]
23. Suryanarayana, C. Mechanical alloying and milling. *Prog. Mater. Sci.* **2001**, *46*, 1–184. [CrossRef]
24. Liu, C.; Shih, S. Kinetics of the reaction of iron blast furnace slag/hydrated lime sorbents with SO_2 at low temperatures: Effects of the presence of CO_2, O_2, and NOX. *Ind. Eng. Chem.* **2009**, *48*, 8335–8340. [CrossRef]
25. Betancur, J.; Barrero, D.; Greneche, F.; Goya, F. El efecto del contenido de agua en la magnetita y en las propiedades estructurales de la goethita. *J. Alloy. Comp.* **2004**, *369*, 247–251. [CrossRef]
26. Gallagher, P.; Warne, S. Thermogravimetry and thermal decomposition of siderite. *Thermochim. Acta* **1980**, *43*, 253–267. [CrossRef]

27. Patterson, J.; Levi, J. Relevance of carbonate minerals in the processing of Australian oil shales. *Fuel* **1991**, *70*, 1252–1259. [CrossRef]

28. Hodkiewicks, J. *Characterizing Carbon Materials with Raman Spectroscopy*; Thermo Fisher Scientific: Madison, WI, USA, 2010.

29. Reitch, S.; Thomsen, C. Raman spectroscopy of graphite. *R. Soc.* **2004**, *362*, 2271–2278. [CrossRef]

30. Wang, Y.; Alsmeyer, S.; McCreery, S. Raman Spectroscopy of Carbon Materials: Structural Basis of Observed Spectra. *Chem. Mater.* **1990**, *2*, 557–563. [CrossRef]

31. Kummar, S. The effect of elevated pressure, temperature and particles morphology on the carbon dioxide capture using zinc oxide. *J. CO_2 Util.* **2014**, *8*, 60–66. [CrossRef]

32. Hassanzadeh, A.; Abbasian, J. Regenerable MgO-based sorbents for high-temperature CO_2 removal from syngas: 1. Sorbent development, evaluation, and reaction modeling. *Fuel* **2010**, *89*, 1287–1297. [CrossRef]

33. Sun, J.; Yang, Y.; Guo, Y.; Xu, Y.; Li, W.; Zhao, A.; Liu, W.; Lu, P. Stabilized CO_2 capture performance of wet mechanically activated dolomite. *Fuel* **2018**, *222*, 334–342. [CrossRef]

34. Ounoughene, G.; Buskens, E.; Santos, R.; Cizer, O.; Van Gerven, T. Solvochemical carbonation of lime using ethanol: Mechanism and enhancement for direct atmospheric CO_2 capture. *J. CO_2 Util.* **2018**, *26*, 143–151. [CrossRef]

35. Benitez, M.; Valverde, J.; Perejon, A.; Sanchez, P.; Perez, L. Effect of milling mechanism on the CO_2 capture performance of limestone in the Calcium Looping process. *Chem. Eng. J.* **2018**, *346*, 549–556. [CrossRef]

36. Alkaç, D.; Atalay, U. Kinetics of thermal decomposition of Hekimhan–Deveci siderite ore samples. *Int. J. Miner. Process.* **2008**, *87*, 120–128. [CrossRef]

37. Rigopoulos, I.; Vasiliades, M.; Ioannou, I.; Esfathiou, A.; Godelistsas, A.; Kyratsy, T. Enhancing the rate of ex situ mineral carbonation in dunites via ball milling. *Adv. Powder Technol.* **2016**, *27*, 360–371. [CrossRef]

38. Rigopoulos, I.; Delimitis, A.; Ioannou, I.; Esftathiou, A.; Kyratsy, T. Effect of ball milling on the carbon sequestration efficiency of serpentinized peridotites. *Miner. Eng.* **2018**, *120*, 66–74. [CrossRef]

39. Criado, J.; Gonzalez, M.; Macias, M. Influence of grinding on both the stability and thermal decomposition mechanical of siderite. *Thermochim. Acta* **1988**, *135*, 219–223. [CrossRef]

40. Bradley, W.; Burst, J.; Graf, D. Crystal chemistry and differential thermal effect of dolomite. *Am. Mineral.* **1953**, *38*, 207–211.

41. Burgio, N.; Lasonna, A.; Magini, M.; Martelli, S.; Padella, F. Mechanical Alloying of the Fe-Zr System. Correlation between Input Energy and End Products. *Il Nuovo Cim. D* **1991**, *13*, 459–476. [CrossRef]

Article

Exergy Analysis and Evaluation of the Different Flowsheeting Configurations for CO$_2$ Capture Plant Using 2-Amino-2-Methyl-1-Propanol (AMP)

Ebuwa Osagie [1],*, Aliyu M. Aliyu [2], Somtochukwu Godfrey Nnabuife [1], Osaze Omoregbe [3] and Victor Etim [1]

[1] School of Water, Energy and Environment, Cranfield University, College Road, Bedford, Bedfordshire MK43 0AL, UK; G.Nnabuife@cranfield.ac.uk (S.G.N.); V.Etim@cranfield.ac.uk (V.E.)

[2] Gas Turbine and Transmissions Research Center, Faculty of Engineering, University of Nottingham, Nottingham NG7 2RD, UK; ali.aliyu@nottingham.ac.uk

[3] Department of Chemical Engineering, University of Birmingham, Edgbaston B15 2TT, UK; OXO850@bham.ac.uk

* Correspondence: ebuwa100@yahoo.com

Received: 27 May 2019; Accepted: 19 June 2019; Published: 24 June 2019

Abstract: This paper presents steady-state simulation and exergy analysis of the 2-amino-2-methyl-1-propanol (AMP)-based post-combustion capture (PCC) plant. Exergy analysis provides the identification of the location, sources of thermodynamic inefficiencies, and magnitude in a thermal system. Furthermore, thermodynamic analysis of different configurations of the process helps to identify opportunities for reducing the steam requirements for each of the configurations. Exergy analysis performed for the AMP-based plant and the different configurations revealed that the rich split with intercooling configuration gave the highest exergy efficiency of 73.6%, while that of the intercooling and the reference AMP-based plant were 57.3% and 55.8% respectively. Thus, exergy analysis of flowsheeting configurations can lead to significant improvements in plant performance and lead to cost reduction for amine-based CO$_2$ capture technologies.

Keywords: 2-Amino-2-Methyl-1-Propanol; modelling and Simulation; post-combustion capture; exergy analysis; flowsheeting configurations

---

1. Introduction

Natural gas power plants play a vital role in meeting energy demands. Power generation from gas-fired power plants produces lots of emissions, which increases the concentration of greenhouse gases in the atmosphere. Thus, CO$_2$ emissions reduction is a high priority demand, and one of the solutions to this problem is carbon capture and storage (CCS). The main restriction for deploying large-scale CO$_2$ capture systems is that these processes reduce the plant net power output for fixed energy due to the addition of carbon capture plant, thereby increasing the net cost of capture [1]. However, the cost associated with commercial capture plants is about 80% of CCS cost [2], which poses a major setback. The reduction in the power output is as a result of the parasitic load of the capture plant, the load demand comes from the reboiler steam requirements drivers such as pumps, compressors, cooling duty needed for the amine process, etc. leading to an energy penalty. This energy penalty can be reduced in a number of ways, many of which are specific to the capture technology employed. For absorption processes, the total reboiler energy can be lowered by an improved process design of the solvent plant [3,4]. Examples of these improved process design include absorber intercooling, rich split, lean amine flash, vapor recompression, configurations and stripping with flash steam, etc. The total energy consumption can be reduced up to 20% in pilot-scale plant studies for the different configurations compared to the conventional amine plants [5].

Several configurations for minimizing energy consumption have been suggested and studied. Leites et al. [6] modelled the intercooler and varied temperatures between 40–80 °C, the whole liquid was removed from the column at each cooling stage and pumped to 1.1 bar to overcome pressure drop. It was concluded that cooling to 40 °C was found to have the maximum effect on reboiler duty and also minimization of additional equipment. Karimi et al. [7] investigated the intercooling effect in CO_2 capture energy consumption, the optimal location for intercooling was about 1/4th to 1/5th of the height of the column from the bottom which brought about 2.84% savings in reboiler energy and it was concluded that intercooling is an option for reducing energy consumption. Aroonwilas and Veawab [8] modelled an intercooler configuration which has been integrated with an amine process to evaluate the energy savings effect as a result of enhanced working capacity. The methodology involved the withdrawal of all the liquid at 1/5th of the column height from the bottom and cooled at 45 °C by varying the lean loading between 0.12–0.25 mol CO_2/mol monoethanolamine (MEA). With the intercooling, a reduction of 10% in the solvent required led to energy savings in the stripper reboiler and also, it was concluded that lean loading above 0.18 mol CO_2/mol MEA had a minimal effect on reboiler duty. Reddy et al. [9] modelled lean amine flash configuration which generates additional steam by flashing hot lean amine leaving the stripper. Results showed 11% reduction in reboiler steam, 16% reduction in cooling water and 6% in stripper diameter. It was observed that hot lean amine temperature was lowered from 120 °C to 103 °C by the flash; this low temperature increases the energy consumption in the stripper bringing about the additional steam generation and improved working capacity. Eisenberg and Johnson [10] modelled a rich split configuration and this resulted in 7.1% savings in reboiler duty over the reference case. It was concluded that for loadings greater than 0.15 mol CO_2/mol MEA, a clear benefit was obtained. But it was later observed that increasing the packing height for a lean loading of 0.2 mol CO_2/mol MEA, for 30% of the cold solvent split, a reboiler duty of 97.8 kW was required, which is about 10.3% higher than the reference case.

Cousins et al. [11] reviewed fifteen amine process configurations (multi-component columns, inter-stage temperature control, heat-integrated stripping column, split flow process, vapor recompression, matrix stripping, heat integration, etc.). The configurations which involved both experimental and simulation-based methods were evaluated with different solvents, and different operating conditions (temperature, pressure and feed composition). It was, therefore, difficult to compare the energy savings on a fair basis. Thus, it was concluded that the configurations considered reduced the energy consumption, but increased the plant complexity. Also, configurations with less additional equipment (e.g., vapor recompression, etc.) gave higher efficiencies than those with more equipment. Ahn et al. [12] evaluated ten different configurations capture plants, this included the multiple alterations (absorber intercooling combined with condensate evaporation and lean amine flash) which were novel in the study using 30 wt% MEA to capture 90% CO_2, reboiler duty savings was maximized by simultaneous application of previous strategies. The comparison was based on total energy consumption (thermal and electrical energy used), the multiple strategies achieved a greater reduction in the energy requirement reducing steam consumption by up to 37% when compared to the simple absorber/stripper configurations. Sharma et al. [13] reviewed and assessed the advantages of fourteen different flow sheeting configurations. The comparison was based on cooling duty and equivalent work. Results showed pump-around was more beneficial than intercooling, while intercooling with rich split was found to be the most beneficial based on additional equipment, and the equivalent energy consumption was 12.9% reduction over the base case. Lars et al. [14] compared different configurations; vapor recompression with split stream gave the best reduction of 11% compared to the conventional. Liang et al. [15] studied five different flow sheeting configurations, the new innovation was the combination of split flow with overhead exchanger and improved split flow with vapor recompression. These innovations decreased equivalent work by 17.21% and 17.52% respectively. Jung et al. [16] suggested a new combination; rich vapor recompression and cold solvent split. Results showed that reboiler heat was reduced from 3.44 MJ/kg CO_2 to 2.75 MJ/kg CO_2. All of these configurations presented above (summarised in Table 1) have achieved the aim of reducing the energy consumption for the MEA capture plant compared to the conventional flowsheet.

Table 1. Summary of representative past works on flowsheeting configurations for CO_2 capture plants.

Author(s)	Detail of Study	Solvent	Exergy Analysis
Leites et al. [6]	Evaluated intercooler at 40–80 °C, operation at 40 °C gave a maximum effect on reboiler duty using monoethanolamine (MEA)	MEA	No
Karimi et al. [7]	Evaluated intercooling effect. The optimal location for intercooling was 1/4th to 1/5th of the height of the column from the bottom, nearly 3% savings in reboiler energy.	MEA	No
Aroonwilas and Veawab [8]	Evaluated intercooling design. Removal of liquid done at 1/5th of column height from the bottom. Cooling at 45 °C resulted in 10% energy savings in the reboiler.	MEA	No
Reddy et al. [9]	Evaluated lean amine flash configuration. Hot lean amine temperature reduced from 120 to 103 °C, the low temperature, however, increases the energy consumption in the stripper.	MEA	No
Eisenberg and Johnson [10]	Evaluated rich split configuration resulting in 7.1% reboiler duty savings over reference case.	MEA	No
Cousins et al. [11]	Reviewed fifteen flowsheeting configurations, these include multi-component columns, inter-stage temperature control, heat-integrated stripping column, split flow process, vapor recompression, matrix stripping, heat integration. Results showed significant energy savings.	MEA	No
Ahn et al. [12]	Evaluated ten different configurations capture plants, this included the multiple alterations (absorber intercooling combined with condensate evaporation and lean amine flash). The multiple strategies achieved a greater reduction in the energy requirement reducing steam consumption by up to 37% when compared to the simple absorber/stripper configurations.	MEA	No
Sharma et al. [13]	Reviewed and assessed the advantages of fourteen different flow sheeting configurations. Results showed pump-around was more beneficial than intercooling.	MEA	No
Lars et al. [14]	Evaluated and compared different configurations; vapor recompression with split stream gave the best reduction of 11% compared to the conventional.	MEA	No
Liang et al. [15]	Five different flow sheeting configurations studied, the new innovation was the combination of split flow with overhead exchanger and improved split flow with vapor recompression These innovations decreased equivalent work by 17.21% and 17.52% respectively	MEA	No
Jung et al. [16]	Evaluated rich vapor recompression and cold solvent split. Results showed that reboiler heat was reduced from 3.44 MJ/kg CO_2 to 2.75 MJ/kg CO_2.	MEA	No
Geuzebroek [17], Lara et al. [18], Olaleye et al. [19]	Exergy analysis of CO_2 capture plant	MEA	Yes
Valenti et al. [20]	Exergy analysis of CO_2 capture plant	Ammonia	Yes

Also, rate-based modelling of CO_2 absorption in a packed column using AMP solutions for the capture plant has been carried out in the literature. Alatiqi et al. [21] used a rate-based model in simulating CO_2 absorption in AMP, MEA, and diethanolamine (DEA) solution. AMP was used to compare the absorption of CO_2 in MEA and DEA solutions. Aboudheir et al. [22] used a rate-based model in simulating the absorption of CO_2 using AMP solutions. Results were validated with experimental plant data. Gabrielsen et al. [23,24] carried out an experimental study using AMP solution and this was used as validation for the simulation of a rate-based model for CO_2 capture in a structured packed column. Afkhamipour and Mofarahi [25] compared rate-based and equilibrium-based models simulation results of a packed column using AMP solution. The rate-based models gave a better prediction of the concentration and temperature profiles than the equilibrium based. Dash et al. [26] explored the benefits of using blended solvents AMP with Piperazine (PZ). These studies presented above have all worked on models for the AMP capture process. There are limited studies on the different configurations using the AMP solvent. Kvamsdal et al. [27] presented a simulated model for the Cesar 1 (AMP + PZ) involving modifications such as intercooling and vapor recompression. The comparison was made with the MEA process using the same modifications. Results showed that the MEA process had lower energy requirements as compared to Cesar 1. Energy consumption accounts for about 25% of total cost thus, the AMP solvent, which has more favourable operating parameters as compared to the MEA solvent as shown in studies [23,28,29] will further minimize energy requirements which will reduce cost. It is therefore important to develop these configurations and utilize the energy savings provided.

Furthermore, exergy analysis which identifies where exergy is destroyed is carried out. The destruction of exergy in a process is proportional to the entropy generation in it; which accounts for the inefficiencies due to irreversibility [30]. Exergy analysis of capture plants using MEA and ammonia solvents have been carried out by few authors [17–20] as shown above in Table 1, to investigate the effects of the associated losses. However, studies to analyse where the losses occur in the AMP-based CO_2 capture plant is lacking. This study includes (i) steady-state rate-based simulation and conventional exergy analysis of the AMP-based PCC process (ii) evaluation of exergy destruction and efficiency in the AMP capture system, (iii) exergy analysis of the different flowsheeting configurations with AMP solvent.

2. Modelling Framework

2.1. Model Description of the Capture Plant

The AMP-based process model was developed using the operating parameters in Aspen Plus® software Version (V) 8.4 (Aspen Technology, Inc., Bedford, MA, USA, released in 2013), and consists of an absorber and a stripper column, with a cross heat exchanger and a pump, all connected in a closed loop cycle as described in studies [26,31]. The validation of the capture plant model with experimental data is presented in the literature [26,28].

2.2. Exergy Analysis

Exergy is defined as the maximum theoretical work that is obtained from a system when its state is brought to the reference state [30]. Exergy analysis is a method employed in the evaluation of the use of energy [32]. Exergy gives the identification of the location, the magnitude and the causes of thermodynamic inefficiencies in a thermal system. In this section, the conventional exergy approach is used to evaluate the exergy destruction and potential for improvement of the CO_2 capture plant. The values of the exergy reference temperature and pressure, which are default parameters in Aspen Plus® V8.4 simulation tool are 298.15 K and 1.013 bar respectively, and these are used in the simulations. A theoretical process in which the thermodynamic reversibility requires that all the process driving forces such as pressure, temperature and chemical potential differences be zero at all points and times [6] leads to producing a maximum amount of high energy consumption [19]. These losses can be reduced by several methods that are based on the second law of thermodynamics such as the counteraction, quasi-static method and the driving force method [6]. In this study, the driving force method is used to reduce the exergy destruction, leading to a reduction in energy consumption in the AMP-based PCC process and using the same absorbent throughout.

The Aspen Plus® V8.4 exergy estimation property set (EXERGYMS) is used in calculating the methods for physical and chemical exergies of the material and heat flows for each component, using the individual streamflow in the AMP-based capture plant. Furthermore, in other to determine the exergy of the reactions containing electrolytes, the thermodynamic properties of the ionic species of AMP were retrieved from the Aspen Plus® databank, V8.4 (Aspen Technology, Inc., Bedford, MA, USA, released in 2013). The standard Gibbs free energy of formation of AMP in the water at infinite dilution (DGAQFM) values used in this study are based on an estimate given in studies [33,34]. Gibbs free energy data which is called from Aspen database is used in the estimation of Gibbs free energy using the empirical relation in Aspen Plus® V8.4 software. The DGAQFM values of -1.628054×10^8 J/kmol and 4.574×10^8 J/kmol were obtained for AMPH$^+$ (AMP protonation) and AMPCOO$^-$ (AMP carbamate formation) respectively. Table 2 below shows the exergy destruction and efficiencies for the equipment in the AMP-based PCC process.

Table 2. Conventional exergy analysis of the AMP-based CO_2 capture plant.

	$E_{fuel(n)}$ (Watts)	$E_{product(n)}$ (Watts)	$E_{destruction(n)}$ (Watts)	E_n (%)
Absorber	45.19	27.02	18.17	59.79
Stripper	5085.56	4224.59	860.97	83.07
Pump	393.27	389.29	3.98	98.99
Cooler	1037.07	307.0	730.06	29.60
Heat exchanger	4174.62	1037.07	3137.55	24.84
Total	10,735.7	5984.98	4750.72	55.75

The equations below are used in evaluating the individual component and the total exergy destruction rate within a component. Thus, the exergy balance [35] for the whole system are given in Equations (1)–(3), while the exergetic performance of the AMP-based capture plant is given in Table 2. Equation 1 which is the fuel exergy of each component is as follows:

$$E_{fuel(total)} = E_{product(total)} + E_{Destruction(total)} + E_{Losses(total)} \tag{1}$$

While for the exergy efficiency of each component (n) which accounts for the thermodynamic losses is given as shown in Equation (2):

$$E_{(n)} = E_{product(n)} / E_{fuel(n)} \tag{2}$$

And the exergy destruction ratio of the nth component is presented in Equation (3):

$$X_{d(n)} = E_{Destruction(n)} / E_{fuel(total)} \tag{3}$$

Table 2, shows the exergy destruction and efficiency for the sub components in the AMP-based capture plant. As observed, the absorber and stripper components had exergy efficiency of 59.8 and 88.5%, respectively, while the heat exchanger gave the lowest exergy efficiency of 25%, these values are close in range with the literature [19].

3. Flowsheeting Configurations

The amine flow sheeting configurations are set up for the capture plant. The reference plant is the standard AMP-based capture plant configuration, as described in Section 2.

3.1. Intercooling Configuration

Absorption of CO_2 from the gas streams is mostly done between temperature 40–60 °C, this is because rates of CO_2 absorption for a 30 wt% amine solvent are highest in this temperature range (11). In other to control the temperature in the absorber so as to reach a higher rich CO_2 loading (high absorption capacity), inter stage cooling as shown in Figure 1 is required.

In this configuration, the exothermic nature of CO_2 absorption in amine present in the absorber leads to a temperature bulge and this impacts absorption negatively [36]. At the location of the T-bulge, the solvent is being extracted, cooled to 40 °C and returned to the absorber which leads to the enhancement of absorption driving force [11]. A lower temperature leads to a reduction in the absorption rates such as chemical kinetics, diffusivities, etc., while an increase in temperature favours the absorption capacity. These two operations compete with each other in the absorber. The temperature in the absorber can be controlled by adjusting the flue gas, the lean solvent temperature and flowrate coming into the absorber. Thus, this balances the temperature at either end of the absorber. With this modification, a reduction in solvent circulation rate which leads to higher absorption capacity is achieved. Hence, this process configuration enables the control of temperature within the absorber and is capable of enhancing CO_2 recovery [11], which is very effective in reducing the energy requirement of a CO_2 capture plant [7]. Results obtained are given in Table 3.

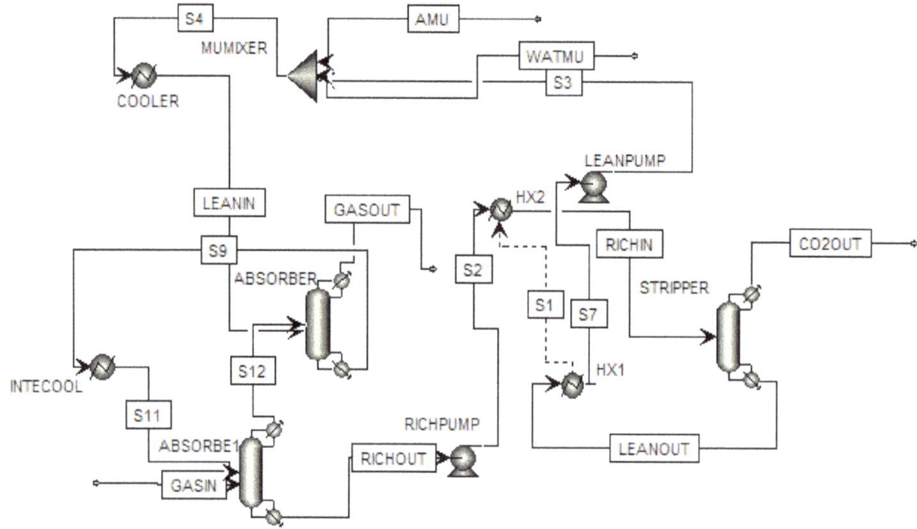

Figure 1. Intercooling Configuration [11].

Table 3. Results for intercooling configuration.

	Rich Loading (mol/mol)	Absorber (ABS) Capacity (mol/mol)	Reboiler Duty (kW)	ABS Temp (°C)
Reference	0.388	0.123	4.77	63.07
AMP-cooled	0.503	0.173	3.60	47.11

As shown in Table 3, lower reboiler duty and higher absorption capacity are achieved for AMP when compared with the reference; this is one of the main benefits of intercooling. This occurs due to the higher rich loading obtained from the intercooled model which leads to a lower circulation rate. Also, literature studies [37] has proven AMP to have a higher loading capacity and lower heat of reaction due to the formation of bicarbonates. The reference plant has a higher temperature bulge (63 °C) in the absorber compared to AMP-cooled (47 °C), thus having a higher heat of reaction which leads to an increment in the reboiler duty. With this intercooling, a gain of 24.5% savings in reboiler duty for AMP-cooled over the reference is achieved.

3.2. Lean Amine Flash Configuration

As shown in Figure 2, the stripper design comprises additional equipment such as a pump, compressor, and the flash drum. Additional steam is generated by flashing hot lean amine exiting the stripper close to ambient pressure, followed by the compressing of the gas stream up to the stripper pressure and re-introducing it into the stripper column [38,39]. In the flash drum present, more CO_2 is desorbed by reducing the pressure in the flash drum and lean loading in the stripper out is further reduced. Thus, additional steam is generated for the AMP-based process since the gas stream is compressed at a higher pressure. Results for the lean amine flash configuration is given in Table 4.

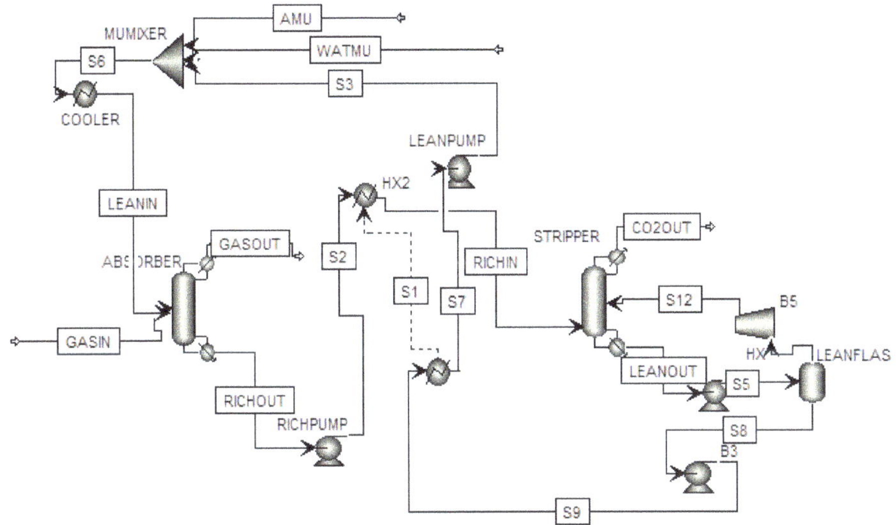

Figure 2. Lean amine flash modification [12].

Table 4. Results for lean amine flash configuration.

	Loading of Lean Out (mol/mol)	Reboiler Duty (kW)	Cross-Heat Exchanger Duty (kW)
Reference	0.265	4.77	17.46
AMP-lean amine flash	0.252	2.79	10.92

Table 4 shows that the loading out of the stripper is further reduced and the solvent working capacity is increased for the AMP-based process with amine flash. For the AMP lean amine flash, compressing at a higher pressure, flashed vapor is heated at a higher temperature of over 140 °C, cross heat exchanger duty is further reduced and a higher stripping efficiency is achieved leading to a savings of 20.92% in reboiler duty of AMP compared to the reference. As a result of the flash, which helps to obtain saturated steam before feeding it into the stripper, hot rich amine temperature is reduced from 140 °C to 113 °C, thus leading to an increase in consumption of energy, for the reference plant as compared to the AMP lean-amine flash.

3.3. Rich Split Configuration

The process in Figure 3 involves the splitting of the rich stream where the split entering the top of the stripping column stays unheated. It has the capability of pre-stripping the cold rich solvent entering at the top of the stripping column; this can be attained due to the vapor released from the rich solvent steam which moves up the stripping column. This helps to thermally regenerate less solvent, thereby reducing regeneration energies [11]. Hence, this configuration process is beneficial for lean loadings above 0.15 mol/mol and this reduces the energy required for regeneration [11].

In the configuration obtained in a study by Cousins et al. [11], 30% of the cold rich solvent was split to the top of the column with a condensate packing height of 1.12 m and a minimum reboiler duty of 97.8 kW was obtained which was about 10.3% higher savings than the reference case. The reason for this is that the reboiler duty achieves a combination of four functions: (i) providing sensible heat to the rich solvent to increase its temperature to the specified reboiler temperature, in which some heat is attained in the lean/rich heat exchanger [11]. (ii) evaporating water in the reboiler, which acts as the stripping agent, aiding the CO_2 removal from the solvent. Thus, steam released will replace steam generated in the reboiler, this is because the generation of steam within the column reduces the operating CO_2 partial pressure below that of the partial pressure in the column, enabling stripping to occur. (iii) Providing heat to reverse the absorption reaction in the absorber, which is in theory, equal to that released due to the exothermic reactions and (iv) providing heat to liberate dissolved CO_2 out of the solvent. Depending on the function which is most dominant under a given set of conditions, the reboiler duty will adjust accordingly to maintain the required stripping rate. The study also revealed that the possibility of obtaining a higher CO_2 flashing will allow the further release of CO_2 in the upper stages of the stripping column and give additional benefits [11].

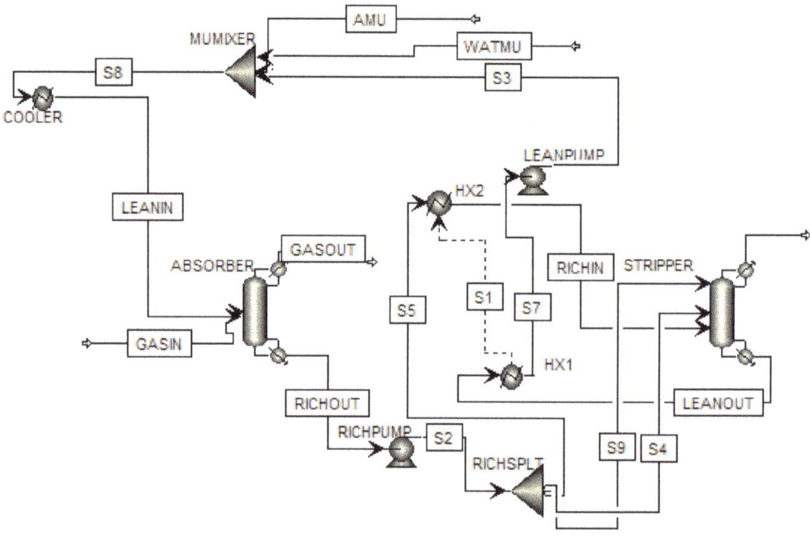

Figure 3. Rich split modification [11].

It should be noted however that the efficiency of the lean/rich heat exchanger will have a significant effect on the results of this process modification. The objective of the lean/rich exchanger is energy conservation. The energy available from the lean amine stream is transferred to the rich amine prior to introducing the rich amine to the stripper. This energy transfer results in a decreased energy requirement for the stripper as observed in the results presented in Table 5.

Table 5. Results for rich split configuration.

	Split-Fraction (%)	Reboiler Duty (kW)	Cross Heat Exchanger Flowrate (kg/s)	Cross Heat Exchanger Duty (kW)
Reference	0	4.77	0.10089	17.46
AMP-rich Split	30	4.7	0.09989	17.30

During the operation considered here, the solvent split fraction was found to have a significant effect on the temperature approach achieved through the lean/rich heat exchanger. As more of the cold rich solvent is split to the top of the column, the lower flowrate through the heat exchanger means that the hot rich solvent can be raised to higher temperatures. The vapor fraction in the hot rich solvent will increase, providing more steam for pre-stripping, which reduces the reboiler duty, as shown in Table 4. Thus, the reboiler duty slightly reduces from 4.77 kW to 4.73 kW for rich split configuration as compared to the reference plant.

3.4. Vapor Recompression

This process modification as shown in Figure 4 involves the extraction of steam from a part of the stripping column, which is recompressed and re-introduced into the regenerator. This operation turns mechanical energy into thermal energy to provide more stripping steam [11]. Hence, this configuration works by providing an additional source of stripping steam for the column, as this lowers the thermal input required by the stripper. The vapor is compressed to five times the operating pressure of the stripping column, before being separated with the condensate recycled back to the stripper [11]. Although this process configuration reduces the reboiler duty, it leads to a corresponding increase in the power requirement due to the addition of the compression stages [11]. Thus, in this study, one expansion stage is used to make comparison easier.

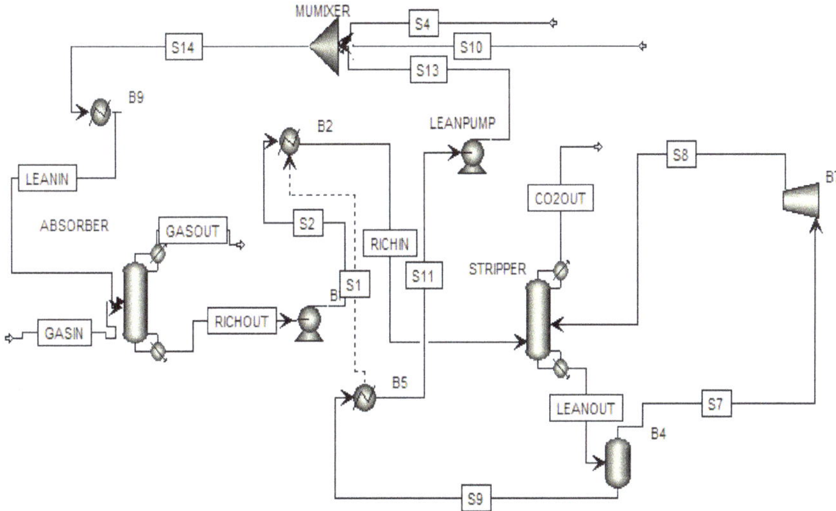

Figure 4. Vapor recompression modification [11].

As shown in Table 6, 59.5% savings in reboiler duty is obtained for the AMP-based plant compared to the reference case. For the vapor recompression configuration, additional stripping steam is generated, this is as a result of the higher pressure which leads to a higher temperature and enables the lean solvent flash pressure drop to increase. Also observed is the reduced heat exchanger duty obtained for the vapor recompression flowsheet, this is because lean solvent temperature increases which lead to a higher stripping efficiency.

Table 6. Results for vapor recompression.

	Reboiler Duty (kW)	Cross-Heat Exchanger Duty (kW)
Reference plant	4.77	17.46
AMP Vap-Recompression	1.93	5.69

3.5. Rich Split with Intercooling

As mentioned earlier, splitting the rich stream by 20–30% as recommended in the literature [11], can increase the absorption capacity which brings about the energy savings for the stripper design. In addition, the rich split modification requires minimal additional equipment. Furthermore, cooling at different stages in the absorber column has a significant effect in reducing the reboiler duty. Thus, with multiple alterations, reboiler duty can be further reduced [40]. Also, since the highest and lowest savings in the reboiler duty for intercooling and rich split configurations respectively, were obtained in this study, a combination of rich split and intercooling configurations will be necessary, to observe if any benefit can be obtained, and to enable further reduction in reboiler duty as shown in Figure 5.

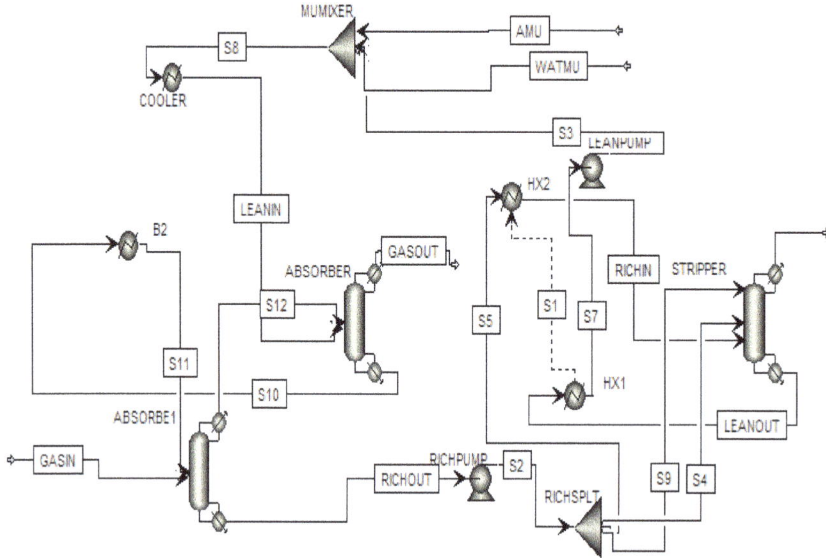

Figure 5. Rich split with intercooling configuration.

Simulation results are reported in Table 7 below, the multiple measures taken (rich split with intercooling) for the AMP-based process led to the energy savings of about 88.5% higher than the reference case. Thus, comparing the five different flowsheeting configurations, the rich split with intercooling configuration gave the best performance. This is in accordance with the literature [13].

Table 7. Results for rich split and intercooling

	Reboiler Duty (kW)	Cross-Heat Exchanger Flowrate (kg/s)	Cross-Heat Exchanger Duty (W)
Reference	4.77	0.10089	17.46
AMP Rich split	0.55	2.55×10^{-4}	5.30

4. Exergy Analysis for the Flowsheeting Configurations

Table 8 below shows the total exergy analysis and performance evaluation of the different flowsheeting configurations. Results show that the rich split with intercooling and vapor recompressions configurations gave the best and worst exergetic performances respectively as compared to the other configurations. Also, as compared to the reference plant which is presented in Table 2, two configurations (rich split with intercooling and intercooling alone and gave higher exergy efficiency than the reference case. While lean amine flash, rich split alone and vapor recompressions gave lower exergy efficiencies as compared to the reference plant. A reason for the low efficiencies could be that, the more efficient the configurations (based on the reboiler duty), the less outlet for exergy losses. Furthermore, Figure 6, shows the efficiency pecentage and the amount of exergy for the different configurations. As clearly seen, the rich split with interccoling configuration, gave the highest exergy efficiency amd the lowest exergy destruction.

Table 8. Summary of Exergy Analysis for the Different Configurations.

Configurations	E_{fuel} (Watts)	$E_{product}$ (Watts)	$E_{destruction}$ (Watts)	E_{eff} (%)
Intercooling	5261.33	3015.67	2245.66	57.32
Rich split	10,515.44	5851.37	4664.07	55.65
Rich slit + intercooling	780.11	573.87	206.23	73.56
Lean amine flash	6360.78	3189.95	3170.83	50.15
Vapor recompression	14,767.7	4197.07	10,570.65	28.42

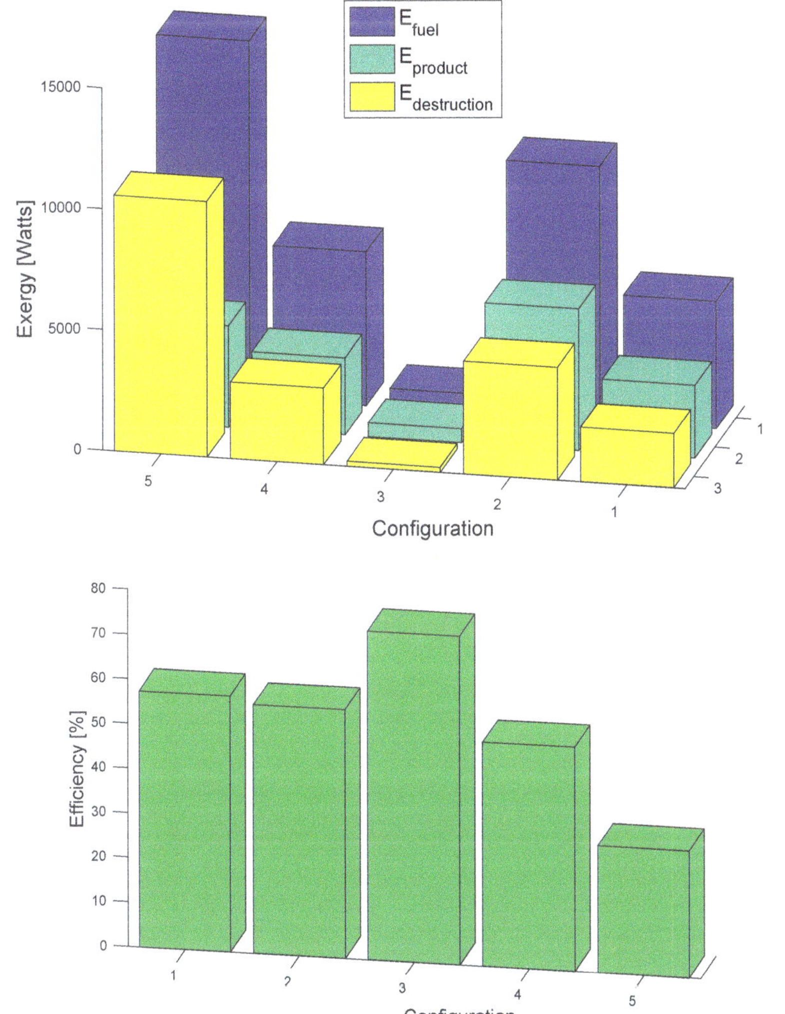

Figure 6. Summary Results of Exergy Analysis for the Different Configurations. For the horizontal "Configuration" axis, 1. Intercooling 2. Rich Split 3. Rich Split with Intercooling 4. Lean Amine Flash 5. Vapor Recompression.

5. Conclusions

In this study, the exergy analysis of the AMP PCC process and its flowsheeting configurations have been evaluated. The operating parameters for the rate-based AMP model present in Aspen Plus® software were used to describe the PCC process. The conventional exergy analysis performed provides an evaluation of energy consumption the CO_2 capture plant from the thermodynamic point of view, and also evaluates the reduction of the exergy destruction. The pump and stripper subsystems of the AMP-based capture plant had the highest exergy destruction, and the cross-heat exchanger subsystem gave the lowest exergy destruction performance.

Several configurations were proposed in the literature to reduce energy requirements in the amine-based CO_2 capture plant, these configurations at atmospheric pressure have been simulated in Aspen Plus® software. Flowsheeting configurations considered in this study include intercooling, lean-amine flash, rich split, vapor recompression and the rich split with intercooling configurations. Results show that the combination configuration (rich split with intercooling) had the highest savings (88.5%) in reboiler duty as compared to the reference AMP-based plant, and the other flowsheeting configurations. Furthermore, exergy analysis performed showed that the rich split with intercooling configuration had the highest exergy efficiency of 74%, followed by the intercooling configuration with 57% exergy efficiency, and that of the reference AMP plant was obtained to be 56%. The other configurations considered in the study had exergy efficiencies lower than that of the reference plant. This study has shown that some of the flowsheeting configurations can reduce the heat required for regeneration, and others can both reduce reboiler duty and at the same time increase the exergy efficiency. Thus, the flowsheeting configurations have significant improvements in the plant performance and may lead to cost reduction for the amine-based CO_2 capture technology. Although the additional equipment for each configuration may incur extra cost, economic analysis is therefore required to ascertain if any cost benefits can be obtained with flowsheeting configurations for the AMP-based PCC process.

Author Contributions: E.O.—methodoogy and writing original draft; A.M.A.—supervision, review and editing; S.G.N.—writing, review and editing; O.O.—review and editing; V.E.—review and editing.

Funding: This research received no external funding.

Conflicts of Interest: The authors declare no conflict of interest.

References

1. Chalmers, H.; Gibbins, J. Initial evaluation of the impact of post-combustion capture of carbon dioxide on supercritical pulverised coal power plant part load performance. *Fuel* **2007**, *86*, 2109–2123. [CrossRef]
2. Rubin, E.S.; Chen, C.; Rao, A.B. Cost and performance of fossil fuel power plants with CO_2 capture and storage. *Energy Policy* **2007**, *35*, 4444–4454. [CrossRef]
3. Davison, J. Performance and costs of power plants with capture and storage of CO_2. *Energy* **2007**, *32*, 1163–1176. [CrossRef]
4. Jassim, M.S.; Rochelle, G.T. Innovative absorber/stripper configurations for CO_2 capture by aqueous monoethanolamine. *Ind. Eng. Chem. Res.* **2006**, *45*, 2465–2472. [CrossRef]
5. Reddy, S.; Scherffius, J.; Freguia, S.; Roberts, C. Fluor's Econamine FG Plus SM Technology. In Proceedings of the Second Annual Conference on Carbon Sequestration, Alexandria, VA, USA, 6 May 2003.
6. Leites, I.L.; Sama, D.A.; Lior, N. The theory and practice of energy saving in the chemical industry: Some methods for reducing thermodynamic irreversibility in chemical technology processes. *Energy* **2003**, *28*, 55–97. [CrossRef]
7. Karimi, M.; Hillestad, M.; Svendsen, H.F. Investigation of intercooling effect in CO_2 capture energy consumption. *Energy Procedia* **2011**, *4*, 1601–1607. [CrossRef]
8. Aroonwilas, A.; Veawab, A. Integration of CO_2 capture unit using single-and blended-amines into supercritical coal-fired power plants: Implications for emission and energy management. *Int. J. Greenh. Gas Control.* **2007**, *1*, 143–150. [CrossRef]

9. Reddy, S.; Scherffius, J.; Gilmartin, J.; Freguia, S. Improved split flow apparatus. Patent No. WO 2004005818 A2, 2004.

10. Eisenberg, B.; Johnson, R.R. Amine Regeneration Process. U.S. Patent 4,152,217A1, 1 May 1979.

11. Cousins, A.; Wardhaugh, L.T.; Feron, P.H. Preliminary analysis of process flow sheet modifications for energy efficient CO_2 capture from flue gases using chemical absorption. *Chem. Eng. Res. Des.* **2011**, *89*, 1237–1251. [CrossRef]

12. Ahn, H.; Luberti, M.; Liu, Z.; Brandani, S. Process configuration studies of the amine capture process for coal-fired power plants. *Int. J. Greenh. Gas Control.* **2013**, *16*, 29–40. [CrossRef]

13. Sharma, M.; Quadii, A.; Khalilpouri, R. Modeling and analysis of process configurations for solvent-based post-combustion carbon capture. *Asia-Pac. J. Chem. Eng.* **2015**, *10*, 764–780. [CrossRef]

14. Lars, E.O.; Terje, B.; Christian, B.; Sven, K.B.; Marius, F. Optimization of configurations for amine based CO_2 absorption using Aspen HYSYS. 7th Trondheim CCS Conference, TCCS-7. *Energy Procedia* **2014**, *51*, 224–233.

15. Liang, Z.; Gao, H.; Rongwong, W.; Na, Y. Comparative studies of stripper overhead vapor integration-based configurations for post-combustion CO_2 capture. *Int. J. Greenh. Gas Control.* **2015**, *34*, 75–84. [CrossRef]

16. Jaeheum, J.; Yeong, S.J.; Ung, L.; Youngsub, L.; Chonghun, H. New Configuration of the CO_2 Capture Process Using Aqueous Monoethanolamine for Coal-Fired Power Plants. *Ind. Eng. Chem. Res.* **2015**, *54*, 3865–3878.

17. Gabrielsen, F.H. Exergy analysis of alkanolamine-based CO_2 removal unit with Aspen Plus. *Energy* **2004**, *29*, 1241–1248. [CrossRef]

18. Lara, Y.; Martinez, A.; Lisbona, P.; Bolea, I.; Gonzalez, A.; Romeo, L.M. Using the second law of thermodynamic to improve CO_2 capture systems. *Energy Procedia* **2011**, *4*, 1043–1050. [CrossRef]

19. Olaleye, A.K.; Wang, M.; Kelsall, G. Steady state simulation and exergy analysis of supercritical coal-fired power plant with CO_2 capture. *Fuel* **2015**, *151*, 57–72. [CrossRef]

20. Valenti, G.; Bonalumi, D.; Macchi, E. Energy and exergy analyses for the carbon capture with the Chilled Ammonia Process (CAP). *Energy Procedia* **2009**, *1*, 1059–1066. [CrossRef]

21. Alatiqi, I.; Sabri, M.F.; Bouhamra, W.; Alper, E. Steady-state rate based modelling for CO_2 amine absorption—Desorption systems. *Gas Sep. Purif.* **1994**, *8*, 3–11. [CrossRef]

22. Aboudheir, A.; Tontiwachwuthikul, P.; Idem, R. Rigorous model for predicting the behavior of CO_2 absorption into AMP in packed-bed absorption columns. *Ind. Eng. Chem. Res.* **2006**, *45*, 2553–2557. [CrossRef]

23. Gabrielsen, J.; Svendsen, H.F.; Michelsen, M.L.; Stenby, E.H.; Kontogeorgis, G.M. Experimental validation of a rate-based model for CO_2 capture using an AMP solution. *Chem. Eng. Sci.* **2007**, *62*, 2397–2413. [CrossRef]

24. Gabrielsen, J.; Michelsen, M.L.; Stenby, E.H.; Kontogeorgis, G.M. Modeling of CO_2 absorber using and AMP solution. *AIChE J.* **2006**, *52*, 3443–3451. [CrossRef]

25. Afkhamipour, M.; Mofarahi, M. Comparison of rate-based and equilibrium-stage models of a packed column for post-combustion CO_2 capture using 2-amino-2-methyl-1-propanol (AMP) solution. *Int. J. Greenh. Gas Control* **2013**, *15*, 186–199. [CrossRef]

26. Dash, S.K.; Samanta, A.N.; Bandyopadhyay, S.S. Simulation and parametric study of post-combustion CO_2 capture process using (AMP + PZ) blended solvent. *Int. J. Greenh. Gas Control* **2014**, *21*, 130–139. [CrossRef]

27. Kvamsdal, H.M.; Chikukwa, A.; Hillestad, M.; Zakeri, A.; Einbu, A. A comparison of different parameter correlation models and the validation of an MEA-based absorber model. *Energy Procedia* **2011**, *4*, 1526–1533. [CrossRef]

28. Osagie, E.; Biliyok, C.; Di Lorenzo, G.; Hanak, D.; Manovic, V. Techno-economic evaluation of the 2-amino-2-methyl-1-propanol (AMP) process for CO_2 capture from natural gas combined cycle power plant. *Int. J. Greenh. Gas Control* **2018**, *70*, 45–56. [CrossRef]

29. MacDowell, N.; Florin, N.; Buchard, A.; Hallett, J.; Galindo, A.; Jackson, G. An overview of CO_2 capture technologies. *Energy Environ. Sci.* **2010**, *3*, 1645. [CrossRef]

30. Mahamud, R.; Khan, M.M.K.; Rasul, M.G.; Leinster, M.G. Exergy Analysis and Efficiency Improvement of a Coal Fired Thermal Power Plant in Queensland. In *Thermal Power Plants—Advanced Applications*; IntechOpen Limited: London, UK, 2013.

31. Aspen Technology. *Rate-Based Model of the CO_2 Capture Process by AMP Using Aspen Plus*; Aspen Technology, Inc.: Cambridge, MA, USA, 2008.

32. Tsatsaronis, G.; Cziesla, F. *Strengths-and-Limitations-of-Exergy-Analysis*; Institute of Energy Engineering: Berlin, Germany, 1999; Volume 69, pp. 93–100.

33. Sherman, B.J. Thermodynamic and Mass Transfer Modelling of Aqueous Hindered Amines for Carbon Dioxide Capture. Ph.D. Dissertation, University of Texas, Austin, TX, USA, 2016.
34. Herrmann, J. Conversion of Experimental Data for Process Modelling of Novel Amine CO_2 Capture Solvents. Master's Thesis, Norwegian University of Science and Technology, Trondheim, Norway, 2014.
35. Yang, Y. Comprehensive exergy-based evaluation and parametric study of a coal-fired ultra-supercritical power plant. *Appl. Energy* **2013**, *112*, 1087–1099. [CrossRef]
36. Kvamsdal, H.M.; Rochelle, G.T. Effects of the temperature bulge in CO_2 absorption from flue gas by aqueous monoethanolamine. *Ind. Eng. Chem. Res.* **2008**, *47*, 867–875. [CrossRef]
37. Anthony, J.L. Carbon dioxide: Generation and capture. In *Energy and Environmental Impacts Related to Sustainability Seminar*; Kansas State University: Manhattan, KS, USA, 2006.
38. Woodhouse, S.; Rushfeldt, P. Improved Absorbent Regeneration. Patent No. WO2008/063079 A2, 29 May 2008.
39. Reddy, S.; Gilmartin, J.; Francuz, V. Integrated Compressor/Stripper Configurations and Methods. U.S. Patent 2009/0205946 A1, 20 August 2009.
40. Le Moullec, Y.; Kanniche, M. Screening of flowsheet modifications for an efficient monoethanolamine (MEA) based post-combustion CO_2 capture. *Int. J. Greenh. Gas Control.* **2011**, *5*, 727–740. [CrossRef]

Article

Reaction Kinetics of Carbon Dioxide in Aqueous Blends of N-Methyldiethanolamine and L-Arginine Using the Stopped-Flow Technique

Nafis Mahmud [1], Abdelbaki Benamor [1,*], Mustafa Nasser [1], Muftah H. El-Naas [1] and Paitoon Tontiwachwuthikul [2]

1 Gas Processing Centre, College of Engineering, Qatar University, Doha 2713, Qatar;
   n.mahmud@qu.edu.qa (N.M.); m.nasser@qu.edu.qa (M.N.); muftah@qu.edu.qa (M.H.E.-N.)
2 Clean Energy Technology Research Institute (CETRI), Faculty of Engineering, University of Regina,
   3737 Wascana Parkway, Regina, SK S4S 0A2, Canada; paitoon@uregina.ca
* Correspondence: benamor.abdelbaki@qu.edu.qa; Tel.: +974-4403-4381; Fax: +974-4403-4371

Received: 7 January 2019; Accepted: 29 January 2019; Published: 6 February 2019

Abstract: Reduction of carbon dioxide emission from natural and industrial flue gases is paramount to help mitigate its effect on global warming. Efforts are continuously deployed worldwide to develop efficient technologies for CO_2 capture. The use of environment friendly amino acids as rate promoters in the present amine systems has attracted the attention of many researchers recently. In this work, the reaction kinetics of carbon dioxide with blends of N-methyldiethanolamine and L-Arginine was investigated using stopped flow technique. The experiments were performed over a temperature range of 293 to 313 K and solution concentration up to one molar of different amino acid/amine ratios. The overall reaction rate constant (k_{ov}) was found to increase with increasing temperature and amine concentration as well as with increased proportion of L-Arginine concentration in the mixture. The experimental data were fitted to the zwitterion and termolecular mechanisms using a nonlinear regression technique with an average absolute deviation (AAD) of 7.6% and 8.0%, respectively. A comparative study of the promoting effect of L-Arginine with that of the effect of Glycine and DEA in MDEA blends showed that MDEA-Arginine blend exhibits faster reaction rate with CO_2 with respect to MDEA-DEA blend, while the case was converse when compared to the MDEA-Glycine blend.

Keywords: Reaction; kinetics; carbon dioxide; N-methyldiethanolamine; L-Arginine; stopped flow technique

---

1. Introduction

The rapid growth of world economies associated with increased fossil fuel consumption for energy needs resulted in the generation of large amounts of greenhouse gases accumulated in the atmosphere. Carbon dioxide (CO_2) is a major contributor to greenhouse gases accountable to the observed climate change and associated environmental problems. Reducing CO_2 concentration in the atmosphere to an acceptable level is necessary for future generation's well-being. Different options are available to capture CO_2; however, amine based reactive solvents is one of the most mature and successful technology used in the industry, especially from large point sources, such as natural gas treatment units and power generation plants [1–6]. Large variants of amine based solvents are available in the market, many of which contain proprietary additives to enhance their absorption performances. Amine based solvents are known by their high absorption capacities and their ability to selectively absorb CO_2/H_2S from natural and flue gases. Conventional amine solvents, such as primary monoethanolamine (MEA), 2-amino-2-methyl-1-propanol (AMP), secondary diethanolamine (DEA), tertiary amine N-methyldiethanolamine (MDEA) and polyamines (such as piperazine (PZ), 2-(2-aminoethylamino) ethanol (AEEA) are efficient for capturing CO_2 from various industrial processes and are still the choice in the industry because of the well-known absorption-regeneration process. However, several drawbacks like low absorption rate, periodic solvent make up to compensate for solvent losses, high regeneration energy requirement and severe equipment corrosion are still associated with their use [7–11].

Liquid tertiary amines, such as MDEA have higher theoretical sorption capacity with a ratio of 1:1 mol [12] but the reaction rate is much slower. To overcome this drawback, blended amines have been suggested [7]. To take advantage of their high loading capacity, low degradation rate and low energy for regeneration, tertiary amines are mixed with faster reacting primary/secondary amines or piperazine to develop new solvents with better CO_2 capture performance such as high absorption and cyclic capacity, fast reaction kinetics, low corrosion, degradation and less heat duty requirement [13,14].

Amino acids, usually called alkaline salts of amines, have recently drawn attention to CO_2 capture due to their exceptional properties [15]. The structure of amino acids consists of two important functional groups, namely amine (-NH$_2$) and carboxylic acid (-COOH) or a sulfonic acid group [16]. Their salt nature makes their volatilities negligible which results in low solvent losses [17]. Their low environmental impact and high biodegradability [18] make them more environmentally friendly [19]. In addition, amino acids have high resistance to oxidative degradation making them a right choice for CO_2 capture from flue gases containing large amounts of oxygen [20]. However, at high concentration or at high CO_2 loading, they tend to precipitate resulting in lower mass transfer [21], which is a major drawback.

Nevertheless, several studies has reported on CO_2 capture using amino acids [22–25]. Siemens developed an amino acid based process and claim it has a reduced energy consumption of about 73% compared to the conventional MEA process [26]. Aqueous solutions of sodium glycinate were proposed for CO_2 absorption [27,28]. Shen et al. [29,30] used potassium salts of lysine and Histidine for CO_2 absorption and concluded that histidine reactivity towards CO_2 was comparable to that of MEA. Portugal et al. [31] used potassium glycinate and potassium threonate for CO_2 absorption purposes [32]. Huang et al. [33] and Wei et al. [34] determined the reaction rate constant of taurate carbamate formation during the absorption of CO_2 into CO_2-free and CO_2-loaded taurate solutions using a wetted-wall column at a temperature range of 293–353 K. The properties necessary for mass transfer evaluation, such us density, viscosity, CO_2 diffusivity, N_2O solubility were reported for several amino acids under different conditions [35–40].

In a previous study on reaction kinetics of amino acids with CO_2, it was observed, that L-Arginine, an amino acid, showed faster reaction rate compared to that of Glycine and Sarcosine when used as a single solvent [41]. Another study on the promoting effect of Glycine in MDEA blends showed that the reaction rate of MDEA with CO_2 could be significantly increased by the presence of an amino acid promoter [42]. However, the promoting effect of L-Arginine in MDEA blends remain unknown. In this work, the reaction kinetics of CO_2 with aqueous mixtures of MDEA and L-Arginine were determined using the stopped flow technique. The temperature was varied from 298 to 313 K and the amine total concentration was varied from 0.25 to 1 mol of different proportions of Arg/MDEA. Our findings provide a new insight to the use of Arg as rate promoter for CO_2 capture blended tertiary amines. The molecular structure of MDEA and L-Arginine are shown in the Figure 1.

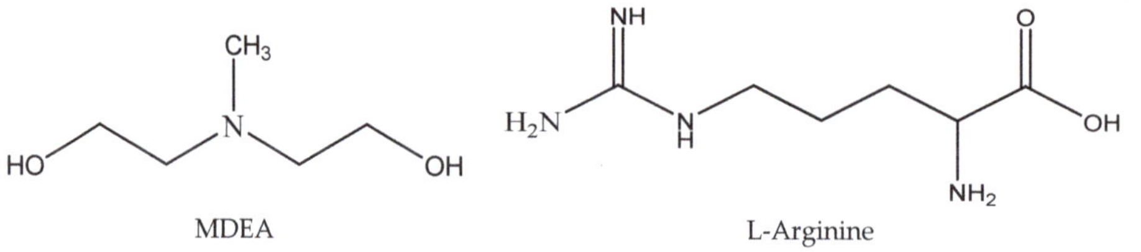

Figure 1. Molecular Structure of MDEA (N-methyldiethanolamine) and L-Arginine.

2. Reaction Models

2.1. Reaction of CO$_2$ with MDEA

It is widely accepted that the reactions of CO$_2$ with primary amines results in formation of a carbamate and a bicarbonate products. However, in case of tertiary amines, only bicarbonates are formed during the reaction with CO$_2$. Therefore, MDEA being a tertiary amine will also not form any carbamates and their reaction of CO$_2$ in aqueous solution is as follows [21]:

$$CO_2 - H_3CN(C_2H_4OH)_2 + H_2O \xrightarrow{k_{MDEA}} H_3CNH^+(C_2H_4OH)_2 + HCO_3^- \tag{1}$$

Its pseudo-first-order reaction rate is:

$$r_{CO_2-MDEA} = -k_{MDEA}[CO_2][MDEA] \tag{2}$$

In addition to this reaction, the formation of bicarbonates in aqueous systems may be considered:

$$CO_2 + OH^- \xrightarrow{k_{OH}} HCO_3^- \tag{3}$$

Its rate of reaction was given as [43]:

$$r_{CO_2-OH} = -k_{OH}[CO_2][OH^-] \tag{4}$$

2.2. Reaction of CO$_2$ with Amino Acid

2.2.1. Zwitterion Mechanism

This mechanism was first coined by Caplow to comprehend the reactions of primary or secondary amines with CO$_2$ [44]. Amino acid [NHR$_1$R$_2$COO$^-$] has a molecular structure similar to that of primary or secondary amines. Its reaction pathway with CO$_2$ is considered to be similar to that of CO$_2$-amines and usually yields the formation of carbamate ion through two successive steps: (i) the formation of an intermediate zwitterion according to Reaction 5, (ii) proton removal by any base present in the solution according to Reaction 6 [22].

$$CO_2 + NHR_1R_2COO^- \underset{k_{-1}}{\overset{k_1}{\rightleftharpoons}} {}^-OOCN^+R_1R_1COO^- \tag{5}$$

$$^-OOCNH^+R_1R_2COO^- + B_i \xrightarrow{k_{b,i}} {}^-OOCNR_1R_2COO^- + BH^+ \tag{6}$$

The corresponding reaction rate is given as [20]:

$$r_{CO2-Arg} = -k_2[Arg][CO_2] \left/ \left(1 + \left(k_{-1} \left/ \left(\sum_i k_{b_i}[B_i]\right)\right.\right)\right)\right. \tag{7}$$

where the term 'k_{b_i}' represents the deprotonation rate constant of the zwitterion by any base. The reaction rate constant can be written as:

$$k_{Arg} = -k_2[Arg] \left/ \left(1 + \left(k_{-1} \left/ \left(\sum_i k_{b_i}[B_i]\right)\right.\right)\right)\right. \tag{8}$$

The analysis of this model reveals two asymptotic cases, namely, $1 \gg \frac{k_{-1}}{\sum_i k_{b_i}[B_i]}$ and $1 \ll \frac{k_{-1}}{\sum_i k_{b_i}[B_i]}$. When the formation of the zwitterion carbamate following Reaction 5 is the rate-limiting step. The first case prevails, thus, Equation (7) reduces to:

$$k_{Arg} = -k_2[Arg] \tag{9}$$

In the opposite case, when $1 << \frac{k_{-1}}{\sum_i k_{b_i}[B_i]}$, the proton removal from the zwitterion intermediate according to Reaction 6 is the rate limiting step; Equation (7) then becomes:

$$k_{Arg} = -k_2[Arg]\left(\sum_i k_{b_i}[B_i]\right)\Big/k_{-1} \tag{10}$$

In the latter case, the reaction order dependency on the amino acid concentration varies from one to two. This phenomenon is commonly observed in CO_2 reaction with primary and secondary amines [26,27] and was proved to prevail in other salts of amino acids [28].

2.2.2. Termolecular Mechanism

Crooks and Donnellan [45] proposed an alternative single-step termolecular mechanism, which involves only one-step in the reaction process as shown in the Figure 2 below.

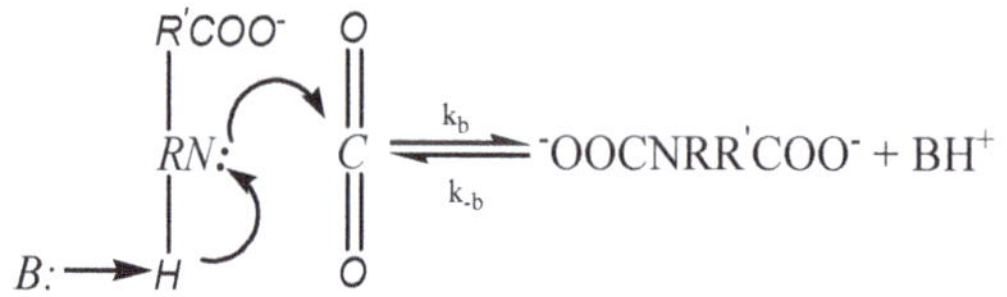

Figure 2. Termolecular Mechanism.

This mechanism was further investigated by Silva and Svendson [46], based on which they suggested that the reaction progresses through the bonding of the CO_2 molecule with the amine, which is stabilized by solvent molecules with hydrogen bonds. This in turn results in the formation of loosely bounded complex. They also added that the carbamate will be formed only when the amine molecule is in the vicinity of zwitterion. An analysis of the rate expression of the termolecular mechanism shows that the reaction of CO_2 with amine is second order with respect to amine. Therefore, in this case, Equation (7) becomes:

$$r_{CO_2} = k_{ov}[CO_2] = [CO_2][Arg]\left\{\sum k_{b_i}[B_i]\right\} \tag{11}$$

Regardless of the mechanism employed, a carbamate and a protonated base are the generally accepted products of the CO_2 reaction with amine.

2.3. Reaction of CO_2 with Mixtures of MDEA and L-Arginine

For blends of MDEA and L-Arginine, the overall reaction rate with CO_2 is considered as the sum of reaction rates of CO_2-MDEA and CO_2-L-Arginine, hence:

$$r_{CO_2} = r_{CO_2-Arg} + r_{CO_2-MDEA} + r_{CO_2-OH} \tag{12}$$

which can be written as:

$$r_{CO_2} = -\left(k_2[Arg]\Big/\left(1+\left(k_{-1}\Big/\sum_i k_{b_i}[B_i]\right)\right)+k_{MDEA}[MDEA]\right)[CO_2] \tag{13}$$

or

$$r_{CO_2} = (k_{Arg}+k_{MDEA}[MDEA])[CO_2] = k_{ov}[CO_2] \tag{14}$$

In case of aqueous solution of L-Arginine, water molecules, 'H$_2$O'; hydroxyl ions, 'OH-' and deprotonated amino acid, 'L-Arginine', act as bases. Therefore, if the reaction is expected to proceed via the zwitterion mechanism, then the based on Equation (8), the term 'k$_{Arg}$' can be defined as follows:

$$k_{Arg} = -\frac{k_2[Arg]}{1 + \left(k_{-1}/\left(k'_{Arg}[Arg] + k'_{OH^-}[OH^-] + k'_{MDEA}[MDEA] + k'_{H_2O}[H_2O]\right)\right)} \tag{15}$$

which can be written as:

$$k_{Arg} = -\frac{[Arg]}{\frac{1}{k_2} + \left(1/\left(\frac{k_2k'_{Arg}}{k_{-1}}[Arg] + \frac{k_2k'_{OH^-}}{k_{-1}}[OH^-] + \frac{k_2k'_{MDEA}}{k_{-1}}[MDEA] + \frac{k_2k'_{H_2O}}{k_{-1}}[H_2O]\right)\right)} \tag{16}$$

By defining new constants, k_a, k_{hyd}, k_b and k_w as $k_a = \frac{k_2 k'_{Arg}}{k_{-1}}$, $k_{hyd} = \frac{k_2 k'_{OH}}{k_{-1}}$, $k_b = \frac{k_2 k'_{MDEA}}{k_{-1}}$ and $k_w = \frac{k_2 k'_{H_2O}}{k_{-1}}$. Equation (13) becomes:

$$k_{Arg} = \frac{[Arg]}{(1/k_2) + \left(1/\left(k_a[Arg] + k_{hyd}[OH^-] + k_b[MDEA] + k_w[H_2O]\right)\right)} \tag{17}$$

However, if the reaction is expected to proceed via the termolecular mechanism, the term, 'k$_{Arg}$' can be redefined as follows:

$$k_{Arg} = [Arg]\left\{k_a[Arg] + k_w[H_2O] + k_{hyd}[OH^-]\right\} \tag{18}$$

3. Materials and Methods

3.1. Materials

Reagents used in this work were analytical grade N-methyldiethanolamine (MDEA) with a mass purity of 99% obtained from Sigma-Aldrich and L-Arginine with purity of 99% purchased from Fluka (St. Louis, MS, USA). All chemicals were used as received without further purification. CO$_2$ solutions were prepared by bubbling analytical grade CO$_2$ (99.99%) for at least half an hour in deionised water. Deionised water was used as solvent throughout the experiments.

3.2. Methods

Pseudo first-order kinetics of CO$_2$ reaction with different aqueous mixtures of L-Arginine in MDEA were measured using stopped-flow apparatus (Hi-Tech Scientific Ltd., and Model SF-61DX2, Bradford-on-Avon, Wiltshire, UK) with a dead time of 1 ms. The apparatus essentially consists of working syringes immersed in water bath, where the reacting solutions are loaded. It also contains a pneumatically controlled drive plate to load the reacting solutions into a mixer and the conductivity of the mixed solution is measured within the conductivity cell. Finally, the mixed solution is collected in a stopping syringe. A schematic diagram of the Stopped-Flow Apparatus has been presented in Figure 3 below.

An external water bath (Alpha RA8, Lauda, Delran, NJ, USA) was used to control the temperature of the sample flow circuits within ±0.10 K. Depending on the applied temperature, the run time of the experiment was varied from 0.05 to 0.2 s. Freshly saturated solutions of CO$_2$ were prepared by bubbling CO$_2$ in deionized water. Concentration of CO$_2$ in the bubbled solution was measured with gas chromatography (GC-6890 from Agilent, Santa Clara, CA, USA) following Shell method®–SMS 2239-04. Fresh water used to dilute the solution in order to maintain the concentration ratio of the amine/amino acid solution 20 times higher than that of the CO$_2$ solution. This was done to ensure that the reaction conditions with respect to [CO$_2$] fall within the pseudo first order regime [47]. The amine/amino acid solutions were also prepared using deionized water. For each run, the CO_2 and amine/amino acid solutions were loaded in two separate syringes. Equal volumes of aqueous solutions of amine/amino acid and CO_2 were mixed in the stopped-flow apparatus. The reaction was monitored by measuring the conductivity change as function of time. The change in the conductivity, Y, with respect to time as described by Knipe et al. [48] was fitted to an exponential equation resembling a first-order kinetics equation:

$$Y = -A . \exp(-k_{ov} . t) + Y_\infty \tag{19}$$

where, 'k_{ov}' is the pseudo first-order reaction rate constant. The averages of three experimental runs were considered to obtain a reproducible and consistent k_{ov} value for the whole range of the tested concentrations and temperatures. The reproducibility error of k_{ov} was found to be less than 3% in all experiments. The experimental results were obtained in the provided 'Kinectic Studio' Software (Bradford-on-Avon, Wiltshire, UK). A screenshot of typical conductivity profile is presented in Figure 4.

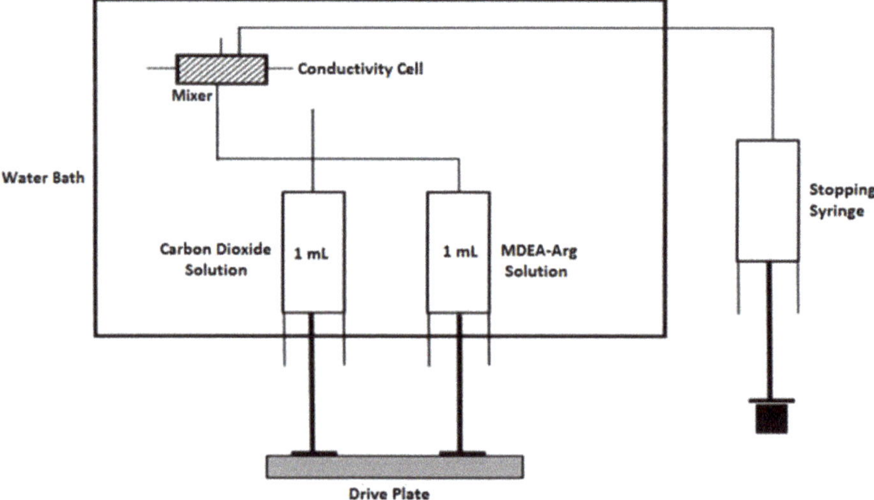

Figure 3. A Schematic Diagram of the Stopped-Flow Apparatus.

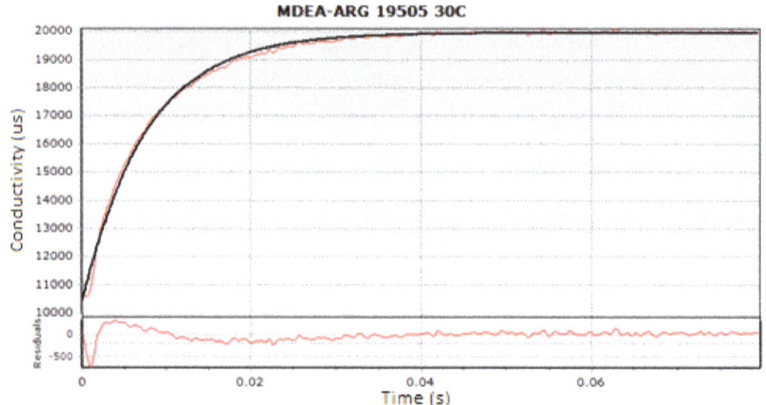

Figure 4. Typical Experimental run for MDEA-Arg at 303 K.

4. Results and Discussion

4.1. Reaction of CO_2 with MDEA and L-Arginine

The obtained pseudo first order rate rate constants, 'k_{ov}', were plotted against temperature for one molar total concentration (see Figure 5). The overall rate constants (k_{ov}) increased with increasing solution temperature as well as with increased [Arg] proportion in the total mixture. Similarly, the plot of the overall rate constants against different Arg/MDEA ratios for a total concentration of 1 mol is shown in Figure 6.

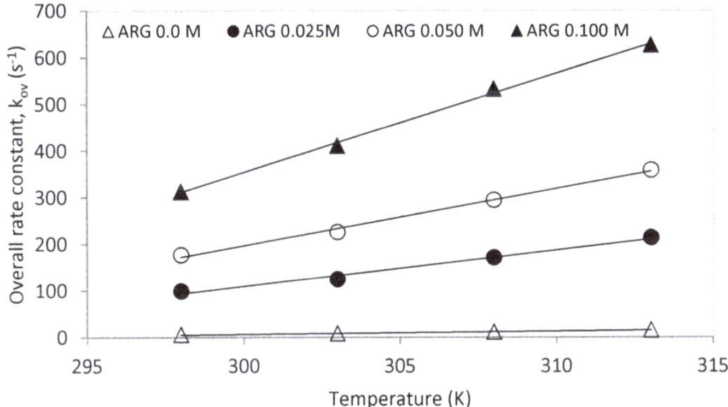

Figure 5. Rate constant, 'k_{ov}', vs. temperature for total 1 M solution.

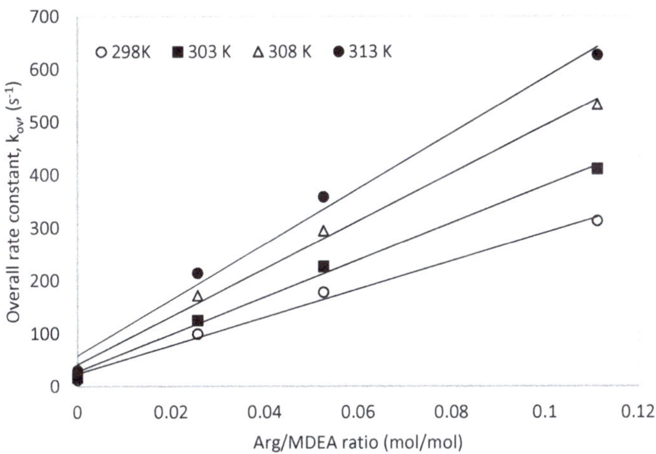

Figure 6. Overall rate constant, 'k_{ov}', vs. different Arg/MDEA ratios for a total 1 M solution.

Upon applying the power law kinetics to plot the overall rate constants against the concentrations of L-Arginine, an average exponent of 0.98 was obtained, which affirms that the pseudo first order regime prevails. Therefore, within the range of the experimental conditions, the reaction can be analysed via the zwitterion mechanism [49]. Additionally, the possibility of using the termolecular mechanism to interpret the obtained data was also verified by plotting k_{ov}/[ARG] against [ARG]. The plots show a satisfactory linear relationship indicating that the termolecular mechanism can also be applied to interpret the obtained experimental kinetics data [49,50]. A typical plot is shown in Figure 7. Hence, obtained experimental kinetics data were analysed using both the zwitterion and termolecular mechanisms.

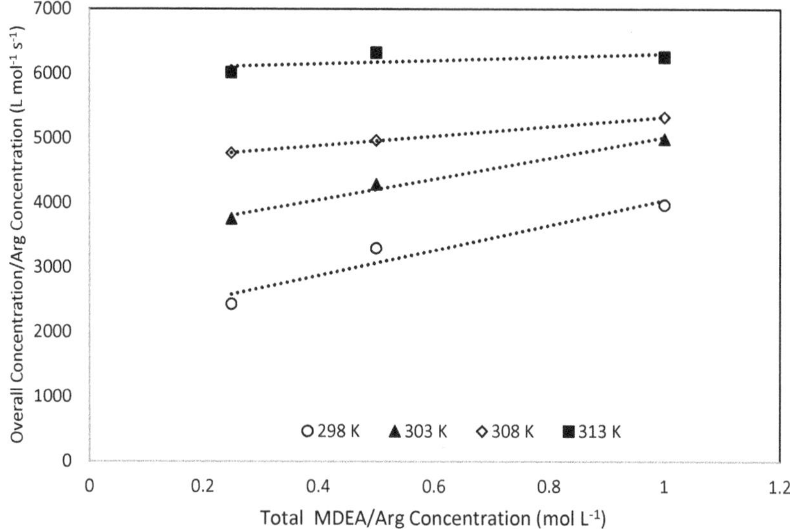

Figure 7. Termolecular mechanism applicability Test for 0.025 M L-Arginine concentration.

4.2. Zwitterion Mechanism

The experimental data of the rate (k_{ov}) of the CO_2 reaction with methyldiethanolamine (MDEA) and L-Arginine (Arg) were obtained by fitting the conductivity-time curves to Equation (19) Zwitterion mechanism was used to interpret the obtained experimental data and the obtained overall rates along with apparent and predicted k_{Arg} rates are presented in Table 1.

Table 1. Rate constants at different temperatures and (MDEA+Arg) concentrations.

ARG	MDEA	OH $\times 10^3$	H_2O	k_{ov}	$k_{Arg\text{-}exp}$	$k_{Arg\text{-}pred}$	AAD%
mol L^{-1}	mol L^{-1}	mol L^{-1}	mol L^{-1}	s^{-1}	s^{-1}	s^{-1}	
				298 K			
0.025	0.975	2.07	49.16	99.17	93.30	80.50	
0.050	0.950	2.07	49.14	177.47	171.70	166.90	
0.100	0.900	2.07	49.08	312.35	306.90	356.00	
0.025	0.475	1.46	52.35	82.40	79.50	70.60	7.60
0.050	0.450	1.46	52.32	152.55	149.80	148.10	
0.100	0.400	1.46	52.26	320.43	318.00	321.80	
0.025	0.225	1.03	53.94	60.75	59.40	65.00	
0.050	0.200	1.03	53.91	125.47	124.30	137.50	
0.075	0.175	1.03	53.88	214.23	213.20	216.90	
				303 K			
0.025	0.975	2.21	49.13	124.65	116.50	104.30	
0.050	0.950	2.21	49.11	226.15	218.20	216.20	
0.100	0.900	2.21	49.05	410.33	402.80	460.40	
0.025	0.475	1.56	52.33	107.15	103.20	91.00	6.30
0.050	0.450	1.56	52.31	195.64	191.90	190.80	
0.100	0.400	1.56	52.25	423.84	420.50	414.30	
0.025	0.225	1.10	53.93	78.32	76.40	83.50	
0.050	0.200	1.10	53.90	165.36	163.70	176.50	
0.075	0.175	1.10	53.88	281.52	280.00	278.40	

Table 1. *Cont.*

ARG	MDEA	OH $\times 10^3$	H$_2$O	k$_{ov}$	k$_{Arg\text{-}exp}$	k$_{Arg\text{-}pred}$	AAD%
mol L^{-1}	mol L^{-1}	mol L^{-1}	mol L^{-1}	s^{-1}	s^{-1}	s^{-1}	
			308 K				
0.025	0.975	2.35	49.11	171.38	160.20	136.10	
0.050	0.950	2.35	49.08	294.32	283.40	281.40	
0.100	0.900	2.35	49.03	532.24	521.90	596.90	
0.025	0.475	1.66	52.32	149.90	144.40	117.60	8.20
0.050	0.450	1.66	52.29	234.18	229.00	246.00	
0.100	0.400	1.66	52.24	496.55	491.90	532.50	
0.025	0.225	1.17	53.92	101.41	98.80	106.90	
0.050	0.200	1.17	53.90	231.21	228.90	225.80	
0.075	0.175	1.17	53.87	357.73	355.70	355.80	
			313 K				
0.025	0.975	2.49	49.09	213.71	198.40	167.80	
0.050	0.950	2.49	49.07	357.97	343.10	347.80	
0.100	0.900	2.49	49.02	625.84	611.70	740.80	
0.025	0.475	1.76	52.31	188.23	180.80	145.90	8.40
0.050	0.450	1.76	52.29	310.72	303.70	306.10	
0.100	0.400	1.76	52.24	632.39	626.10	665.10	
0.025	0.225	1.24	53.92	130.31	126.80	133.50	
0.050	0.200	1.24	53.90	302.60	299.50	282.50	
0.075	0.175	1.24	53.87	451.29	448.50	445.90	
			AAD%				7.60

The rate constant k$_{arg}$ was calculated using Equation (17) by subtracting k$_{MDEA}$ and k$_{OH}$ values from k$_{ov}$ values. The values of k$_{MDEA}$, k$_{OH}$, k$_2$, k$_a$ and k$_w$ were estimated from the previous works [41–43,51]. It is to be noted that, there was a typographical error within the power of the previously reported rate expression for k$_w$ (1.23×10^{12} e$^{-\frac{4364.7}{T}}$ was reported instead of 1.23×10^9 e$^{-\frac{4364.7}{T}}$) [41], this error has been revised in this work and the correct rate expression for k$_w$ along with the other expressions that were used in this work have been listed in Table 2.

Table 2. List of rate expressions used for zwitterion mechanism.

Rate	Equation	References
k$_{MDEA}$ (m^3 kmol^{-2} s^{-1})	$k_2 = 2.56 \times 10^9 e^{-\frac{5922.0}{T}}$	Benamor et al. [42]
k$_{OH}$ (m^3 kmol^{-2} s^{-1})	$k_{OH} = 4.33 \times 10^{10} e^{-\frac{66666.0}{T}}$	Pinsent et al. [43]
k$_2$ (m^3 kmol^{-1} s^{-1})	$k_2 = 2.81 \times 10^{10} e^{-\frac{4482.9}{T}}$	Mahmud et al. [41]
k$_a$ (m^6 kmol^{-2} s^{-1})	$k_a = 7.96 \times 10^{10} e^{-\frac{4603.8}{T}}$	Mahmud et al. [41]
k$_w$ (m^6 kmol^{-2} s^{-1})	$k_w = 1.23 \times 10^9 e^{-\frac{4364.7}{T}}$	Mahmud et al. [41] *

* Corrected value.

Experimental rate constants, 'k$_{Arg}$', data were fitted to Equation (17) to extract the individual blocks of rate constants described in this equation. The concentrations of water molecules [H$_2$O] were calculated by mass balance while those of hydroxyl ions [OH$^-$] were estimated from the relation given by Astarita et al. [52]. The use of this relationship is justified since the CO$_2$ loading in the amine solution was very small as it was verified by Gas Chromatography throughout all experiments.

$$[OH^-] = \sqrt{\frac{K_W}{K_{Pi}}[AM]} \text{ for } \alpha \leq 10^{-3} \tag{20}$$

Using Equation (20), the total $[OH^-]$ was taken to be the sum of $[OH^-]$ ions produced by MDEA and those produced by Arg. The water dissociation constant, 'K_w' and protonation constant, 'K_{pi}', for MDEA and L-Arginine were expressed as a function of temperature according to the following equation:

$$LnK_i = \frac{a_i}{T} + b_i lnT + c_i T + d_i \tag{21}$$

Values of the constants a_i–d_i are given in Table 3.

Table 3. Values of different equilibrium constant used to estimate OH^- in Equation (20).

Parameter	a_i	b_i	c_i	d_i	Validity Range	Source
$K_{p1(MDEA)}$	-8483.95	-13.8328	0	87.39717	293–333 K	[53]
$K_{p2(Arginine)}$	-3268.3	0	0	-9.9729	293–323 K	[41]
K_w	-13445.9	-22.4773	0	140.932	273–498 K	[54]

Applying a nonlinear regression technique using Excel solver, experimental k_{arg} values were fitted to Equation (17) taking into account the species concentrations, H_2O, Arg, OH^- and MDEA previously calculated. Since the rate expressions for the terms k_2, k_a and k_w were already available from previous work [41]. The regression analysis was initially performed to generate the values of the term k_b and k_{OH} only. However, the obtained values for the k_{OH} indicated no catalytic influence on the k_{Arg} values. This can be attributed to the fact that the concentration of the hydroxyl ions is very low compared to other bases in the system, which leads to the conclusion that there is no significant influence of catalytic hydroxyl ions on the kinetics. Furthermore, the reaction between hydroxyl and CO_2 exhibits slower kinetics which has been previously demonstrated by Gou et al. [55]. Hence, the final regression analysis were performed excluding the k_{OH} term. The generated rate constant values for the k_b term are summarized in Table 4.

Table 4. Reaction rate constants at different temperatures using Zwitterion mechanism.

Temperature	298 K	303 K	308 K	313 K
Rate constant, k_b (m^6 $kmol^{-2}$ s^{-1})	2321.0	3127.5	4400.5	5130

Using these generated rate constants, the overall reaction rate constant, 'k_{Arg}', values were predicted using Equation (17) and were plotted against real experimental data in a parietal plot as shown in Figure 8. It is very clear that the adopted rate model along with extracted individual rate constants represent very well the experimental results with an average absolute deviation, AAD, of 7.6%. Figure 8 validated the choice of the kinetics model used to interpret the data of the reaction of CO_2 with mixtures of MDEA and Arg represented in Equation (12). Furthermore, these results confirm the contribution of Arg and MDEA species in the base-catalytic formation of carbamate.

The individual rate constants at different temperatures were plotted as a function of temperature according to Arrhenius equation as shown in Figure 9 and associated parameters are summarized in Table 5. The activation energy (E_a) of each reaction was derived from the Arrhenius plots along with the pre-exponential coefficient of each rate constant.

From Table 5, it can be observed that the activation energy of L-Arginine (39.15 kJ mol^{-1}) is smaller than that of MDEA (49.24 kJ mol^{-1}), which shows that L-Arginine reacts faster with CO_2 than MDEA. In fact, L-Arginine having a molecular structure similar to that of primary amines, have a faster reaction rate compared to tertiary amine MDEA. The E_a for Arg, MDEA and H_2O catalytic carbamate formation showed that the contribution of water to the overall formation of carbamate (36.29 kJ mol^{-1}) is the most important followed by that of L-Arginine (38.28 kJ mol^{-1}), while the contribution of MDEA to this reaction (42.27 kJ mol^{-1}) were found to be the least.

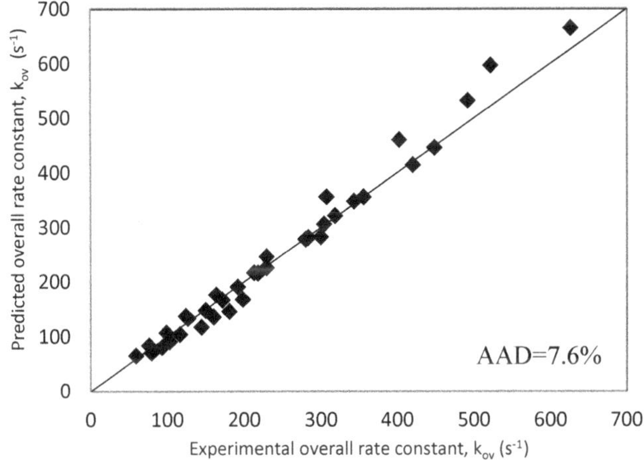

Figure 8. Parity plot of experimental and predicted k_{Arg} values for Zwitterion mechanism.

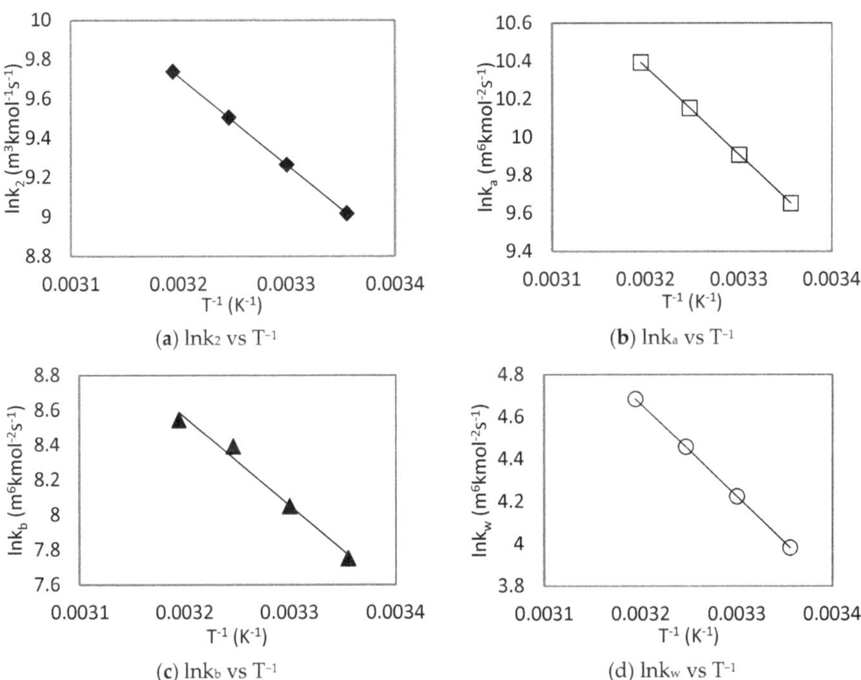

Figure 9. Arrhenius plots of CO_2-MDEA-Arg rate constants using zwitterion mechanism.

Table 5. Summarized kinetics rate constants for CO_2-MDEA-Arg reaction using Zwitterion Mechanism.

Rate	$k_{ov} = k_{MDEA}[MDEA] + k_{OH^-}[OH^-] + \dfrac{[Arg]}{\frac{1}{k_2} + k_a[Arg] + k_b[MDEA] + k_w[H_2O]}$		References
	Lnk_0	E_a (kJ mol^{-1})	Expression	
k_{MDEA} (m^3 kmol^{-2} s^{-1})	21.66	49.24	$k_{MDEA} = 2.56 \times 10^9 e^{-\frac{5922.0}{T}}$	Benamor et al. [42]
k_{OH} (m^3 kmol^{-2} s^{-1})	24.49	554.26	$k_{OH} = 4.33 \times 10^{10} e^{-\frac{66666.0}{T}}$	Pinsent et al. [43]
k_2 (m^3 kmol^{-2} s^{-1})	24.06	37.27	$k_2 = 2.81 \times 10^{10} e^{-\frac{4482.9}{T}}$	Mahmud et al. [41]
k_a (m^6 kmol^{-2} s^{-1})	25.10	38.28	$k_a = 7.96 \times 10^{10} e^{-\frac{4603.8}{T}}$	Mahmud et al. [41]
k_b (m^6 kmol^{-2} s^{-1})	24.83	42.27	$k_b = 6.07 \times 10^{10} e^{-\frac{5083.8}{T}}$	This Work
k_w (m^6 kmol^{-2} s^{-1})	18.63	36.29	$k_w = 1.23 \times 10^9 e^{-\frac{4364.7}{T}}$	Mahmud et al. [41] *

* Corrected expression.

4.3. Termolecular Mechanism

Since the termolecular applicability tests revealed the possibility of applying this mechanism to interpret the experimental data, the kinetics of CO_2-MDEA-Arg were then further investigated via this mechanism. Similar to that of the zwitterion mechanism, the values of k_a and k_w were also estimated from previous work [41] and are listed in Table 6.

Table 6. List of rate expressions used for Termolecular mechanism.

Rate	Equation	References
k_a (m^6 kmol^{-2} s^{-1})	$k_a = 5.72 \times 10^{10} e^{-\frac{4769.00}{T}}$	Mahmud et al. [4]
k_w (m^6 kmol^{-2} s^{-1})	$k_w = 9.41 \times 10^7 e^{-\frac{4365.00}{T}}$	Mahmud et al. [41]

Using the k_a and k_w values, previously obtained k_{Arg} values were then fitted in accordance to the Equation (18). Excel solver was then used to regress the experimental data to obtain the rate expressions. The apparent and predicted k_{Arg} values obtained using the termolecular mechanism are presented in Table 7.

Similar to the results obtained using the zwitterion mechanism, the obtained fitting results showed that hydroxyl ion (k_{hyd}) had a negligible effect for CO_2-MDEA-Arginine reaction using termolecular mechanism. Only L-Arginine, MDEA and water concentrations effects were found to be significant. The natural logarithm of the individual rate constants was plotted against T^{-1} according to Arrhenius equation as shown in Figure 10. The generated rate constant values for the k_b term are summarized in Table 8.

The predicted rate constant values were compared to the experimental ones as via a parity plot shown in Figure 11, which displayed good agreement between both values with an AAD of 8.0%. Since the AAD% is very close to that obtained in case of zwitterion mechanism (7.60%), it can be suggested that termolecular mechanism can be also used to interpret the obtained experimental data. The obtained rate expressions using termolecular are summarized in Table 9.

Table 7. Rate constants at different temperatures and (MDEA+Arg) concentrations.

ARG	MDEA	OH	H_2O	$k_{Arg\text{-}exp}$	$k_{Arg\text{-}pred}$	AAD%
mol L^{-1}	mol L^{-1}	mol L^{-1}	mol L^{-1}	s^{-1}	s^{-1}	
298 K						
0.025	0.975	3.2×10^{-4}	49.16	93.3	79.7	
0.050	0.950	4.5×10^{-4}	49.14	171.8	166.2	
0.100	0.900	6.2×10^{-4}	49.08	306.9	358.9	
0.025	0.475	2.3×10^{-4}	52.35	79.5	70.0	7.8
0.050	0.450	3.1×10^{-4}	52.32	149.8	146.6	
0.100	0.400	4.1×10^{-4}	52.26	318.0	319.9	
0.025	0.225	1.5×10^{-4}	53.94	59.4	65.1	
0.050	0.200	2.1×10^{-4}	53.91	124.3	136.8	
0.075	0.175	2.4×10^{-4}	53.88	213.2	215.2	
303 K						
0.025	0.975	3.0×10^{-4}	49.16	116.5	103.4	
0.050	0.950	4.2×10^{-4}	49.14	218.2	215.3	
0.100	0.900	5.8×10^{-4}	49.08	402.8	465.2	
0.025	0.475	2.1×10^{-4}	52.35	103.2	90.0	6.9
0.050	0.450	2.9×10^{-4}	52.32	191.9	188.7	
0.100	0.400	3.8×10^{-4}	52.26	420.5	411.9	
0.025	0.225	1.4×10^{-4}	53.94	76.4	83.4	
0.050	0.200	1.9×10^{-4}	53.91	163.7	175.4	
0.075	0.175	2.2×10^{-4}	53.88	280.1	276.0	
308 K						
0.025	0.98	2.8×10^{-4}	49.16	160.2	134.7	
0.050	0.95	3.9×10^{-4}	49.14	283.4	280.3	
0.100	0.90	5.4×10^{-4}	49.08	521.9	604.6	
0.025	0.48	2.0×10^{-4}	52.35	144.5	115.8	8.6
0.050	0.45	2.7×10^{-4}	52.32	229.0	242.6	
0.100	0.40	3.6×10^{-4}	52.26	492.0	529.2	
0.025	0.23	1.3×10^{-4}	53.94	98.8	106.4	
0.050	0.20	1.8×10^{-4}	53.91	228.9	223.8	
0.075	0.18	2.1×10^{-4}	53.88	355.7	352.2	
313 K						
0.025	0.98	2.6×10^{-4}	49.16	198.5	164.8	
0.050	0.95	3.6×10^{-4}	49.14	343.1	343.9	
0.100	0.90	5.0×10^{-4}	49.08	611.8	745.2	
0.025	0.48	1.8×10^{-4}	52.35	180.8	143.3	8.8
0.050	0.45	2.5×10^{-4}	52.32	303.7	301.0	
0.100	0.40	3.3×10^{-4}	52.26	626.1	659.5	
0.025	0.23	1.3×10^{-4}	53.94	126.8	132.6	
0.050	0.20	1.7×10^{-4}	53.91	299.5	279.6	
0.075	0.18	1.9×10^{-4}	53.88	448.6	441.0	
Overall AAD%						8.0

Table 8. Reaction rate constants at different temperatures using termolecular mechanism.

Rate Constant	298 K	303 K	308 K	313 K
k_b (m^6 kmol^{-2} s^{-1})	1043.0	1397.0	1927.0	2240.0

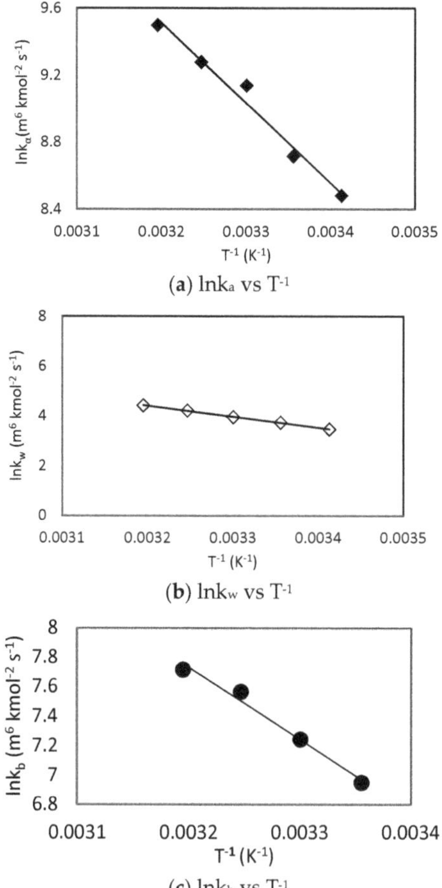

(**a**) lnk_a vs T^{-1}

(**b**) lnk_w vs T^{-1}

(**c**) lnk_b vs T^{-1}

Figure 10. Arrhenius plots of CO_2-MDEA-Arg rate constants using termolecular mechanism.

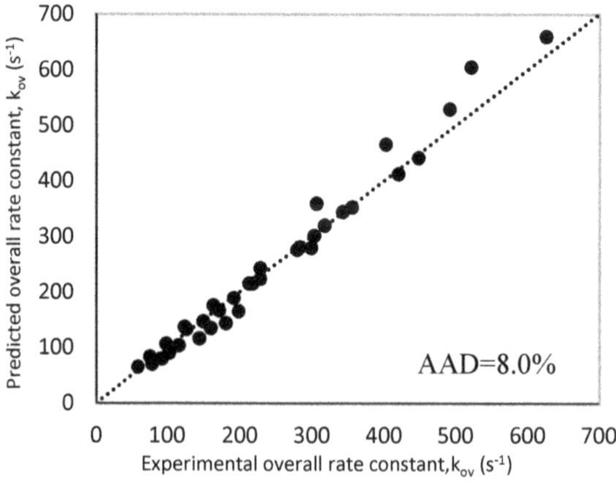

Figure 11. Parity plot of experimental and predicted 'k_{Arg}' values for termolecular mechanism.

Table 9. Summarized kinetics rate constants for CO_2-MDEA-Arg reaction using Termolecular Mechanism.

Rate	$k_{ov} = k_{MDEA}[MDEA] + k_{OH^-}[OH^-] + k_a[Arg] + k_b[MDEA] + k_w[H_2O]$			References
	lnk_0	E_a (kJ mol^{-1})	Expression	
k_{MDEA} (m^3 kmol^{-2} s^{-1})	21.66	49.24	$k_{MDEA} = 2.56 \times 10^9 e^{-\frac{5922.0}{T}}$	Benamor et al. [42]
k_{OH} (m^3 kmol^{-2} s^{-1})	24.49	554.26	$k_{OH} = 4.33 \times 10^{10} e^{-\frac{66666.0}{T}}$	Pinsent et al. [43]
k_a (m^6 kmol^{-2} s^{-1})	24.77	38.28	$k_a = 5.72x \times 10^{10} e^{-\frac{4769.0}{T}}$	Mahmud et al. [41]
k_b (m^6 kmol^{-2} s^{-1})	23.38	40.65	$k_b = 1.42 \times 10^{10} e^{-\frac{4888.6}{T}}$	This Work
k_w (m^6 kmol^{-2} s^{-1})	18.63	36.29	$k_w = 9.41 \times 10^7 e^{-\frac{4365.0}{T}}$	Mahmud et al. [41]

Upon evaluating the obtained rate expression for the 'k_b' term in both mechanisms, it is observed that activation energies in both models are comparable to each other (E_a^Z = 42.27 kJ mol^{-1} and E_a^T = 40.65 kJ mol^{-1}). Furthermore, it is noticed that regardless of the model used catalytic effect of L-Arginine ($E_a^Z = E_a^T$ = 38.28 kJ mol^{-1}) is higher than the catalytic effect of MDEA. Based on this, it can be concluded that the CO_2-MDEA-Arg reactions can be effectively interpreted using both zwitterion and termolecular mechanisms.

5. Comparison with Other Amine Systems

5.1. Comparison with Secondary, Tertiary and Sterically Hindered Amine

The obtained rate constants data for 1M MDEA-Arg (0.9M MDEA + 0.1M Arg) in this work were compared with those of DEA [56], MDEA [42] and AMP [57] as shown in Figure 12. It was observed that the rate constants of MDEA-Arg were much lower than that of secondary amine (DEA) and lower than that of sterically hindered amine (AMP). However, the rate constant of MDEA-Arg mixtures were always higher than those of the tertiary amine (MDEA). This elucidates the effect of L-Arginine presence in the MDEA-Arg blend which can enhance the overall kinetics of the CO_2-MDEA reaction and make it comparable to other secondary and hindered amines. Based on the above analysis, the overall rate constants of these amine systems with CO_2 were found to be in the following order: DEA > AMP > MDEA-Arg > MDEA.

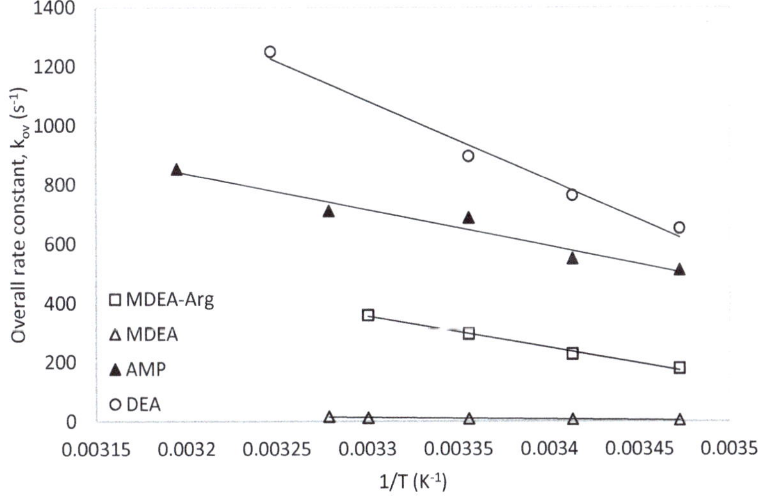

Figure 12. Comparison of the obtained data with MDEA-Arg with other amine systems.

5.2. Comparison of the Promoting Effect of L-Arginine

The promoting effect of L-Arginine investigated in this work was also compared with that of the available data of MDEA-Glycine [42] and MDEA-DEA [58] systems at different temperatures and at the same overall concentration (1 M) as shown in Figure 13. For all three compared systems, the same 0.1 M of the promoter was added. It is observed that the addition of 0.1 M L-Arginine in the MDEA-Arg has resulted in higher overall rate constant compared to the addition of the 0.1 M DEA. However, the addition of 0.1 M Glycine in the MDEA blend has resulted in higher overall rate constant compared to that of 0.1 M L-Arginine. Although the previous study [41] revealed that the kinetics of L-Arginine alone with CO_2 is higher than that of the Glycine, Guo et al. [55] observed that the Glycine at higher pH exhibits faster kinetics. Since the presence of MDEA can increase pH of the solution, it triggers the base form of Glycine to react with the CO_2 resulting in a higher overall rate constant in the MDEA-Glycine blend as observed in the work of Benamor et al. [42]. The presence of MDEA has more significant catalytic effect towards the formation of zwitterion intermediate in MDEA-Glycine (E_a = 24.67 kJ mol^{-1} [42]) compared to that of MDEA-Arg (E_a = 42.27 kJ mol^{-1}). Furthermore, activation energy for the reaction of zwitterion intermediate formation of Glycine in the MDEA-Gly is 22.95 kJ mol^{-1} [42] is also lower than that of L-Arginine (E_a = 37.27 kJ mol^{-1}) in the MDEA-Arg blend. Therefore, based on this analysis, the rate constants of the three blended amine systems with CO_2 were found to be in the following order: MDEA-Gly > MDEA-Arg > MDEA-DEA.

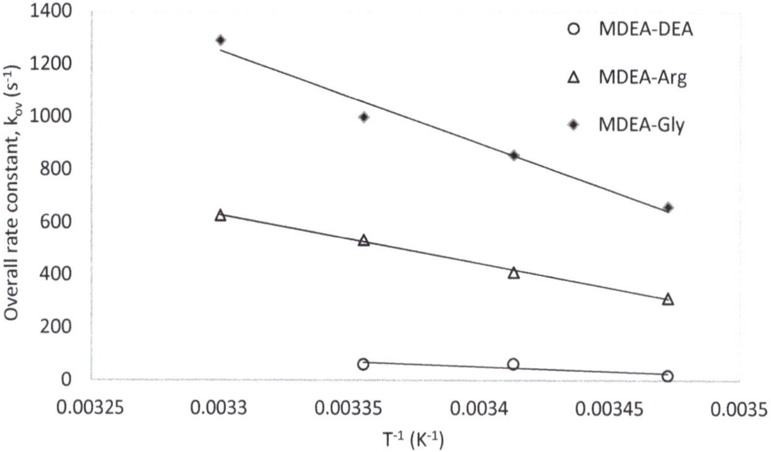

Figure 13. Comparison of MDEA-Arg with other MDEA blends.

6. Conclusions

The kinetics of the reaction of CO_2 with MDEA + Arginine in aqueous solutions was studied using the stopped-flow technique for the first time. The measurements were performed for a concentration range from 0.25 M to 1 M and a temperature range from 298 to 313 K. The rate constants were well correlated by Arrhenius equation type. The activation energies for the rate constants were estimated. Both of the adopted zwitterion and termolecular models were very accurate in representing the experimental data over a range of five different temperatures from 293 to 313 K with an AAD of 7.6% and 8.0%, respectively. The contribution of L-Arginine, MDEA and H_2O to the catalytic carbamate formation pathway was assessed using rate constants generated form the reaction of arginine alone with CO_2. The results showed that the contribution of arginine to the overall formation is more significant followed by the contribution of water in both models, while the contribution of MDEA molecules was found to be the least. Based on the regression results, rate expression for the catalytic formation of zwitterion was to be $k_b = 6.07 \times 10^{10} e^{-\frac{5083.8}{T}}$ for the zwitterion mechanism and $k_b = 1.42 \times 10^{10} e^{-\frac{4888.6}{T}}$ for the termolecular mechanism. A comparison of the obtained overall rate constant with other amine systems revealed the MDEA-Arginine-CO_2 reaction was faster than that of MDEA-CO_2 but slower than that of secondary and sterically hindered amine. A further comparison with MDEA-promotor blends showed that the reaction of MDEA-Arginine with CO_2 is slower than MDEA-Glycine but faster than MDEA-DEA. This was attributed to the fact that the catalytic contribution of L-Arginine for the formation of zwitterion intermediate is lower compared to Glycine in MDEA blends. Furthermore, presence of MDEA can significantly catalyse the formation zwitterion intermediate in MDEA-Glycine blend (E_a = 24.67 kJ mol^{-1} [43]) than that of MDEA-Arg blend (E_a = 42.27 kJ mol^{-1}). Consequently, a faster reaction kinetics was observed in MDEA-Glycine-CO_2 reactions than MDEA-Arg-CO_2 reactions.

Author Contributions: Conceptualization, A.B., N.M., and P.T.; Methodology, A.B. and N.M.; Software, A.B.; Validation, A.B., N.M., M.N., M.H.E.-N. and P.T.; Formal Analysis, A.B., N.M, M.N. and M.H.E.-N.; Investigation, A.B., N.M. and M.H.E.-N.; Resources, A.B.; Data Curation, A.B., N.M. and M.N.; Writing-Original Draft Preparation, N.M., and A.B.; Writing-Review & Editing, N.M., A.B., M.N., M.H.E.-N. and P.T.; Visualization, A.B., M.N. and N.M.; Supervision, A.B. and P.T.; Project Administration, A.B. and M.H.E.-N.; Funding Acquisition, A.B.

Funding: This paper was made possible by an NPRP Grant # 7 - 1154 - 2 – 433 from the Qatar National Research Fund (a member of Qatar Foundation). The statements made herein are solely the responsibility of the authors.

Acknowledgments: The authors thank Ahmed Soliman and Dan Jerry Cortes for providing laboratory support.

Conflicts of Interest: The authors declare no conflict of interest.

Nomenclature

MDEA	N-methyldiethanolamine
Arg	L-Arginine
Gly	Glycine
MEA	Monoethanolamine
DEA	Diethanolamine
AMP	2-Amino-2-Methyl-1-Propanol
$r_{CO_2\text{-}MDEA}$ (l mol^{-1} s^{-1})	Reaction rate of CO_2 with MDEA
$r_{CO_2\text{-}OH}$ (l mol^{-1} s^{-1})	Reaction rate of CO_2 with hydroxyl ions
$r_{CO_2\text{-}OH}$ (l mol^{-1} s^{-1})	Reaction rate of CO_2 with L-Arginine
r_{CO_2} (l mol^{-1} s^{-1})	Reaction rate of CO_2 with MDEA-Arginine
k_{MDEA} (s^{-1})	Overall reaction rate of CO_2 and MDEA
k_{OH} (s^{-1})	Overall reaction rate of CO_2 and Hydroxyl ion
k_{Arg} (s^{-1})	Overall reaction rate of CO_2 and L-Arginine
k_{ov} (s^{-1})	Overall reaction rate with CO_2 with MDEA and L-Arginine
k_1 (m^3 kmol^{-1} s^{-1})	Reaction rate constant of the formation of the intermediate Zwitterion
K_{MDEA} (m^3 kmol^{-1} s^{-1})	Reaction rate constant of CO_2 and MDEA reaction.
K_{OH} (m^3 kmol^{-1} s^{-1})	Reaction rate constant of CO_2 and hydroxyl ion reaction.
k_{-1} (s^{-1})	Reaction rate constant of the consumption of the intermediate Zwitterion
$k_{b,I}$ (s^{-1})	Individual reaction rate constants according to zwitterion mechanism
T (K)	Temperature
t (s)	Time
K_w (mol l^{-1})	Water dissociation constant
K_{Pi} (mol l^{-1})	Protonation constant for MDEA and L-Arginine
$k_{Arg\text{-}exp}$ (s^{-1})	Experimental apparent rate constant of CO_2 and L-Arginine.
$k_{Arg\text{-}pred}$ (s^{-1})	Predicted apparent rate constant of CO_2 and L-Arginine.
E_a (kJ mol^{-1})	Activation energy
E_a^Z (kJ mol^{-1})	Activation energy obtained in zwitterion mechanism
E_a^T (kJ mol^{-1})	Activation energy obtained in termolecular mechanism
k_a (m^6 kmol^{-2} s^{-1})	Catalytic contribution of L-Arginine in the reaction rate according to zwitterion mechanism
k_{hyd} (m^6 kmol^{-2} s^{-1})	Catalytic contribution of hydroxyl ion in the reaction rate according to zwitterion mechanism
k_b (m^6 kmol^{-2} s^{-1})	Catalytic contribution of MDEA in the reaction rate according to zwitterion mechanism

References

1. Tontiwachwuthikul, P.; Idem, R.; Gelowitz, D.; Liang, Z.H.; Supap, T.; Chan, C.W.; Sanpasertparnich, T.; Saiwan, C.; Smithson, H. Recent progress and new development of post-combustion carbon-capture technology using reactive solvents. *Carbon Manag.* **2011**, *2*, 261–263. [CrossRef]
2. Zhou, Q.; Chan, C.W.; Tontiwachiwuthikul, P.; Idem, R.; Gelowitz, D. A statistical analysis of the carbon dioxide capture process. *Int. J. Greenh. Gas Control* **2009**, *3*, 535–544. [CrossRef]
3. Rubin, E.S.; Mantripragada, H.; Marks, A.; Versteeg, P.; Kitchin, J. The outlook for improved carbon capture technology. *Prog. Energy Combust. Sci.* **2012**, *38*, 630–671. [CrossRef]
4. Lee, A.S.; Kitchin, J.R. Chemical and Molecular Descriptors for the Reactivity of Amines with CO_2. *Ind. Eng. Chem. Res.* **2012**, *51*, 13609–13618. [CrossRef]
5. Zhang, Z.; Borhani, T.N.G.; El-Naas, M.H. Chapter 4.5—Carbon Capture. In *Exergetic, Energetic and Environmental Dimensions*; Dincer, I., Colpan, C.O., Kizilkan, O., Eds.; Academic Press: Cambridge, MA, USA, 2018; pp. 997–1016. ISBN 978-0-12-813734-5.
6. Zhang, Z.; Yan, Y.; Chen, Y.; Zhang, L. Investigation of CO_2 absorption in methyldiethanolamine and 2-(1-piperazinyl)-ethylamine using hollow fiber membrane contactors: Part, C. Effect of operating variables. *J. Nat. Gas Sci. Eng.* **2014**, *20*, 58–66. [CrossRef]
7. Chakravarty, T.; Phukan, U.K.; Weiland, R.H. Reaction of Acid Gases with mixtures of amines. *Chem. Eng. Prog.* **1985**, *81*, 32–36.
8. Liang, Z.; Rongwong, W.; Liu, H.; Fu, K.; Gao, H.; Cao, F.; Zhang, R.; Sema, T.; Henni, A.; Sumon, K.; et al. Recent progress and new developments in post-combustion carbon-capture technology with amine based solvents. *Int. J. Greenh. Gas Control* **2015**, *40*, 26–54. [CrossRef]
9. Benamor, A.; Ali, B.S.; Aroua, M.K. Kinetic of CO_2 absorption and carbamate formation in aqueous solutions of diethanolamine. *Korean J. Chem. Eng.* **2008**, *25*, 451–460. [CrossRef]
10. Zhao, X.; Cui, Q.; Wang, B.; Yan, X.; Singh, S.; Zhang, F.; Gao, X.; Li, Y. Recent progress of amine modified sorbents for capturing CO_2 from flue gas. *Chin. J. Chem. Eng.* **2018**, *26*, 2292–2302. [CrossRef]
11. Wang, Q.; Luo, J.; Zhong, Z.; Borgna, A. CO_2 capture by solid adsorbents and their applications: Current status and new trends. *Energy Environ. Sci.* **2011**, *4*, 42–55. [CrossRef]
12. Munoz, D.M.; Portugal, A.F.; Lozano, A.E.; de la Campa, J.G.; de Abajo, J. New liquid absorbents for the removal of CO_2 from gas mixtures. *Energy Environ. Sci.* **2009**, *2*, 883–891. [CrossRef]
13. Brœder, P.; Svendsen, H.F. Capacity and Kinetics of Solvents for Post-Combustion CO_2 Capture. *Energy Procedia* **2012**, *23*, 45–54. [CrossRef]
14. Puxty, G.; Rowland, R.; Allport, A.; Yang, Q.; Bown, M.; Burns, R.; Maeder, M.; Attalla, M. Carbon Dioxide Postcombustion Capture: A Novel Screening Study of the Carbon Dioxide Absorption Performance of 76 Amines. *Environ. Sci. Technol.* **2009**, *43*, 6427–6433. [CrossRef]
15. Zhang, Z.; Li, Y.; Zhang, W.; Wang, J.; Soltanian, M.R.; Olabi, A.G. Effectiveness of amino acid salt solutions in capturing CO_2: A review. *Renew. Sustain. Energy Rev.* **2018**, *98*, 179–188. [CrossRef]
16. Lerche, B.M. *CO_2 Capture from Flue Gas Using Amino Acid Salt Solutions*; Department of Chemical and Biochemical Engineering, Technical University of Denmark: Kongens Lyngby, Denmark, 2012.
17. Goetheer, E.; Nell, L. First pilot results from TNO's solvent development workflow. *Carbon Capture J.* **2009**, *8*, 2–3.
18. Eide-Haugmo, I.; Brakstad, O.G.; Hoff, K.A.; da Silva, E.F.; Svendsen, H.F. Marine biodegradability and ecotoxicity of solvents for CO 2-capture of natural gas. *Int. J. Greenh. Gas Control* **2012**, *9*, 184–192. [CrossRef]
19. Aronu, U.E.; Svendsen, H.F.; Hoff, K.A. Investigation of amine amino acid salts for carbon dioxide absorption. *Int. J. Greenh. Gas Control* **2010**, *4*, 771–775. [CrossRef]
20. Kumar, P.; Hogendoorn, J.; Versteeg, G.; Feron, P. Kinetics of the reaction of CO_2 with aqueous potassium salt of taurine and glycine. *AIChE J.* **2003**, *49*, 203–213. [CrossRef]
21. Kumar, P.; Hogendoorn, J.; Feron, P.; Versteeg, G. Equilibrium solubility of CO_2 in aqueous potassium taurate solutions: Part 1. Crystallization in carbon dioxide loaded aqueous salt solutions of amino acids. *Ind. Eng. Chem. Res.* **2003**, *42*, 2832–2840. [CrossRef]
22. Aronu, U.E.; Hessen, E.T.; Haug-Warberg, T.; Hoff, K.A.; Svendsen, H.F. Vapor–liquid equilibrium in amino acid salt system: Experiments and modeling. *Chem. Eng. Sci.* **2011**, *66*, 2191–2198. [CrossRef]

23. Kumar, P.; Hogendoorn, J.; Timmer, S.; Feron, P.; Versteeg, G. Equilibrium solubility of CO_2 in aqueous potassium taurate solutions: Part 2. Experimental VLE data and model. *Ind. Eng. Chem. Res.* **2003**, *42*, 2841–2852. [CrossRef]
24. Majchrowicz, M.; Brilman, D. Solubility of CO_2 in aqueous potassium l-prolinate solutions—Absorber conditions. *Chem. Eng. Sci.* **2012**, *72*, 35–44. [CrossRef]
25. Portugal, A.F.; Sousa, J.M.; Magalhães, F.D.; Mendes, A. Solubility of carbon dioxide in aqueous solutions of amino acid salts. *Chem. Eng. Sci.* **2009**, *64*, 1993–2002. [CrossRef]
26. Schneider, R.; Schramm, H. Environmentally friendly and economic carbon capture from power plant flue gases. In Proceedings of the 1st Post Combustion Capture Conference, Abu Dhabi, UAE, 17–19 May 2011.
27. Park, S.-W.; Son, Y.-S.; Park, D.-W.; Oh, K.-J. Absorption of carbon dioxide into aqueous solution of sodium glycinate. *Sep. Sci. Technol.* **2008**, *43*, 3003–3019. [CrossRef]
28. Lee, S.; Song, H.-J.; Maken, S.; Park, J.-W. Kinetics of CO_2 absorption in aqueous sodium glycinate solutions. *Ind. Eng. Chem. Res.* **2007**, *46*, 1578–1583. [CrossRef]
29. Shen, S.; Yang, Y.-N.; Bian, Y.; Zhao, Y. Kinetics of CO_2 Absorption into Aqueous Basic Amino Acid Salt: Potassium Salt of Lysine Solution. *Environ. Sci. Technol.* **2016**, *50*, 2054–2063. [CrossRef]
30. Shen, S.; Yang, Y.-n.; Zhao, Y.; Bian, Y. Reaction kinetics of carbon dioxide absorption into aqueous potassium salt of histidine. *Chem. Eng. Sci.* **2016**, *146*, 76–87. [CrossRef]
31. Portugal, A.; Derks, P.; Versteeg, G.; Magalhaes, F.; Mendes, A. Characterization of potassium glycinate for carbon dioxide absorption purposes. *Chem. Eng. Sci.* **2007**, *62*, 6534–6547. [CrossRef]
32. Portugal, A.; Magalhaes, F.; Mendes, A. Carbon dioxide absorption kinetics in potassium threonate. *Chem. Eng. Sci.* **2008**, *63*, 3493–3503. [CrossRef]
33. Hwang, K.-S.; Park, D.-W.; Oh, K.-J.; Kim, S.-S.; Park, S.-W. Chemical Absorption of Carbon Dioxide into Aqueous Solution of Potassium Threonate. *Sep. Sci. Technol.* **2010**, *45*, 497–507. [CrossRef]
34. Wei, C.-C.; Puxty, G.; Feron, P. Amino acid salts for CO_2 capture at flue gas temperatures. *Chem. Eng. Sci.* **2014**, *107*, 218–226. [CrossRef]
35. Aronu, U.E.; Hartono, A.; Svendsen, H.F. Density, viscosity and N2O solubility of aqueous amino acid salt and amine amino acid salt solutions. *J. Chem. Thermodyn.* **2012**, *45*, 90–99. [CrossRef]
36. Kumar, P.S.; Hogendoorn, J.; Feron, P.; Versteeg, G. Density, viscosity, solubility and diffusivity of N2O in aqueous amino acid salt solutions. *J. Chem. Eng. Data* **2001**, *46*, 1357–1361. [CrossRef]
37. Hamborg, E.S.; Niederer, J.P.; Versteeg, G.F. Dissociation constants and thermodynamic properties of amino acids used in CO_2 absorption from (293 to 353) K. *J. Chem. Eng. Data* **2007**, *52*, 2491–2502. [CrossRef]
38. Hamborg, E.S.; van Swaaij, W.P.; Versteeg, G.F. Diffusivities in aqueous solutions of the potassium salt of amino acids. *J. Chem. Eng. Data* **2008**, *53*, 1141–1145. [CrossRef]
39. Van Holst, J.; Versteeg, G.; Brilman, D.; Hogendoorn, J. Kinetic study of CO_2 with various amino acid salts in aqueous solution. *Chem. Eng. Sci.* **2009**, *64*, 59–68. [CrossRef]
40. Dalton, J.B.; Schmidt, C.L. The solubilities of certain amino acids in water, the densities of their solutions at twenty-five degrees and the calculated heats of solution and partial molal volumes. *J. Biol. Chem.* **1933**, *103*, 549–578.
41. Mahmud, N.; Benamor, A.; Nasser, M.S.; Al-Marri, M.J.; Qiblawey, H.; Tontiwachwuthikul, P. Reaction kinetics of carbon dioxide with aqueous solutions of l-Arginine, Glycine & Sarcosine using the stopped flow technique. *Int. J. Greenh. Gas Control* **2017**, *63*, 47–58. [CrossRef]
42. Benamor, A.; Al-Marri, M.J.; Khraisheh, M.; Nasser, M.S.; Tontiwachwuthikul, P. Reaction kinetics of carbon dioxide in aqueous blends of N-methyldiethanolamine and glycine using the stopped flow technique. *J. Nat. Gas Sci. Eng.* **2016**, *33*, 186–195. [CrossRef]
43. Pinsent, B.R.W.; Pearson, L.; Roughton, F.J.W. The kinetics of combination of carbon dioxide with hydroxide ions. *Trans. Faraday Soc.* **1956**, *52*, 1512–1520. [CrossRef]
44. Caplow, M. Kinetics of carbamate formation and breakdown. *J. Am. Chem. Soc.* **1968**, *90*, 6795–6803. [CrossRef]
45. Crooks, J.E.; Donnellan, J.P. Kinetics and mechanism of the reaction between carbon dioxide and amines in aqueous solution. *J. Chem. Soc. Perkin Trans.* **1989**, *2*, 331–333. [CrossRef]
46. Da Silva, E.F.; Svendsen, H.F. Computational chemistry study of reactions, equilibrium and kinetics of chemical CO_2 absorption. *Int. J. Greenh. Gas Control* **2007**, *1*, 151–157. [CrossRef]
47. Ali, S.H.; Merchant, S.Q.; Fahim, M.A. Kinetic study of reactive absorption of some primary amines with carbon dioxide in ethanol solution. *Sep. Purifi. Technol.* **2000**, *18*, 163–175. [CrossRef]

48. Knipe, A.C.; McLean, D.; Tranter, R.L. A fast response conductivity amplifier for chemical kinetics. *J. Phys. E Sci. Instrum.* **1974**, *7*, 586–590. [CrossRef]

49. Alper, E.; Bouhamra, W. Kinetics and mechanisms of reaction between carbon disulphide and morpholine in aqueous solutions. *Chem. Eng. Technol.* **1994**, *17*, 138–140. [CrossRef]

50. Vaidya, P.D.; Kenig, E.Y. Termolecular Kinetic Model for CO_2-Alkanolamine Reactions: An Overview. *Chem. Eng. Technol.* **2010**, *33*, 1577–1581. [CrossRef]

51. Peters, L.; Hussain, A.; Follmann, M.; Melin, T.; Hägg, M.B. CO_2 removal from natural gas by employing amine absorption and membrane technology—A technical and economical analysis. *Chem. Eng. J.* **2011**, *172*, 952–960. [CrossRef]

52. Astaria, G.; Savage, D.W.; Bisio, A. *Gas Treating with Chemical Solvents*; John Wiley & Sons Inc.: New York, NY, USA, 1983; ISBN 978-0471057680.

53. Littel, R.J.; Bos, M.; Knoop, G.J. Dissociation constants of some alkanolamines at 293, 303, 318 and 333 K. *J. Chem. Eng. Data* **1990**, *35*, 276–277. [CrossRef]

54. Edwards, T.; Maurer, G.; Newman, J.; Prausnitz, J. Vapor-liquid equilibria in multicomponent aqueous solutions of volatile weak electrolytes. *AIChE J.* **1978**, *24*, 966–976. [CrossRef]

55. Guo, D.; Thee, H.; Tan, C.Y.; Chen, J.; Fei, W.; Kentish, S.; Stevens, G.W.; da Silva, G. Amino Acids as Carbon Capture Solvents: Chemical Kinetics and Mechanism of the Glycine + CO_2 Reaction. *Energy Fuels* **2013**, *27*, 3898–3904. [CrossRef]

56. Littel, R.J.; Versteeg, G.F.; Van Swaaij, W.P.M. Kinetics of CO_2 with primary and secondary amines in aqueous solutions—II. Influence of temperature on zwitterion formation and deprotonation rates. *Chem. Eng. Sci.* **1992**, *47*, 2037–2045. [CrossRef]

57. Xu, S.; Wang, Y.-W.; Otto, F.D.; Mather, A.E. Kinetics of the reaction of carbon dioxide with 2-amino-2-methyl-1-propanol solutions. *Chem. Eng. Sci.* **1996**, *51*, 841–850. [CrossRef]

58. Benamor, A.; Al-Marri, M.J. Reactive Absorption of CO_2 into Aqueous Mixtures of Methyldiethanolamine and Diethanolamine. *Int. J. Chem. Eng. Appl.* **2014**, *5*, 291–297. [CrossRef]

Article

Investigation of Pore-Formers to Modify Extrusion-Spheronized CaO-Based Pellets for CO_2 Capture

Zonghao Zhang, Shuai Pi, Donglin He, Changlei Qin * and Jingyu Ran *

Key Laboratory of Low-grade Energy Utilization Technologies and Systems of Ministry of Education, School of Energy and Power Engineering, Chongqing University, Chongqing 400044, China; zzh.cqu@foxmail.com (Z.Z.); p.shuai@cqu.edu.cn (S.P.); cqu.hedl@foxmail.com (D.H.)
* Correspondence: c.qin@cqu.edu.cn (C.Q.); ranjy@cqu.edu.cn (J.R.);
  Tel.: +86-6510-3101 (C.Q.); +86-6511-2813 (J.R.)

Received: 4 January 2019; Accepted: 22 January 2019; Published: 24 January 2019

Abstract: The application of circulating fluidized bed technology in calcium looping (CaL) requires that CaO-based sorbents should be manufactured in the form of spherical pellets. However, the pelletization of powdered sorbents is always hampered by the problem that the mechanical strength of sorbents is improved at the cost of loss in CO_2 sorption performance. To promote both the CO_2 sorption and anti-attrition performance, in this work, four kinds of pore-forming materials were screened and utilized to prepare sorbent pellets via the extrusion-spheronization process. In addition, impacts of the additional content of pore-forming material and their particle sizes were also investigated comprehensively. It was found that the addition of 5 wt.% polyethylene possesses the highest CO_2 capture capacity (0.155 g-CO_2/g-sorbent in the 25th cycle) and mechanical performance of 4.0 N after high-temperature calcination, which were about 14% higher and 25% improved, compared to pure calcium hydrate pellets. The smaller particle size of pore-forming material was observed to lead to a better performance in CO_2 sorption, while for mechanical performance, there was an optimal size for the pore-former used.

Keywords: CO_2 capture; calcium looping; chemical sorption; anti-attrition; pore-former particle size

---

1. Introduction

Greenhouse gases, such as CO_2 which is mainly produced from fossil fuel combustion, are believed to be the major contributors to the rise of global temperatures [1]. It is predicted that the total emission amount of CO_2 will increase to 40.2 Gt by 2030 [2]. Therefore, many researchers around the world focus their studies on various technologies to reduce the emission of CO_2, and the capture, utilization, and storage (CCUS) of CO_2 has been considered as the most effective way to solve the problem [3–6]. Capture is the key process in CCUS, and among different capture technologies, calcium looping (CaL) has been demonstrated as having good potential in achieving high-efficiency CO_2 separation with affordable cost [7]. With the development of CaL, some pilot plant projects have already been put into large-scale testing [8–10].

Calcium looping is based on the reversible chemical reaction of $CaO + CO_2 \Leftrightarrow CaCO_3$. Generally, CO_2 is captured in a carbonator at around 650 °C by CaO-based sorbents and released subsequently in a calciner above 900 °C. In this way, CO_2 in the flue gas can be separated and collected in high purity. Though CaL has the significant advantages of low cost and high CO_2 uptake capacity, there are still some obstacles in the commercialization of calcium looping. Firstly, all the natural sorbents will face a severe problem of loss-in-capacity due to sintering at a high temperature [11,12]. Another problem is attrition and fragmentation of sorbents due to its cyclic operation in fluidized bed systems [13–15].

As a result, partial sorbent will be lost and cannot be used repeatedly and economically. Although tremendous efforts have been made in improving the capacity and stability of sorbents [16–30], these materials still face the problem that powders are too small and must be produced in the form of pellets before they can be practically applied in CaL [14,31].

Pelletization is a good way to solve the problem by shaping powdered sorbent to the desired size with suitable anti-attrition property for industrial application. However, pelletization process usually causes the loss of CO_2 uptake capacity of sorbent due to the destruction of the original porous structure, so pore-forming materials are needed in the process to obtain sorbent pellets with balanced mechanical and chemical performance. Sun et al. [27] found that carbide sorbent doped with microcrystalline via extrusion-spheronization could get a carbonation conversion of 52.64% in the 25th cycle on the premise of sacrificing mechanical strength. Firas et al. [32] reported that biomass-derived materials could increase the porosity of pellets and achieve a CO_2 capture capacity of around 30% higher than biomass-free pellets. Therefore, it is important to find suitable pore-forming candidates to improve chemical sorption and minimize the negative impact on mechanical performance.

In order to obtain sorbent pellets with balanced mechanical and chemical performance, in this work, the pore-forming materials, widely used in industry, including polyethylene, polystyrene, graphite, and microcrystalline cellulose were screened and tested by the fabrication of CaO-based sorbent pellets using an extrusion-spheronization method. The effects of different pore-forming materials, their size, and doping ratio were comprehensively investigated and evaluated, and the roles of pore-forming materials in the preparation of sorbent pellets were well understood.

2. Experimental

2.1. Materials

Calcium hydroxide (CH) powder (>95%) with a particle diameter of 30–50 μm was used as the precursor of CaO. Four types of pore-forming materials including: polyethylene (denoted as PE, >99%), polystyrene (PS, >99%), graphite powder (C, >99%), and microcrystalline cellulose (MC, >99%) were screened and tested. For the specific test, PE powder with four different sizes of 6 μm, 12 μm, 30 μm, and 150 μm were utilized in the work.

2.2. Sorbent Pellets Preparation

Sorbent pellets were manufactured using an extrusion-spheronization method. For each sorbent, first, weighted calcium hydroxide and pore-forming material were vigorously dry-blended for 20 min and wet-mixed for 5 min in a stainless steel basin to get the homogeneous mixtures. Then, the wet mixtures were extruded into cylinders with a diameter of 1 mm by a mini-extruder (LEAP E-26, Chongqing, China). After that, the cylinders were cut off and rounded in a spheronizer (LEAP R-120, Chongqing, China) with a rotational speed of 3000 rpm for 25 min. Finally, pellets with a diameter of 0.75–1.25 mm were obtained through sieving and one-day air drying (called fresh pellets). A part of the pellets was calcined in a muffle furnace at 900 °C for 10 min (called pre-calcination). For simplicity, sorbent pellets were named as CH-PMX-Y, where PM refers to the pore-forming material, and X and Y are its initial mass content and particle size, respectively. For example, CH-PE10-12 means the pellet was doped with 10 wt. % PE whose particle size is 12 μm in preparation. The pellets consisting of pure calcium hydroxide is denoted as CH.

2.3. Thermo-Gravimetric Analysis

The CO_2 capture capacity and CaO conversion of the samples were tested in a thermo-gravimetric analyzer (NETZSCH TG209 F3, Selb, Germany). Approximately 15 mg of the sample was placed on a corundum crucible in TGA and heated to 900 °C at a rate of 30 °C/min under a N_2 flow of 85 mL/min, and the temperature was kept for 5 min to remove the moisture completely. Then, the sample was cooled down to 650 °C at a rate of −30 °C/min. Once the temperature was reached, a

CO_2 flow of 15 mL/min was added immediately, and the CO_2 sorption condition was kept for 20 min. The aforementioned process of carbonation and calcination was repeated totally 25 times to investigate the cyclic CO_2 capture performance. Based on the mass data recorded, CaO carbonation conversion (Xn, %) and CO_2 sorption capacity (Cn, g-CO_2/g-sorbent) of the samples were calculated using the following formulas:

$$Cn = \frac{m - m_0}{m_0} \tag{1}$$

$$Xn = \frac{m - m_0}{m_0 \varphi} \times \frac{M_{CaO}}{M_{CO_2}} \times 100\% \tag{2}$$

where m is the maximum mass of CaO-based sorbents in sorption stage and m_0 is minimum mass in calcination stage, φ is the mass content of CaO in the CaO-based sorbent, M_{CaO} and M_{CO_2} are the molar mass of CaO and CO_2, respectively. The CO_2 sortion capacity uncertainty is around ± 0.002 g-CO_2/g-sorbent based on repeated tests.

2.4. Pellets Impact Crushing Test

Impact crushing of the pellet was carried out using an apparatus built by ourselves, as shown in Figure 1, according to the literature [33–35]. A high-pressure air bottle with a volumetric flow meter was used to control the velocity of air flow, and two valves were used to feed samples without the escaping of gas and particles. Particles were accelerated in an educator (1.2 m in length and 10 mm in I.D) by air at a speed of 18 m/s and impacted a stainless target (inclined by 60° with respect to the vertical direction) in the collection chamber. To filter the entrained fine particle (<12 μm), a sintered porous metal plate was installed on the top of the chamber.

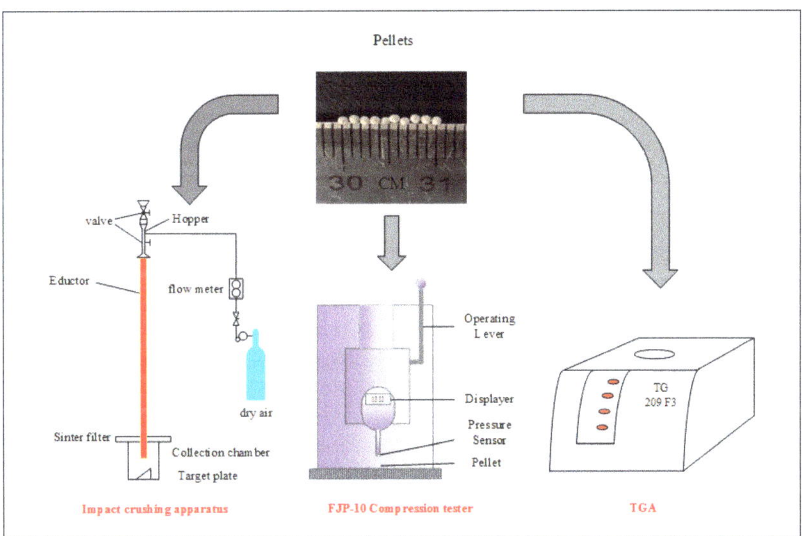

Figure 1. Schematic diagram of experimental devices.

In each test, 0.5 g samples were loaded, and after impact crushing residual particles were collected and screened through sieving into 7 size ranges: 750~1250 μm, 500~750 μm, 375~500 μm, 187.5~375 μm, 75~187.5 μm, 12~75 μm, and <12 μm. In the work, particles with a diameter smaller than 187.5 μm are regarded as completely crushed particles.

2.5. Compressive Strength Test

The compressive strength of pellets was tested using a precision digital compression tester (SHIMPO FGP-10, Kyoto, Japan), and each sample was tested for 10 times. The maximum crushing force was obtained by slowly increasing the pressure until the particle was crushed. Compression strength was evaluated by the average crushing force and the error bar was calculated.

2.6. Characterization of Sorbents

Surface morphology of sorbents was captured using a field emission scanning electron microscopy (FESEM, SU8020, Hitachi, Tokyo, Japan). To determine the specific Brunauer–Emmett–Teller (BET) surface, Barrett–Joyner–Halenda (BJH) pore volume and pore size distribution of selected samples, N_2 adsorption/desorption analysis was measured at approximately $-196\ ^{\circ}C$ using a Micromeritics TriStar II 3020 instrument after outgassing under vacuum for 18 h at 200 $^{\circ}C$.

3. Results and Discussion

3.1. Decomposition of Pore-Forming Materials

To understand thermal properties of pore-forming materials, they were first tested in the TGA at a constant ramping rate of 30 $^{\circ}C$/min to 900 $^{\circ}C$ followed by isothermal calcination for 5 min, and the results are shown in Figure 2. It was seen that under the atmosphere of N_2, MC, PS, and PE started to pyrolyze at 300–400 $^{\circ}C$ with a quick weight loss until they were completely decomposed. In contrast, there was almost no decomposition of C in N_2 flow. When the atmosphere was switched to a flow containing 15 vol. % O_2 balanced with N_2, C was observed to start burning at 600 $^{\circ}C$ until it was burnt out at around 900 $^{\circ}C$. Based on the decomposition characteristics, it can be concluded that all pore-forming materials would be burnt into gases to create pores without any solid residues left in the prepared sorbent pellets.

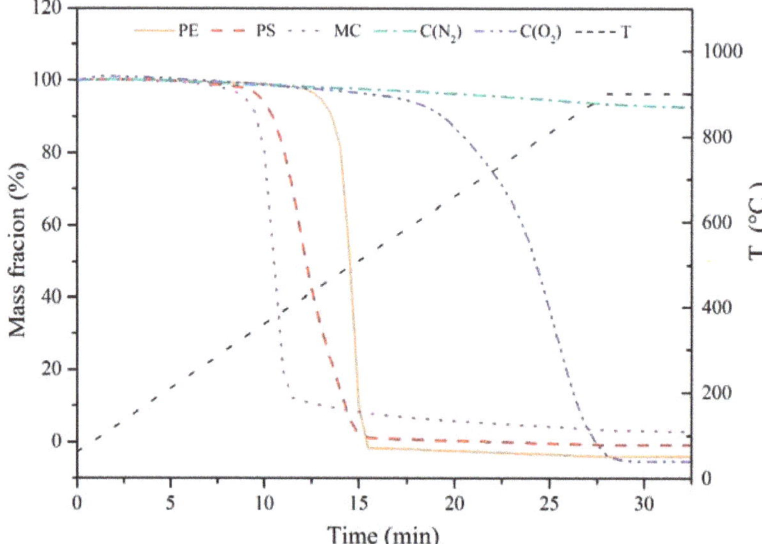

Figure 2. Decomposition property of pore-forming materials at a constant ramping rate of 30 $^{\circ}C$/min to 900 $^{\circ}C$ under the atmosphere of N_2 or N_2/O_2.

3.2. Characterization of Sorbent Pellets

To see the effect of pore-forming materials on the structure of sorbent pellets, cross-sectional images were captured of the calcined CH and CH-PE5 using the field emission scanning electron microscopy, and the results are depicted in Figure 3. It is very clear that grains and pores in CH-PE5 are much smaller and their distributions are more uniform than those in the sample of CH. These differences can be attributed to the positive pore-forming effect of PE on sorbents microstructure during its decomposition to gaseous phases. The roles of pore-formers could be further understood from the results of the N_2 adsorption/desorption test. Table 1 summarizes the specific surface area, BJH average pore width, and pore volume of CH and CH-PE5. It is seen that CH-PE5 pellets have a BET surface area of 11.66 m^2/g, BJH cumulative pore volume of 0.0436 cm^3/g, and an average pore diameter of 15.2 nm, which are all superior to those of CH. Figure 4 indicates that the higher pore volume of CH-PE5 is mainly correlated to pores of size in the range of 2–40 nm, which also contributed to the higher specific surface area in comparison with CH. Based on these results, it is reasonable to predict that the better microstructure of CH-PE5 would lead to an easier diffusion of CO_2 within the sorbent and the following carbonation reaction.

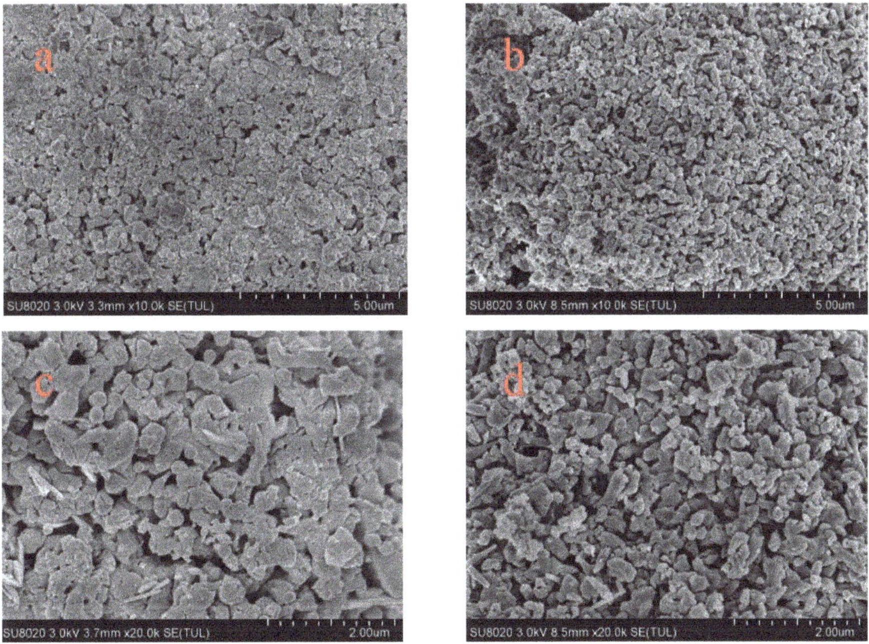

Figure 3. Cross-sectional images of calcined (**a**,**c**) CH (Ca(OH)$_2$), and (**b**,**d**) CH-PE5 (Ca(OH)$_2$ doped with 5 wt. % PE).

Table 1. Microstructures of sorbent pellets after calcination at 900 °C.

Sorbents	BET (Brunauer–Emmett–Teller) Surface Area (m^2/g)	Pore Volume (cm^3/g)	Pore Size (nm)
CH	7.86 ± 0.01	0.0415 ± 0.0001	21.09 ± 0.01
CH-PE5	11.66 ± 0.01	0.0436 ± 0.0001	15.21 ± 0.01

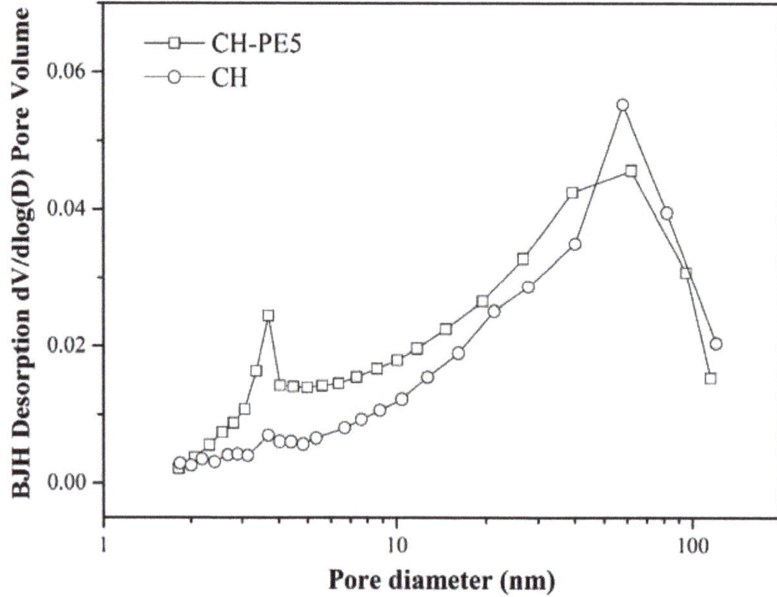

Figure 4. Pore size distribution of CH-PE5 and CH pellets.

3.3. Effect of Various Pore-Forming Materials

To understand the impact of different pore-forming materials, CO_2 sorption/desorption performance, compressive strength, and anti-impact crushing capacity of prepared sorbent pellets were tested and evaluated in this section. Pore-forming materials with a particle size of 12 μm were used, and their contents were kept at the same of 5 wt. %.

3.3.1. Sorption/Desorption Performance

Cyclic CO_2 uptake capacity and CaO conversion of sorbent pellets prepared with different pore-formers were presented in Figure 5. It shows that CH possesses a CO_2 capture around 0.556 g-CO_2/g-sorbents initially, but the value quickly decreased to 0.136 g-CO_2/g-sorbent after 25 cycles. For sorbent pellets with the addition of pore-forming materials in preparation, CH-PE5 shows the highest CO_2 capture capacity of 0.574 g-CO_2/g-sorbent in the first cycle and 0.155 g-CO_2/g-sorbent after 25 cycles, which are both higher than those of CH. This is in accordance with the aforementioned characterization results, demonstrating that CH-PE5 modified inner structure of sorbents during the pyrolysis of PE, resulted in a shifting of pores to smaller sizes of 2–40 nm (see Figure 4), and an increase of surface area and pore volume (see Table 1). It is well known that a bigger pore volume means a smaller CO_2 transfer resistance, and a higher specific surface area could supply more available sites in pellets to react with CO_2 [32,34]. Thus, CO_2 sorption performance of CH-PE5 was improved. The CH-C5 shows a similar CO_2 uptake capacity, which is only slightly lower than CH-PE5 but higher than CH. For CH-PS5 and CH-MC5, no improvement in sorption performance was observed compared to CH. The CaO conversion, the other aspect of sorption property, shares the same trend with the CO_2 uptake capacity.

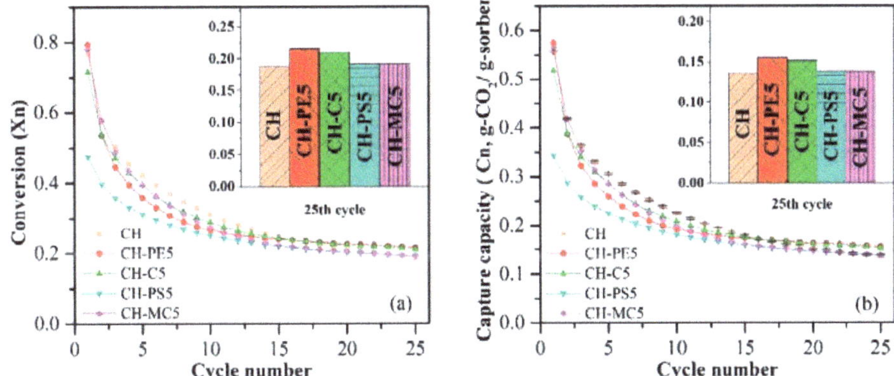

Figure 5. (a) Carbonation conversion and (b) CO_2 capture capacity of sorbent pellets prepared with different pore-forming materials. Testing conditions: calcination at 900 °C for 5 min in N_2, and carbonation at 650 °C for 20 min in 15 vol. % CO_2.

3.3.2. Mechanical Performance

Figure 6 plots the compressive strength of sorbent pellets (with/without calcination) prepared from different pore-forming materials. Without the calcination, CH pellets show a compressive strength of 6.7 N, while CH-PE5 possesses a value of 7.6 N, the highest among all fresh pellets. After calcination, the mechanical strength of all pellets experiences a significant decline. Among them, CH shows the lowest compressive strength of 3.2 N, almost a half loss comparing to that without calcination. By contrast, the addition of PS leads to about 30% improvement of the compressive strength to 4.2 N, and calcined pellets of CH-PE5, CH-C5, and CH-MC5 demonstrated a compressive strength of 4.0 N, 4.0 N, and 3.2 N, respectively. Since the compressive strength of calcined pellets is more meaningful in CaL application, it is concluded that the addition of pore-formers PE, PS, and C could modify pore structure of sorbents in the way that benefits the mechanical strength of pellets.

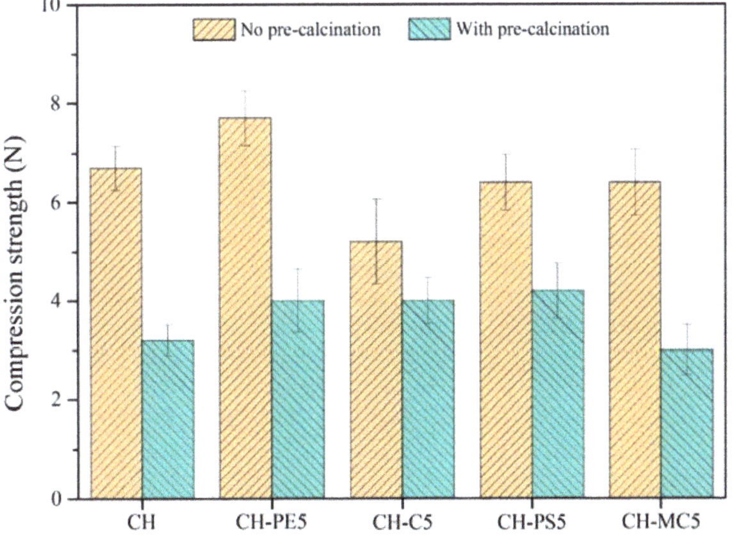

Figure 6. Compressive strength of sorbent pellets (at a mean diameter of 1 mm) with different pore-forming materials.

The results of impact crushing test are shown in Figure 7. By comparing particle size distribution after impact crushing, attrition resistance of diverse sorbent pellets could be evaluated. Obviously, the anti-attrition ability of fresh pellets is much better than calcined pellets, which, however, is more meaningful in the practical CaL application. That is our focus in the following work. It is seen that when particles smaller than 187.5 μm are regarded as completely crushed materials, the mass loss fraction of CH pellets reach to nearly 24.4% from the cumulative percentage results. In contrast, the value for CH-PE5% and CH-C5 was only around 5.9% and 5.2%, respectively. Meanwhile, the mass fraction of residual particles within a diameter range of 750–1250 μm is 11.2% for CH-PE5 from the mass fraction results, which is the highest. These results indicate that the addition of a small amount pore-formers enhances the mechanical strength of sorbent pellets. It is possible that the pores formed during the pyrolysis of pore-formers at a relatively low temperature are able to act as channels for the release of CO_2 in the decomposition of $CaCO_3$, avoiding the formation of additional cracks, and is beneficial for keeping the mechanical strength.

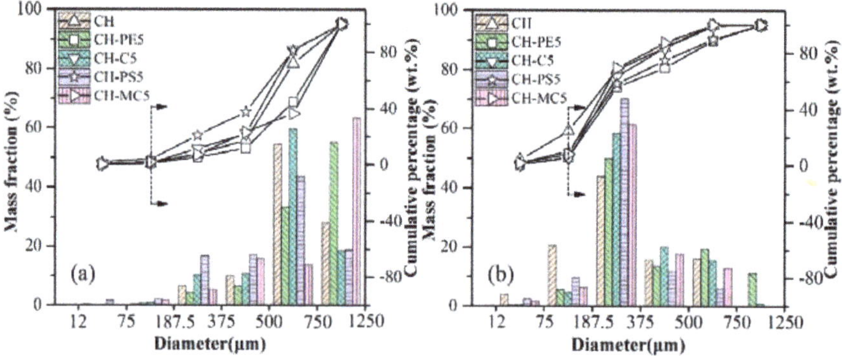

Figure 7. Impact crushing resulted in particle size distribution of sorbent pellets with different pore-forming materials: (**a**) fresh, (**b**) after calcination at 900 °C.

3.4. Effect of Addition Content of PE

The above results indicate that the addition of PE resulted in the best improvement in both the CO_2 sorption and the mechanical performance among all pore-formers, thus we optimized its addition content from 2.5% to 20% in this part, while keeping the particle size of PE at 12 μm.

3.4.1. Sorption/Desorption Performance

Figure 8 shows that CO_2 sorption capacity of all CH-PE samples were improved under the conditions studied. A tendency of first-increase-then-decrease was observed with the increasing addition of PE, and the peak appears when 5 wt. % PE was utilized. Theoretically, with the increasing addition of pore-forming materials, CO_2 sorption performance should keep going up. However, it never happened in our experiment. The reason could be that the most suitable pore size range for long cyclic CO_2 capture is 20–50 nm [35]. However, the addition of too much PE would result in pores that are out of the aforementioned optimum range and lead to a loss in CO_2 capacity in a long cycle.

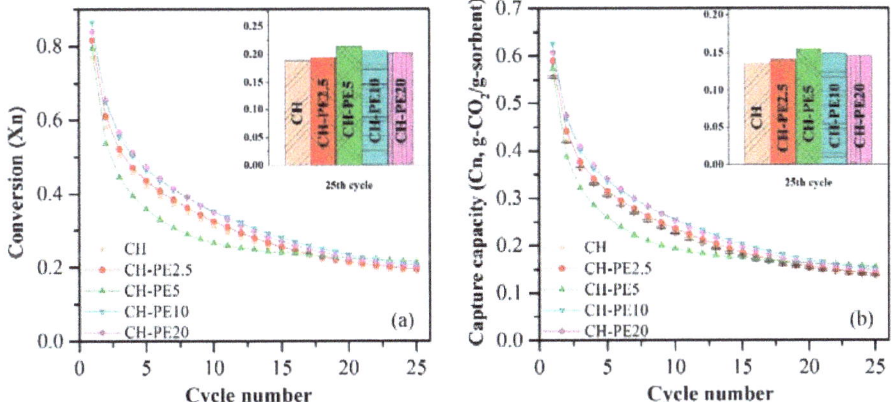

Figure 8. (**a**) Carbonation conversion and (**b**) CO_2 capture capacity of sorbent pellets with different addition of PE. Testing conditions: calcination at 900 °C for 5 min in N_2, and carbonation at 650 °C for 20 min in 15 vol. % CO_2.

3.4.2. Mechanical Performance

The compressive strength of pellets with different addition of PE is illustrated in Figure 9. Similar to the variation of chemical sorption, the compressive strength of calcined CH-PE goes up with the increasing addition of PE, especially for 2.5 wt. % and 5 wt. %, and then slightly declines. It is also clear that pellets without calcination have a much better mechanical strength. The results of impact crushing test for different CH-PE sorbent pellets are summarized in Figure 10. It shows that calcined CH-PE2.5 and CH-PE5 pellets have a much better attrition resistance, whose complete mass loss is 6.9% and 5.9%, respectively, from the cumulative percentage results. In contrast, the value for CH-PE10 and CH-PE20 pellets is 22.1% and 31.1%, which are similar to CH. In addition, CH-PE5% has the biggest mass fraction of residual particle whose size ranges in 750–1250 μm. So, it could be concluded that, within a small addition of PE, the mechanical strength could be significantly enhanced. However, when further increasing the amount of PE, the cavities and gases released in PE pyrolysis are too much to hold a favorable structure for the good mechanical strength of sorbent pellets.

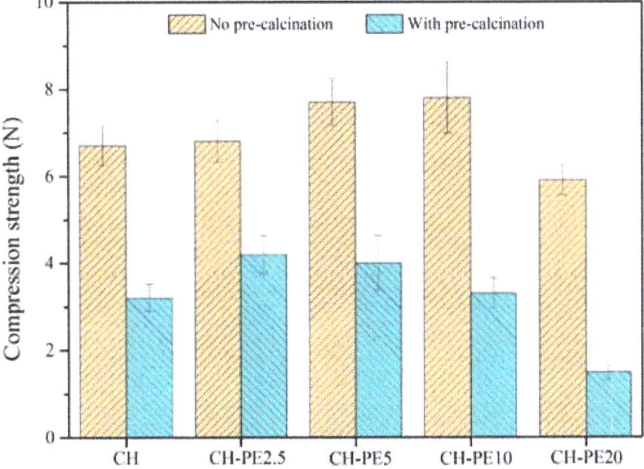

Figure 9. Compressive strength of synthetic sorbent pellets (a mean diameter of 1 mm) with different addition of PE.

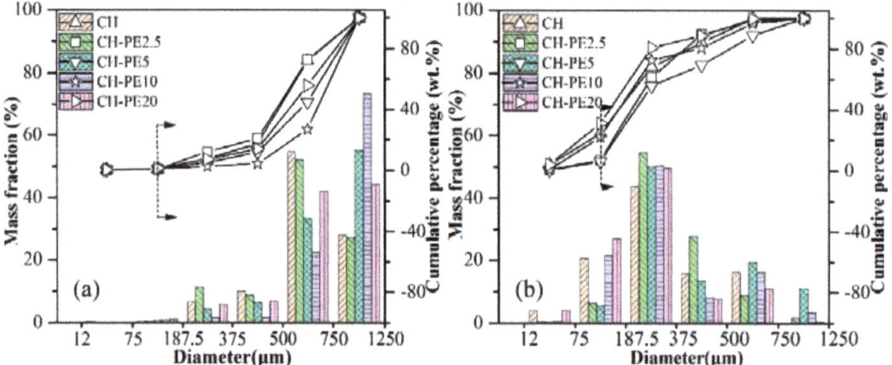

Figure 10. Impact crushing resulted in particle size distribution of sorbent pellets with different additions of PE: (**a**) fresh, (**b**) after calcination at 900 °C.

3.5. Effect of Pore-Former Particle Size

Particle size of the pore-forming material could possibly change the performance of sorbent pellets. However, few experiments were reported in this field deeply. Here, PE with varying particle sizes from 6 μm to 150 μm were selected and tested while its addition content was kept at 5%.

3.5.1. Sorption/Desorption Performance

The CO_2 sorption performance of sorbent pellets with different particle sizes of PE are presented in Figure 11. It is evident that sorbent pellets with a smaller particle size of PE have a better performance in CO_2 capturing. The CH-PE5-6 pellet had the highest CO_2 uptake capacity of 0.157 g-CO_2/g-sorbent which was 6.8% higher than CH-PE5-150 pellets, followed by CH-PE5-12 and CH-PE5-50. It can be concluded that the chemical performance was inversely proportional to the particle size of pore-former utilized. It is very likely that with the same content of PE added, the smaller particle size could lead to the formation of a more uniform distribution of smaller pores, in other words, a bigger specific surface area, and bigger pore volume. Thus, there is more "activated space" for the CO_2 to react with CaO.

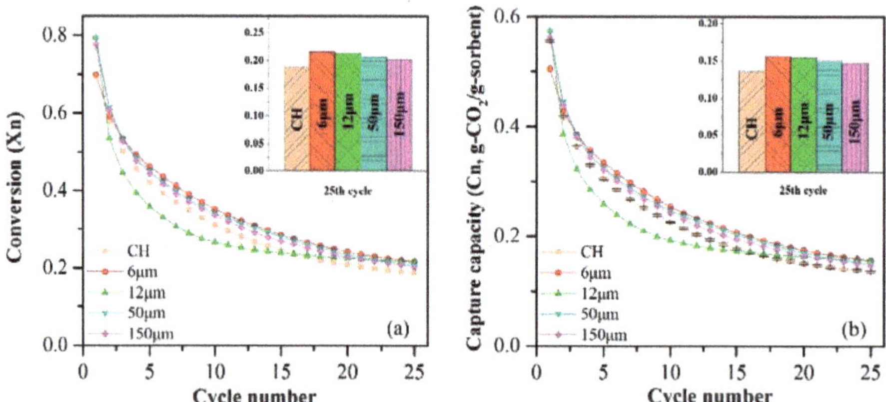

Figure 11. (**a**) Carbonation conversion and (**b**) CO_2 capture capacity of synthetic pellets with different particle sizes of PE. Testing conditions: calcination at 900 °C for 5 min in N_2, and carbonation at 650 °C for 20 min in 15 vol. % CO_2.

3.5.2. Mechanical Performance

Compressive strength of synthetic sorbent pellets with different particle sizes of PE are shown in Figure 12. A tendency of first-increase-then-decrease compressive strength was observed with the increasing particle size of PE, for both fresh and calcined pellets. The CH-PE5-12 possessed the biggest compressive strength among all calcined pellets while CH-PE5-150 was the worst. The ratio of the mechanical strength value for pellets after calcination to fresh pellet was also calculated in order to understand the influence of different particle sizes. The value is 55.9%, 51.9%, 23.9%, and 18.9% for CH-PE5-6, CH-PE5-12, CH-PE5-50, and CH-PE5-150, respectively, showing that bigger particle size of the pore-former has a negative effect in maintaining the compressive strength of pellets during high-temperature calcination.

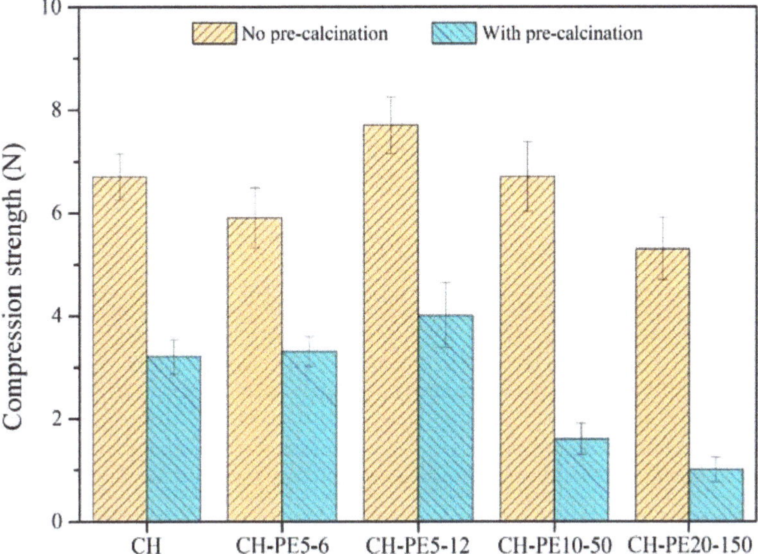

Figure 12. Compressive strength of synthetic sorbent pellets (a mean diameter of 1 mm) with different particle size of PE.

Figure 13 summarizes impacting test results for sorbent pellets with different particle sizes of PE. For the various fresh pellets, their particle distribution after impacting test is very similar, and the complete mass loss is around 1% from the cumulative percentage results. For calcined ones, due to the significant loss in mechanical strength of pellets added with bigger PE (50 μm, 150 μm), lots of them broke up during the pre-calcination process. Thus, we did not present results in the diagram. But conclusion can also be reached that with the smaller particle size of pore-former added, the better mechanical strength sorbent pellets have, which is also consistent with the conclusion obtained in the compression test.

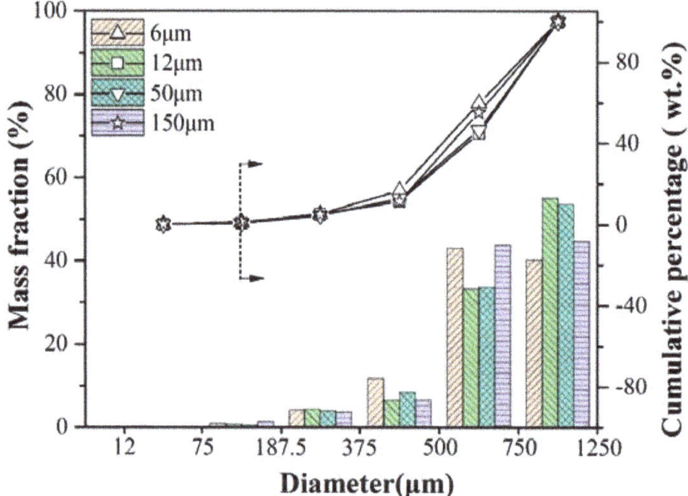

Figure 13. Impact crushing resulted in particle size distribution of fresh sorbent pellets with different particle size of PE.

4. Conclusions

This work studied CO_2 uptake and anti-attrition performance of CaO-based pellets synthesized via extrusion-spheronization with the addition of a small amount of pore-forming materials. Four kinds of pore-forming materials with varying additional content and particle size were investigated in this work. It was found that the addition of PE had a positive effect on enhancing the CO_2 sorption capacity meanwhile maintaining a relatively high mechanical strength, compared to pure $Ca(OH)_2$ pellets. The reason is micro-structures (i.e., pore distribution, surface area, and pore volume) of synthetic sorbents were modified in a way promoting the diffusion and subsequent CO_2 reaction with sorbents. After 25 typical cycles, pellets with 5% PE captured 14% more CO_2 and possessed 25% higher mechanical strength than pure $Ca(OH)_2$ pellets. The lesser or greater addition of PE did not bring further performance enhancement. The particle size of pore-formers was also observed to affect the performance of prepared sorbent pellets, and the smaller ones led to a better chemical performance in CO_2 sorption. In contrast, there was an optimal size of PE (12 µm) for the mechanical strength of sorbent pellets.

Author Contributions: Conceptualization, C.Q. and Z.Z.; investigation, Z.Z.; resources, C.Q.; data curation, S.P. and D.H.; writing—original draft preparation, Z.Z.; writing—review and editing, C.Q.; supervision, C.Q. and J.R..; project administration, C.Q and J.R.; funding acquisition, C.Q. and J.R.

Funding: This research was funded by the National Key R&D Program of China, grant number 2017YFB0603300; National Natural Science Foundation of China, grant number 51606018; Chongqing Basic Science and Advanced Technology Research Program, grant number cstc2017jcyjAX0324; the Innovation Program for Chongqing Overseas Returnees, grant number cx2017021; Fundamental Research Funds for the Central Universities, grant number 2018CDJDDL0005; Key Industrial Generic Technology Innovation Project of Chongqing, grant number cstc2016zdcy-ztzx20006.

Acknowledgments: The authors are grateful for financial support from the National Key R&D Program of China (No. 2017YFB0603300), National Natural Science Foundation of China (No. 51606018), Chongqing Basic Science and Advanced Technology Research Program (No. cstc2017jcyjAX0324), the Innovation Program for Chongqing Overseas Returnees (No. cx2017021), Fundamental Research Funds for the Central Universities (No. 2018CDJDDL0005), and Key Industrial Generic Technology Innovation Project of Chongqing (No. cstc2016zdcy-ztzx20006).

Conflicts of Interest: The authors declare no conflict of interest.

References

1. Pachauri, K.; Meyer, A. Climate Change 2014: Synthesis Report. *Environ. Policy Collect.* **2014**, *27*, 408.
2. International Energy Agency. *World Energy Outlook*; International Energy Agency: Paris, France, 2012.
3. Liu, W.Q.; An, H.; Qin, C.L.; Yin, J.J.; Wang, G.X.; Feng, B.; Xu, M.H. Performance Enhancement of Calcium Oxide Sorbents for Cyclic CO_2 Capture—A Review. *Energy Fuels* **2012**, *26*, 2751–2767. [CrossRef]
4. Frank, Z. Energy and Material Balance of CO_2 Capture from Ambient Air. *Environ. Sci. Technol.* **2007**, *41*, 7558–7563.
5. Zhang, Z.; Li, Y.; Zhang, W.; Wang, J.; Soltanian, M.R.; Olabi, A.G. Effectiveness of amino acid salt solutions in capturing CO_2: A review. *Renew. Sustain. Energy Rev.* **2018**, *98*, 179–188. [CrossRef]
6. Zhang, Z.; Cai, J.; Chen, F.; Li, H.; Zhang, W.; Qi, W. Progress in enhancement of CO_2 absorption by nanofluids: A mini review of mechanisms and current status. *Renew. Energy* **2018**, *118*, 527–535. [CrossRef]
7. Qin, C.; He, D.; Zhang, Z.; Tan, L.; Ran, J. The consecutive calcination sulfation in calcium looping for CO_2 capture Particle modeling and behaviour investigation. *Chem. Eng. J.* **2018**, *334*, 2238–2249. [CrossRef]
8. Alonso, M.; Rodríguez, N.; González, B.; Grasa, G.; Murillo, R.; Abanades, J.C. Carbon dioxide capture from combustion flue gases with a calcium oxide chemical loop. Experimental results and process development. *Int J Grennh Gas Con.* **2010**, *4*, 167–173. [CrossRef]
9. Lu, D.Y.; Hughes, R.W.; Anthony, E.J. Ca-based sorbent looping combustion for CO_2 capture in pilot-scale dual fluidized beds. *Fuel Process. Technol.* **2008**, *89*, 1386–1395. [CrossRef]
10. Ströhle, J.; Junk, M.; Kremer, J.; Galloy, A.; Epple, B. Carbonate looping experiments in a 1 MW th pilot plant and model validation. *Fuel* **2014**, *127*, 13–22. [CrossRef]
11. Qin, C.; Yin, J.; Feng, B.; Ran, J.; Li, Z.; Manovic, V. Modelling of the calcination behaviour of a uniformly-distributed $CuO/CaCO_3$ particle in Ca–Cu chemical looping. *Appl. Energy* **2016**, *164*, 400–410. [CrossRef]
12. Qin, C.; Bo, F.; Yin, J.; Ran, J.; Li, Z.; Manovic, V. Matching of kinetics of $CaCO_3$ decomposition and CuO reduction with CH_4 in Ca–Cu chemical looping. *Chem. Eng. J.* **2015**, *262*, 665–675. [CrossRef]
13. Materic, V.; Hyland, M.; Jones, M.I.; Holt, R. Investigation of the friability of Ca looping sorbents during and after hydration based reactivation. *Fuel* **2014**, *127*, 70–77. [CrossRef]
14. Scala, F.; Salatino, P.; Boerefijn, R.; Ghadiri, M. Attrition of sorbents during fluidized bed calcination and sulphation. *Powder Technol.* **2000**, *107*, 153–167. [CrossRef]
15. Chen, H.; Zhao, C.; Yang, Y. Enhancement of attrition resistance and cyclic CO_2 capture of calcium-based sorbent pellets. *Fuel Process. Technol.* **2013**, *116*, 116–122. [CrossRef]
16. Liu, W.; Low, N.W.; Feng, B.; Wang, G.; Diniz da Costa, J.C. Calcium precursors for the production of CaO sorbents for multicycle CO_2 capture. *Environ. Sci. Technol.* **2009**, *44*, 841–847. [CrossRef] [PubMed]
17. Martavaltzi, C.S.; Lemonidou, A.A. Development of new CaO based sorbent materials for CO_2 removal at high temperature. *Microporous Mesoporous Mater.* **2008**, *110*, 119–127. [CrossRef]
18. Wang, Y.; Zhu, Y.; Wu, S. A new nano CaO-based CO_2 adsorbent prepared using an adsorption phase technique. *Chem. Eng. J.* **2013**, *218*, 39–45. [CrossRef]
19. Xu, Y.; Luo, C.; Zheng, Y.; Ding, H.; Zhou, D.; Zhang, L. Natural Calcium-Based Sorbents Doped with Sea Salt for Cyclic CO_2 Capture. *Chem. Eng. Technol.* **2017**, *40*, 522–528. [CrossRef]
20. Al-Jeboori, M.J.; Fennell, P.S.; Nguyen, M.; Ke, F. Effects of Different Dopants and Doping Procedures on the Reactivity of CaO-based Sorbents for CO_2 Capture. *Energy Fuels* **2012**, *26*, 6584–6594. [CrossRef]
21. Manovic, V.; Fennell, P.S.; Al-Jeboori, M.J.; Anthony, E.J. Steam-Enhanced Calcium Looping Cycles with Calcium Aluminate Pellets Doped with Bromides. *Ind. Eng. Chem. Res.* **2013**, *52*, 7677–7683. [CrossRef]
22. Wang, Y.; Lin, S.; Suzuki, Y. Limestone Calcination with CO_2 Capture (II): Decomposition in CO_2/Steam and CO_2/N_2 Atmospheres. *Energy Fuels* **2008**, *22*, 2326–2331. [CrossRef]
23. Yu, Z.; Duan, L.; Su, C.; Li, Y.; Anthony, E.J. Effect of steam hydration on reactivity and strength of cement-supported calcium sorbents for CO_2 capture. *Greenh. Gases Sci. Technol.* **2017**, *7*, 915–926. [CrossRef]
24. García-Labiano, F.; Abad, A.; De Diego, L.; Gayan, P.; Adanez, J. Calcination of calcium-based sorbents at pressure in a broad range of CO_2 concentrations. *Chem. Eng. Sci.* **2002**, *57*, 2381–2393. [CrossRef]
25. Fennell, P.S.; Pacciani, R.; Dennis, J.S.; Davidson, J.F.; Hayhurst, A.N. The Effects of Repeated Cycles of Calcination and Carbonation on a Variety of Different Limestones, as Measured in a Hot Fluidized Bed of Sand. *Energy Fuels* **2007**, *21*, 2072–2081. [CrossRef]

26. Lu, H.; Smirniotis, P.G.; Ernst, F.O.; Pratsinis, S.E. Nanostructured Ca-based sorbents with high CO_2 uptake efficiency. *Chem. Eng. Sci.* **2009**, *64*, 1936–1943. [CrossRef]

27. Sun, J.; Liu, W.; Hu, Y.; Wu, J.; Li, M.; Yang, X.; Wang, W.; Xu, M. Enhanced performance of extruded–spheronized carbide slag pellets for high temperature CO_2 capture. *Chem. Eng. J.* **2016**, *285*, 293–303. [CrossRef]

28. He, D.; Qin, C.; Manovic, V.; Ran, J.; Bo, F. Study on the interaction between CaO-based sorbents and coal ash in calcium looping process. *Fuel Process. Technol.* **2017**, *156*, 339–347. [CrossRef]

29. Xu, Y.; Luo, C.; Zheng, Y.; Ding, H.; Wang, Q.; Shen, Q.; Li, X.; Zhang, L. Characteristics and performance of CaO-based high temperature CO_2 sorbents derived from sol-gel process with different supports. *RSC Adv.* **2016**, *6*, 79285–79296. [CrossRef]

30. Xu, Y.; Ding, H.; Luo, C.; Zheng, Y.; Xu, Y.; Li, X.; Zhang, Z.; Shen, C.; Zhang, L. Porous spherical calcium-based sorbents prepared by a bamboo templating method for cyclic CO_2 capture. *Fuel* **2018**, *219*, 94–102. [CrossRef]

31. Manovic, V.; Wu, Y.H.; He, I.; Anthony, E.J. Spray Water Reactivation/Pelletization of Spent CaO-based Sorbent from Calcium Looping Cycles. *Environ. Sci. Technol.* **2012**, *46*, 12720–12725. [CrossRef]

32. Ridha, F.N.; Wu, Y.; Manovic, V.; Macchi, A.; Anthony, E.J. Enhanced CO_2 capture by biomass-templated $Ca(OH)_2$-based pellets. *Chem. Eng. J.* **2015**, *274*, 69–75. [CrossRef]

33. Duan, L.; Yu, Z.; Erans, M.; Li, Y.; Manovic, V.; Anthony, E.J. Attrition study of cement-supported biomass-activated calcium sorbents for CO_2 capture. *Ind. Eng. Chem. Res.* **2016**, *55*, 9476–9484. [CrossRef]

34. Scala, F.; Montagnaro, F.; Salatino, P. Attrition of Limestone by Impact Loading in Fluidized Beds. *Energy Fuels* **2007**, *21*, 2566–2572. [CrossRef]

35. Scala, F.; Salatino, P. Flue gas desulfurization under simulated oxyfiring fluidized bed combustion conditions: The influence of limestone attrition and fragmentation. *Chem. Eng. Sci.* **2010**, *65*, 556–561. [CrossRef]

Article

Optimal Design of a Carbon Dioxide Separation Process with Market Uncertainty and Waste Reduction

Juan Pablo Gutierrez [1,*], Eleonora Erdmann [1] and Davide Manca [2]

[1] Instituto de Investigaciones para la Industria Química (INIQUI, CONICET-UNSa), Facultad de Ingeniería, Universidad Nacional de Salta. Av. Bolivia 5150, Salta 4400, Argentina; eleonora@unsa.edu.ar

[2] PSE-Lab, CMIC Department, Politecnico di Milano, P.zza Leonardo da Vinci 32, 20133 Milan, Italy; davide.manca@polimi.it

* Correspondence: gutierrezjp@unsa.edu.ar

Received: 11 March 2019; Accepted: 27 May 2019; Published: 5 June 2019

Abstract: The aim of this work is to optimize the conceptual design of an amine-based carbon dioxide (CO_2) separation process for Enhanced Oil Recovery (EOR). A systematic approach is applied to predict the economic profitability of the system while reducing the environmental impacts. Firstly, we model the process with UniSim and determine the governing degrees of freedom (DoF) through a sensitivity analysis. Then, we proceed with the formulation of the economic problem, where the employment of econometric models allows us to predict the highest dynamic economic potential (DEP). In the second part, we apply the Waste Reduction (WAR) algorithm to quantify the environmental risks of the studied process. This method is based on the minimization of the potential environmental indicator (PEI) by using the generalization of the Waste Reduction algorithm. Results show that the CO_2 separation plant is promising in terms of economic revenues. However, the PEI value indicates that the higher the profitability, the larger the environmental risk. The optimal value of the DEP corresponds to 0.0274 kmol/h and 60 °C, with a plant capacity according to the mole flow rate of the produced acid gas. In addition, the highest environmental risk is observed at the upper bounds of the DoF.

Keywords: optimal conceptual design; market prediction; economic uncertainty; environmental impact; carbon dioxide separation

---

1. Introduction

Several stages exist to recover the original pressure of mature oil and gas wells. Among those already applied, the Enhanced Oil Recovery (EOR) with carbon dioxide (CO_2) proved to be a mid-term solution to increase the oil production to its original levels while capturing thousands of tonnes of CO_2 [1,2].

Haszeldine [3] states that the first injections of carbon dioxide into the microscopic pores of sedimentary rocks date from the early 1970s. Successful cases of CO_2-EOR have been reported in the United States, United Kingdom, Norway, and Canada by Wright et al. [4] and Mumford et al. [5]. The injection of CO_2 was also evaluated in the reservoirs of Argentina, a region where EOR pilot experiences were barely intended. Although the results provided good revenues, the CO_2-EOR in the region remains unmaterialized after more than twenty years since first being discussed [6].

The main problem related to this procedure is the large and continuous amount of CO_2 necessary to start the EOR injection [7]. In this regard, Herzog [8] reports that the common sources for large amounts of CO_2 correspond to the acid gas coming from natural gas processing.

Kwak et al. [9] compare different technologies for CO_2 separation from natural gas. Based on simulation and economic studies, they conclude that chemical absorption with methyldiethanolamine (MDEA) is the least expensive and most feasible option to separate carbon dioxide. Moreover,

Leung et al. [10] note the amine processes' high efficiency, large amounts of acid gas as a side product, and the possibility to regenerate the solvent. Other comparable processes include separation with polymeric membranes, cryogenic separation, physical solvents, and hybrid technologies.

Another task when evaluating CO_2-EOR possibilities is the large dependence of oil and gas companies upon economic conditions and countries' institutional frameworks [11]. For instance, Ponzo et al. [12] state that changing market structures influence the long-term evolution of gas quotations and, consequently, the development of gas fields. Moreover, interdependency among variations of time with technical, operative, and economic conditions has been assigned to perform economic evaluation by Manolas et al. [13]. Classically, the interaction between the operating aspects and economic revenues during the definition of a process is first estimated according to the conventional conceptual design [14]. Conceptual process design (CD) consists of the selection of proper operation units, their sequences, and the recycling structure needed to obtain a specified product [15]. However, Sepiacci et al. [16] explain that this conventional method is no longer representative when considering market uncertainty, demand and offer fluctuations, and the price instability of commodities and utilities. Then, Manca and Grana [17] introduced the benefits of dynamic conceptual design (DCD). Based on CD and the economic potentials (EP) presented by [14], DCD takes into account the dynamic features of price/cost fluctuations within a given time horizon.

Indeed, the process design of chemical industries are considered complete when performing the environmental risk analysis of new process systems. Currently, there is a great deal of interest in the development of methods that can be used to minimize the generation of pollution, and there are numerous efforts underway in this area [18]. Specifically, this interest has increased with the world's awareness of CO_2 emissions and made process engineering adopt practices to mitigate the effects of climate change [19].

For the above reasons, we apply the concept of DCD to obtain and condition CO_2 for EOR purposes. As can be anticipated, we focus our study to establish the conceptual design of the process in the context of market instability and future uncertainties. CO_2 for EOR is obtained from a natural gas sweetening design that uses MDEA as solvent; the specifications for the produced CO_2 include a 95 mol% concentration of the acid gas, compressed at 6500 kPa [20].

An optimization problem is formulated with the aim of minimizing the Dynamic Economic Potential (DEP) of the design. In this sense, Mores et al. [21] state that two degrees of freedom (DoF) govern the optimization problem of the CO_2 MDEA absorption—the recycled amine flow rate and its temperature. However, we extend the analysis to prove that the variable most affecting the energy demands of the plant is the water makeup of the amine solution, and thus more proper DoF.

Then, we analyze the historical prices of products and raw materials by using statistical tools. We present natural gas prices as references to estimate the evolution of the rest of the involved components by using numerical correlations. Linear Regression Models (such as AutoRegressive model with an eXternal input, ARX) are applied to interpret the behavior of past quotations. We switch the contribution of these economic models into econometrics to make them capable of predicting quotations and generating future market scenarios.

On the other hand, we perform an assessment to find the pair of DoF that reduce the environmental potential index. The method is adapted from the Waste-Reduction algorithm applied to chemical processes presented by Young et al. [22]. The Waste Reduction (WAR) algorithm has been developed to describe the flow and the generation of potential environmental impact through a chemical process.

2. Process Description

The purpose of a natural gas sweetening process is to remove the acid gases from a sour natural gas stream. Due to the high selectivity of the solvent, the by-product of this process is a high-purity CO_2 material stream that, after conditioning, can be used as an EOR fluid.

The regular process of natural gas sweetening to obtain CO_2 is divided into two parts [23]. In the first stage, which consists of an absorber column, the natural acid is put in countercurrent contact with a descending MDEA aqueous solution—a so-called lean amine [24]. Fouad and Berrouk [25] and Kazemi et al. [26] indicate that low temperatures and high pressures favor the exothermic reaction that occurs in the unit. After contact, the aqueous solution of amine is pressurized, heated, and sent to the regeneration stage [27]. This second stage consists of a distillation column where the acid gas is removed from the amine solution due to an external heat contribution. Different studies have been performed in order to optimize the energy requirements of the regeneration column [28–30]. The liquid from the regenerator column is cooled and pumped back to the absorption stage [31,32]. Water and MDEA are placed in the stream from the bottom of the column to the absorption tower to compensate for leaks within the operation. Meanwhile, the high-purity CO_2 from the top of the regenerator is sent to a series of four centrifugal compressors to considerably increase the pressure. Original well pressures are required to dispose of the CO_2 as an injection fluid; in this case the value remains over 6500 kPa. The 4-stage compression design includes intercooling units and intermediate separation stages [33].

3. Methods

3.1. Simulation Base Case

A process of CO_2 absorption and compression is modeled by using UniSim [34]. Natural gas at a value of 2500 mm^3/d is assumed as the plant's capacity. The conditions of the plant are those reported in the work of Gutierrez et al. [19]: sour natural gas at 35 °C and 6178 kPa with 93 and 4 mol% of CH_4 and CO_2, respectively. Also, the conditions of the lean amine are reproduced. We consider 21,000 kmol/h of an aqueous MDEA solution (38 wt%), at 42 °C and 9610 kPa. A 24-tray absorption column operates at the pressure of the inlet gas. Rich MDEA from the bottom of the absorber is flashed at 441 kPa, heated up to 90 °C, and then sent to regeneration. The regeneration column consists of 24 trays and operates at 90 °C and 443 kPa. To provide the column with an external heat, we assume a reboiler unit using natural gas as fuel. Recycled MDEA is pumped and cooled, first exchanging heat with the rich amine, and then with a cooler so that it reproduces the temperature of absorption.

A 4-stage compression system is employed to increase the pressure of the produced CO_2 up to 6865 kPa [33]. Figure 1 shows the simulation of (**a**) the CO_2 separation plant and (**b**) the compression sector to produce the high-purity CO_2 stream.

Muhammad and GadelHak [35] explain that the main variables affecting CO_2 absorption are solvent flow rate and the absorber temperature, this last through the cooling of the lean amine stream.

As we anticipated, two streams conform the solvent inlet flow stream, one corresponding to a pure MDEA stream and the other connected to the makeup water. Generally speaking, two independent variables are related to the same degree of freedom, so in this study we determine whether there is a strong dependency between the main energy requirements and the independent variables. Similar to Torres-Ortega et al. [36], we perform a sensitivity analysis to evaluate suitable ranges of variation for the decision variables along the optimization.

3.2. Predictive Concept Design

This section provides the dynamic approach to the economic assessment for the CO_2 conditioning plant. Econometrics models (EM) are employed to simulate and evaluate future trajectories of prices and costs.

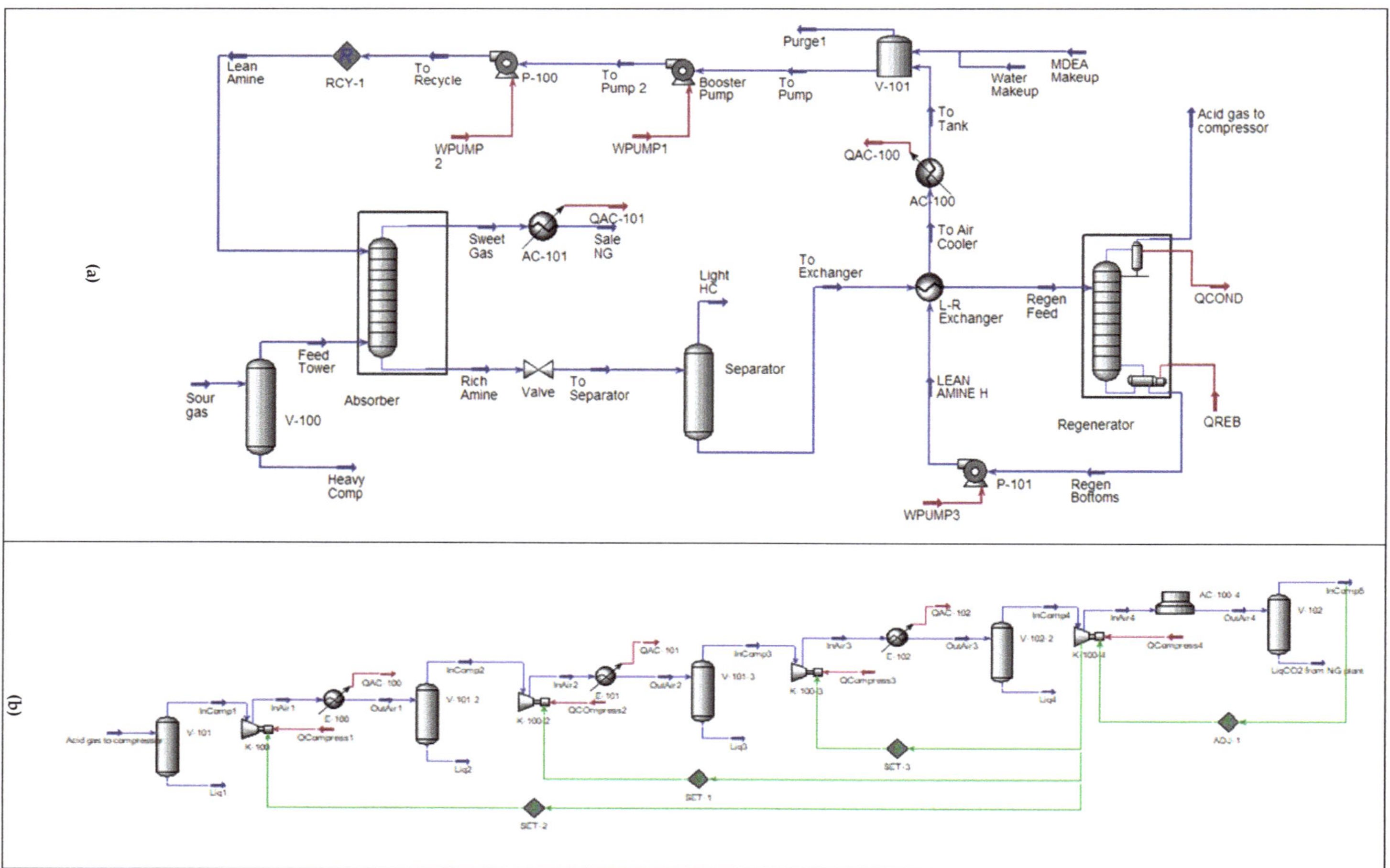

Figure 1. Simulation model in UniSim: (**a**) absorption sector and (**b**) compression sector.

3.2.1. Development of Econometric Models

The first step while performing EM is the selection of a reference component (RC). Manca [37] employs RC historical quotations to estimate the economic dynamics of all commodities and utilities in the process he analyzed. Moreover, Manca [38] suggests that an RC must be representative of the sector where the plant operates, with the availability of frequent data and updated price evolution.

A good RC for the industry of Oil and Gas is crude oil (CO) [39]. CO, and also the evolution of natural gas (NG), quotations are traced daily for EIA [40]. However, the prices of natural gas produced in the basins of Argentina are also indicated monthly by the Ministry of Energy [41]. In this study, we perform the EM for both CO and NG as potential candidates for reference components.

A structural auto regression model is applied to separately autocorrelate both West Texas Intermediate (WTI) crude oil and US natural gas prices [42]. For both potential candidates, we analyze monthly quotations from July, 2007 to July, 2017 (the last available date). To correlate the historical values of the quotations, we use a similar methodology as the one used by Zhou et al. [43] regarding the coefficients of the Pearson equation (Equation (1)). Pearson coefficients (PCs) measure the strength and direction of the linear relationship between two random variables [44]. In this case, both variables represent the monthly quotations of the RC, but differ in one period:

$$r_k = \frac{\sum_{t=k+1}\left(Y_t - \overline{Y}\right)\left(Y_{t-k} - \overline{Y}\right)}{\sum_{t=1}^{n}\left(Y_t - \overline{Y}\right)^2} \tag{1}$$

where r_k denotes the PC for a particular period. $\sum_{t=k+1}\left(Y_t - \overline{Y}\right)\left(Y_{t-k} - \overline{Y}\right)$ is the covariance of the quotations (Y_t) with respect to one-period of the previous quotations (Y_{t-k}), and $\sum_{t=1}^{n}\left(Y_t - \overline{Y}\right)^2$ is the squared of the standard deviation. r_k varies from -1 to 1 and, in general, the higher the correlation coefficient, the stronger the relationship is [45]. Dancey and Reidy [46] state that if r_k ranges from 0.7 to 0.9, the strength of correlation is high, and quite enough to determine the size of the correlation. This characteristic can be visualized when plotting the coefficient versus the time lag between the quotations.

3.2.2. Formulation of the Economic Optimization

Once the EM are identified, it is viable to run the grid-search optimization according to the regular process conceptual design (PCD). In the optimization problem, we determine the set of DoF that maximizes the cumulated value of the Dynamic Economic Potential of order four (DEP4), Equations (2)–(4).

$$(Cumulated)_i = \sum_{j=1}^{N} DEP4_{j,i}; \; i = 1, \ldots, I \tag{2}$$

$$DEP4_i\left(\frac{USD}{y}\right) = \sum_{j=1}^{N} Revenues_{j,i} \cdot nHpY - \frac{CAPEX}{N/12} \tag{3}$$

$$Revenues_{j,i}\left(\frac{USD}{y}\right) = \sum_{p=1}^{NP} C_{p,j,i} \cdot F_P - \sum_{r=1}^{NR} C_{r,j,i} \cdot F_r - OPEX_{j,i} \tag{4}$$

where $DEP4$ is the fourth-level economic potential calculated for the $i-th$ economic scenario. j, i are the subscripts for a specific month and scenario, respectively; $nHpY$ is the number of working hours per year. N stands for the number of months to perform the economic assessment. NP, NR, F_P, and F_r represent the number of products and reactants, their flow rates, and C their costs. The $CAPEX$ term is estimated according to the empirical equations reported by Douglas [14]. Six main units are considered for the calculation: absorber and distillation columns, MDEA heat exchanger, and two air coolers.

The $OPEX$ term considers a price trajectory for each raw material, by-product, and utility, for the $i-th$ scenario. The main contributors of the $OPEX$ are two air coolers, a condenser, reboiler fuel, and the total power required for the acid gas compressors (Gutierrez et al. [19]). The material and energy balances required to calculate the *OPEX* are taken from the steady-state simulation of the process.

The goal of the optimization is to determine the combination of DoF that maximizes the value of $(Cumulated)_i$, with respect to a set of generated scenarios, where the assessment becomes probabilistic. To obtain a high-purity CO_2 material stream, Gutierrez et al. [19] use a limit value of 2 mol% in the gas coming from the top of the absorber, so we consider the molar fraction of CO_2 as a restriction for the stated problem.

3.3. Waste Reduction Algorithm

We employ the Waste Reduction (WAR) algorithm to describe the flow and the generation potential environmental impact through the process under study [22]. The general methodology of the WAR algorithm defines Potential Environmental Impact (PEI) indexes to characterize the generation of the potential impact in a process, divided into eight categories.

The first four categories evaluate, globally, the environmental friendliness of a process: human toxicity potential by ingestion (HTPI), human toxicity potential by exposure (both dermal and inhalation) (HTPE), terrestrial toxicity potential (TTP), and aquatic toxicity potential (ATP).

On the other hand, the other four are related to the toxicological aspects of the involved chemicals within the process: global warming potential (GWP), ozone depletion potential (ODP), photochemical oxidation potential (PCOP), and acidification potential (AP).

The potential environmental impacts are calculated from stream mass flow rates, stream composition, and a relative potential environmental impact score for each chemical present in the separation process [18].

According to the notation of Young and Cabezas [47], the output PEI to the chemical process can be rewritten as Equation (5):

$$i_{out}^{(cp)} = \sum_{w=1}^{N} \dot{M}_w^{(out)} \sum_i x_{i,w}\psi_i \tag{5}$$

$$\psi_k = \sum_l \alpha_l \psi_{i,l}^s \tag{6}$$

where $\dot{M}_w^{(out)}$ is the output mass flow rate of stream w, $x_{i,w}$ is the mass fraction of the chemical i in the stream is w, and ψ_i is the overall PEI for the chemical k. ψ_i can be calculated from Equation (6). In Equation (6), $\psi_{i,l}^s$ is the normalized specific PEI of chemical k for the impact category l, and α_l is the relative weighing factor of impact category l [39,47]. A unitary value is assigned to α, to illustrate the case where the eight categories have the same importance in our evaluation [48]. Normalized impact scores are obtained from the WAR algorithm add-in included in the latest release of the CAPE-OPEN to CAPE-OPEN (COCO) Simulation Environment, available from http://www.cocosimulator.org/ [49].

4. Results

4.1. Simulation Output

Figure 2 reports the evolution of the main energy requirements, according to the variation of the independent variables. For this case, we present the reboiler requirement versus (a) the flow rate of the make-up of MDEA, (b) the make-up of water, and (c) the temperature of the recycled amine.

Figure 3 reports the evolution of the condenser energy requirement, according to the variation of the chosen independent variables. Again, we present the condenser requirement versus (a) the flow rate of the make-up of MDEA, (b) the make-up of water, and (c) the temperature of the recycled amine.

Figure 4 shows the evolution of the total compressor power (kW), respect to the variation of the same independent variables. We present the compressor power demand versus (a) the flow rate of the make-up of MDEA, (b) the make-up of water, and (c) the temperature of the recycled amine.

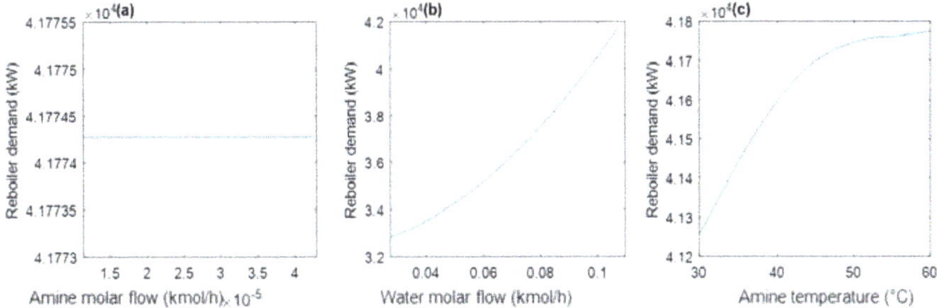

Figure 2. Variation of reboiler energy requirements (kW) versus (**a**) the make-up of amine molar flow (kmol/h), (**b**) water make-up (kmol/h), and (**c**) the recycled methyldiethanolamine (MDEA) temperature (°C).

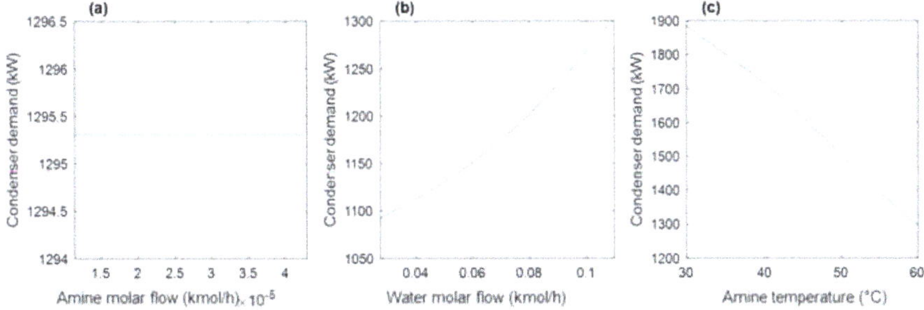

Figure 3. Variation of condenser energy requirement (kW) versus (**a**) the make-up of amine molar flow (kmol/h), (**b**) water make-up (kmol/h), and (**c**) the recycled MDEA temperature (°C).

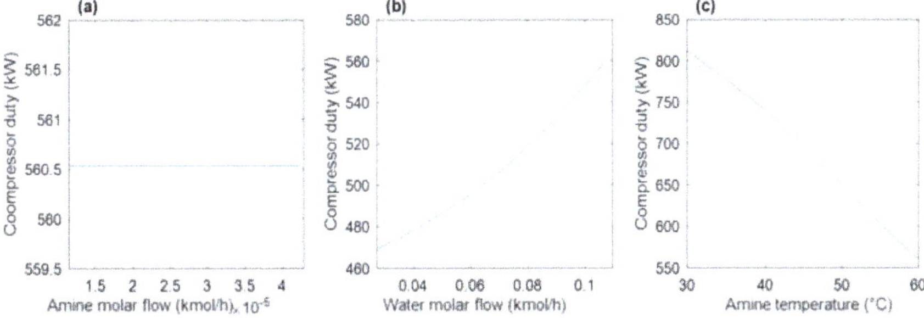

Figure 4. Variation of compressor power demand (kW) versus (**a**) the make-up of amine molar flow (kmol/h), (**b**) water make-up (kmol/h), and (**c**) the recycled MDEA temperature (°C).

Figure 5 shows the evolution of the cooling system requirements (kW), with respect to the variation of the available variables. We present the energy demand of the coolers (AC-100 and AC-101) versus (a) the flow rate of the make-up of MDEA, (b) the make-up of water, and (c) the temperature of the recycled amine.

Figures 2–5 expose a remarkable dependency between the main energy consumptions and the temperature of the recycled MDEA. Moreover, it was illustrated that the energy requirements strongly depend on the flow rate of the water make up. On the other hand, the variation of the MDEA flow rate proves to not alter the energy requirement of the reboiler, condenser, compressors, and the air-coolers. With this analysis, it is demonstrated that the proper DoF, representing the reduction of the recycle MDEA flow rate, corresponds to the water makeup of the process. Previous articles state that the decision variable is the recycled aqueous amine flowrate, but it is demonstrated here that the variable of most impact is the water make-up to conform to that flowrate. Thus, for the objective functions in this work, the decision variables are the water mole flow and the temperature of the recycled amine.

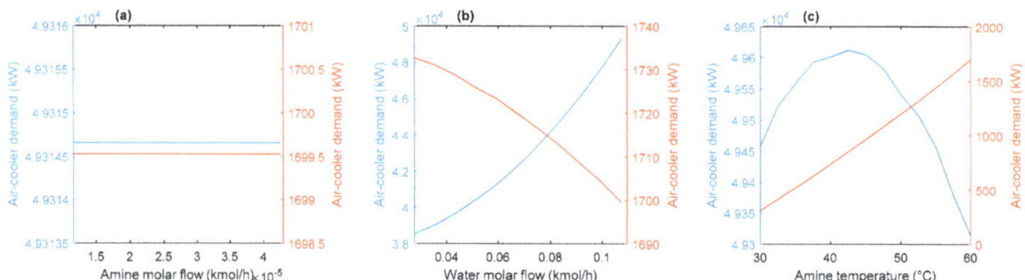

Figure 5. Variation of air-coolers energy demand (kW) versus (**a**) the make-up of amine molar flow (kmol/h), (**b**) water make-up (kmol/h), and (**c**) the recycled MDEA temperature (°C). Blue: cooling system of the recycled amine; orange: cooling system of the sweet natural gas.

4.2. Economic Scenarios

Figure 6 shows the autocorrelograms (PC versus lag time) of (a) CO and (b) NG.

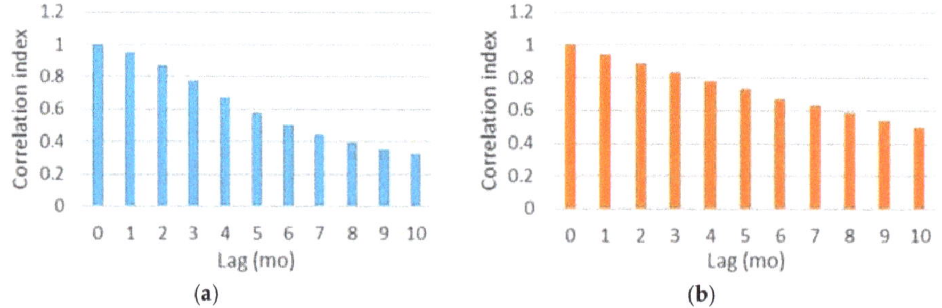

Figure 6. Autocorrelograms of (**a**) crude oil (CO) and (**b**) natural gas (NG).

By analyzing the autocorrelograms shown in Figure 6, one can deduce that the CO quotation at the month $k + 1$ depends mostly on the two previous quotations.

4.2.1. Correlation

In this subsection, we evaluate the relationship among all involved commodities with respect to the potential RC. In Figure 7, we expose the correlation between (a) CO_2, (b) MDEA, and (c) Electric Energy (EE) with respect to crude oil quotations. It can be seen that correlation values change in the range of [−1, 1]. If the two sets are perfectly correlated (e.g., are the same set), the correlation index is 1. On the contrary, if they are anti-correlated (e.g., the two sets have opposite trends), it is −1.

Figure 8 exposes the correlation between the same components and natural gas quotations. CO_2 quotations were estimated in accordance with the work presented by Cook [50]. It can be seen that values of correlation between the set of quotations present higher values compared to the ones obtained by correlating the crude oil. Then, NG is selected as a reference component and econometric model, as presented in Section 4.2.2.

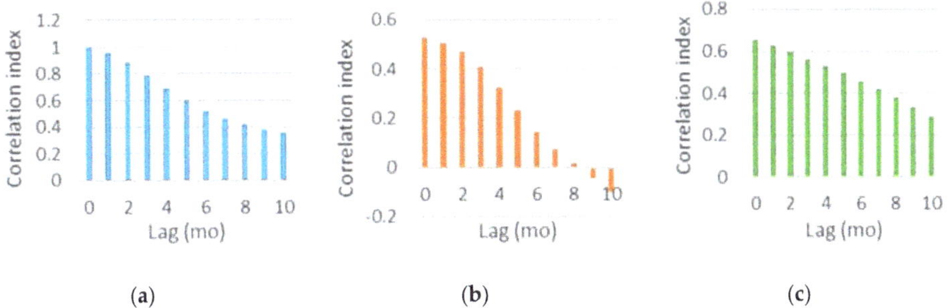

Figure 7. Correlation between CO and (**a**) CO_2, (**b**) MDEA, and (**c**) electricity quotations.

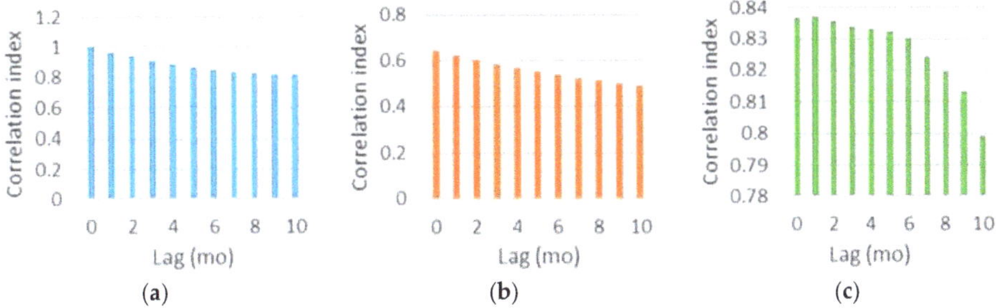

Figure 8. Correlation between NG and (**a**) CO_2, (**b**) MDEA, and (**c**) electricity quotations.

4.2.2. Econometric Models

From Figures 7 and 8, we observe better correlation indexes when comparing to the NG quotations. Then, the EM of NG as RC becomes the one expressed through Equation (5).

Where $P_{NG,k+1}$ is the monthly quotation of NG. σ and $\overline{P}$ are the standard deviation of the prices and the average of relative errors, respectively. *rand* is a stochastic function normally distributed, and A, B, and C are adaptive parameters calculated with linear regression, minimizing the square error between real quotations and those predicted by the model [51].

Manca [38] reports EM for toluene, benzene, propylene, and cumene prices based on a dedicated (auto)correlogram analysis. According to our correlation indexes, we elaborate the EM for the CO_2 conditioning process. Table 1 presents Autoregressive Distributed Lag (ADL) models for estimating each quotation evolution, without the stochastic factor.

$$P_{NG,k+1} = \left(A + B{\cdot}P_{NG,k} + C{\cdot}P_{NG,k-1}\right){\cdot}\left(1 + rand{\cdot}\sigma_{NG} + \overline{P}_{NG}\right). \tag{7}$$

Table 1. ADL EM for NG, CO_2, and MDEA prices, without the stochastic factor.

Component	Model
CO_2	$P_{CO_2,k+1} = A + B{\cdot}P_{NG,k+1} + C{\cdot}P_{CO_2,k} + D{\cdot}P_{CO_2,k-1}$
MDEA	$P_{MDEA,k+1} = A + B{\cdot}P_{NG,k+1} + C{\cdot}P_{MDEA,k} + D{\cdot}P_{MDEA,k-1}$

To simply the forecast EE quotations, we adopt previous monthly prices of the Ministry of Energy [41]. Similar to Manca [52], the EM for EE is based on (auto)correlograms and the economic dependency of the EE to NG. From these observations, it is feasible to apply the model represented by Equation (8):

$$P_{EE,k+1} = A + B{\cdot}P_{NG,k} + C{\cdot}P_{EE,k} \tag{8}$$

where the price of EE ($P_{EE,k+1}$) is estimated employing previous quotations of NG and EE. Table 2 reports the adaptive coefficients, including the models of NG, CO_2, MDEA, and EE.

Table 2. Adaptive parameters of ADL EM of NG, CO_2, and MDEA.

Component	A	B	C	D	σ	$\overline{P}$
NG	0.0362	−0.0285	1.2205	-	0.1918	0.0705
CO_2	0.0033	0.0078	1.4167	−0.4870	0.0606	0.0074
MDEA	0.1124	0.9731	0	−0.0171	0.0126	0.0002

We use the EM of CO_2, MDEA and EE to generate a set of random economic scenarios. Figure 9 shows eight predicted trajectories from the EM of NG, MDEA, CO_2, and EE, during a time horizon of 120 months, in different random colors. It shows a probabilistic approach, based on a distribution of multiple viable economic scenarios.

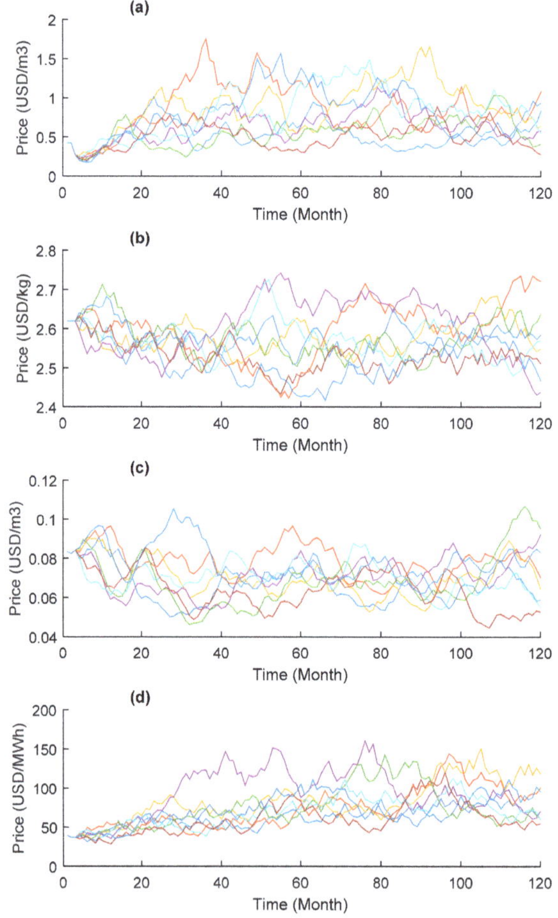

Figure 9. Random price trajectories, for (**a**) NG expressed, (**b**) MDEA, (**c**) CO_2, and (**d**) EE.

Particularly for the case of the Electric Energy (Figure 9d), we present a brief predictive model where A = 2.98, B = 1.316, and C = 0.81 (Sepiacci et al. [16]), in Equation (8). The predictive nature of this model is given by its dependency with the forecasted prices of NG. Other reported models associate the EE prices with the crude oil quotations, but those forecasts are also of random variability [53]. The Electric Energy has a great impact as a process utility because of the type of its cooling equipment and compression system. Although the prices of the utility vary periodically with the time domain, we assume this simplified behavior for the scope of this article.

Each colorful line corresponds to random trajectories generated from the econometric models of Equations (6)–(8). The prices of each item can follow one of the colorful trajectories within the time horizon.

4.3. Optimal Economic and Environmental Friendly Design

Results concerning the optimal design of the CO_2 separation plant are shown in this section.

4.3.1. DEP4 Cumulated

Figure 10 shows, on the y-axis, the value of the cumulated DEP4 (USD), and on the x-axis it shows the time series of market quotations. Each bar represents the value of the DEP4 calculated by considering the market quotations of the corresponding month based on the specific plant configuration that maximizes the DEP4 value.

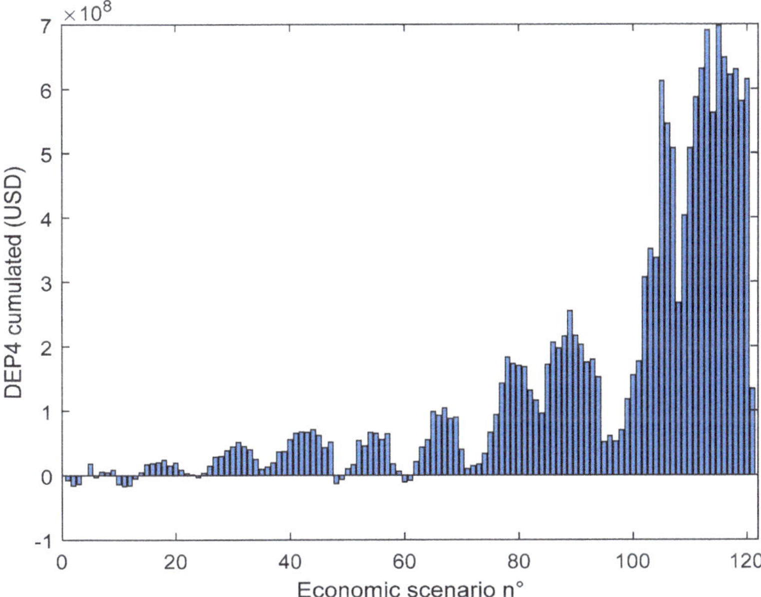

Figure 10. Fluctuation of the DEP4 (USD) according to the number of the generated scenario.

The generated models are used to produce a set of economic scenarios that are distributed according to the modeled fluctuations of quotations and the stochastic contribution of the reference component. The cornerstone of this methodology is symbolized by the number of scenarios that are called for quantifying a set of different scenarios subject to the price/cost trajectories obtained by the econometric models through their stochastic contribution (Random). Therefore, it refers to a probabilistic concept of PCD that is grounded in the distribution of possible economic scenarios for this specific process. A necessary condition for economic sustainability is that the DEPs are positive.

It can be seen that DEP4 varies, even to negative values, during the time domain. Each bar of the graph represents the higher value of EP4, corresponding to the best combination of the DoF, at one particular month. In general, the economic potential fluctuation strongly depends on the price volatility of raw materials and final products. Where positive, the obtained DEP4 is of an eight-power magnitude, which demonstrates the economic potential of the plant in accordance with the predictive models.

4.3.2. Economic Optimal

Figure 11 illustrates the trend of the cumulated DEP4 as a function of the DoF, the water flow make-up, and the temperature of the recycled amine. The presented surface represents the maximization of Equation (4), where a total capital expenditure of 1.44×10^7 USD is estimated from the calculation. As previously stated, the DEP4 is not represented by a single value but by a distribution of values, one for each scenario. In order to have a simple representation of the economic objective function, we present the average value of the cumulated DEP4. The results of Equation (4) show that the average of the cumulated DEP4 reaches eight order values.

The configuration yielding the maximum value of the cumulated DEP4 corresponds to a temperature equal to 60 °C for the MDEA to recycle and value the water amine flow rate equal to 0.0274 kmol/h.

Based on this experience, high temperatures of MDEA imply that the conversion of the absorption reaction is increased and, consequently, the produced CO_2 is increased. Interestingly, an increment of the water flow rate proves that the MDEA concentration of 38 wt% can be modified to obtain a better performance in terms of the economical aspect of this process. At the same value of temperature, 60 °C, and 0.1074 kmol/h, the cumulated DEP4 is equal to 1.06×10^8 USD. The order of magnitude of this DEP4 is even higher than the one obtained by Sepiacci et al. [16], who obtained a six-order DEP4 while applying this methodology in a petrochemical process.

4.3.3. Minimal Environmental Risks

Figure 12 shows the behavior of the PEI. In this case, the highest environmental risk is observed at the upper bounds of the DoF.

A probabilistic approach to future scenarios is concerned to find the combination of decisive DoF that maximizes the indicator of economic sustainability. Similarly, the potential environmental risk is also evaluated. Results show that this CO_2 separation design is promising, although the PEI indicates that the higher the profitability, the larger the environmental risk is. The environmental risk appears at high values of water make-up flow and recycle amine temperatures. This situation may be explained by the toxicological aspects of the involved chemicals within the process—an increase in the power of the cooling stage and modification of the reboiler combustion parameters.

(a)

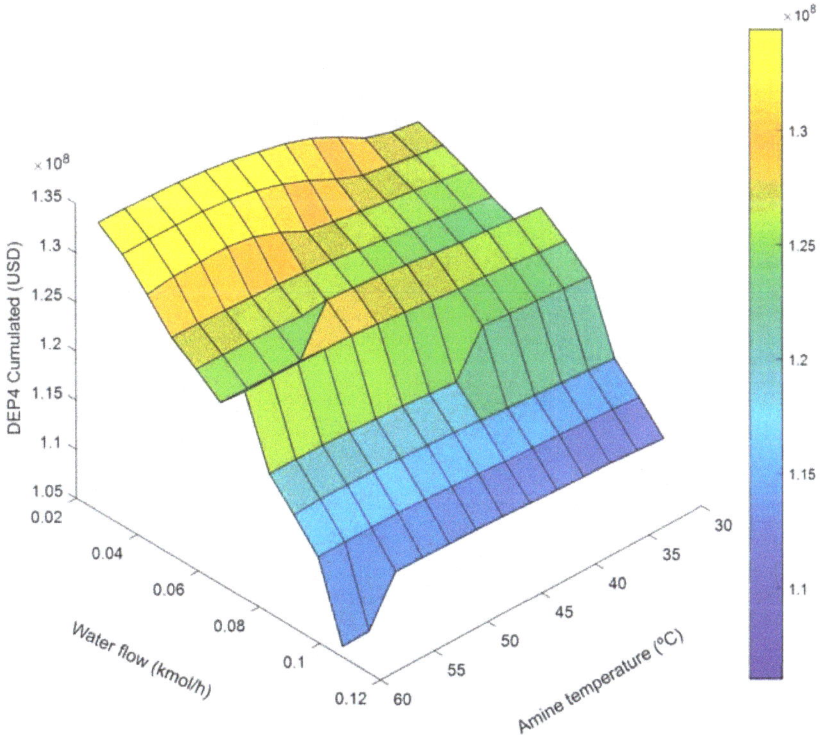

(b)

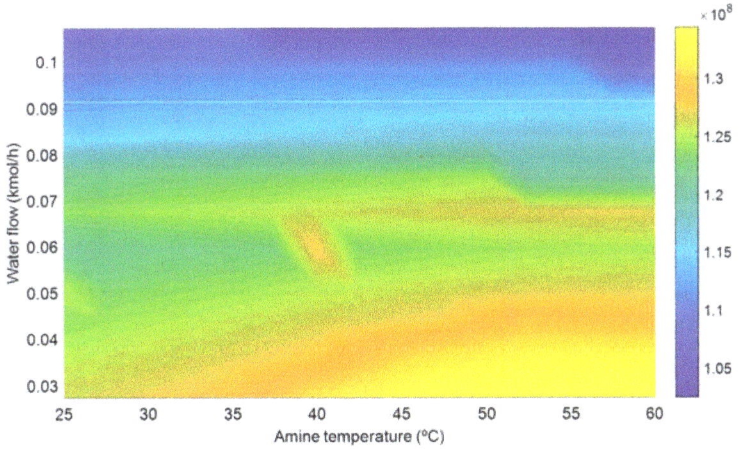

Figure 11. (**a**,**b**). Average cumulated DEP4 (USD) function with respect to water amine molar flow rate (kmol/h) and recycle MDEA temperature (°C), based on the PCD method.

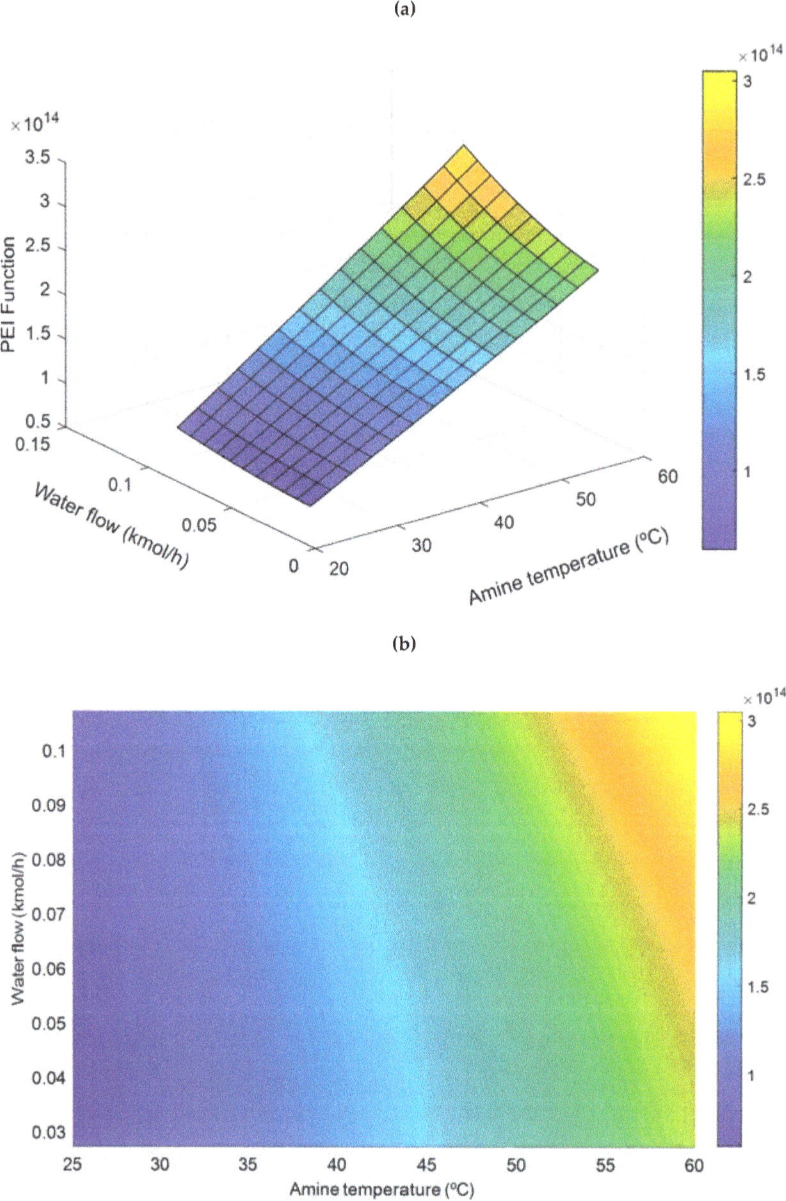

Figure 12. (**a**,**b**). PEI function with respect to water amine molar flow rate (kmol/h) and recycle MDEA temperature (°C), based on Waste Reduction.

5. Conclusions and Future Developments

This paper evaluates the process to obtain and condition CO_2 to be used as an EOR fluid, in the Argentine Basin of Neuquén. We focus the study on the evaluation of economic aspects in a context of market variability and price uncertainties. PCD methodology is adopted to achieve the aim of the article. With this technique, a probabilistic approach to future scenarios is used to find the combination of decisive DoF that maximizes the indicator of economic sustainability. According to the results, the implementation of the plant at this stage of the study is feasible and suggests promising values for revenues and economic profitability.

The results of this preliminary study are promising. The economic potential of the four order is proven to be high, with a magnitude of eight order in USD/y. Further, the statistical indexes prove that the plant is profitable within 12 years of the process time's life. Finally, the conditions of the plant maximizing the EP are identified—a recycle amine flow of 0.0274 kmol/h at 60 °C proved to be an optimal combination of the decision variables. In respect to the 'green' risks, it is demonstrated that the higher the upper bounds of the DoF, the higher the environmental risk is.

The evaluation of DoF and their impact on the energy requirements of the plant have led to a notable conclusion—the decision variable affecting the consumer is the water makeup of the plant. Thus, a new perspective for authors working with a similar process is presented in this paper.

Future work can extend the limits of this methodology and include a higher number of DoFs, such as the ones related to the regeneration of the column, which is rarely discussed in the bibliography. In addition, the economic potential evaluation can be extended with heat integration coming from the pinch technology.

The last important aspect to be noted is that the CO_2 was historically considered to be a by-product, and in the past, it was a common practice to flare it. However, the recuperation and condition of this gas, and the installation of a proper plant operating at proper conditions, might be the starting point for implementing the technology of EOR in the region, taking into account volatile market scenarios.

Author Contributions: Conceptualization, J.P.G., E.E. and D.M.; Methodology, J.P.G., E.E. and D.M.; Validation, J.P.G., E.E. and D.M.; Investigation, J.P.G., E.E. and D.M.; Resources, E.E. and D.M.; Writing-Original Draft Preparation, J.P.G. and E.E.; Writing-Review & Editing, J.P.G., E.E. and D.M.; Supervision, E.E. and D.M.; Funding Acquisition, J.P.G., E.E. and D.M.

Funding: This publication has been produced with the funding of the ERASMUS MUNDUS (Action 2 Strand 1) SUSTAIN-T Program, under the coordination of Politecnico di Milano, Italy. The authors also acknowledge the funding of CONICET (Grant 2222016000218900) and the Universidad Nacional de Salta (CIUNSa Projects 2253/0, 2465, and 2645), Argentina.

Conflicts of Interest: The authors declare no conflict of interest.

References

1. Roussanaly, S.; Grimstad, A.-A. The Economic Value of CO_2 for EOR Applications. *Energy Procedia* **2014**, *63*, 7836–7843. [CrossRef]
2. Yang, H.; Xu, Z.; Fan, M.; Gupta, R.; Slimane, R.B.; Bland, A.E.; Wright, I. Progress in carbon dioxide separation and capture: A review. *J. Environ. Sci.* **2008**, *20*, 14–27. [CrossRef]
3. Haszeldine, R.S. Carbon capture and storage: How green can black be? *Science* **2009**, *325*, 1647–1652. [CrossRef] [PubMed]
4. Wright, I.W.; Lee, A.; Middleton, P.; Lowe, C.; Imbus, S.W.; Miracca, I. CO_2 Capture Project: Initial Results. In Proceedings of the SPE International Conference on Health, Safety, and Environment in Oil and Gas Exploration and Production, Society of Petroleum Engineers, Calgary, AB, Canada, 29–31 March 2004.
5. Mumford, K.A.; Wu, Y.; Smith, K.H.; Stevens, G.W. Review of solvent based carbon-dioxide capture technologies. *Front. Chem. Sci. Eng.* **2015**, *9*, 125–141. [CrossRef]
6. Brush, R.M.; Davitt, H.J.; Aimar, O.B.; Arguello, J.; Whiteside, J.M. Immiscible CO_2 flooding for increased oil recovery and reduced emissions. In Proceedings of the SPE/DOE Improved Oil Recovery Symposium, Society of Petroleum Engineers, Tulsa, Oklahoma, 3–5 April 2000.
7. Mazzetti, M.J.; Skagestad, R.; Mathisen, A.; Eldrup, N.H. CO_2 from natural gas sweetening to kick-start EOR in the North Sea. *Energy Procedia* **2014**, *63*, 7280–7289. [CrossRef]
8. Herzog, H.J. Scaling up carbon dioxide capture and storage: From megatons to gigatons. *Energy Econ.* **2011**, *33*, 597–604. [CrossRef]
9. Kwak, D.-H.; Yun, D.; Binns, M.; Yeo, Y.-K.; Kim, J.-K. Conceptual process design of CO_2 recovery plants for enhanced oil recovery applications. *Ind. Eng. Chem. Res.* **2014**, *53*, 14385–14396. [CrossRef]

10. Leung, D.Y.; Caramanna, G.; Maroto-Valer, M.M. An overview of current status of carbon dioxide capture and storage technologies. *Renew. Sustain. Energy Rev.* **2014**, *39*, 426–443. [CrossRef]

11. Chávez-Rodríguez, M.; Varela, D.; Rodrigues, F.; Salvagno, J.B.; Köberle, A.C.; Vasquez-Arroyo, E.; Raineri, R.; Rabinovich, G. The role of LNG and unconventional gas in the future natural gas markets of Argentina and Chile. *J. Nat. Gas Sci. Eng.* **2017**, *45*, 584–598. [CrossRef]

12. Ponzo, R.; Dyner, I.; Arango, S.; Larsen, E.R. Regulation and development of the Argentinean gas market. *Energy Policy* **2011**, *39*, 1070–1079. [CrossRef]

13. Manolas, D.A.; Frangopoulos, C.A.; Gialamas, T.P.; Tsahalis, D.T. Operation optimization of an industrial cogeneration system by a genetic algorithm. *Energy Convers. Manag.* **1997**, *38*, 1625–1636. [CrossRef]

14. Douglas, J.M. *Conceptual Design of Chemical Processes*; McGraw-Hill: New York, NY, USA, 1988; Volume 1110.

15. Harmsen, G. Industrial best practices of conceptual process design. *Chem. Eng. Process. Process Intensif.* **2004**, *43*, 671–675. [CrossRef]

16. Sepiacci, P.; Depetri, V.; Manca, D. A systematic approach to the optimal design of chemical plants with waste reduction and market uncertainty. *Comput. Chem. Eng.* **2017**, *102*, 96–109. [CrossRef]

17. Manca, D.; Grana, R. Dynamic conceptual design of industrial processes. *Comput. Chem. Eng.* **2010**, *34*, 656–667. [CrossRef]

18. Cabezas, H.; Bare, J.C.; Mallick, S.K. Pollution prevention with chemical process simulators: The generalized waste reduction (WAR) algorithm—Full version. *Comput. Chem. Eng.* **1999**, *23*, 623–634. [CrossRef]

19. Gutierrez, J.P.; Ruiz, E.L.A.; Erdmann, E. Energy requirements, GHG emissions and investment costs in natural gas sweetening processes. *J. Nat. Gas Sci. Eng.* **2017**, *38*, 187–194. [CrossRef]

20. Gallo, G.; Erdmann, E. Potencialidad el EOR con CO_2 en reservorios de baja permeabilidad de la cuenca Neuquina. In *Congreso de Produccion y Desarrollo de Reservas*; Instituto Argentino del Petroleo y Gas: Buenos Aires, Argentina, 2016.

21. Mores, P.; Scenna, N.; Mussati, S. Post-combustion CO_2 capture process: Equilibrium stage mathematical model of the chemical absorption of CO_2 into monoethanolamine (MEA) aqueous solution. *Chem. Eng. Res. Des.* **2011**, *89*, 1587–1599. [CrossRef]

22. Young, D.; Scharp, R.; Cabezas, H. The waste reduction (WAR) algorithm: Environmental impacts, energy consumption, and engineering economics. *Waste Manag.* **2000**, *20*, 605–615. [CrossRef]

23. Erdmann, E.; Ruiz, L.A.; Martínez, J.; Gutierrez, J.P.; Tarifa, E. Endulzamiento de gas natural con aminas. Simulación del proceso y análisis de sensibilidad paramétrico. *Avances en Ciencias e Ingeniería*. **2012**, *3*, 89–101.

24. Green, D.W.; Perry, R.H. *Chemical Engineers' Handbook*; McGraw-Hill: New York, NY, USA, 1973.

25. Fouad, W.A.; Berrouk, A.S. Using mixed tertiary amines for gas sweetening energy requirement reduction. *J. Nat. Gas Sci. Eng.* **2013**, *11*, 12–17. [CrossRef]

26. Kazemi, A.; Malayeri, M.; kharaji, A.G.; Shariati, A. Feasibility study, simulation and economical evaluation of natural gas sweetening processes—Part 1: A case study on a low capacity plant in iran. *J. Nat. Gas Sci. Eng.* **2014**, *20*, 16–22. [CrossRef]

27. Gutierrez, J.P.; Benitez, L.A.; Ale Ruiz, E.L.; Erdmann, E. A sensitivity analysis and a comparison of two simulators performance for the process of natural gas sweetening. *J. Nat. Gas Sci. Eng.* **2016**, *31*, 800–807. [CrossRef]

28. Al-Lagtah, N.M.; Al-Habsi, S.; Onaizi, S.A. Optimization and performance improvement of Lekhwair natural gas sweetening plant using Aspen HYSYS. *J. Nat. Gas Sci. Eng.* **2015**, *26*, 367–381. [CrossRef]

29. Kvamsdal, H.; Jakobsen, J.; Hoff, K. Dynamic modeling and simulation of a CO_2 absorber column for post-combustion CO_2 capture. *Chem. Eng. Process. Process Intensif.* **2009**, *48*, 135–144. [CrossRef]

30. Prölss, K.; Tummescheit, H.; Velut, S.; Åkesson, J. Dynamic model of a post-combustion absorption unit for use in a non-linear model predictive control scheme. *Energy Procedia* **2011**, *4*, 2620–2627. [CrossRef]

31. Behroozsarand, A.; Zamaniyan, A. Multiobjective optimization scheme for industrial synthesis gas sweetening plant in GTL process. *J. Nat. Gas Chem.* **2011**, *20*, 99–109. [CrossRef]

32. Øi, L.E.; Bråthen, T.; Berg, C.; Brekne, S.K.; Flatin, M.; Johnsen, R.; Moen, I.G.; Thomassen, E. Optimization of configurations for amine based CO_2 absorption using Aspen HYSYS. *Energy Procedia* **2014**, *51*, 224–233. [CrossRef]

33. Gutierrez, J.P.; Erdmann, E.; Manca, D. Multi-objective optimization of a CO_2-EOR process from the sustainability criteria. In *28th European Symposium on Computer Aided Process Engineering*; Elsevier: Graz, Austria, 2018.

34. Honeywell. *UniSim Design*; Honeywell International Inc.: Charlotte, NC, USA, 2016.

35. Muhammad, A.; GadelHak, Y. Correlating the additional amine sweetening cost to acid gases load in natural gas using Aspen Hysys. *J. Nat. Gas Sci. Eng.* **2014**, *17*, 119–130. [CrossRef]

36. Torres-Ortega, C.E.; Segovia-Hernández, J.G.; Gómez-Castro, F.I.; Hernández, S.; Bonilla-Petriciolet, A.; Rong, B.-G.; Errico, M. Design, optimization and controllability of an alternative process based on extractive distillation for an ethane–carbon dioxide mixture. *Chem. Eng. Process. Process Intensif.* **2013**, *74*, 55–68. [CrossRef]

37. Manca, D. A methodology to forecast the price of commodities. In *Computer Aided Chemical Engineering*; Karimi, I.A., Srinivasan, R., Eds.; Elsevier: Amsterdam, The Netherlands, 2012; pp. 1306–1310.

38. Manca, D. Modeling the commodity fluctuations of OPEX terms. *Comput. Chem. Eng.* **2013**, *57*, 3–9. [CrossRef]

39. Sepiacci, P.; Manca, D. Economic assessment of chemical plants supported by environmental and social sustainability. *Chem. Eng. Trans.* **2015**, *43*, 2209–2214.

40. EIA. US Energy Information Administration. 2018. Available online: https://www.eia.gov/ (accessed on 1 June 2018).

41. Ministry-of-Energy. Reporte de Produccion. 2017; Presidencia de la Nacion Argentina. Available online: https://www.se.gob.ar/ (accessed on 1 June 2018).

42. Wiggins, S.; Etienne, X.L. Turbulent times: Uncovering the origins of US natural gas price fluctuations since deregulation. *Energy Econ.* **2017**, *64*, 196–205. [CrossRef]

43. Zhou, H.; Deng, Z.; Xia, Y.; Fu, M. A new sampling method in particle filter based on Pearson correlation coefficient. *Neurocomputing* **2016**, *216*, 208–215. [CrossRef]

44. Lee Rodgers, J.; Nicewander, W.A. Thirteen ways to look at the correlation coefficient. *Am. Stat.* **1988**, *42*, 59–66. [CrossRef]

45. Mohamed Salleh, F.H.; Arif, S.M.; Zainudin, S.; Firdaus-Raih, M. Reconstructing gene regulatory networks from knock-out data using Gaussian Noise Model and Pearson Correlation Coefficient. *Comput. Biol. Chem.* **2015**, *59*, 3–14. [CrossRef] [PubMed]

46. Dancey, C.P.; Reidy, J. *Statistics without Maths for Psychology*; Pearson Education: London, UK, 2007.

47. Young, D.; Cabezas, H. Designing sustainable processes with simulation: The waste reduction (WAR) algorithm. *Comput. Chem. Eng.* **1999**, *23*, 1477–1491. [CrossRef]

48. Marticorena, A.A.; Mandagarán, B.A.; Campanella, E.A. Análisis del Impacto Ambiental de la Recuperación de Metanol en la Producción de Biodiesel usando el Algoritmo de Reducción de Desechos WAR. *Inf. Tecnol.* **2010**, *21*, 23–30. [CrossRef]

49. Barrett, W.M.; van Baten, J.; Martin, T. Implementation of the waste reduction (WAR) algorithm utilizing flowsheet monitoring. *Comput. Chem. Eng.* **2011**, *35*, 2680–2686. [CrossRef]

50. Cook, B. EORI's Economic scoping model. In Proceedings of the 8th Annual EORI Casper CO_2 Conference, Enhanced Oil Recovery Institute, University of Wyoming, Laramie, WY, USA, 9 July 2014.

51. Mazzetto, F.; Ortiz-Gutiérrez, R.A.; Manca, D.; Bezzo, F. Strategic design of bioethanol supply chains including commodity market dynamics. *Ind. Eng. Chem. Res.* **2013**, *52*, 10305–10316. [CrossRef]

52. Manca, D. Price model of electrical energy for PSE applications. *Comput. Chem. Eng.* **2016**, *84*, 208–216. [CrossRef]

53. Manca, D. A methodology to forecast the price of electric energy. In *Computer Aided Chemical Engineering*; Elsevier: Amsterdam, The Netherlands, 2013; pp. 679–684.

Article

Regeneration of Sodium Hydroxide from a Biogas Upgrading Unit through the Synthesis of Precipitated Calcium Carbonate: An Experimental Influence Study of Reaction Parameters

Francisco M. Baena-Moreno [1,2,*], Mónica Rodríguez-Galán [1], Fernando Vega [1], T. R. Reina [2], Luis F. Vilches [1] and Benito Navarrete [1]

1   Chemical and Environmental Engineering Department, Technical School of Engineering, University of Seville, C/Camino de los Descubrimientos s/n, 41092 Sevilla, Spain; mrgmonica@us.es (M.R.-G.); fvega1@us.es (F.V.); luisvilches@us.es (L.F.V.); bnavarrete@us.es (B.N.)
2   Department of Chemical and Process Engineering, University of Surrey, Guildford GU2 7XH, UK; t.ramirezreina@surrey.ac.uk
*   Correspondence: fbaena2@us.es; Tel.: +34-680-252-606

Received: 11 October 2018; Accepted: 22 October 2018; Published: 24 October 2018

Abstract: This article presents a regeneration method of a sodium hydroxide (NaOH) solution from a biogas upgrading unit through calcium carbonate ($CaCO_3$) precipitation as a valuable by-product, as an alternative to the elevated energy consumption employed via the physical regeneration process. The purpose of this work was to study the main parameters that may affect NaOH regeneration using an aqueous sodium carbonate (Na_2CO_3) solution and calcium hydroxide ($Ca(OH)_2$) as reactive agent for regeneration and carbonate slurry production, in order to outperform the regeneration efficiencies reported in earlier works. Moreover, Raman spectroscopy and Scanning Electron Microscopy (SEM) were employed to characterize the solid obtained. The studied parameters were reaction time, reaction temperature, and molar ratio between $Ca(OH)_2$ and Na_2CO_3. In addition, the influence of small quantities of NaOH at the beginning of the precipitation process was studied. The results indicate that regeneration efficiencies between 53%–97% can be obtained varying the main parameters mentioned above, and also both Raman spectroscopy and SEM images reveal the formation of a carbonate phase in the obtained solid. These results confirmed the technical feasibility of this biogas upgrading process through $CaCO_3$ production.

Keywords: carbon capture and utilization; biogas upgrading; calcium carbonate precipitation; chemical absorption

---

1. Introduction

Climate change is one of the major problems that has plagued humanity in recent times, consisting of a significant and lasting modification of local and global patterns of climate on the planet. The frequency and intensity of meteorological phenomena such as rainfall, hurricanes, storms, decreasing extent of ice, rising sea level and, above all, the increasing average temperature of the Earth's atmosphere are the main evidences found by scientists that corroborate climate change [1,2]. According to the Intergovernmental Panel on Climate Change (IPCC) [1], the main origin is the anthropogenic emissions of so-called greenhouse gases (GHG), due to the use of fossil fuels such as coal, oil and natural gas for the production of electricity, transportation or industrial uses, CO_2 being the most relevant among the greenhouse gases. For this reason, the use of renewables energies which reduce CO_2 emissions could be found as one of the fields most investigated in the last decade [3–11].

One of the most promising renewable energy sources is biomass [12]. Biomethane is obtained by upgrading biogas produced from anaerobic digestion of different types of biomass. There are several ways to use this biomethane as an energy resource, which depend on technical, economic and legislative factors of each country [13]. Biomethane is an improved biogas from landfills, farms, sewage treatment plants, agriculture or other sources [14]. For use, biomethane must be submitted to a process called upgrading, which separates the undesired compounds (mainly CO_2) and adapts its composition to the standards set by the legislation corresponding to that suitable for a fuel gas [15]. CO_2 content in biogas as produced in anaerobic digestion varies between 35%–45% [16,17]. To remove CO_2, very diverse techniques have been studied: Pressure Swing Adsorption (PSA) [18,19], Water Scrubbing (WS) [20,21], Organic Physical Scrubbing (OPS) [22,23], Chemical Absorption Scrubbing (CAS) [13,24], Membrane Separation (MS) [25,26] and Cryogenic Separation (CS) [27,28]. CAS is one of the most promising technique [29,30], both with amines (monoethanolamine (MEA) or Piperazine (PZ)) and caustic solvents (NaOH or potassium hydroxide (KOH)). Previous studies have reported high capture yields and selectivity to CO_2, achieving similar capture efficiencies from 90% to 99% [31–35]. In the case of amine solvents, a high regeneration efficiency of the solvent can be obtained via physical regeneration, with an acceptable energy consumption [34]. However, compared with the previous solvents explained, when employing caustic solvents, an elevated energy consumption is necessary in order to regenerate the solvent physically, which makes these solvents less usable than the conventional MEA [35]. In this reaction, Na_2CO_3 or potassium carbonate (K_2CO_3) is obtained as a consequence of the absorption step.

$$2NaOH/KOH(aq) + CO_2(g) \rightarrow +Na_2CO_3/K_2CO_3 + H_2O \tag{1}$$

An alternative path for CO_2 utilization that avoids the energy penalty in the regeneration stage of the solvent would be the synthesis and separation of chemicals based on calcium (Ca^+) by the precipitation processes into the solvent solution, as for example in $Ca(OH)_2$ or residues with high Ca^+ content. This alternative is very attractive from an economic point of view in order to drastically reduce costs of CO_2 capture and valorize CO_2 as a commercial by-product. An interesting by-product to be taken into account when alkaline hydroxides are used as solvents is $CaCO_3$. $CaCO_3$ can be produced through chemical reaction with $Ca(OH)_2$ and precipitated as a solid [33], according to the next reaction:

$$Na_2CO_3/K_2CO_3(aq) + Ca(OH)_2(s) \rightarrow 2NaOH/KOH(aq) + CaCO_3(s) \tag{2}$$

The type of $CaCO_3$ obtained as a by-product is called Precipitated Calcium Carbonate (PCC). PCC is consumed in huge quantities and in variated applications for different industrial sectors, such as a filler for plastic materials, paper, foods, printing ink and medical necessities [36]. This synergy process between biogas upgrading, CO_2 capture and PCC production is shown in Figure 1.

This process is much less energy intensive than physical regeneration previously studied by various authors [35,37,38], making the process economically attractive. Many researches focused on the carbonation of residues for storing CO_2, as for instance steel slags [39], air pollution control residues [32,33], argon oxygen decarburization slags [23], incineration bottom ash [40] or basic oxygen furnace slags [41]. Baciocchi et al. [32,33] proposed an application of the process above explained using both NaOH and KOH as solvents, and Air Pollution Control residues as Ca^+ sources, focusing on the amount of CO_2 that could be definitely stored by this residues [32,42]. However, their results showed a non-valuable by-product from a commercial point of view and solvent regeneration efficiencies from 50% to 60% due to the employment of the residues. Therefore, the purpose of this work was to study the main parameters that may affect NaOH regeneration using an aqueous Na_2CO_3 solution and $Ca(OH)_2$ as a reactive for regeneration and carbonate slurry production, in order to achieve better regeneration efficiencies than previous works and obtaining PCC as a valuable by-product. With this, the foundations for future works are laid, in which valuable by-products would be obtained that would allow to achieve a more economical and sustainable process for carbon capture and utilization.

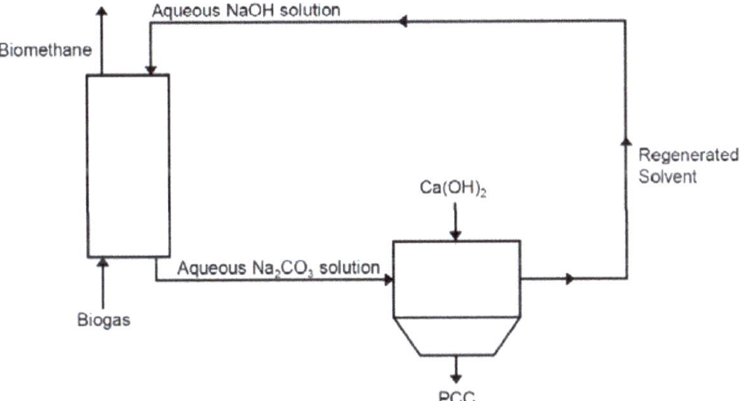

Figure 1. Biogas upgrading through Precipitated Calcium Carbonate (PCC) production.

2. Materials and Methods

2.1. Materials

Chemical compounds used in the experiments ($Ca(OH)_2$, Na_2CO_3, $CaCO_3$, $NaOH$) were provided by PanReac-AppliChem (Barcelona, Spain) (pure-grade or pharma-grade, 99% purity).

2.2. Regeneration Experiments

In general, the regeneration experiments were carried out following the methodology exposed below, which will be explained in greater depth later. First, the solutions of the reactants were prepared, at the same time that the instruments needed for the precipitation reaction were tuned. After these steps, the reaction was produced, which, once finished, was filtered and separated quickly for analyzing. The main parameters considered for results were NaOH regeneration efficiency and carbonate phase reached. The three most important variables studied in these experiments were the reaction time, the reaction temperature and the molar ratio between $Ca(OH)_2$ and Na_2CO_3, which according to previous works may have a considerable effect on the regeneration efficiency of the process [32,33]. The matrix of experiments carried out can be found in Table 1. In order to study how each parameter affects by itself, a standard value was set for each of them, according to the bibliography for similar studies [31–33,35], and later they were varied one by one. The standard value for temperature reaction was set at 50 °C, molar ratio at 1.2 mol Ca/Na_2CO_3 and reaction time at 30 min. Furthermore, the influence of an initial addition of NaOH in the Na_2CO_3 solution was tested, in order to analyze the effect on NaOH regeneration.

Lab scale batch precipitation experiments were carried out in a 600 mL beaker placed in a water bath to control temperature tests. The value of the pH gives some clues of the compounds that may be present in the solution. Therefore, to check that the NaOH regeneration reaction had the desired effect, the pH was measured and checked to be in the range 12–14, which is characteristic for hydroxides solutions [31]. During each whole experiment time, the solutions were stirred by an electromagnetic magnet at a constant speed of 1000 rpm. For temperature and pH measuremsent, a thermometer and a pH-meter by Trison Instrument (BANDELIN electronic GmbH & Co. KG, Berlin, Germany) were employed. Measures were continually carried out and recorded in a data logger. Reproducibility checks were conducted resulting in an overall experimental error of ±2% for the regeneration efficiency calculations. The first steps of the procedures followed were to prepare both

Na_2CO_3 and $Ca(OH)_2$ solutions. In the case of Na_2CO_3, the aqueous solution was set at 20 g/100 mL according to the basis typical values expected after the absorption step [21,28], while the concentration of the $Ca(OH)_2$ solution was stoichiometrically calculated for each test, as will be explained later. At the beginning of each experiment, a 200 mL distilled water slurry of $Ca(OH)_2$ was poured into the beaker placed. After 15 min, 200 mL of Na_2CO_3 aqueous solution was added to start the reaction time. At the end of each experiment, the solution was vacuum filtered immediately and 50 mL sample was taken to determine the concentration of NaOH by inductively coupled plasma atomic emission spectroscopy. The solid obtained by filtration was dried at 105 °C to ensure a carbonate phase which was characterized by means of Scanning Electron Microscopy (SEM) and Raman spectroscopy.

Table 1. Matrix of the experiments carried out.

TEST	TIME (MIN)	TEMPERATURE (°C)	MOLAR RATIO	NaOH INITIAL (M)
Standard	30	50	1.2	0
1	5	50	1.2	0
2	15	50	1.2	0
3	45	50	1.2	0
4	60	50	1.2	0
5	90	50	1.2	0
6	120	50	1.2	0
7	30	30	1.2	0
8	30	35	1.2	0
9	30	40	1.2	0
10	30	45	1.2	0
11	30	50	1.2	0
12	30	55	1.2	0
13	30	60	1.2	0
14	30	65	1.2	0
15	30	70	1.2	0
16	30	50	0.7	0
17	30	50	0.8	0
18	30	50	0.9	0
19	30	50	1	0
20	30	50	1.1	0
21	30	50	1.2	0
22	30	50	1.3	0
23	30	50	1.4	0
24	30	50	1.5	0
25	5	50	1.2	1
26	15	50	1.2	1
27	30	50	1.2	1
28	45	50	1.2	1
29	60	50	1.2	1
30	90	50	1.2	1
31	120	50	1.2	1

Raman measurements of the powders samples were recorded using a Thermo DXR2 spectrometer (Thermo Fisher Scientific, Waltham, MA, USA) equipped with a Leica DMLM microscope (Thermo Fisher Scientific, Waltham, MA, USA). The wavelength of applied excitation line was 532 nm ion laser and 50× objective of 8-mm optical was used to focus the depolarized laser beam on a sport of about 3 μm in diameter.

A JEOL JSM6400 (JEOL Ltd., Tokyo, Japan) operated at 20 KV equipped with energy dispersive X-ray spectroscopy (EDX) and a wavelength dispersive X-ray spectroscopy (WDS) systems was used for the microstructural/chemical characterization (SEM with EDS and WDS).

3. Results

This section reports the experimental results of the different tests carried out. The results of NaOH regeneration efficiency are presented, with reference to every parameter studied. NaOH regeneration efficiency is defined has follow:

$$\text{NaOH regeneration efficiency}(\%) = \frac{\text{NaOH regenerated}}{\text{Maximum NaOH to regenerate}} \times 100$$

As it has been set previously, NaOH regenerated was determined by inductively coupled plasma atomic emission spectroscopy, while the maximum NaOH to be regenerated can be easily stochiometrically calculated from the concentration of the Na_2CO_3 initial solution.

Then, some Raman spectroscopies of PCC are shown to demonstrate the carbonate phase reached, which are accompanied by some SEM images that contribute to verify the results predicted by Raman.

3.1. NaOH Regeneration

Figures 2–4 show the regeneration efficiency curves of the filtered solutions resulting from regeneration experiments carried out with Ca(OH)$_2$ at different reaction times, temperatures and molar ratios, respectively.

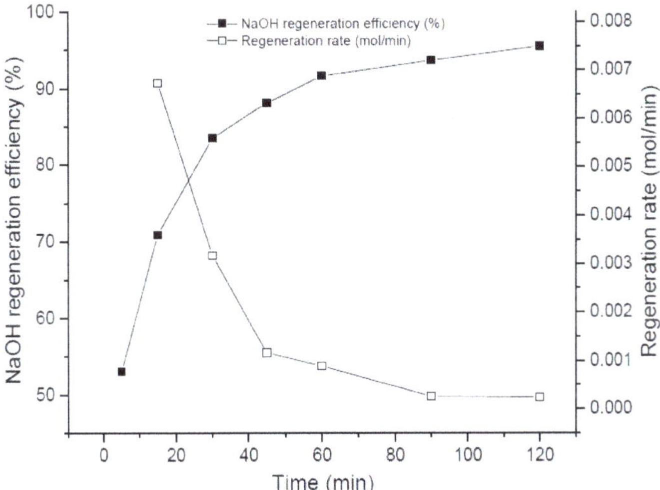

Figure 2. Evolution of NaOH regeneration with time at fixed Temperature 50 °C and R = 1.2.

3.1.1. Reaction Time Effect

Figure 2 shows the effect of the reaction time in the regeneration phenomenon. As depicted in the plot, NaOH regeneration efficiency varies from 53% to 83% approximately from 5 min to 30 min of reaction time, and later, the slope of the curve changes drastically, passing through 91% regeneration efficiency at 60 min, until achieving a 95% of NaOH regeneration at 120 min. This means, in fact, that in a hypothetical real reactor, a duplication of its volume will be necessary to achieve an increase of 4% approximately (from 30 min to 60 min). As can be seen, from 60 min to 120 min, less than 0.001 mol NaOH is regenerated per minute. Thus, to operate at 120 min residence time is not worthy from a plant design point of view. The intersection of the two curves (regeneration rate and regeneration efficiency) indicates an interesting and very likely optimum operational point where a fair balance between both tendencies can be reached.

3.1.2. Temperature Influence

As for the temperature influence, Figure 3 shows the effect of different temperatures in the regeneration studies. Temperature effect reflects a linear trend showcasing a direct correlation between NaOH regeneration efficiency and process temperature. Indeed, in the best case scenario (at 70 °C) it can reach 97% of NaOH regeneration efficiency. Normalizing the regeneration capacity by the incremental temperature (empty symbols in the Figure) maximum is obtained at around 50 °C which somehow indicates that the increment in temperature has a stronger impact on the regeneration efficiency in the low-medium temperature range. This is an important result to be highlighted from an energy consumption perspective, as a temperature of 50 °C could be easily achieved through low-cost and/or renewable energy sources such as solar.

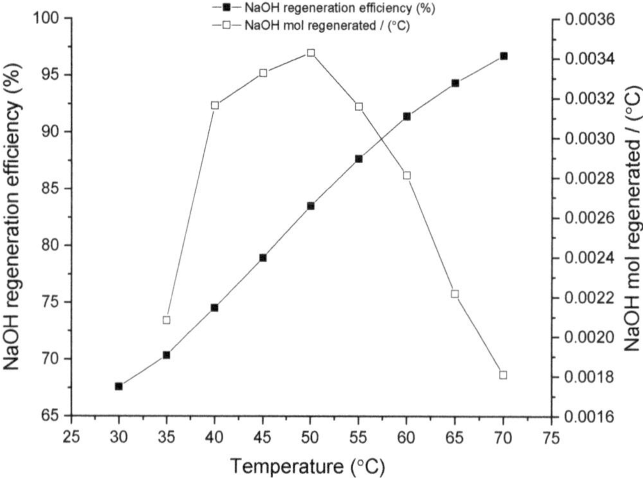

Figure 3. Influence of temperature on NaOH regeneration. Experiments carried out at R = 1.2 and t = 30 min.

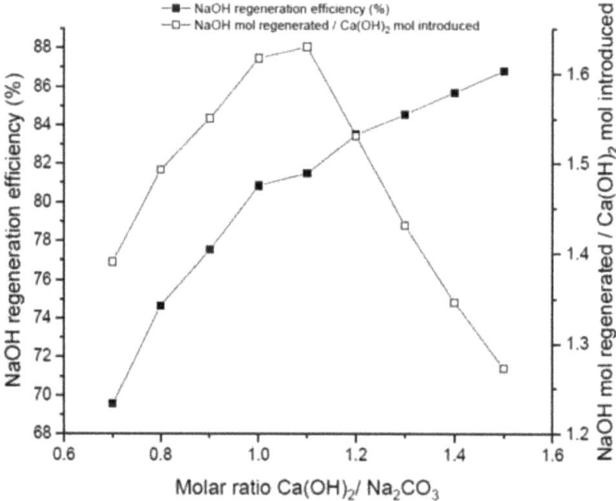

Figure 4. Influence of molar ratio (R) in the regeneration experiments. Runs conducted at t = 30 min and 50 °C.

3.1.3. Molar Ratio Influence

Molar ratio inlet carbonate/precipitant agent is another important parameter to consider in the regeneration process. As can be seen in Figure 4, NaOH regeneration efficiency is favored by an increase of the molar ratio. Nevertheless, this increase of molar ratio promotes a higher quantity of Ca^{2+} ions that should be removed before recirculating the absorbent to the absorption tower in order to prevent accumulation of $Ca(OH)_2$ which eventually may lead to fouling phenomenon in the tower. In parallel, as can be observed in Figure 4, NaOH mol regeneration per mol of $Ca(OH)_2$ introduced decreases upon increasing the molar ratio above a threshold value of R = 1.1. This value set an optimum operational point beyond which no further benefits are envisaged from the process point of view.

3.1.4. Effect of NaOH Spark in the Regeneration Efficiency

The addition of small quantities of NaOH at the beginning of the precipitation process may promote the recovery of NaOH (initially entering the precipitation reactor in the form of Na_2CO_3). Also, it should be taken into account that, in a real industrial plant, 100% conversion from NaOH to Na_2CO_3 would hardly be reached in the absorption stage. In this sense, small amounts of NaOH as "sparking species" were added to investigate its effect on the process. The impact exerted by the addition of an initial concentration of NaOH (1 M) in the Na_2CO_3 solution in terms of NaOH regeneration efficiency is reported in Figure 5. Analyzing this Figure, it may be noted that the NaOH regeneration was slower than the results obtained without an initial NaOH concentration. It seems that the presence of alkaline compounds do not benefit the regeneration process—a fact that can be related to the poorer solubility of $Ca(OH)_2$ due to the ion common effect as previously observed elsewhere [32,36,43].

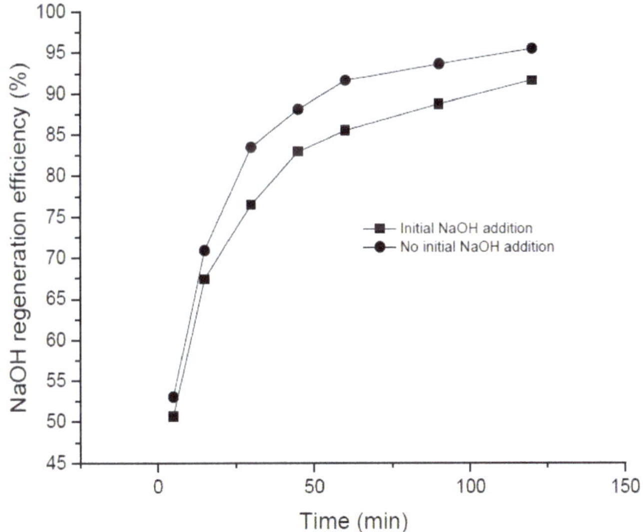

Figure 5. Comparison curves of NaOH regeneration efficiency with and without initial addition of NaOH.

3.2. Physicochemical Characterization of the PCC

Aiming to determine the purity of the carbonates obtained during the recovery process, a combined Raman-SEM study was conducted on selected samples. Figure 6 shows the Raman spectra of the recovered carbonated after 30 min of reaction at 50 °C using a R = 1.2 in comparison with standards samples of pure $CaCO_3$ and pure $Ca(OH)_2$. $CaCO_3$ typically presents a monoclinic structure belonging to the P21/c group [44]. The main characteristic band of $CaCO_3$ polymorphs is a strong and narrow feature which appears at around 1100 cm^{-1}. Also, another band ca. 700 cm^{-1} is typically ascribed to this type of structure [44]. As can be seen, these two peaks are presented in PCC spectra, confirming the successful precipitation process. In fact, the spectrum of our PCC sample resembles that of the $CaCO_3$ standard as shown in Figure 6. Nevertheless, it must be highlighted that a certain amount of $Ca(OH)_2$ remains present in our solid sample as intended by the Raman vibration mode at ca. 400 cm^{-1} which matches well with the most intense band on the $Ca(OH)_2$ standard. In fact, these data correlate well with the regeneration efficiency data discussed above where 100% regeneration is never reached. In this sense, Raman experiments indicate that despite the fact that the regeneration process is highly effective, there is still some room for further improvements.

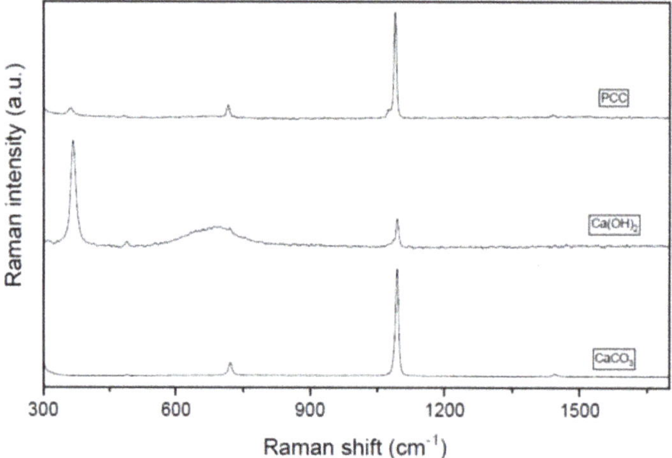

Figure 6. Raman spectra of the PCC obtained (time = 30 min, T = 50 °C, R = 1.2) and the $Ca(OH)_2$ and $CaCO_3$ standards.

Scanning Electron Microscopy images are useful to gain further insights on the samples structures. Selected SEM images of different sections of the sample studied by Raman are presented in Figure 7. SEM images again confirmed the presence of $CaCO_3$ with the typical morphology of calcite as previously observed by Altiner et al. [45]. In the case of SEM, it is hard to distinguish between $CaCO_3$ and $Ca(OH)_2$—especially when the amount of $Ca(OH)_2$ is just a minor contribution in the overall sample composition. In general terms, our SEM study confirms that the successful carbonate precipitation is in good agreement with the regeneration efficiency studies and the Raman results.

Figure 7. SEM images of the PCC obtained (time = 30 min, T = 50 °C, R = 1.2).

4. Conclusions

The results obtained from this lab scale work have confirmed the technical feasibility of this biogas upgrading process through PCC production. In general, the majority of the tests have shown better regeneration efficiencies than previous studies identified in the first section of this work (53%–97% vs. 50%–60%) [32,33]. The multiple reaction parameters have a different impact on the overall process performance. For instance, it was identified that the ideal reaction time would be around 30–60 min for T = 50 °C and R = 1.2, leading to compact reactor units. As for the temperature effect, a maximum NaOH mol regenerated per grade is reached at 50 °C for t = 30 min and R = 1.2; this would be an advisable value for a real process since it could be reached easily through the employment of a renewable energy source. The molar ratio $Ca(OH)_2/Na_2CO_3$ also influences the process, 1.1 being an ideal ratio to be implemented for realistic operations, for t = 30 min and T = 50 °C. This result has been chosen taking into account the maximum in the curve of NaOH mol regeneration per mol of $Ca(OH)_2$. The presence of small quantities of NaOH do not benefit the regeneration process and in fact it produces a decrease in the regeneration efficiency due to the ion common effect; this would suggest an effort to get the maximum percentage of NaOH conversion to Na_2CO_3 in the absorption stage.

Raman and SEM studies confirm the large majority presence of $CaCO_3$ on the recovered material. Interestingly, although the obtained solid is mainly composed by calcite type $CaCO_3$, some traces of $Ca(OH)_2$ are still present. This opens some room for further research to improve the regeneration process.

Author Contributions: Conceptualization, F.M.B.-M., L.F.V. and B.N.; Methodology, F.M.B.-M., M.R.-G., F.V. and T.R.R.; Data curation, F.M.B.-M. and T.R.R.; Writing—original draft preparation, F.M.B.-M., T.R.R., M.R.-G., B.N.; Writing—review and editing, F.M.B.-M., T.R.R., M.R.-G., B.N.; Supervision, T.R.R., M.R.-G., B.N.; Funding acquisition, B.N., L.F.V. and T.R.R.

Funding: This work was supported by University of Seville through its V PPIT-US. Financial support for this work was also provided by the EPSRC grant EP/R512904/1 as well as the Royal Society Research Grant RSGR1180353. This work was also partially sponsored by the CO2Chem UK through the EPSRC grant EP/P026435/1.

Conflicts of Interest: The authors declare no conflict of interest.

References

1. IPCC. *Global Warming of 1.5 °C. Summary for Policymakers*; Intergovernmental Panel on Climate Change: Incheon, Korea, 2018.
2. International Energy Agency. *World Energy Outlook Special Report: Energy and Climate Change*; International Energy Agency: Paris, France, 2015.
3. Decardi-Nelson, B.; Liu, S.; Liu, J. Improving Flexibility and Energy Efficiency of Post-Combustion CO_2 Capture Plants Using Economic Model Predictive Control. *Processes* **2018**, *6*, 135. [CrossRef]
4. Tollkötter, A.; Kockmann, N. Absorption and Chemisorption of Small Levitated Single Bubbles in Aqueous Solutions. *Processes* **2014**, *2*, 200–215. [CrossRef]
5. Adams, T.A.; Hoseinzade, L.; Madabhushi, P.B.; Okeke, I.J. Comparison of CO_2 Capture Approaches for Fossil-Based Power Generation: Review and Meta-Study. *Processes* **2017**, *5*, 44. [CrossRef]
6. Taimoor, A.A.; Al-shahrani, S.; Muhammad, A. Ionic Liquid (1-Butyl-3-Metylimidazolium Methane Sulphonate) Corrosion and Energy Analysis for High Pressure CO_2 Absorption Process. *Processes* **2018**, *5*, 45. [CrossRef]
7. Li, J.; Ahmed, R.; Li, X. Thermodynamic Modeling of CO_2-N_2-O_2-Brine-Carbonates in Conditions from Surface to High Temperature and Pressure. *Energies* **2018**, *11*, 2627. [CrossRef]
8. Hu, J.; Galvita, V.; Poelman, H.; Marin, G. Advanced Chemical Looping Materials for CO_2 Utilization: A Review. *Materials* **2018**, *11*, 1187. [CrossRef] [PubMed]
9. Zhang, Z.; Yan, Y.; Zhang, L.; Chen, Y.; Ju, S. CFD investigation of CO_2 capture by methyldiethanolamine and 2-(1-piperazinyl)-ethylamine in membranes: Part B. Effect of membrane properties. *J. Nat. Gas Sci. Eng.* **2014**, *19*, 311–316. [CrossRef]
10. Zhang, Z.; Cai, J.; Chen, F.; Li, H.; Zhang, W.; Qi, W. Progress in enhancement of CO_2 absorption by nanofluids: A mini review of mechanisms and current status. *Renew. Energy* **2018**, *118*, 527–535. [CrossRef]
11. Zhang, Z.; Li, H.; Chang, H.; Pan, Z.; Luo, X. Machine learning predictive framework for CO_2 thermodynamic properties in solution. *J. CO2 Util.* **2018**, *26*, 152–159. [CrossRef]
12. Patel, D.; Kellici, S.; Saha, B. Green Process Engineering as the Key to Future Processes. *Processes* **2014**, *10*, 311–332. [CrossRef]
13. Zhang, Z.; Li, Y.; Zhang, W.; Wang, J.; Soltaninan, M.R.; Olabie, A.G. Effectiveness of amino acid salt solutions in capturing CO_2: A review. *Renew. Sustain. Energy Rev.* **2018**, *98*, 179–188. [CrossRef]
14. Wheeler, P.; Holm-Nielsen, J.B.; Jaatinen, T.; Wellinger, A.; Lindberg, A.; Pettigrew, A. *Biogas Upgrading and Utilisation*; IEA Bioenergy: Paris, France, 1999; pp. 3–20.

15. Petersson, A.; Wellinger, A. *Biogas Upgrading Technologies—Developments and Innovations*; IEA Bioenergy: Paris, France, 2009; Volume 20. [CrossRef]
16. Zhou, K.; Chaemchuen, S.; Verpoort, F. Alternative materials in technologies for Biogas upgrading via CO_2 capture. *Renew. Sustain. Energy Rev.* **2017**, *79*, 1414–1441. [CrossRef]
17. Kadam, R.; Panwar, N.L. Recent advancement in biogas enrichment and its applications. *Renew. Sustain. Energy Rev.* **2017**, *73*, 892–903. [CrossRef]
18. Alonso-Vicario, A.; Ochoa-Gómez, J.R.; Gil-Río, S.; Gómez-Jiménez-Aberasturi, O.; Ramírez-López, C.A.; Torrecilla-Soria, J.; Domínguez, A. Purification and upgrading of biogas by pressure swing adsorption on synthetic and natural zeolites. *Microporous Mesoporous Mater.* **2010**, *134*, 100–107. [CrossRef]
19. Kim, Y.J.; Nam, Y.S.; Kang, Y.T. Study on a numerical model and PSA (pressure swing adsorption) process experiment for CH_4/CO_2 separation from biogas. *Energy* **2015**, *91*, 732–741. [CrossRef]
20. Nie, H.; Jiang, H.; Chong, D.; Wu, Q.; Xu, C.; Zhou, H. Comparison of water scrubbing and propylene carbonate absorption for biogas upgrading process. *Energy Fuels* **2013**, *27*, 3239–3245. [CrossRef]
21. Budzianowski, W.M.; Wylock, C.E.; Marciniak, P.A. Power requirements of biogas upgrading by water scrubbing and biomethane compression: Comparative analysis of various plant configurations. *Energy Convers. Manag.* **2017**, *141*, 2–19. [CrossRef]
22. Niesner, J.; Jecha, D.; Stehlík, P. Biogas upgrading technologies: State of art review in European region. *Chem. Eng. Trans.* **2013**, *35*, 517–522. [CrossRef]
23. Ozturk, B.; Demirciyeva, F. Comparison of biogas upgrading performances of different mixed matrix membranes. *Chem. Eng. J.* **2013**, *222*, 209–217. [CrossRef]
24. Cousins, A.; Wardhaugh, L.T.; Feron, P.H.M. A survey of process flow sheet modifications for energy efficient CO_2 capture from flue gases using chemical absorption. *Int. J. Greenh. Gas Control Int. J.* **2011**, *5*, 605–619. [CrossRef]
25. Chen, X.Y.; Vinh-Thang, H.; Ramirez, A.A.; Rodrigue, D.; Kaliaguine, S. Membrane gas separation technologies for biogas upgrading. *RSC Adv.* **2015**, *5*, 24399–24448. [CrossRef]
26. Zhang, Z.; Yan, Y.; Zhang, L.; Chen, Y.; Ran, J.; Pu, G.; Qin, C. Theoretical Study on CO_2 Absorption from Biogas by Membrane Contactors: Effect of Operating Parameters. *Ind. Eng. Chem. Res.* **2014**, *53*, 14075–14083. [CrossRef]
27. Tuinier, M.J.; Van Sint Annaland, M. Biogas purification using cryogenic packed-bed technology. *Ind. Eng. Chem. Res.* **2012**, *51*. [CrossRef]
28. Chiesa, P.; Campanari, S.; Manzolini, G. CO_2 cryogenic separation from combined cycles integrated with molten carbonate fuel cells. *Int. J. Hydrogen Energy* **2011**, *36*, 10355–10365. [CrossRef]
29. Ryckebosch, E.; Drouillon, M.; Vervaeren, H. Techniques for transformation of biogas to biomethane. *Biomass Bioenergy* **2011**, *35*, 1633–1645. [CrossRef]
30. Tippayawong, N.; Thanompongchart, P. Biogas quality upgrade by simultaneous removal of CO_2 and H_2S in a packed column reactor. *Energy* **2010**, *35*, 4531–4535. [CrossRef]
31. Baciocchi, R.; Costa, G.; Gavasci, R.; Lombardi, L.; Zingaretti, D. Regeneration of a spent alkaline solution from a biogas upgrading unit by carbonation of APC residues. *Chem. Eng. J.* **2012**, *179*, 63–71. [CrossRef]
32. Baciocchi, R.; Corti, A.; Costa, G.; Lombardi, L.; Zingaretti, D. Storage of carbon dioxide captured in a pilot-scale biogas upgrading plant by accelerated carbonation of industrial residues. *Energy Procedia* **2011**, *4*, 4985–4992. [CrossRef]
33. Baciocchi, R.; Carnevale, E.; Costa, G.; Gavasci, R.; Lombardi, L.; Olivieri, T.; Zanchi, L.; Zingaretti, D. Performance of a biogas upgrading process based on alkali absorption with regeneration using air pollution control residues. *Waste Manag.* **2013**, *33*, 2694–2705. [CrossRef] [PubMed]
34. Vega, F.; Cano, M.; Gallego, M.; Camino, S.; Camino, J.A.; Navarrete, B. Evaluation of MEA 5 M performance at different CO_2 concentrations of flue gas tested at a CO_2 capture lab-scale plant. *Energy Procedia* **2017**, *114*, 6222–6228. [CrossRef]
35. Leonzio, G. Upgrading of biogas to bio-methane with chemical absorption process: Simulation and environmental impact. *J. Clean. Prod.* **2016**, *131*, 364–375. [CrossRef]
36. Ahn, J.W.; Kim, J.H.; Park, H.S.; Kim, J.A.; Han, C.; Kim, H. Synthesis of single phase aragonite precipitated calcium carbonate in $Ca(OH)_2$-Na_2CO_3-NaOH reaction system. *Korean J. Chem. Eng.* **2005**, *22*, 852–856. [CrossRef]

37. Yeh, J.; Pennline, H.; Resnik, K. Study of CO_2 absorption and desorption in a packed column. *Energy Fuels* **2001**, *15*, 272–278. [CrossRef]

38. Rao, A.B.; Rubin, E.S. A Technical, Economic, and Environmental Assessment of Amine-Based CO_2 Capture Technology for Power Plant Greenhouse Gas Control. *Environ. Sci. Technol.* **2002**, *36*, 4467–4475. [CrossRef] [PubMed]

39. Librandi, P.; Costa, G.; De Souza, A.C.B.; Stendardo, S.; Luna, A.S.; Baciocchi, R. Carbonation of Steel Slag: Testing of the Wet Route in a Pilot-scale Reactor. *Energy Procedia* **2017**, *114*, 5381–5392. [CrossRef]

40. Morone, M.; Costa, G.; Polettini, A.; Pomi, R.; Baciocchi, R. Valorization of steel slag by a combined carbonation and granulation treatment. *Miner. Eng.* **2014**, *59*, 82–90. [CrossRef]

41. Santos, R.M.; Knops, P.C.M.; Rijnsburger, K.L.; Chiang, Y.W. CO_2 Energy Reactor—Integrated Mineral Carbonation: Perspectives on Lab-Scale Investigation and Products Valorization. *Front. Energy Res.* **2016**. [CrossRef]

42. Baciocchi, R.; Carnevale, E.; Corti, A.; Costa, G.; Lombardi, L.; Olivieri, T.; Zanchi, L.; Zingaretti, D. Innovative process for biogas upgrading with CO_2 storage: Results from pilot plant operation. *Biomass Bioenergy* **2013**, *53*, 128–137. [CrossRef]

43. Konno, H.; Nanri, Y.; Kitamura, M. Crystallization of aragonite in the causticizing reaction. *Powder Technol.* **2002**, *123*, 33–39. [CrossRef]

44. Dandeu, A.; Humbert, B.; Carteret, C.; Muhr, H.; Plasari, E.; Bossoutrot, J.M. Raman spectroscopy—A powerful tool for the quantitative determination of the composition of polymorph mixtures: Application to $CaCO_3$ polymorph mixtures. *Chem. Eng. Technol.* **2006**, *29*, 221–225. [CrossRef]

45. Altiner, M.; Yildirim, M. Production of precipitated calcium carbonate particles with different morphologies from dolomite ore in the presence of various hydroxide additives. *Physicochem. Probl. Miner. Process* **2017**, *53*, 413–426. [CrossRef]

Article

Effect of Physical and Mechanical Activation on the Physicochemical Structure of Coal-Based Activated Carbons for SO$_2$ Adsorption

Dongdong Liu [1] , **Zhengkai Hao** [1], **Xiaoman Zhao** [1], **Rui Su** [1], **Weizhi Feng** [1], **Song Li** [1] **and Boyin Jia** [2,*]

[1]   College of Engineering and Technology, Jilin Agricultural University, Changchun 130118, China; liudongdong@jlau.edu.cn (D.L.); hzk980924@163.com (Z.H.); zhaoxiaoman0930@163.com (X.Z.); gaojihui0809@163.com (R.S.); fengweizhijlau@163.com (W.F.); shangguansong@aliyun.com (S.L.)

[2]   College of Animal Science and Technology, Jilin Agricultural University, Changchun 130118, China

*   Correspondence: jiaboyin@jlau.edu.cn

Received: 9 September 2019; Accepted: 26 September 2019; Published: 5 October 2019

Abstract: The SO$_2$ adsorption efficiency of activated carbons (ACs) is clearly dependent on its physicochemical structure. Related to this, the effect of physical and mechanical activation on the physicochemical structure of coal-based ACs has been investigated in this work. In the stage of CO$_2$ activation, the rapid decrease of the defective structure and the growth of aromatic layers accompanied by the dehydrogenation of aromatic rings result in the ordered conversion of the microstructure and severe carbon losses on the surfaces of Char-PA, while the oxygen content of Char-PA, including C=O (39.6%), C–O (27.3%), O–C=O (18.4%) and chemisorbed O (or H$_2$O) (14.7%), is increased to 4.03%. Char-PA presents a relatively low S$_{BET}$ value (414.78 m^2/g) owing to the high value of Non-V_{mic} (58.33%). In the subsequent mechanical activation from 12 to 48 h under N$_2$ and dry ice, the strong mechanical collision caused by ball-milling can destroy the closely arranged crystalline layers and the collapse of mesopores and macropores, resulting in the disordered conversion of the microstructure and the formation of a defective structure, and a sustained increase in the S$_{BET}$ value from 715.89 to 1259.74 m^2/g can be found with the prolonging of the ball-milling time. There is a gradual increase in the oxygen content from 6.79 to 9.48% for Char-PA-CO$_2$-12/48 obtained by ball-milling under CO$_2$. Remarkably, the varieties of physicochemical parameters of Char-PA-CO$_2$-12/48 are more obvious than those of Char-PA-N$_2$-12/48 under the same ball-milling time, which is related to the stronger solid-gas reactions caused by the mechanical collision under dry ice. Finally, the results of the SO$_2$ adsorption test of typical samples indicate that Char-PA-CO$_2$-48 with a desirable physicochemical structure can maintain 100% efficiency within 30 min and that its SO$_2$ adsorption capacity can reach 138.5 mg/g at the end of the experiment. After the 10th cycle of thermal regeneration, Char-PA-CO$_2$-48 still has a strong adsorptive capacity (81.2 mg/g).

Keywords: activated carbons; physical activation; mechanical activation; physicochemical structure; SO$_2$ adsorption

---

1. Introduction

For a long time, SO$_2$ emission from large coal-fired power plants has seriously polluted the environment and threatened human health [1]. The traditional wet desulfurization technology using calcium-based absorbent is unable to satisfy sustainable development, owing to its ecological destruction and the production of massive low-value by-products [2]. Dry desulfurization technology using porous carbon materials (such as activated carbons, ACs) as adsorbents has a better application prospect owing to its low price and ability to produce valuable by-products (such as sulfuric acid) [3].

ACs are usually prepared via a physical activation method and chemical activation method. In the process of chemical activation, the substantial water/acid is consumed to remove a large number of residual reagents (such as KOH [4,5], K_2CO_3 [6,7], $ZnCl_2$ [8], and H_3PO_4 [9]), which not only increases the preparation costs but also causes environmental pollution. Thus, it is highly desirable to apply physical activation as a green and low-cost method for the preparation of ACs. In the process of physical activation, activation agents including steam, CO_2 and their mixtures can partially etch the carbon network to produce some porosity and functional groups, and some researchers [10–14] have found that CO_2 activation can make it easier to generate pores inside the particles than steam activation can. In addition, the preparation of ACs using coal as raw material can also meet desulfurization requirements in coal-fired power plants. Furthermore, the desulfurization performance of ACs is closely related to its physicochemical structure, such as a lot of active sites and a high specific surface area (S_{BET}) in the presence of the hierarchical pore [15,16]. In the process of physical activation, the pore development follows a branched model. First, the micropore is formed on the surface of particles at the initial stage, after which the successive diffusion of activated agents from surface to core helps the formation of a new micropore; meanwhile, the formation of mesopores and macropores originates from the enlargement of the original micropores [17,18]; this branched model inevitably leads to a low specific surface area (S_{BET}) and a high carbon loss on the surface of the particles, even under various activation conditions (such as the activation temperature, activation time and activated gas species, etc.) [13,14,19,20]. In addition, the active sites of porous carbon materials usually include oxygen/nitrogen-containing functional groups and defects at the edge of the aromatic layers. Zhu et al. [13,14] have found that the surface chemical properties of ACs are also significantly changed to form some oxygen-containing functional groups with the carbon loss at the stage of CO_2 activation. Some researchers and our previous work [21–23] have found that the dehydrogenation of aromatic rings accompanied by the rapid decrease of the defect structure helps the vertical condensation and the transverse growth of aromatic layers during the physical activation, thus resulting in the removal of a large number of active sites. In summary, it is difficult to obtain the ideal ACs with a desirable porosity and more active sites in the stage of physical activation.

A ball-milling technique is a novel method to synthesize materials by mechanical activation without producing hazardous products (which can destroy the chemical bonds of the macromolecular structure), finally resulting in the molecular rearrangement, amorphization and recrystallization of the crystal structure [24,25]. Ong et al. [26] have found that the specific surface area (S_{BET}) and the oxygen content of the sample increase rapidly from 6 m^2/g to 450 m^2/g and from 3.6% to 5.0% within 12 h of milling in the oxidizing atmosphere. Zhang et al. [27] also found that the size distribution of the sample ends up being narrower and its average particle size decreases with the increase of the milling time from 0.25 to 8 h because of the high collision strength between the agate balls during the ball-milling process. Salver-Disma et al. [28] have demonstrated that the mechanical milling of natural graphite is one possibility for producing disordered carbons with large intercalation capacities. The milled graphite contains large amounts of defects and present anisotropies [29]. Nevertheless, information regarding the changes in the physicochemical structure of coal-based ACs during physical and mechanical activation is still limited.

In this work, a precursor with a stable carbon-based framework obtained by pyrolysis could be treated first by CO_2 activation, which ensures the formation of the original pores (such as some micropores and the hierarchical pores) and some oxygen-containing functional groups. In order to further increase the specific surface area and quantities of the active sites, the samples mentioned above continued to be activated via the ball-milling method under different times (12 h and 48 h, respectively) and different atmospheres (dry ice or N_2) to further increase the specific surface area and quantities of the active sites. In addition, the physicochemical structure of all the samples were measured by a D/max-rb X-ray diffractometer (XRD), Raman spectroscopy, Nitrogen adsorption, X-ray photoelectron spectroscopy (XPS), transmission electron microscope (TEM) and high-resolution scanning electron microscope (SEM). Finally, to verify the application potentials of the ACs with the ideal physicochemical structure, an SO_2 removal test was performed to further explore the relationship between the physicochemical structure and SO_2 adsorption of ACs.

2. Materials and Methods

2.1. Materials

In this work, Jixi bituminous coal with particle sizes of 200–350 μm was obtained from the northeast of China. In order to eliminate the interference of minerals in the raw material, Jixi bituminous coal was treated sequentially using 30 wt % hydrofluoric acid (HF) and 5 mol·L^{-1} hydrochloric acid (HCl) following the steps in the literature [30]. Then, the acid-treated samples were washed using deionized water and dried in an oven at 90 °C for 12 h. The proximate analysis and elemental analysis of the acid-treated samples (JX) were given in Table 1.

Table 1. The proximate analyses and elemental analyses of JX (wt %).

V_{ad}	FC_{ad}	A_{ad}	M_{ad}	C_{daf}	H_{daf}	O_{daf} *	N_{daf}	S_{daf}
39.66	56.60	0.12	3.62	74.81	19.49	4.01	1.31	0.38

* By difference; ad (air-dried basis): the coal in dry air was used as a benchmark; daf (dry ash free basis): the remaining component after the removal of water and ash in coal was used as a benchmark; *V*: volatile; *FC*: fixed carbon; *A*: ash; *M*: moisture; *C*: carbon element; *H*: hydrogen element; *O*: oxygen element; *N*: nitrogen element; *S*: sulfur element.

2.2. Experimental Process

2.2.1. CO_2 activation

10 g of JX were heated to 900 °C at a constant rate of 5 °C/min in an argon atmosphere flow of 600 mL/min and held for 60 min, and this sample was marked as Char. Then, argon atmosphere was converted to CO_2 (99.999%) at 600 mL/min and held for 60 min, before being finally cooled down to room temperature under an argon atmosphere and being marked as Char-PA.

2.2.2. Mechanical activation

Char-PA was prepared by a planetary ball mill (Pulverisette 6, FRITSCH, Idar-Oberstein, Germany,) by applying a rotation speed of 400 rpm in dry ice or N_2 atmosphere for 12 h and 48 h, respectively. Additionally, a ball-to-powder weight ratio of 15:1 and 12 stainless steel balls with a diameter of 10 mm were used in the process of ball-milling, and these treated samples were marked as Char-PA-N_2/CO_2-different activation times. In addition, thermal annealing of a typical sample was processed at 800 °C for 60 min in 5% H_2/Ar atmosphere; this treated sample was marked as Char-PA-N_2/CO_2-different activation times-H_2.

2.3. Measurement Analysis

The surface topography of the samples was obtained via a scanning electron microscope (SEM, Quanta 200, FEI, Oregon, OR, USA) at 200 kV. The microstructure of the samples was tested by high-resolution transmission electron microscope (HRTEM, Tecnai G2 F30, FEI, Oregon, OR, USA) at 300 kV. The crystal information of the samples was received by a D/max-rb X-ray diffractometer (XRD, D8 ADVANCE, Brooke, Karlsruhe, Germany) at a fixed scanning speed of 3°/min from 5° to 85°. The hybrid carbon information of the samples was received by Raman spectroscopy at a stable scanning scope from 1000–1800 cm^{-1} using a 532 nm wavelength laser. The elemental composition, chemical state and relative concentration on the surface of the samples were obtained by X-ray photoelectron spectroscopy (XPS, K-Alpha, Thermo Fisher Scientific, Waltham, MA, USA) with an Al Kα X-ray at 14 kV and 6 mA [31]. The pore parameters of the samples were received by a micromeritics adsorption apparatus (ASAP2020, Micromeritics, Norcross, GA, USA) at 77 K and a relative pressure (P/P_0) range from 10^{-7} to 1 [32]. The vacuum degassing pretreatment of the tested samples was carried out at 473 K for 12 h. Moreover, the specific surface area (S_{BET}) of the samples was calculated using a BET model in a relative pressure range of 0.05–0.2 [33]; the total pore volume (V_{tot}) caused by the adsorption value of liquid nitrogen at a relative pressure of 0.98 was obtained using the *t*-plot method [34]; the micropore volume (V_{mic}) of the samples was calculated using the Horvath–Kawazoe (HK) method [35]; the nonlocal density functional theory (NLDFT) was used to obtain the pore size distribution of the micropore and mesopore, and the relative pressure range was 10^{-7}~0.9 [13].

2.4. SO$_2$ Adsorption Test

A fixed-bed experimental system was used to investigate the SO$_2$ adsorption performance of ACs. This system included four parts: the gas distribution system, the fixed-bed reactor, the heating and insulation system, and the gas analysis system, as shown in Figure 1. First, 250 mL of deionized water were added to the humidifying tank, which was placed in a constant temperature water bath to maintain it at 90 °C. Simulated flue gas was prepared by mixing a certain amount of N_2 with SO_2, O_2 and other gases. The water vapor content in the simulated flue gas was controlled by adjusting the N_2 flow rate.

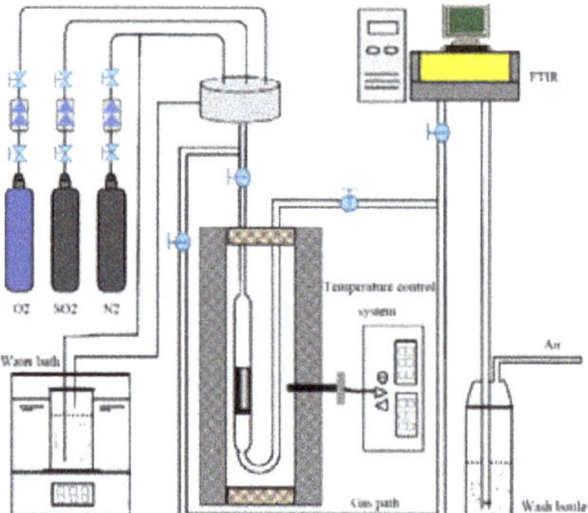

Figure 1. Schematic figure of the fixed-bed reactor system for SO$_2$ adsorption.

The experimental process is as follows: 5 g of tested sample were placed in a glass tube reactor that consisted of a glass tube and sand core. Furthermore, this sand core not only supported AC particles, but also played the role of current sharing. The filling height and diameter of the adsorbent in the glass tube reactor were 25 mm and 20 mm, respectively. The SO$_2$ adsorption test was performed at 80 °C for 210 min. The gas volumetric composition used in the experiments was: SO_2, 1500 ppm; O_2, 5%; water vapor, 10%; and N_2, balance, total flow rate 250 mL·min^{-1}. The SO$_2$ concentrations at the entrance and exit were measured by an on-line Fourier transform infrared gas analyzer (FTIR, Dx4000, Gasmet, Vantaa, Finland) for calculating the SO$_2$ removal efficiency and the change of the removal rate with time. According to the integral area and reaction time on the removal curve, the SO$_2$ removal capacity of the coal-based ACs was obtained [36]. In addition, the SO$_2$ removal efficiency, SO$_2$ removal rate and cumulative sulphur capacity were calculated with the following formula:

SO$_2$ removal efficiency (*DeSO$_2$*):

$$DeSO_2(\%) = \frac{SO_2(in) - SO_2(out)}{SO_2(in)} \times 100\% \tag{1}$$

In the formula, SO_2 (*in*) and SO_2 (*out*) are the SO$_2$ concentrations at the reactor inlet and outlet measured by FTIR, respectively.

SO_2 removal rate (RSO_2):

$$RSO_2(\text{mg} \cdot \text{g}^{-1} \cdot \text{min}^{-1}) = \frac{DeSO_2 \cdot SO_2(in) \cdot 0.3 \cdot 64}{22400} \tag{2}$$

In the formula, $DeSO_2$ is the SO_2 removal efficiency and SO_2 (in) is the SO_2 concentration at the reactor inlet measured by FTIR,

The accumulated sulfur capacity (ASO_2) refers to the cumulative removal capacity of SO_2 from samples varying with time and is obtained by integrating the removal rate of SO_2 with the time.

$$ASO_2(\text{mg} \cdot \text{g}^{-1}) = \int_0^t RSO_2 dt \tag{3}$$

2.5. Thermal Regeneration

For the thermal regeneration, desulfurized ACs were placed in a glass tube, and afterwards were settled in a vertical furnace (Figure 1). The ACs were heated in a flow of nitrogen (200 mL/min), at a heating rate of 10 °C min^{-1} and maintained at 400 °C for 30 min, before being cooled for 30 min while the nitrogen purge was continued.

3. Results and Discussion

3.1. Microstructure and Surface Morphology Analysis by HRTEM and SEM

Figure 2 shows several HRTEM and SEM images of samples produced under pyrolysis and different activation conditions, including (a) HRTEM and (h) SEM of Char; (b) HRTEM and (i) SEM of Char-PA; (c) HRTEM and (j) SEM of Char-PA-N$_2$-12; (d) HRTEM and (k) SEM of Char-PA-N$_2$-48; (e) HRTEM and (l) SEM of Char-PA-CO$_2$-12; (f) HRTEM and (m) SEM of Char-PA-CO$_2$-48; (g) HRTEM and (n) SEM of Char-PA-CO$_2$-48H$_2$.There are some crystalline layers with different orientations near small quantities of amorphous carbon for Char in Figure 2a. In addition, a smooth surface and compact texture of Char can be found in Figure 2h. He et al. [37] found that metaplast material that was formed via the combination of transferable hydrogen and aliphatic hydrocarbons during pyrolysis can not only promote the ordered arrangement of aromatic layers but also reshape the particle surface. After CO$_2$ activation, a large amount of long and multi-layer crystallite with a consistent orientation can be found in Char-PA, as shown in Figure 2b; the highly ordered crystalline layers of Char-PA can hinder the further diffusion of activated gas into the interior of particles, leading to severe carbon losses on the particle surfaces, as shown in Figure 2i. After 12 h of high-energy ball milling, some short and thin crystallite layers of Char-PA-N$_2$/CO$_2$-12 with a granular structure and rough surface can be found in Figure 2c,e,j,l. Furthermore, as the time of ball milling increases from 12 h to 48 h, blurred boundaries and a disordered arrangement of crystalline layers of Char-PA-N$_2$/CO$_2$-48 with a rougher surface and some smaller particles are also found in Figure 2d,f,k,m. The above results show that the mechanical collision caused by ball milling can significantly promote the disordered conversion of the microstructure and improve the surface morphology of particles with the increase of the ball-milling times from 12 to 48 h. In the process of mechanical activation, the high-energy ball milling can reduce the reaction temperature of the solid state and cause the increase of the local temperature on the material [24,25]. Based on this fact, a strong mechanical collision under dry ice can rapidly promote the solid-gas reactions between CO$_2$ and the carbon matrix, as a result of which Char-PA-CO$_2$-12/48 shows more obvious changes in its microcrystalline structure and surface morphology than Char-PA-N$_2$-12/48 under the same ball-milling time.

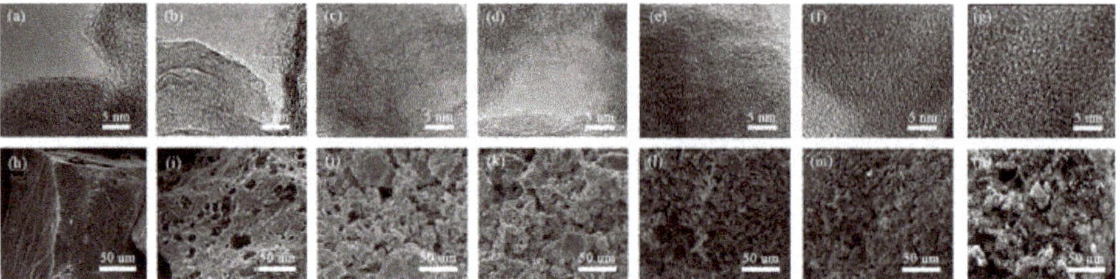

Figure 2. The HRTEM and SEM images of the samples produced at different treatment conditions. (**a**) HRTEM and (**h**) SEM of Char; (**b**) HRTEM and (**i**) SEM of Char-PA; (**c**) HRTEM and (**j**) SEM of Char-PA-N$_2$-12; (**d**) HRTEM and (**k**) SEM of Char-PA-N$_2$-48; (**e**) HRTEM and (**l**) SEM of Char-PA-CO$_2$-12; (**f**) HRTEM and (**m**) SEM of Char-PA-CO$_2$-48; (**g**) HRTEM and (**n**) SEM of Char-PA-CO$_2$-48H$_2$.

3.2. Crystal Structure Analysis by XRD

The XRD profiles of the samples produced under pyrolysis and different activation conditions are given in Figure 3. There are two obvious broad diffraction peaks at $2\theta = 16$–$32°$ and 36–$52°$ in all of the samples; the information on two diffraction peaks (such as positions and half-peak width) can be obtained using the peak fitting treatment, as shown in Figure 4. Some crystal parameters, including the interlayer distance (d_{002}), stacking height (L_c), and the size (L_a) and numbers (N) of the aromatic layers, are calculated via the following formulas [38]:

$$d_{002} = \frac{\lambda}{2\sin\theta} \tag{4}$$

$$L_c = \frac{0.89\lambda}{\beta\cos\theta} \tag{5}$$

$$L_a = \frac{1.84\lambda}{\beta\cos\theta} \tag{6}$$

$$N = \frac{L_c}{d_{002}} \tag{7}$$

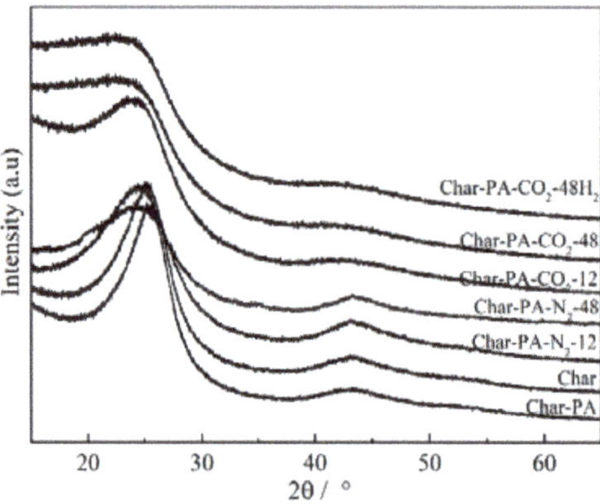

Figure 3. The XRD profiles from the samples produced at different treatment conditions.

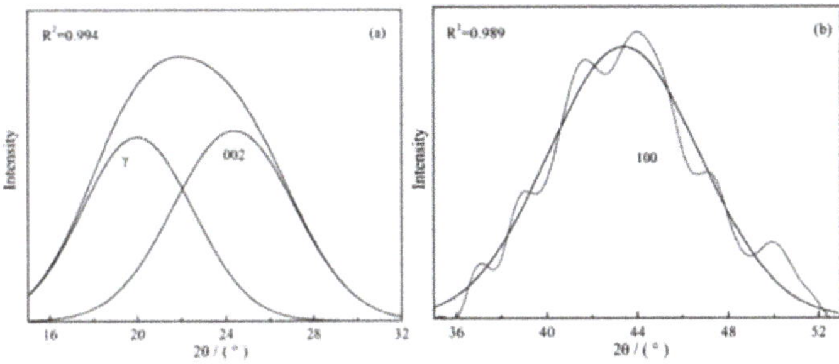

Figure 4. The fitting curve of the peaks for Char-PA in the 2θ range (**a**) 15–32° and (**b**) 35–53°.

In the above formulas, λ: the wavelength of the X-ray, $\lambda = 1.54$ Å; θ: peaks' positions (°); and β: half peak width. The results of the crystal parameters of all the samples are given in Table 2.

Table 2. The crystal parameters of the samples produced at different treatment conditions.

Samples	La (Å)	Lc (Å)	d_{002} (Å)	$N = Lc/d_{002}$
Char	24.81	13.48	3.50	3.85
Char-PA	25.73	13.79	3.45	3.99
Char-PA-N$_2$-12	24.31	13.15	3.58	3.67
Char-PA-N$_2$-48	23.19	12.43	3.72	3.34
Char-PA-CO$_2$-12	23.91	12.86	3.69	3.48
Char-PA-CO$_2$-48	22.28	11.79	3.96	2.98
Char-PA-CO$_2$-48H$_2$	22.75	12.97	4.05	3.20

First, the rapid increase in the numbers (N), stacking height (L_c) and size (L_a) of the aromatic layers and the obvious reduction in the layer spacing (d_{002}) can be found in the Char-PA during CO$_2$ activation. The longitudinal condensation and transversal growth of the aromatic layers are related to the rapid consumption of the side chains and bridge bonds within the aromatic layers and defective structure on the edge of the aromatic layers [14,39]. Then, with the ball-milling time increases from 12 to 48 h, there is a sustained decrease in the L_a, L_c and N values and a persistent increase in the d_{002} value for Char-PA-CO$_2$/N$_2$-12/48, indicating the disordered conversion of the microcrystalline structure. These changes are closely related to the breakdown and distortion of the aromatic layers caused by a strong mechanical collision, which can further destroy the parallelism of the layer and the constancy of the interlayer spacing. Finally, the changes of the crystal parameters of Char-PA-CO$_2$-12/48 are more obvious than those of Char-PA-N$_2$-12/48 under the same ball-milling time, which may be related to the fact that the local high temperature caused by ball milling promotes CO$_2$ etching on the aromatic layers. In this process, the more defective structures on the edge of the aromatic layers can also be formed as active sites.

3.3. Carbon Structure Analysis by Raman

The Raman spectra of the samples produced under pyrolysis and different activation conditions are shown in Figure 5. There are two obvious broad diffraction peaks at 1230–1450 cm^{-1} (D peak) and 1450–1580 cm^{-1} (G peak) in all of the samples. Some researchers [40,41] have found that the widening of the D and G peaks is serious for incomplete graphitized materials in Raman spectra, indicating the existence of many sp^2-hybridized structures and sp^2–sp^3 hybridized structures. Thus, it is necessary to resolve overlapped peaks using a fitting treatment, including for the D$_1$ peak (1300 cm^{-1}), D$_3$ peak (1520 cm^{-1}), D$_4$ peak (1200 cm^{-1}) and G peak (1550 cm^{-1}). Figure 6 shows the fitting curve of the Raman spectrum for Char-PA-N$_2$-48. Furthermore, the D$_1$ peak is represented as the defective sp^2 bonding carbon atoms; the D$_3$ peak is represented as the amorphous sp^2 bonding carbon atoms;

the D_4 peak is represented as the sp^2–sp^3 bonding carbon atoms; the G peak is represented as the crystalline sp^2 bonding carbon atoms; furthermore, the relative quantities of different hybridized structures is as follows: (1) A_{D1}/A_G represents the relative quantity of the big aromatic rings, including C–C between aromatic rings and aromatics with no fewer than 6 rings with a defective structure; (2) A_{D3}/A_G represents the relative quantity of the small aromatic rings including aromatics with 3–5 rings and the semi-circle breathing of aromatic rings; and (3) A_{D4}/A_G represents the relative quantity of the cross-linking structure, including $C_{aromatic}$–C_{alkyl}, aromatic (aliphatic) ethers and C–C on hydroaromatic rings [42]. The results of the different hybrid carbons in the form of area ratios of the samples are shown in Table 3.

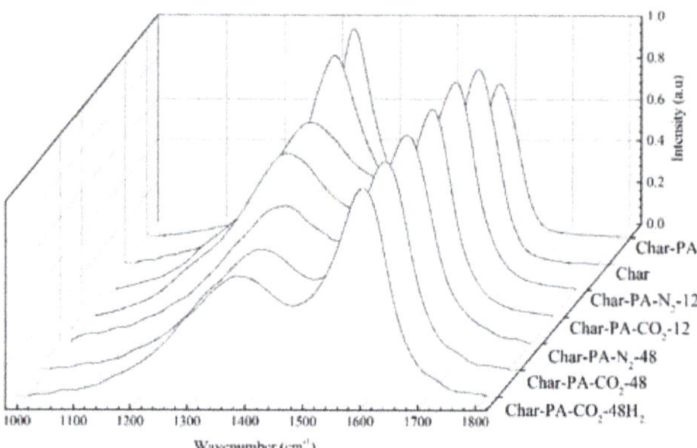

Figure 5. The Raman spectra from the samples produced at different treatment conditions.

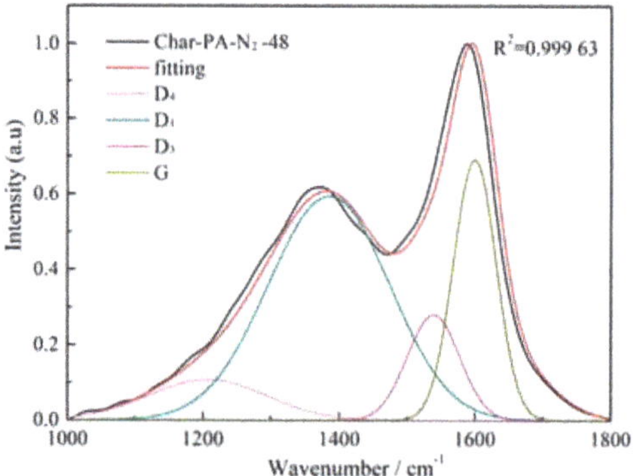

Figure 6. The fitting curve of the Raman spectrum for Char-PA-N$_2$-48.

Table 3. Hybrid carbon in the form of the area ratio of the samples produced at different treatment conditions.

Samples	A_{D1}/A_G	A_{D3}/A_G	A_{D4}/A_G
Char	3.179	1.947	0.589
Char-PA	1.460	0.479	0.187
Char-PA-N$_2$-12	2.014	0.481	0.385
Char-PA-N$_2$-48	3.311	0.495	0.597
Char-PA-CO$_2$-12	2.187	0.486	0.418
CH-PA-CO$_2$-48	3.785	0.501	0.657
Char-PA-CO$_2$-48H$_2$	3.785	0.501	0.657

First, the values of A_{D1}/A_G, A_{D3}/A_G and A_{D4}/A_G of Char-PA obviously decrease when compared to those of Char. In the process of CO_2 activation, the defective structure at the edge of the aromatic layers and some small aromatic rings are consumed preferentially [43].The consumption of the defective structure can promote the dehydrogenation of the big aromatic rings; and the removal of the small aromatic rings can help the inner reactivation of the big aromatic rings accompanied by the breakdown of the cross-linking structure [44]; these changes finally lead to a substantial increase in the quantities of crystalline sp^2 bonding carbon atoms, which further promotes the stability of the carbon structure. Then, with the increase of the ball-milling time from 12 to 48 h, the values of A_{D1}/A_G and A_{D4}/A_G of Char-PA-N_2/CO_2-12/48 increase gradually; but there is a slight increase in the value of A_{D3}/A_G. It can be inferred that a strong mechanical collision caused by the ball milling has partly destroyed the crystalline sp^2 bonding carbon structure, decomposing it into the big aromatic rings, which is accompanied by the formation of new crosslinking structures. In particular, the variety of the hybrid carbon parameters of Char-PA-CO_2-12/48 is more obvious than that of Char-PA-N_2-12/48 under the same ball-milling time. Some oxygen atoms from dry ice are easily bonded and fixed to the carbon matrix in the form of cross-linking bonds (such as -COO- and -O-), and some oxygen-containing heterocycles under a local high temperature are caused by a strong mechanical collision; furthermore, the presence of O-containing structures can also promote the reorganization of aromatic fragments to form more big aromatic rings.

3.4. Pore Structure Analysis by N_2 Adsorption

The N_2 adsorption isotherm and pore-size distribution of the samples under pyrolysis and different activation conditions are given in Figure 7, and the corresponding pore parameters are shown in Table 4.

Table 4. The pore structure parameters of the samples at different treatment conditions.

Samples	S_{BET} (m^2/g) [a]	V_t (m^3/g) [b]	V_{mic} (m^3/g) [c]	Non-V_{mic} (%) [d]
Char	48.45	0.048	0.034	29.17
Char-PA	414.78	0.24	0.10	58.33
Char-PA-N_2-12	715.89	0.33	0.22	33.33
Char-PA-N_2-48	919.74	0.38	0.30	21.05
Char-PA-CO_2-12	859.14	0.36	0.26	27.78
Char-PA-CO_2-48	1259.74	0.42	0.34	19.056
Char-PA-CO_2-48H_2	1214.85	0.38	0.31	18.42

[a] Specific surface area determined by the BET method for P/P_0 from 0.05 to 0.24. [b] Total pore volume calculated at $P/P0 \frac{1}{4} 0.98$. [c] Volume of micropores (< 2 nm) calculated by the t-plot method. [d] V_t minus V_{mic} (>2 nm).

First, the N_2 adsorption isotherm of Char is attributed to a type I according to the IUPAC classification, and its N_2 adsorption capacity is very small, showing small amounts of pores. The metaplast formed by the combination of transferable hydrogen with free radicals during pyrolysis can plug the pores, thus leading to an S_{BET} value of 48.45 m$^2 \cdot$g^{-1}, V_{mic} value of 0.034 m$^3 \cdot$g^{-1} and non-V_{mic} value of 29.17% for Char, with a narrow size distribution of less than 2 nm. Then, the N_2 adsorption isotherm of Char-PA exhibits a typical characteristic of type IV, with the increase of the relative pressure from 0 to 1. This isotherm has begun to branch, and a hysteresis loop has also been formed with the increase of the relative pressure, indicating the formation of hierarchical pores. In the process of CO_2 activation, some micropores can be formed at the initial stage of activation, after which these micropores can further be enlarged into mesopores and macropores with the prolongation of the activation time; finally, these mesopores and macropores, as channels, can promote the diffusion of activated gas to help the production of new micropores [17,18]. Therefore, the N_2 adsorption capacities of Char-PA are higher than those of Char with the increase of the relative pressure, presenting an S_{BET} value of 414.78 m$^2 \cdot$g^{-1} and V_{mic} value of 0.1 m$^3 \cdot$g^{-1}. However, the non-V_{mic} value of 58.33% of

Char-PA with a wide pore distribution indicates the rapid development of mesopores and macropores instead of that of micropores during CO_2 activation.

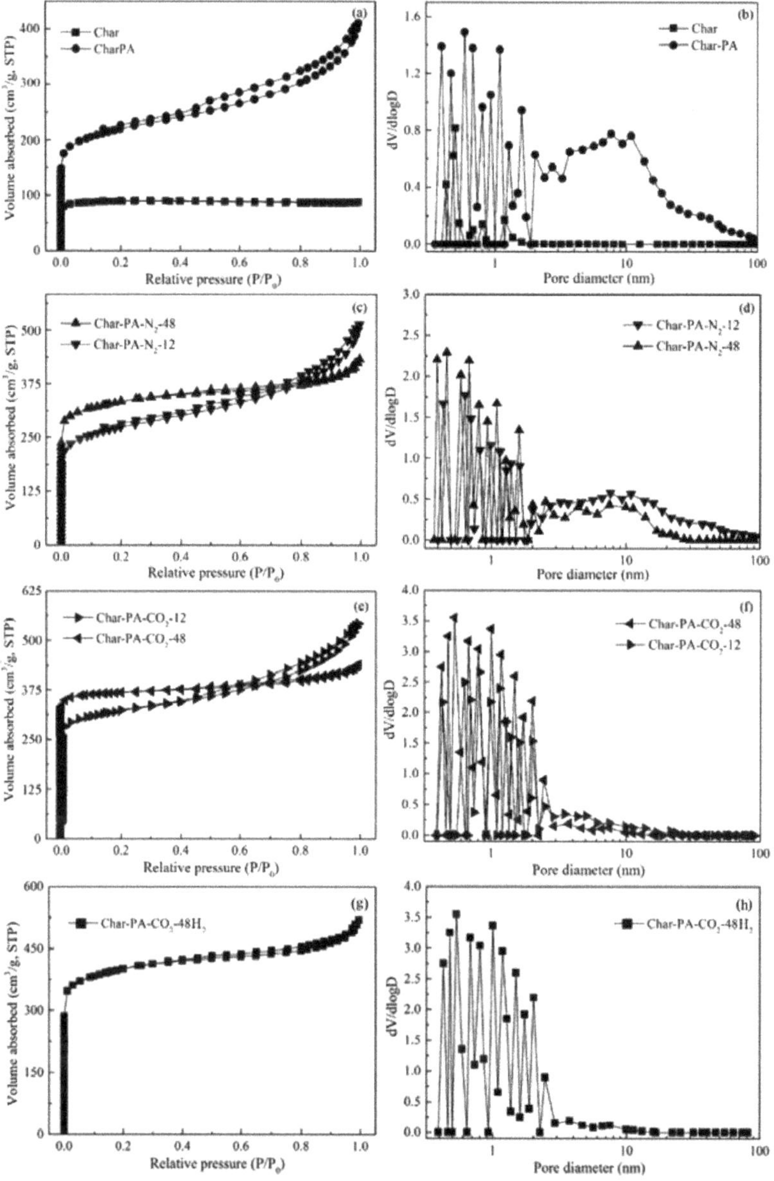

Figure 7. (**a**,**c**,**e**,**g**) The N_2 adsorption isotherm and (**b**,**d**,**f**,**h**) pore-size distribution of the samples at different treatment conditions.

After 12 h of high-energy ball milling, the N_2 adsorption isotherms of Char-PA-CO_2/N_2-12 exhibit the typical characteristic of type IV, with an increase of the relative pressure from 0 to 1. With the increase of the ball-milling time from 12 to 48 h, the flat development of adsorption isotherms and a small hysteresis loop can also be found for Char-PA-CO_2/N_2-48, with high N_2 adsorption capacities at a low pressure and presenting a rapid increase of the S_{BET} and V_{mic} values and an obvious decrease of the non-V_{mic} value of Char-PA-CO_2/N_2-12/48, accompanied by a gradual narrowing of the pore size distribution. These changes indicate a sustained formation of new micropores and a rapid decrease of mesopores and macropores during the ball milling, which are related to the fact that the collapse of the mesopores and macropores caused by the strong mechanical collision brings about the production of new micropores. Furthermore, a stronger solid-gas reaction caused by the mechanical collision under dry ice can favor the development of the porous structure. Therefore, the variety of pore structures of Char-PA-CO_2-12/48 is more obvious than that of Char-PA-N_2-12/48 under the same ball-milling time.

3.5. Surface Chemical Structure Analysis of XPS

Figure 8 shows the broad scanning energy spectrum of all the samples, determined by XPS in the range of 10–1200 eV binding energy to obtain the strongest peak for most elements. There are two obvious peaks (C1s and O1s peaks) in Char-PA and Char-PA-N_2/CO_2-12/48, indicating the dominant position of C and O in the element composition; however, the disappearance of the O1s peak of Char is related to the release of the oxygen elements in the form of small molecules (such as CO, CO_2) during pyrolysis. With the increase of the ball-milling time from 12 to 48 h, there is a slight increase in the oxygen content of Char-PA-N_2-12/48 from 4.09% to 4.11%. A ball milling under dry ice can rapidly promote the combination of surface unsaturated carbon atoms with CO_2 to fix a large number of O atoms in the form of oxygen-containing functional groups; thus, the oxygen content of Char-PA-CO_2-12/48 increases rapidly from 6.79% to 9.48%.

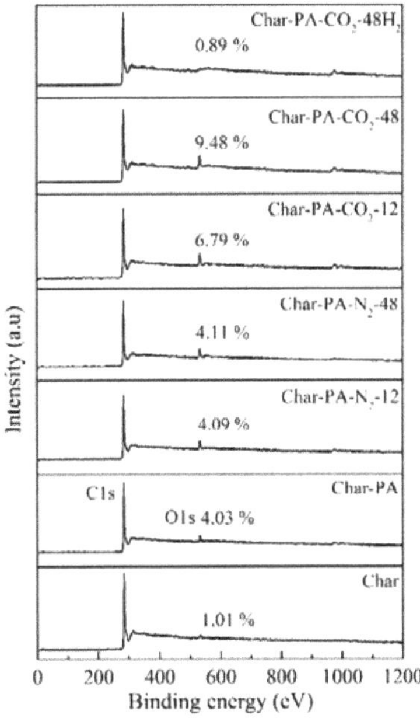

Figure 8. The survey XPS spectra of the samples at different treatment conditions.

In carbon materials, surface oxygen-containing functional groups are the most important functional groups that have been identified as affecting the surface chemical properties of carbon materials. In order to further explore the types and contents of oxygen functional groups on the surface of Char-PA and Char-PA-N_2/CO_2-12/48 quantitatively, the O1s peak of five samples is fitted and analyzed according to different binding energies, as follows: carboxyl or ester carbon (C=O) at 530.9 eV, phenolic hydroxyl or ether (C–O) at 532.4 eV, carboxyl or ester carbon (O–C=O) at 533.8 eV and chemisorbed O_2 (or H_2O) at 535.2 eV. The results of the curve fitting of the five samples are given in Table 5. In addition, the XPS spectra of Char and Char-PA-CO_2-48H_2 are not treated further using the fitting method due to a minor oxygen content. After CO_2 activation, the carbonyl and quinone group (C=O) of Char-PA accounts for the most part (39.6%); its phenol and ether group (C–O) takes second place (27.3%), and the carboxyl group or ester group (O–C=O) and chemisorbed O (or H_2O) come last (18.4% and 14.7%) out of all the oxygen-containing functional groups. After ball milling under N_2, the proportions of C=O, C–O and O–C=O of Char-PA-N_2-12/48 are gradually consistent with each other with the prolongation of the ball-milling time. After ball milling under dry ice, the proportion of O–C=O of Char-PA-CO_2-12/48 increases and its proportion of C–O, C=O decreases gradually, indicating the existence of oxidation phenomena. The above results illustrate that the content and type of oxygen-containing groups in the samples can be controlled by the ball-milling treatment in different atmospheres.

Table 5. The results of the curve fitting of Char-PA and Char-PA-N_2/CO_2-12/48 in O1s from the XPS spectra.

Samples	C–O (%)	C=O (%)	O–C=O (%)	Chemisorbed O (or H_2O) (%)
Char-PA	27.3	39.6	18.4	14.7
Char-PA-N_2-12	29.5	35.4	25.6	9.5
Char-PA-N_2-48	30.4	31.7	29.7	8.2
Char-PA-CO_2-12	26.7	29.5	36.1	7.7
Char-PA-CO_2-48	23.9	26.4	41.2	8.5

3.6. Study of SO_2 Adsorption

In order to further explore the effect of the physicochemical structure of ACs on SO_2 adsorption, an SO_2 adsorption test of Char-PA, Char-PA-N_2-48, Char-PA-CO_2-48 and Char-PA-CO_2-48-H_2 from a simulated flue gas is performed at 80 °C for 210 min. In these tested samples, Char-PA-CO_2-48-H_2 is obtained via a thermal annealing treatment of Char-PA-CO_2-48 at 800 °C for 1 h in 5% H_2/Ar atmosphere, and the results of the physicochemical structure and corresponding parameters are given in Figures 2–8 and Tables 2–4. It can be found that the thermal annealing process has removed almost all of oxygen-containing functional groups (the oxygen content is only 0.89% in Figure 8) but cannot further change its porosity, microstructure and surface morphology. The results of the SO_2 removal of Char-PA, Char-PA-N_2-48, Char-PA-CO_2-48 and Char-PA-CO_2-48-H_2 are shown in Figure 9.

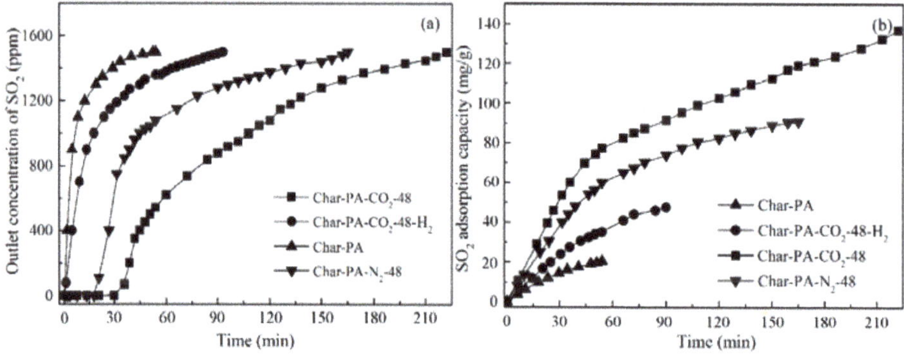

Figure 9. SO_2 removal of typical ACs: (**a**) SO_2 breakthrough curve and (**b**) SO_2 adsorption quantity.

First, the efficient adsorption of Char-PA is mainly presented within only 60 min, while the SO_2 concentrations of the gas outlet arrive quickly 1500 ppm; this result indicates that Char-PA has already been penetrated by SO_2. In addition, Char-PA can only achieve 21.2 mg/g at the end of the experiment. Therefore, Char-PA, with undeveloped pores and fewer activated sites, can only maintain an efficient SO_2 adsorption and conversion at the initial stage. Then, the SO_2 adsorption of Char-PA-CO_2-48 can be maintained at 100% within 30 min. After that, there is a slow increase from 70 to 1500 ppm in the SO_2 concentrations of the gas outlet, and its SO_2 adsorption capacity can reach 138.5 mg/g at the end of the experiment. Remarkably, Char-PA-N_2-48, with relatively developed pores and some active sites, also has a relatively high SO_2 adsorptive capacity (92.2 mg/g) at the end of the experiment, and its SO_2 adsorption curve is similar to the curve of Char-PA-CO_2-48. In the presence of O_2 and H_2O, the SO_2 removal of porous carbon is a multi-step heterogeneous reaction. First, SO_2, O_2 and H_2O can be rapidly absorbed by a lot of micropores, inside of which the oxidation and hydration of SO_2 with O_2 and H_2O is further catalyzed by active sites to form H_2SO_4 [45]; after that, the active sites can continue to migrate H_2SO_4 from micropores to meso-/macro-pores to release the microporous space [46]. Therefore, Char-PA-CO_2-48, with a large quantity of micropores and active sites, presents a high adsorption capacity as compared to that of samples in the previous literatures [47–49]. In addition, Char-PA-CO_2-48-H_2, with a large number of micropores and fewer active sites, has a limited adsorptive capacity (48.7 mg/g) at the end of the experiment, and its SO_2 adsorption curve is similar to the curve of Char-PA. In the desulphurization process of Char-PA-CO_2-48-H_2, SO_2 can be adsorbed effectively by its micropores in the initial stage, but the adsorbed SO_2 within the micropores cannot be further catalyzed and migrated into the mesopores and macropores due to its limited active sites.

In order to measure the cyclic desulfurization performance of the prepared adsorbent, a thermal regeneration of Char-PA-CO_2-48 is performed at 400 °C for 30 min in N_2 atmosphere, and the corresponding result is given in Figure 10. The SO_2 removal capacities of Char-PA-CO_2-48 exhibit a general decreasing trend from 133.3 mg/g in the first-time desulfurization to 81.2 mg/g in the 10th cycle. Pi et al. [47] found that the pore structure and chemically active sites were damaged during a long-time (30 min), high-temperature (400 °C) and repeated thermal-treatment process, thus leading to rapidly decreased SO_2 removal capacities.

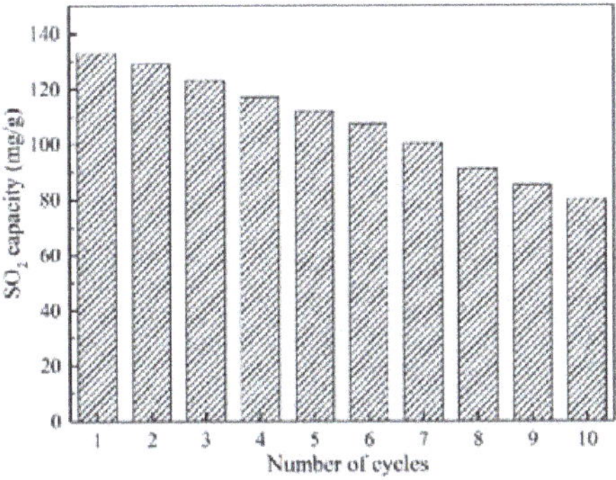

Figure 10. SO_2 removal capacities vs. cycling number of Char-PA-CO_2-48

4. Conclusions

The effect of physical and mechanical activation on the physicochemical structure of coal-based activated carbons (ACs) for SO_2 adsorption has been investigated in this work. Char, using Jixi bituminous coal as raw materials obtained by pyrolysis, is activated sequentially via physical and mechanical methods. The results of the physicochemical structure of a series of AC samples indicate that a substantial reduction in the defective structure at the edge of the aromatic layers and the rapid growth of the aromatic layers accompanied by the dehydrogenation of the aromatic rings result in the order transformation of microstructures of Char-PA and its severe carbon losses on the particle surfaces in the stage of CO_2 activation. Furthermore, the oxygen content of Char-PA is increased to 4.03%, and the proportions of the different oxygen-containing functional groups in Char-PA are as follows: C=O (39.6%), C–O (27.3%), O–C=O (18.4%) and chemisorbed O (or H_2O) (14.7%). The pore development of Char-PA follows a hierarchical model, leading to a relatively low S_{BET} value (414.78 m^2/g) and a high value of Non-V_{mic} (58.33%). Char-PA with undeveloped pores and fewer activated sites can only maintain an efficient SO_2 adsorption and conversion within 60 min and achieve 21.2 mg/g at the end of the experiment. In the subsequent mechanical activation under N_2 and dry ice from 12 to 48 h, the strong mechanical collision can improve the surface morphology and destroy the parallelism of the aromatic layers and the constancy of the interlayer spacing, resulting in the disordered conversion of the microstructure and the formation of more defective structures with the prolonging of the ball-milling time. In addition, the collapse of mesopores and macropores caused by a strong ball milling facilitates the formation of more micropores, leading to a sustained increase in the S_{BET} value from 715.89 to 1259.74 m^2/g and of the micropore volume from 0.22 to 0.34 m^3/g, as well as a sustained decrease in Non-V_{mic} from 33.33 to 19.05% with the prolonging of the ball-milling time. However, the oxygen content of Char-PA-N_2-12/48 increases slowly from 4.09 to 4.11%, presenting a similar distribution proportion, whereas the oxygen content of Char-PA-CO_2-12/48 increases rapidly from 6.79 to 9.48%, presenting an increased proportion of O–C=O and a decreased proportion of C–O, C=O. It is worth noting that the varieties of physicochemical parameters of Char-PA-CO_2-12/48 are more obvious than those of Char-PA-N_2-12/48 under the same ball-milling time, which is related to the strong solid-gas reactions between CO_2 and the carbon matrix caused by the mechanical collision under dry ice. The desulfurization efficiency of Char-PA-CO_2-48 with a desirable physicochemical structure can be maintained at 100% within 30 min and reached 138.5 mg/g. Char-PA-N_2-48 has a similar structure to Char-PA-CO_2-48, thus presenting a relatively high SO_2 adsorptive capacity (92.2 mg/g). Char-PA-CO_2-48-H_2, with fewer active sites obtained by the thermal annealing treatment, has a limited adsorptive capacity (48.7 mg/g) at the end of the experiment. After the 10th cycle of thermal regeneration, Char-PA-CO_2-48 still has a strong adsorptive capacity (81.2 mg/g).

Author Contributions: D.L., S.L. and W.F. conceived and designed the experiments; X.Z., R.S. and Z.H. carried out the experiments; D.L. wrote the paper; D.L. and B.J. reviewed the paper.

Funding: This research was funded by National Natural Science Foundation of China, grant number 51806080, and Scientific Research Fund Project of Jilin Agricultural University, grant number 201801, and Jilin Province Education Department Science and Technology Program during the Thirteenth Five-year Plan Period, grant number JJKH20190940KJ.

Conflicts of Interest: The authors declare no conflict of interest.

References

1. Liu, D.D.; Gao, J.H.; Cao, Q.X.; Wu, S.H.; Qin, Y.K. Improvement of activated carbon from Jixi bituminous coal by air peroxidation. *Energy Fuels* **2017**, *31*, 1406–1415. [CrossRef]
2. Sumathi, S.; Bhatia, K.S.; Lee, K.T.; Mohamed, A.R. Adsorption isotherm models and properties of SO_2 and NO removal by palm shell activated carbon supported with cerium (Ce/PSAC). *Chem. Eng. J.* **2010**, *162*, 194–200. [CrossRef]
3. Bahamon, D.; Vega, L.F. Systematic evaluation of materials for post-combustion CO_2 capture in a temperature swing adsorption process. *Chem. Eng. J.* **2016**, *284*, 438–447. [CrossRef]
4. He, X.J.; Zhang, H.B.; Zhang, H. Direct synthesis of 3D hollow porous graphene balls from coal tar pitch for high performance supercapacitors. *J. Mater. Chem. A* **2014**, *46*, 19633–19640. [CrossRef]
5. Parimal, C.B.; Aola, S.; Rituparna, K.; Mridushmita, B.; Chubaakum, P.; Dipak, S. Activated carbon synthesized from biomass material using single-step KOH activation for adsorption of fluoride: Experimental and theoretical investigation. *Korean J. Chem. Eng.* **2019**, *36*, 551–562. [CrossRef]
6. Li, Y.T.; Pi, Y.T.; Lu, L.M. Hierarchical porous active carbon from fallen leaves by synergy of K_2CO_3, and their supercapacitor performance. *J. Power Sources* **2015**, *299*, 519–528. [CrossRef]

7. Foo, K.Y.; Hameed, B.H. Preparation and characterization of activated carbon from sunflower seed oil residue via microwave assisted K_2CO_3 activation. *Bioresour. Technol.* **2010**, *102*, 9794–9799. [CrossRef]

8. Ma, G.F.; Yang, Q.; Sun, K.J. Nitrogen-doped porous carbon derived from biomass waste for high-performance supercapacitor. *Bioresour. Technol.* **2015**, *197*, 137–142. [CrossRef]

9. Elmouwahidi, A.; Bailón-García, E.; Pérez-Cadenas, A.F. Activated carbons from KOH and H_3PO_4-activation of olive residues and its application as supercapacitor electrodes. *Electrochim. Acta* **2017**, *229*, 219–228. [CrossRef]

10. Rodríguez-Reinoso, F.; Molina-Sabio, M.; González, M.T. The use of steam and CO_2 as activating agents in the preparation of activated carbons. *Carbon* **1995**, *33*, 15–23. [CrossRef]

11. Yang, L.; Huang, T.; Jiang, X.; Jiang, W.J. Effect of steam and CO2activation on characteristics and desulfurization performance of pyrolusite modified activated carbon. *Adsorption* **2016**, *22*, 1099–1107. [CrossRef]

12. Valente Nabais, J.M.; Nunes, P.; Carrott, P.J.M.; Carrott, M.M.L.R.; García, A.M.; Díaz-Díezb, M.A. Production of activated carbons from coffee endocarp by CO_2 and steam activation. *Fuel Process. Technol.* **2008**, *89*, 262–268. [CrossRef]

13. Zhu, Y.W.; Gao, J.H.; Li, Y.; Sun, F. Preparation of activated carbons for SO_2 adsorption by CO_2 and steam activation. *J. Taiwan Inst. Chem. Eng.* **2012**, *43*, 112–119. [CrossRef]

14. Zhu, Y.W.; Gao, J.H.; Li, Y.; Sun, F.; Qin, Y.K. Preparation and characterization of activated carbons for SO_2 adsorption from Taixi anthracite by physical activation with steam. *Korean J. Chem. Eng.* **2011**, *28*, 2344–2350. [CrossRef]

15. Shu, S.; Guo, J.X.; Liu, X.L.; Wang, X.J.; Yin, H.Q.; Luo, D.M. Effects of pore sizes and oxygen-containing functional groups on desulfurization activity of Fe/NAC prepared by ultrasonic-assisted impregnation. *Appl. Surf. Sci.* **2016**, *360*, 684–692. [CrossRef]

16. Davini, P. Adsorption and desorption of SO_2 on active carbon: The effect of surface basic groups. *Carbon* **1990**, *28*, 565–571. [CrossRef]

17. Shiratori, N.; Lee, K.J.; Miyawaki, J.; Hong, S.H.; Mochida, I.; An, B.; Yokogawa, K.; Jang, J.; Yoon, S.H. Pore structure analysis of activated carbon fiber by microdomain-based model. *Langmuir* **2009**, *25*, 7631–7637. [CrossRef]

18. Liu, D.D.; Gao, J.H.; Wu, S.H.; Qin, Y.K. Effect of char structures caused by varying the amount of $FeCl_3$ on the pore development during activation. *RSC Adv.* **2016**, *6*, 87478–87485. [CrossRef]

19. Srinivas, G.; Yue, L.; Neal, S.; Taner, Y.; Zheng, X.G. Design of hyperporous graphene networks and their application in solid-amine based carbon capture systems. *J. Mater. Chem. A* **2017**, *5*, 17833–17840. [CrossRef]

20. Hu, B.; Wang, K.; Wu, L.H.; Yu, S.H.; Antonietti, M.; Titirici, M.M. Engineering carbon materials from the hydrothermal carbonization process of biomass. *Adv. Mater.* **2010**, *22*, 813–828. [CrossRef]

21. Wigmans, T. Industrial aspects of production and use of activated carbons. *Carbon* **1989**, *27*, 13–22. [CrossRef]

22. Dahn, J.R.; Xing, W.; Gao, Y. The "falling cards model" for the structure of microporous carbons. *Carbon* **1997**, *35*, 825–830. [CrossRef]

23. Liu, D.D.; Jia, B.Y.; Liu, X.; Zhao, B.J.; Gao, J.H.; Cao, Q.X.; Wu, S.H.; Qin, Y.K. Effects of oxygen functional groups and $FeCl_3$ on the evolution of physico-chemical structure in activated carbon obtained from Jixi bituminous coal. *RSC Adv.* **2018**, *8*, 8569–8579. [CrossRef]

24. Delogu, F.; Gorrasi, G.; Sorrentino, A. Fabrication of polymer nanocomposites via ball milling: Present status and future perspectives. *Prog. Mater. Sci.* **2017**, *86*, 75–126. [CrossRef]

25. He, S.H.; Qin, Y.B.; Walid, E.; Li, L.; Cui, J.; Ma, Y. Effect of ball-milling on the physicochemical properties of maize starch. *Biotechnol. Rep.* **2014**, *3*, 54–59. [CrossRef]

26. Ong, T.S.; Yang, H. Effect of atmosphere on the mechanical milling of natural graphite. *Carbon* **2000**, *38*, 2077–2085. [CrossRef]

27. Zhang, L.M.; Zhang, R.; Zhan, L.; Qiao, W.M.; Liang, X.Y.; Ling, L.C. Efiect of ball-milling technology on pore structure and electrochemical properties of activated carbon. *J. Shanghai Univ. (Engl. Ed.)* **2008**, *12*, 372–376. [CrossRef]

28. Salver-Disma, F.; Lenain, C.; Beaudoin, B.; Aymard, L.; Tarascon, J.M. Unique effect of mechanical milling on the lithium intercalation properties of different carbons. *Solid State Ion.* **1997**, *98*, 145–158. [CrossRef]

29. Salver-Disma, F.; Tarascon, J.M.; Clinard, C.; Rouzaud, J.N. Transmission electron microscopy studies on carbon materials prepared by mechanical milling. *Carbon* **1999**, *37*, 1941–1959. [CrossRef]

30. Liu, D.D.; Jia, B.Y.; Li, S.; Dong, L.J.; Gao, J.H.; Qin, Y.K. Effect of pyrolysis conditions on the improvement of the physicochemical structure of activated carbon obtained from Jixi bituminous coal. *Asia Pac. J. Chem. Eng.* **2019**, *14*, 1–12. [CrossRef]

31. Pietrzak, R. XPS study and physico-chemical properties of nitrogen-enriched microporous activated carbon from high volatile bituminous coal. *Fuel* **2009**, *88*, 1871–1877. [CrossRef]

32. Gong, X.Z.; Guo, Z.C.; Wang, Z. Variation of char structure during anthracite pyrolysis catalyzed by Fe_2O_3 and its influence on char combustion reactivity. *Energy Fuels* **2009**, *23*, 4547–4552. [CrossRef]

33. Belhachemi, M.; Rios, R.V.; Addoun, F.; Silvestre-Albero, J.; Sepulveda-Escribano, A.; Rodrìguez-Reinoso, F. Preparation of activated carbon from date pits: Effect of the activation agent and liquid phase oxidation. *J. Anal. Appl. Pyrolysis* **2009**, *86*, 168–172. [CrossRef]

34. Yang, K.B.; Peng, J.H.; Xia, H.Y.; Zhang, L.B.; Srinivasakannan, C.; Guo, S.H. Textural characteristics of activated carbon by single step CO_2 activation from coconut shells. *J. Taiwan Inst. Chem. Eng.* **2010**, *41*, 367–372. [CrossRef]

35. Heras, F.; Alonso-Morales, N.; Jimenez-Cordero, D.; Gilarranz, M.A.; Rodriguez, J.J. Granular mesoporous activated carbons from waste tires by cyclic oxygen chemisorption-desorption. *Ind. Eng. Chem. Res.* **2012**, *51*, 2609–2614. [CrossRef]

36. Karatepe, N.; Orbak, I.; Yavuz, R.; Özyuğuran, A. Sulfur dioxide adsorption by activated carbons having different textural and chemical properties. *Fuel* **2008**, *87*, 3207–3215. [CrossRef]

37. He, X.F.; Jin, L.J.; Wang, D.; Zhao, Y.P.; Zhu, S.W.; Hu, H.Q. Integrated process of coal pyrolysis with CO_2 reforming of methane by dielectric barrier discharge plasma. *Energy Fuels* **2011**, *25*, 4036–4042. [CrossRef]

38. Li, W.; Zhu, Y.M. Structural characteristics of coal vitrinite during pyrolysis. *Energy Fuels* **2014**, *28*, 3645–3654. [CrossRef]

39. Davini, P. SO_2 adsorption by activated carbons with various burnoffs obtained from a bituminous coal. *Carbon* **2001**, *39*, 1387–1393. [CrossRef]

40. Li, T.; Zhang, L.; Li, D. Effects of gasification atmosphere and temperature on char structural evolution during the gasification of collie sub-bituminous coal. *Fuel* **2014**, *117*, 1190–1195. [CrossRef]

41. Sasezky, A.; Muckenhuber, H.; Grothe, H. Raman microspectroscopy of soot and related carbonaceous materials: Spectral analysis and structural information. *Carbon* **2005**, *43*, 1731–1742. [CrossRef]

42. Sathe, C.; Pang, Y.; Li, C.Z. Effects of heating rate and ion-exchangeable cations on the pyrolysis yields from a Victorian brown coal. *Energy Fuels* **1999**, *13*, 748–755. [CrossRef]

43. Guo, X.; Tay, H.L.; Li, C.Z. Changes in char structure during the gasification of a Victorian brown coal in steam and oxygen at 800 °C. *Energy Fuels* **2008**, *22*, 4034–4038. [CrossRef]

44. Li, Y.; Yang, H.P.; Hu, J.H.; Wang, X.H.; Chen, H.P. Effect of catalysts on the reactivity and structure evolution of char in petroleum coke steam gasification. *Fuel* **2014**, *117*, 1174–1180. [CrossRef]

45. Gaur, V.; Asthana, R.; Verma, N. Removal of SO_2 by activated carbon fibers inthe presence of O_2 and H_2O. *Carbon* **2006**, *44*, 46–60. [CrossRef]

46. Sun, F.; Gao, J.; Liu, X.; Tang, X.; Wu, S. A systematic investigation of SO_2 removal dynamics by coal-based activated cokes: The synergic enhancement effect of hierarchical pore configuration and gas components. *Appl. Surf. Sci.* **2015**, *357*, 1895–1901. [CrossRef]

47. Pi, X.X.; Sun, F.; Gao, J.H.; Zhu, Y.W.; Wang, L.J.; Qu, Z.B.; Liu, H.; Zhao, G.B. Microwave irradiation induced high-efficiency regeneration for desulfurized activated coke: A comparative study with conventional thermal regeneration. *Energy Fuels* **2017**, *31*, 9693–9702. [CrossRef]

48. Guo, Y.Y.; Li, Y.R.; Zhu, T.Y.; Ye, M.; Wang, X. Adsorption of SO_2 and chlorobenzene on activated carbon. *Adsorption* **2013**, *19*, 1109–1116. [CrossRef]

49. Guo, Y.Y.; Li, Y.R.; Zhu, T.Y.; Ye, M. Effects of concentration and adsorption product on the adsorption of SO_2 and NO on activated carbon. *Energy Fuels* **2013**, *27*, 360–366. [CrossRef]

Article

Catalytic Effect of NaCl on the Improvement of the Physicochemical Structure of Coal-Based Activated Carbons for SO$_2$ Adsorption

Dongdong Liu [1], Rui Su [1], Zhengkai Hao [1], Xiaoman Zhao [1], Boyin Jia [2,*] and Liangjie Dong [1]

[1] College of Engineering and Technology, Jilin Agricultural University, Changchun 130118, China; liudongdong@jlau.edu.cn (D.L.); gaojihui0809@163.com (R.S.); hzk980924@163.com (Z.H.); zhaoxiaoman0930@163.com (X.Z.); jiaboyinjby@163.com (L.D.)

[2] College of Animal Science and Technology, Jilin Agricultural University, Changchun 130118, China

* Correspondence: jiaboyin@jlau.edu.cn

Received: 25 April 2019; Accepted: 28 May 2019; Published: 5 June 2019

Abstract: The utilization of coal-based activated carbons focuses on improving the physicochemical structure for achieving high-capacity. Herein, the catalytic effect of NaCl (1 and 3 wt%) in the presence of oxygen functional groups on the improvement of the physicochemical structure of coal-based activated carbons is studied in this work. A large quantity of Na can be retained in 1NaJXO and 3NaJXO with the presence of oxygen functional groups to promote further its catalytic characteristics during pyrolysis, resulting in the disordered transformation of the carbon structure. In addition, the development of micropores is mainly affected by the distribution and movement of Na catalyst, whereas the growth of mesopores is mainly influenced by the evolution of oxygen functional groups. Then, the active sites of 3NaJXO-800 can no longer be consumed preferentially in the presence of Na catalyst during subsequent CO$_2$ activation to facilitate the sustained disordered conversion of the microstructure and the rapid development of the micropores, resulting in the obvious high S$_{BET}$ value as activation proceeds. And the high S$_{BET}$/burn-off ratio value (41.48 m^2·g^{-1}/%) of 3NaJXO-800 with a high value of S$_{BET}$ (1995.35 m^2·g^{-1}) at a low burn-off value (48.1%) can be obtained, presenting the high efficiency of pore formation. Finally, the SO$_2$ adsorption efficiency of 3NaJXO-800-48.1 maintains at 100% within 90 min. After 180 min, 3NaJXO-800-48.1 still presents a high adsorptive capacity (140.2 mg/g). It is observed that a large micropore volume in the case of hierarchical pore structure necessarily assures optimal adsorption of SO$_2$.

Keywords: activated carbons; catalytic activation; physicochemical structure; SO$_2$ adsorption

---

1. Introduction

Recently, gas sorption, storage and separation in porous nanocarbons and metal–organic frameworks have received increasing attention. In particular, the tunable porosity, surface area and functionality of the lightweight and stable graphene-based materials open up great scope for those applications [1]. Activated carbons (ACs) as an adsorbent material is a promising choice to achieve the gas pollutants (such as SO$_2$, NO$_x$) adsorption [2]. In the case of ACs, the gas pollutants are adsorbed and catalyzed in abundant active sites within micropores, then can be further migrated and stored in developed mesopores [3],so the effective removal of gas pollutants is closely related to the physicochemical structure of ACs, including more amounts of active sites, a hierarchical distribution of pores and a high specific surface area (S$_{BET}$).Among some synthesis methods (such as physical or chemical activation [4], soft/hard template [5,6], hydrothermal carbonization and self-assembly [7,8]) and sources of raw materials (such as coal, biomass, wastes, MOF and ordered mesoporous carbons), the traditional physical activation using H$_2$O, CO$_2$, flue gas, or their mixtures as activation agents can partially etch the carbon-based framework to obtain desirable porosity and more active sites over other methods, in addition, the coal as the most suitable raw materials instead of other materials can satisfy the industrial application of ACs [9]. Therefore, coal-based ACs produced by physical activation process meets the requirements of economic and environmental friendliness.

In order to improve the physicochemical structure of coal-based ACs, some researchers concentrate mainly on various activation modes (such as the single activation and mixing activation) [10,11]. However, the obtained products present the low S_{BET} values between 600 and 1000 m^2/g at the high burn-off value of approximately around 60–85% and the rapid consumption of active sites during activation, finally resulting in a higher cost [12–16]. Thus, it is difficult to obtain the ideal AC only by adjusting the activation conditions during activation. Our previous research found that the number of initial pores and active sites of chars produced by pyrolysis and the disordered conversion of carbon structure of chars during activation have important effects on the ideal AC production [17]. Furthermore, more initial pores can promote rapidly the diffusion of the activated gas into the particles' interior to avoid external loss of quality effectively, resulting in the high efficiency of pore formation. In addition, a lot of active sites and the disordered conversion of carbon structure at high activation temperatures can further promote etching of carbon-based framework to obtain desirable porosity and more new active sites.

Alkali metals (such as Na and K) in raw coal play an efficient catalytic role in gasification which is similar to physical activation reaction, and the corresponding catalytic mechanism has been investigated by some researchers [18–22]. Alkali metal not only can fundamentally change the reaction path between activated gas and active sites, but also can accelerate the reaction between the active gas and coal matrix by providing catalytic active sites, finally, the increase of new active sites and the disordered conversion of carbon structure during gasification. Importantly, most of the works in the literature reported that large amounts of alkali metals have been released into the gas phase during the temperature range of 300–900 °C, resulting in the absence of a large number of catalyst in the subsequent activation stage [23–27]. However, some studies also reported that the introduction of oxygen functional groups play an important role in the movement and catalytic effect of alkali metals [28–30]. Alkali metals (M) can be fixed inside char during pyrolysis as intermediates (such as C-O-M and -COOM) form which act as catalytic active sites to react with activated gas at the activation stage. In addition, in our previous study [31] and Francisco et al. [32] found that more active sites in chars can be produced by the introduction of oxygen functional groups during pyrolysis.

In this work, we systematically investigated the catalytic effect of alkali metal under the introduction of oxygen functional groups regarding the improve of the physicochemical structure of Jixi bituminous coal-based activated carbons. A series of samples were prepared by loading various amounts of the NaCl (1 wt% and 3 wt%) and/or the subsequent pre-oxidation in the air at 200 °C for 48 h. To verify the application potentials of the ideal activated carbons with developed pore structure, SO_2 removal tests also were conducted by portable FTIR. The characteristic parameters of all samples were determined by scanning electron microscopy (SEM), nitrogen adsorption, X-ray diffraction (XRD) and Raman spectroscopy.

2. Materials and Methods

2.1. Materials

Jixi bituminous coal was collected from the southeast of Heilongjiang province in China, and acted as the source material for the preparation of ACs. Different particle sizes of 250–380 μm from the Jixi bituminous coal were obtained through crushing and sieving. Importantly, the raw materials were demineralized sequentially using 6 mol·L^{-1} HCl and 40 wt% HF [33]. Afterwards, the sample was treated with deionized water and dried in an oven at 80 °C overnight. After that, the proximate and ultimate analysis of a demineralized sample (denoted as JX) were shown in Table 1.

Table 1. Proximate and ultimate analyses of JX (wt%).

V_{ad}	FC_{ad}	A_{ad}	M_{ad}	C_{daf}	H_{daf}	O_{daf} *	N_{daf}	S_{daf}
39.66	56.60	0.12	3.62	74.81	19.49	4.01	1.31	0.38

* By difference; ad (air-dried basis): The coal in dry air was used as a benchmark; daf (dry ash free basis): The remaining component after the removal of water and ash in coal was used as a benchmark.

Na-loaded samples were prepared by liquid impregnation. A known amount of NaCl powder was first added into a beaker and dissolved by deionized water under magnetic stirring. NaCl powder was obtained from Kemiou, Tianjin, China. A pre-weighed amount of JX was then added into the beaker to make a coal-water slurry. The coal-water slurry was dried at 80 °C with magnetic stirring of 300 r/min, until the water was completely evaporated. The mass content of Na in the mineral-loaded coal samples is controlled at 1 wt% and 3 wt%. The Na-loaded samples were denoted as 1NaJX and 3NaJX. In addition, JX was oxidized in air at 200 °C for 48 h, the oxidized sample was marked as JXO. Then, a predetermined amount of NaCl powder (0.03 g and 0.09 g) and 3 g of JXO were mixed using above liquid impregnation to prepare the Na-loaded oxidized samples and they were marked as 1NaJXO and 3NaJXO.

2.2. Experimental Process

The 3 g of samples were heated at a constant rate of 10 °C/min in nitrogen (99.999%) flow of 300 mL/min by the three-stage fixed-bed reactor. Thermal upgrading stopped when the final temperature reached 300, 400, 500, 600, 700, 800, 900 and 1000 °C then maintained for 10 min, and after that, the samples were quickly cooled in a nitrogen atmosphere and was marked as JX, JXO, 1NaJX, 3NaJX, 1NaJXO and 3NaJXO- pyrolysis temperature. After that, the atmosphere was switched to CO_2 (99.999%) at the same flow rate for a certain time to produce ACs with different porous structures, which were marked as the samples-pyrolysis time—burn-off value. In order to eliminate the interference of Na-based compounds in chars for the results of XRD and Raman. The 10 g of testing samples, including Na-based compounds were washed with 300 mL of 0.2 mol L−1 HCl at 60 °C for 24 h using magnetic stirring, then the residual acid-soluble inorganic salts on the surface of coal particles were removed by filtering and washing with 300 mL deionized water twice, respectively. In addition, the dissolved Na element in the residual liquid was quantified by an ICP-AES. This testing process was repeated three times, and the test results were averaged.

2.3. Measurement Analysis

The visualized results of surface topography of samples were obtained by SEM (Quanta 200, FEI, Hillsboro, OR, USA) a t 200 kV. The crystal parameters of the samples were obtained by a D/max-rb X-ray diffractometer (XRD, D8 ADVANCE, Brooke, Karlsruhe, Germany) in the 20 range from 10° to 80°, and the scanning rate of XRD was set at a stable value of 3°/min. The different hybrid carbon structures of the samples were obtained by Raman spectroscopy via a 532 nm wavelength laser, and the scanning scope was determined from 1000–1800 cm^{-1}. The information of pore structure was obtained by a micromeritics adsorption apparatus (ASAP2020, Micromeritics, Norcross, GA, USA) in a relative pressure (P/P$_0$) range from 10^{-7} to 1, and the analysis temperature was set at 77 K [34]. The vacuum degassing process for samples was performed before the test and analysis experiment, and the temperature and time were set to 473 K and 12 h. Furthermore, the specific parameters of pore structure, such as the specific surface area (S_{BET}), the micropore area (S_{mic}), and the micropore volume (V_{mic}), were calculated using the following formulas: Brunauer–Emmett–Teller (BET) equation, the t-plot method, Horvath–Kawazoe (HK) method, and nonlocal density functional theory (NLDFT), respectively [35]. In addition, the parameter of the total pore volume (V_t) was obtained at 0.98 relative pressure. Elemental analysis (EA) was performed using an analyzer (Vario MACRO cube, Elementar, Langenselbold, Germany) for determination of the total carbon and oxygen content of the bulk samples.

The quantification of sodium was obtained by an ICP-AES (Optima 5300 DV, PerkinElmer, Boston, MA, USA).

2.4. The SO₂ Adsorption Test

The SO_2 adsorption experiments were carried out in a fixed bed reactor using an on-line Fourier transform infrared gas analyzer (Dx4000, Gasmet, Vantaa, Finland) to continuously monitor the SO_2 concentrations. The experimental system consists of a tubular reactor (20 mm diameter), placed in a vertical furnace, with a system of valves and mass flow controllers in order to select the flow and the composition of the inlet gas, as shown in Figure S1. In each typical running, 2.5 g of the sample was put into the tubular reactor at 80 °C within 120 min. The gas volumetric composition used in experiments was: SO_2, 1500 ppm; O_2, 5%; water vapor, 10%; N_2, balance, total flow rate 200 mL·min^{-1}. SO_2 removal efficiency and rate versus time were evaluated by detected concentrations of SO_2 at the inlet and outlet in real time through the gas analyzer. The SO_2 removal capacity of coal-based activated carbons was calculated by the integrating area above the removal curves and reaction time [36].

3. Results and Discussion

3.1. Sodium Release of NaCl Loaded Chars at Pyrolysis

Figure 1 shows the retention of Na (%) in different chars during pyrolysis. With the increase of pyrolysis temperature, the retention of Na in the char all decreases.

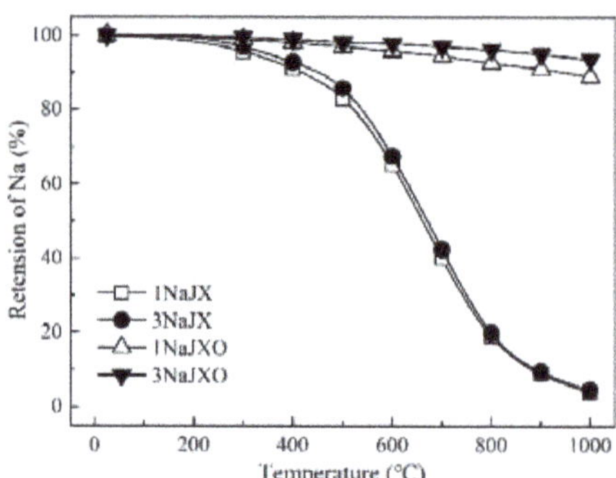

Figure 1. The retention of Na (%) in different chars at pyrolysis.

The trends of the retention of Na in 1NaJX and 3NaJX are similar, especially when the pyrolysis temperature is higher than 500 °C, the retention of Na rapidly decrease from 85 to 5%, indicating the release of Na from chars cannot be changed only by adding the amount of catalyst. Li et al. 26 found that the repeated fracture and recombination processes are presented between alkali metal and coal/coke system (CM), resulting in the production of more stable chemical bonds as the increase of pyrolysis temperature, the entire reaction process can be expressed as follows:

$$(CM-Na) = (-CM) + Na \tag{1}$$

$$(-CM) = (-CM') + gas \tag{2}$$

$$(-CM') + Na = (-CM'-Na) \tag{3}$$

In addition, it is noteworthy that large quantities of Na are retained in 1NaJXO and 3NaJXO during pyrolysis. Some intermediates (such as -O-Na and −COONa) are formed to fix Na within chars in the temperature range from 300 °C to 600 °C, resulting in the release of a small amount of sodium. Then these intermediates with low thermal stability (such as −COONa) may further be dissociated into the volatile matter to produce more active sites, promoting the re-bonding process between Na and carbon matrix (CM), the reaction process can be expressed as follows [37]:

$$(-COO\text{-}Na) + (-CM) \rightarrow (CM-Na) + CO_2 \tag{4}$$

However, other intermediates with high thermal stability $(-O-Na)$ can continue to stabilize alkali metal Na within chars even at a higher temperature.

3.2. Carbon Structure Analysis of Chars at Pyrolysis

The Raman spectra of the different chars are shown in Figure S2. To obtain more information about the hybrid carbon structure of chars, each Raman spectrum was treated further into five band areas by the fitting method [38]. Importantly, the ratios of the different band areas, such as A_{D1}/A_G, A_{D3}/A_G, A_{D4}/A_G, and A_{D1}/A_{D3}, represent the different types of hybrid carbo as follow in turn: The defect degree of the microcrystalline structure; the amorphous carbon; the relative quantity of cross-linking bonds; the ratio between big rings relative to small fused rings in chars [39,40]. The parameters of the carbon structure of carbonized chars are given in Figure 2.

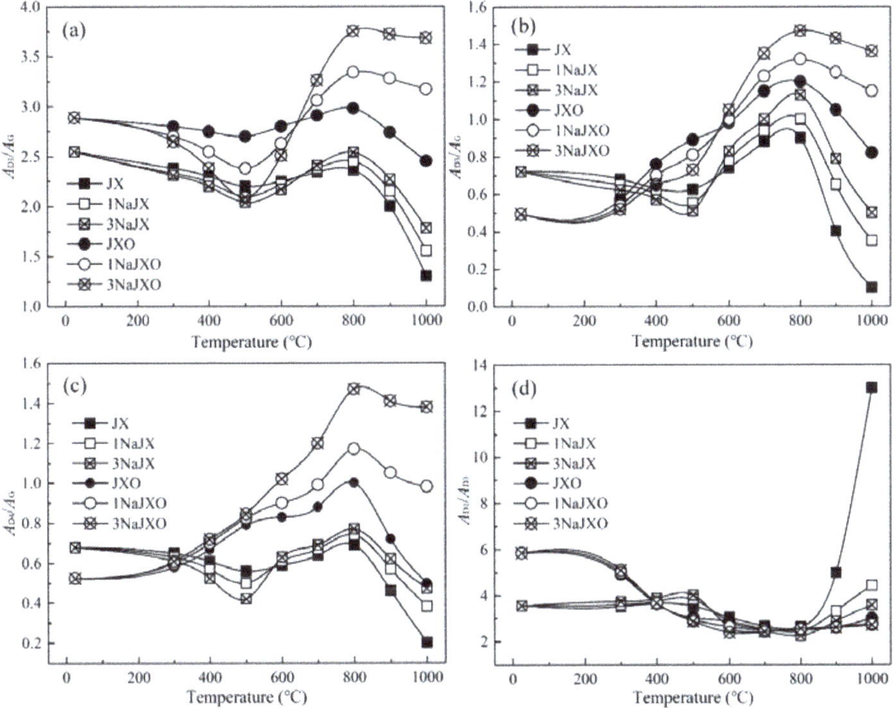

Figure 2. Raman data from chars at pyrolysis (**a**) A_{D1}/A_G; (**b**) A_{D3}/A_G; (**c**) A_{D4}/A_G; (**d**) A_{D1}/A_{D3}.

First of all, the values of A_{D1}/A_G, A_{D3}/A_G and A_{D4}/A_G of JX, 1NaJX and 3NaJX decrease obviously with the increase of Na loading during temperature rising from 25 to 500 °C, whereas the A_{D1}/A_{D3} value increases gradually. These results indicate that the splitting of big aromatic rings (A_{D1}/A_G) and the remove of a large number of amorphous sp^2 bonding carbon atoms (A_{D3}/A_G) as volatile matter at the beginning of pyrolysis result in the increase in the A_{D1}/A_{D3} value. Moreover, these reactions can be further strengthened under the presence of Na. The value of A_{D3}/A_G of 1NaJX and 3NaJX continues to decrease in this stage, due to the release of more organic components caused by the catalytic decomposition of Na. Alternatively, the value of A_{D1}/A_G decreases slightly and A_{D1}/A_{D3} decrease quickly, whereas the values of A_{D3}/A_G and A_{D4}/A_G of JXO increase obviously. The cross-linking reaction of oxygen-containing functional groups not only promotes the formation of cross-linking bonds (such as -COO- and -O-) and more small aromatic rings (such as some oxygen-containing heterocycles), but hinders the intense decomposition of aromatic structure. However, the values of A_{D1}/A_G and A_{D1}/A_{D3} decrease obviously and the A_{D3}/A_G and A_{D4}/A_G of 1NaJXO and 3NaJXO increase slightly as compared with that of JXO. The formation of some intermediates can fix Na within chars to promote further its catalytic decomposition and reduce the number of cross-linking bonds.

Then, the values of A_{D1}/A_G, A_{D3}/A_G and A_{D4}/A_G of JX, 1NaJX and 3NaJX increase obviously with the increase of Na loading during temperature rising from 500 to 800 °C, whereas the A_{D1}/A_{D3} value decreases gradually. The increase of some aromatic rings originates from the formation of more new cross-linking bonds at this stage. Especially, there is no significant change in all parameters for JX from 700 to 800 °C, where the process of cross-linking reaction can be shortened to hinder the production of sp^2-sp^3 bonding carbon atoms; the amorphous sp^2 bonding carbon atoms and transformation of small aromatic rings to form big aromatic rings, these results indicate the—graphitization conversion of microcrystalline. In addition, a large amount of Na is released as volatile, the remaining Na is bonded to carbon matrix (CM) to form the stable chemical bonds (CM-Na) at high temperature during which more free radical fragments have been produced. Thus, the cross-linking reaction can be strengthened continuously accompanied by the combination of free radical fragments, resulting in an obvious increase in all parameters of 1NaJX and 3NaJX. The values of A_{D1}/A_{D3}, A_{D1}/A_G, A_{D3}/A_G and A_{D4}/A_G of JXO increase at this stage. The break of oxygen-containing structures with low thermal stability not only helps the production of new cross-linking bonds with high thermal stability to form furthermore small aromatic rings, but also promotes the transformation of small aromatic rings to big aromatic rings. However, the value of A_{D1}/A_{D3} of 1NaJXO and 3NaJXO decreases, and the values of A_{D1}/A_G, A_{D3}/A_G and A_{D4}/A_G continue to increase than that of JXO. The retention of more Na caused by the existence of oxygen-containing structures enhances further the catalytic and cross-linking reaction, resulting in the higher activity of coal chars.

Finally, the values of A_{D1}/A_G, A_{D3}/A_G and A_{D4}/A_G of JX decrease obviously, whereas the A_{D1}/A_{D3} value increases from 800 to 1000 °C, and the change range of the corresponding parameters of 1NaJX, 3NaJX, JXO, 1NaJXO and 3NaJXO reduces gradually. These results indicate the relevant reactions (including the break of cross-linking bonds; the transformation of the isolated sp^2 structure and the amorphous sp^2 bonding carbon atoms into the crystalline sp^2 structure) all are promoted at higher pyrolysis temperature, presenting the more ordered conversion of carbon structure with lower reactivity. In addition, the change of A_{D1}/A_{D3} value indicates that the amorphous sp^2 bonding carbon atoms are more easily transformed into the crystalline sp^2 structure (G peak). In particular, Na can promote further the stability of some oxygen-containing structures to hinder the graphitization conversion of the carbon structure.

3.3. Crystal Structure Analysis of Coal Chars at Pyrolysis

The XRD profiles of the different chars are shown in Figure S3. Some important structure feature information of aromatic layers (such as layer distance (d_{002}), stacking height (L_c) and width (L_a)) can be obtained through the fitting treatment of two obvious broad diffraction peaks at $2\theta = 24°$–$27°$ and $41°$–$44°$ in all samples [41]. The results of themicrocrystalline structure are shown in Figure 3.

First of all, the La value of JX, 1NaJX and 3NaJX decreases and the Lc value first decreases and then increases, and the d_{002} first increases and then decreases from 25 to 500 °C. These changes may be related to the break of chemical bonds and the release of organic fragments (such as $\bullet C_n H_{2n+1}$, $\bullet OC_n H_{2n+1}$, and substituted benzene) [41]. The break of chemical bonds and the slow release of organic fragments facilitate the production of the metaplast material, resulting in the movement, the orientation adjustment and the stacking of aromatic layers (namely the presence of a plastic behavior) for JX. However, the addition of NaCl accelerates the depolymerization of aromatic structure unit to produce more volatile matters and the smaller the aromatic structure, which weakens the production of the metaplast material, the movement and the stacking of aromatic layers [42]. Alternatively, the values of L_a, L_c and d_{002} of JXO, 1NaJXO and 3NaJXO increase. The growth and stacking of aromatic layers are promoted by cross-linking reaction of oxygen-containing functional groups, and the significant reductions of organic fragments during the pre-oxidation stage is beneficial to the increase of the layer distance, whereas the formation of some intermediates can reduce the number of cross-linking bonds.

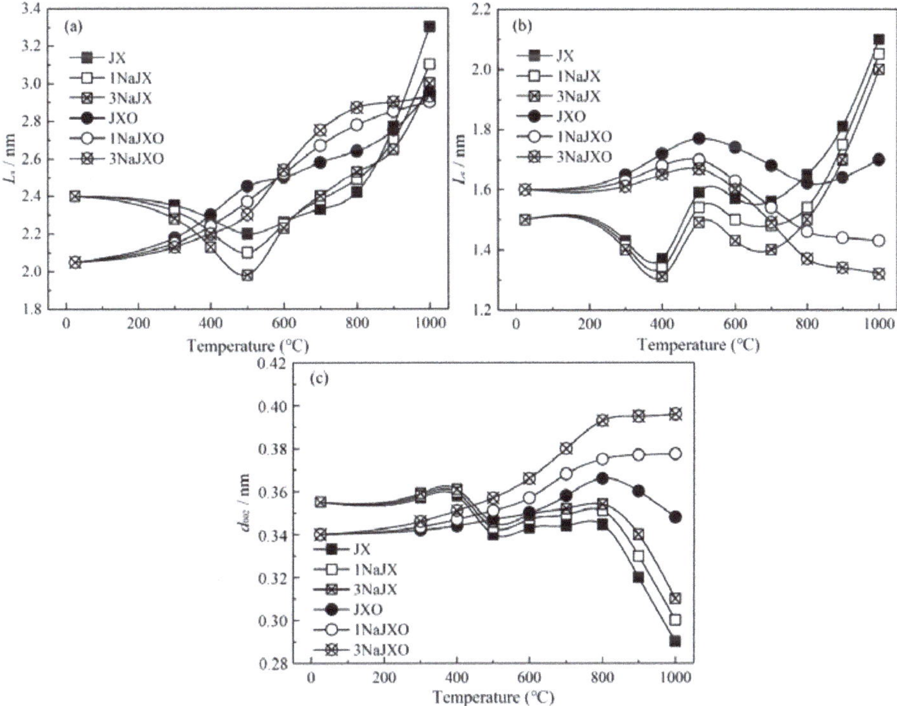

Figure 3. XRD data from chars at pyrolysis (**a**) L_a; (**b**) L_c; (**c**) d_{002}.

Then, the values of L_a and d_{002} of JX,1NaJX and 3NaJX increase and L_c value decrease first and then increase obviously from 500 to 800 °C, these changes may be related to the competition between the break and production of cross-linking bonds. Some disordered aromatic structure units are formed by cross-linking reaction of aromatic layers, resulting in the disordered array and connection of aromatic layers. The formation of stable chemical bonds (CM-Na) at a high temperature cannot only enlarge the spacing of aromatic layer, but produces more organic fragments to increase the amount of cross-linking bonds [43]. However, the presence of more ordered aromatic structure units facilitates the break of chemical bonds instead of the cross-linking reaction of aromatic layers, presenting a graphitization tendency. Alternatively, the values of L_a and d_{002} increase of JXO, 1NaJXO and 3NaJXO, and L_c value decrease with the increase of Na loading. The break of cross-linking bonds with low thermal stability promotes the depolymerization of aromatic layers, and the production of new cross-linking bonds with high thermal stability helps the formation of more aromatic rings with disordered structure. The retention of a large number of Na enhances further the break and formation of cross-linking bonds with different thermal stability, presenting the obvious non-graphitized tendency.

Finally, the values of L_a, L_c and N of JX, 1NaJX, 3NaJX and JXO increase and the d_{002} value decreases rapidly from 800 to 1000 °C. These changes indicate the microcrystalline structure has transformed into a highly ordered graphite-like structure, and the ordered stacking and rapid growth of aromatic layers are presented at higher pyrolysis temperature. The variation of these parameters can be weakened with the increase of Na loading or the existence of oxygen-containing structures, but it is not easy to hinder the ordered condensation of aromatic layers at a high temperature. Remarkably, the values of L_a and d_{002} of 1NaJXO and 3NaJXO increase gently and the values of L_c gradually decrease, these results indicate that the coexistence of oxygen-containing structures and Na can help further the disordered transformation of aromatic structures at high temperature.

3.4. Specific Surface Area (S_{BET}) Analysis of Coal Chars at Pyrolysis

The value of S_{BET} and V_t of chars is shown in Table 2. First of all, the S_{BET} value of JX increases first slightly and then decreases rapidly from 25 to 500 °C. The formation of pores is related to the release of volatile matter at the beginning of pyrolysis, but the increase of pore volume is limited. With the increase of pyrolysis temperature, a large number of metaplast materials are formed to block the pores of chars. However, the release of more volatile matter and the formation of fewer metaplast materials are performed with the increase of Na loading, leading to the relatively developed pore structure of 1NaJX and 3NaJX compared to that of JX. Alternatively, the S_{BET} value of JXO, 1NaJXO and 3NaJXO increases continuously with the increase of Na loading. The depolymerization and recombination of aromatic structure promote the release of volatile matter and disordered stacking of aromatic layers, and there are no metaplast materials to block the pores for oxidized coal, resulting in the development of the pore. In addition, the depolymerization and cross-linking recombination of aromatic structure can be strengthened under the existence of Na, therefore, the pores structure of 1NaJXO and 3JXNaO can be developed further.

Table 2. The value of S_{BET} (m^2/g) of chars at pyrolysis.

°C S_{BET} Samples	JX	1NaJX	3NaJX	JXO	1NaJXO	3NaJXO
25	21	21	21	54	54	54
300	25	29	33	67	78	83
400	35	40	47	75	87	94
500	8	12	14	84	100	114
600	16.4	19	23	97	115	126
700	27.3	35.6	38.6	111	134	149
800	42	50	55	133	156	167
900	29	32	34	110	150	162
1000	18	24	27	84	141	155

Then, the S_{BET} value of JX, 1NaJX and 3NaJX increases from 500 to 800 °C. The formation of cross-linking bonds between aromatic layers at this condensation stage promotes the production of pores. However, the pores have never been fully developed, due to blockage of metaplast materials. Alternatively, the S_{BET} value of JXO, 1NaJXO and 3NaJXO increases obviously with the increase of Na. The break of oxygen-containing functional groups and oxygen heterocycles promotes the formation of cross-linking bonds and the release of volatile matter, resulting in the disordered condensation of aromatic structure and the formation of more pores. These processes can be strengthened further with the increase of Na loading.

Finally, the S_{BET} value of JX, 1NaJX, 3NaJX, JXO decreases obviously from 800 to 1000 °C. This result is related to the collapse and expansion of microporous to form mes- or marc-opore and the further collapse of mesopore and macropore can be presented at higher pyrolysis temperature. However, it is difficult to prevent the collapse of pores only in the presence of Na or oxygen-containing functional groups. Remarkably, the S_{BET} value of 1NaJXO and 3NaJXO indicates that the coexistence of oxygen-containing structures and Na can stabilize and develop the three-dimensional spatial structure of aromatic layers, thus resulting in the sustained development of porosity.

3.5. Crystal Structure Analysis of Typical Chars During Activation

3NaJXO-800 is the most suitable precursor for subsequent activation, due to its disordered carbon structure, abundant initial pores and active sites. JX-800 and JXO-800 are used as the contrastive precursor. Therefore, the change of the physicochemical structure for these typical chars during activation can be analyzed in detail. The XRD profiles and crystal parameters of JX-800, JXO-800 and 3NaJXO-800 at different burn offs are shown in Figure S4 and Table 3.

Table 3. XRD data of typical chars at different burn offs.

Samples	La (A)	Lc (A)	d_{002} (A)	$N = Lc/d_{002}$
JX-800	24.20	16.50	3.45	4.78
JX-800-19.5	24.66	16.25	3.38	4.81
JX-800-29.1	26.07	17.06	3.30	5.17
JX-800-51.3	27.64	18.27	3.11	5.87
JXO-800	26.40	16.20	3.66	4.43
JXO-800-19.7	26.13	16.01	3.71	4.31
JXO-800-29.4	26.04	15.77	3.78	4.17
JXO-800-47.2	26.98	16.51	3.58	4.61
3NaJXO-800	28.70	13.70	3.93	3.48
3NaJXO-800-19.3	27.55	13.49	4.01	3.36
3NaJXO-800-29.2	27.36	13.11	4.15	3.16
3NaJXO-800-48.1	27.18	12.77	4.23	3.02

First of all, there is a sustained increase in the values of L_a and N of JX-800 and decrease in the d_{002} value, however, the L_c value first decreases from 0 to 19.5% and then increases from 19.5 to 51.3%. These changes indicate that the microcrystalline of JX-800 always develops towards a highly ordered structure during activation. At the beginning of activation, the reaction between activated gas and the active sites and some sandwich materials in the longitudinal aromatic layers results in a decrease in the values of d_{002} and L_c. With an increase in carbon loss, the aromatic layers with the more ordered orientation rapidly begin to condense and distort that promotes the thickness and size of the microcrystalline.

Alternatively, the values of L_c, L_a and N of JXO-800 first decrease and then increase, whereas d_{002} value first increases and then decreases with an increase in burn offs. These changes indicate that the existence of oxygen-containing structure of JXO-800 may hinder the ordered transformation of aromatic layers at the beginning of activation. More active sites (including the defects and the oxygen-containing side chains and bridge of basic unit in aromatic layers) are removed by activated gas to strengthen the etching of carbon-based framework, thus resulting in the decrease of the thickness and size of the microcrystalline structure. With an increase in carbon loss, the more active sites of JXO-800 have been consumed, therefore the horizontal and longitudinal condensation of aromatic layers have been presented; moreover, the rapid reduction of lamellar spacing also indicates the highly ordered transformation of aromatic structures.

Finally, there is a sustained decrease in values of L_a, L_c and N of 3NaJXO-800 and the increase on d_{002} value during activation, indicating that the addition of Na catalyst has promoted a continuous disordered conversion of aromatic structures. The distortion of the longitudinal aromatic structure can accompany with its catalytic cracking, thus resulting in an obvious decrease in the L_c value. Na bonded and fixed in carbon matrix may destroy the parallelism of the layer and the constancy of the interlayer spacing, thus increasing the interlayer spacing; however, Na-based compounds can further accelerate the etching of aromatic layer instead of their ordered condensation and growth.

3.6. Carbon Structure Analysis of Typical Chars During Activation

The Raman spectra and corresponding parameters and of JX-800, JXO-800 and 3NaJXO-800 at different burn offs are shown in Figure S5 and Table 4.

Table 4. Raman data of typical chars at different burn offs.

Samples	A_{D1}/A_G	A_{D3}/A_G	A_{D4}/A_G	A_{D1}/A_{D3}
JX-800	2.36	0.91	0.69	2.60
JX-800-19.5	2.92	0.83	0.58	3.52
JX-800-29.1	3.68	0.78	0.39	4.72
JX-800-51.3	1.81	0.53	0.11	3.42
JXO-800	2.98	1.21	1.01	2.48
JXO-800-19.7	2.87	1.28	1.19	2.24
JXO-800-29.4	2.80	1.32	1.23	2.12
JXO-800-47.2	2.71	1.01	0.99	2.68
3NaJXO-800	3.75	1.47	1.50	2.55
3NaJXO-800-19.3	3.70	1.55	1.57	2.39
3NaJXO-800-29.2	3.61	1.64	1.69	2.20
3NaJXO-800-48.1	3.53	1.79	1.74	1.97

There is a sustained decrease in the values of A_{D3}/A_G and A_{D4}/A_G, and the values of A_{D1}/A_G and A_{D1}/A_{D3} first increase at the low burn offs and then decrease at the high burn offs. At the beginning of activation, the active sites are consumed by activated gas preferentially, resulting in a decrease in the A_{D3}/A_G and A_{D4}/A_G values. In addition, the growth and conversion of aromatic ring and its conversion into big aromatic ring structures promote the increase in the A_{D1}/A_{D3} and A_{D1}/A_G values. With an increase in burn-offs, the interior of aromatic structure can be activated by the continuous penetration of the activated gas to further induce the condensation of the aromatic ring [42]. Alternatively, there is a sustained decrease in A_{D1}/A_G value of JXO-800, whereas the values of A_{D3}/A_G and A_{D4}/A_G first increases and then decreases, and the A_{D1}/A_{D3} value first decreases and then increases. At the beginning of activation, the existence of more active sites promotes further the etching of carbon structure, thus hindering its growth. In addition, the existence of more oxygen-containing structure facilitates the production of new cross-linking bonds and small aromatic ring, but the reaction path between active sites and active gas still can't be changed. With an increase in burn-offs, consistent reduction of oxygen-containing structure and self-consumption of the small aromatic ring and its conversion into a big aromatic ring are presented to promote the formation of more crystalline sp^2 structure. Finally, there is a sustained decrease in the values of A_{D1}/A_{D3} and A_{D1}/A_G of 3NaJXO-800 and the increase in the values of A_{D3}/A_G and A_{D4}/A_G. It can be inferred that the presence of Na catalyst can change the reaction pathways between the carbon structure and activated gas. More concretely, the active sites can no longer be consumed with activated gas preferentially, and the big aromatic rings would begin to decompose into small aromatic rings and the Na catalyst also hinder the formation of the crystalline sp^2 structure. In other words, a large number of broken fragments are produced by the catalytic effect of Na, resulting in the formation of newer cross-linking structure. Moreover, the presence of oxygen-containing structures is conducive to the reorganization of aromatic fragments.

3.7. Pore Structure Development of Typical Chars During Activation

To analyze the pore development of JX-800, JXO-800 and 3NaXO-800 at different burn-off values during activation, N_2 adsorption isotherms and parameters of porous structure are shown in Figure 4 and Table 5.

First of all, the S_{BET} value of 101.78 $m^2{\cdot}g^{-1}$, V_{mic} value of 0.06 $m^3{\cdot}g^{-1}$ and non-V_{mic} value of 6.25% of JX-800-19.5 are obtained, showing the development of micropores at the beginning of activation. These values increase gradually with an increase in burn-offs from 19.5% to 51.3%, that are related to the enlargement of micropores into mesopores and the production of some new micropores. Remarkably, the rapid increase of non-V_{mic} value indicates the more obvious development of mesopore rather than that of new micropores. At a higher burn offs, an S_{BET}/burn-off ratio value of 13.23 $m^2\ g^{-1}$/% of JX-800-51.3 is obtained. Many macropores from 2 μm to 35 μm on the particle surfaces can be found in JX-800-51.3 from Figure 5a, indicating the severe carbon losses on the particle surfaces during activation. This result may be related with the ordered conversion of the aromatic structure of JX-800 with less initial pores can hinder the penetration of activated gas into the interior of char structure during activation, thus resulting in the occurrence of more reactions on the particle surfaces rather than in the interior to decrease the production of the pores.

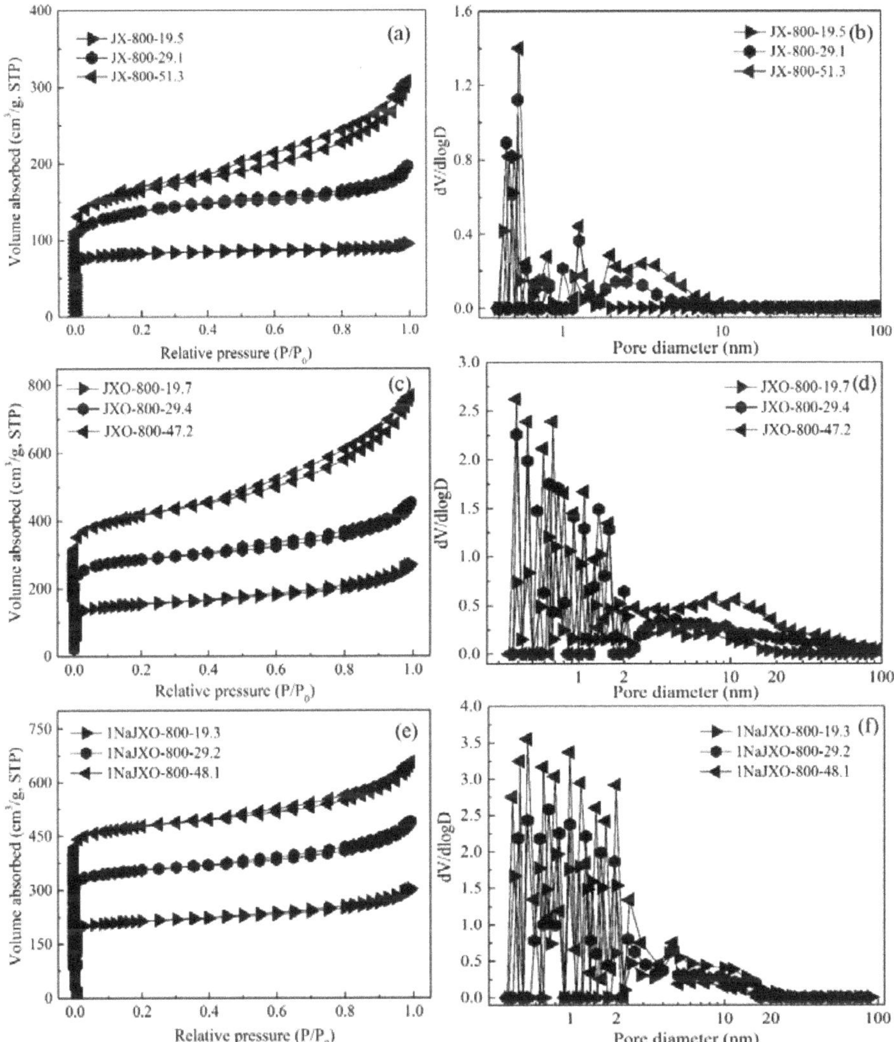

Figure 4. N_2 adsorption isotherms (**a**), (**c**) (**e**) and pore-size distributions (**b**), (**d**) (**f**) of activated carbons at different burn offs.

Then, the increase in V_t, V_{mic} and S_{BET} values and the decrease in the non-V_{mic} value of JXO-800 with the increase of burn offs from 0 to 29.4% show an obvious growth of micropores. These changes are related to the initial pores of JXO-800 act as channels to promote the diffusion of activated gas and more active sites produced by pyrolysis can strengthen the etching of carbon structure. All pore parameters of JXO-800 increase gradually with the increase of burn offs from 29.4 to 47.2%, indicating the formation of new micropores slows down; however, a rapid development of mesopores mainly results from the widening of the pores. The oxygen functional groups of JXO-800 as active sites are consumed gradually with continuous activation, resulting in limited micropores development. In particular, an S_{BET}/burn-off ratio value of 22.99 m^2 g^{-1}/% of JXO-800-47.2 is obtained. No severe carbon losses and macropores on the particle surfaces of JXO-800-47.2 are found from Figure 5b, these changes are related to the penetration of activated gas into the interior of the particle during activation.

Table 5. Pore structure parameters of typical chars at different burn offs.

Samples	S_{BET} (m^2/g)	V_t (m^3/g)	V_{mic} (m^3/g)	Non-V_{mic} (%)	S_{EBT}/Burn offs (m^2 g^{-1}/%)
JX-800	42	0.029	0.020	3.1	-
JX-800-19.5	101.78	0.064	0.060	6.25	5.22
JX-800-29.1	315.78	0.165	0.131	20.61	10.85
JX-800-51.3	678.90	0.285	0.175	38.59	13.23
JXO-800	133	0.101	0.059	41.58	-
JXO-800-19.7	352.78	0.189	0.132	30.16	17.91
JXO-800-29.4	648.21	0.297	0.247	16.84	22.05
JXO-800-47.2	1085.53	0.356	0.278	21.91	22.99
3NaJXO-800	167	0.199	0.132	33.67	-
3NaJXO-800-19.3	659.35	0.310	0.245	20.97	34.16
3NaJXO-800-29.2	1156.57	0.378	0.310	17.99	39.61
3NaJXO-800-48.1	1995.35	0.481	0.421	12.47	41.48

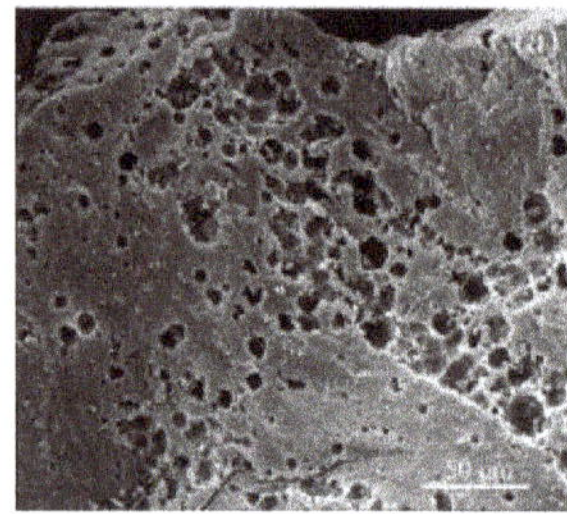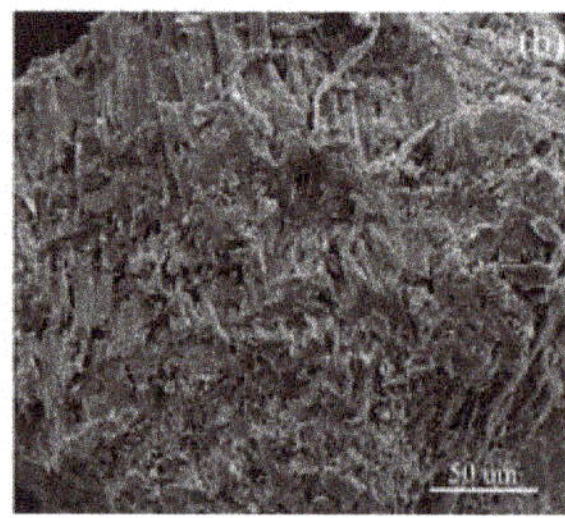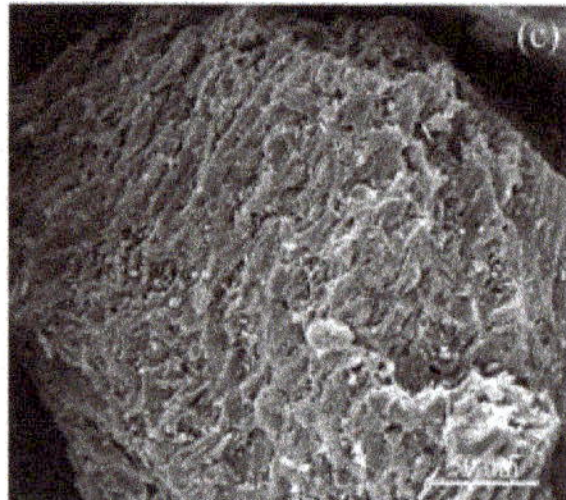

Figure 5. SEM images of typical chars under final burn-off values (**a**) JX-800-51.3; (**b**) JXO-800-47.2; (**c**) 3NaJXO-800-48.1.

Finally, the rapid increase in V_t, V_{mic} and S_{BET} values and the rapid decrease in non-V_{mic} value of 3NaJXO-800 during the whole stage of activation are shown in Table 5. Along with the gradual consumption of oxygen functional groups, the disordered conversion of carbon structure and more active sites in the presence of Na-based catalysts facilitate a sustained formation of more micropores. Although the catalysts might have moved and agglomerated on the particle surfaces with an increase of burn-off at high temperature activation, the simultaneous existence of oxygen functional groups and Na-based catalyst can constantly enhance the development of micropores. Importantly, am S_{BET}/burn-off ratio value of 41.48 m^2 g^{-1}/% of 3NaJXO-800-48.1 is obtained. Moreover, the surface morphology of 3NaJXO-800-48.1 is similar to that of JXO-800-47.2 from Figure 5c, indicating no severe carbon losses on the particle surfaces of 3NaJXO-800-48.1.

3.8. Study of SO$_2$ Adsorption

The SO$_2$ adsorption test from a simulated flue gas in which JX-800-51.3, JXO-800-47.2 and 3NaJXO-800-48.1 as testing samples is carried out under 80 °C, the result of SO$_2$ removal of the samples is shown in Figure 6.

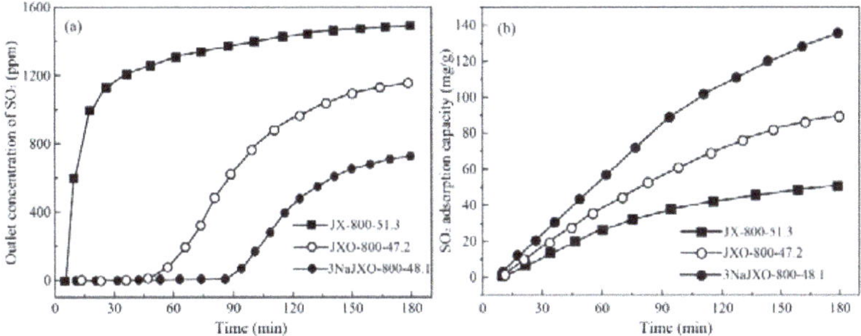

Figure 6. SO_2 removal of typical activated carbon (**a**) SO_2 breakthrough curve; (**b**) SO_2 adsorption quantity.

The SO_2 adsorption capacities of JX-800-51.3 are mainly exhibited within only 30 min, after that, its desulfurization performance is obviously reduced from 30 to 180 min. The SO_2 concentrations of gas outlet for JX-800-51.3 have achieved 1200 ppm about 30 min, showing that it has been penetrated basically by SO_2. The SO_2 adsorption capacity of JX-800-51.3 can only achieve 50.2 mg/g. Then SO_2 adsorption efficiency of 3NaJXO-800-48.1 maintains at 100% within 90 min. The SO_2 concentrations of gas outlet for 3NaJXO-800-48.1 slowly increase from 90 to 180 min and only achieve 700 ppm at 180 min, indicating the adsorptive capacity of 3NaJXO-800-48.1 with 140.2 mg/g is still strong. However, the SO_2 adsorption capacities of JXO-800-47.2 are presented between JX-800-51.3 and 3NaJXO-800-48.1. In the case of AC, the hierarchical structure (micro- and mesopores) is critical to the SO_2 removal process, the adsorption and catalysis processes of SO_2 are performed within the micropores, and the developed mesopores promote the migration and storage of produced sulfuric acid [3]. In addition, a high specific surface area (S_{BET}), which is related to the degree of well-developed pores, promotes desulfurization [44]. In addition, Zhu et al. [16,17] found that with the increase of burnout rate, the amount of basic functional groups has a good positive correlation with micropore. When the micropore originates from the microcrystalline structure etched by activated gas, they have a better linear relationship. The more active sites and high micropore volume of 3NaJXO-800-48.1 promote the adsorption and catalysis processes of SO_2 within the micropores, then a well-developed mesopore is conducive to the migration and storage of sulfuric acid to release consistently the active sites as adsorption sites within the micropores, which ensures the persistent adsorption capacity. However, the SO_2 adsorption capacities of JX-800-51.3 has reached saturation quickly in the initial stage, due to its less active sites and low micropores value, then undeveloped mesopore of JX-800-51.3 is unable to meet the storage of more sulfuric acid to release consistently the active sites, thus presenting a low SO_2 adsorption capacity.

4. Conclusions

A catalytic effect of NaCl (1 and 3 wt%) in presence of oxygen functional groups (created by air pre-oxidation at 200 °C for 48 h) has provided control of the physicochemical structure of Jixi bituminous coal-based ACs for high efficiency of SO_2 adsorption. In the phase of pyrolysis, a large number of Na catalyst can be fixed first in the interior of chars by the oxygen functional groups in the form of some intermediates (such as -O-Na and -COONa), then Na can be re-bonded with carbon matrix (CM-Na) at high temperatures, during which the catalytic cracking characteristics of Na catalyst plays a more important role, finally resulting in the disordered conversion of microstructure and the number of more active sites. Moreover, Na catalyst also helps the development of micropores; however, the evolution of oxygen functional groups mainly facilitates the production of mesopores. In the phase of activation, the reaction pathway of active sites and CO_2 was changed by the presence of Na catalyst, leading to a consistent disordered conversion of the microstructure and the production of new active sites of 3NaXO-800. With an increase in burn offs, the existence of Na catalyst facilitates the etching of the carbon structure to develop continuously the micropores rather than only widening of the pores to form mesopore and macropore. Finally, the S_{BET} values (1995.35 $m^2 \cdot g^{-1}$) of 3NaXO-800-48.1 with the high S_{BET}/burn-off ratio values of 41.48 $m^2 \cdot g^{-1}$/%. is obtained, presenting a persistent high adsorption efficiency (100%) within 90 min and a high SO_2 adsorption capacity with 140.2 mg/g after 180 min.

Supplementary Materials: The following are available online at http://www.mdpi.com/2227-9717/7/6/338/s1, Figure S1: Schematic figure of the fixed bed reactor system for SO_2 adsorption, Figure S2: Raman spectra from chars produced by pyrolysis, Figure S3: XRD profiles from chars produced by pyrolysis, Figure S4: XRD profiles of coal chars at different burn offs during activation (**a**) JX-800; (**b**) JXO-800; (**c**) 3NaJXO-800; Figure S5: Raman spectra of coal chars at different burn offs during activation (**a**) JX-800; (**b**) JXO-800; (**c**) 3NaJXO-800.

Author Contributions: D.L., B.J. and L.D. conceived and designed the experiments; X.Z., R.S. and Z.H. carried out the experiments; D.L. wrote the paper; D.L. and B.J. reviewed the paper.

Funding: This research was funded by National Natural Science Foundation of China, grant number 51806080, and Scientific Research Fund Project of Jilin Agricultural University, grant number 201801, and Jilin Province Education Department Science and Technology Program during the Thirteenth Five-year Plan Period, grant number JJKH20190940KJ.

Conflicts of Interest: The authors declare no conflict of interest.

References

1. Srinivas, G.; Zheng, X.G. Graphene-based materials: Synthesis and gas sorption, storage and separation. *Prog. Mater. Sci.* **2015**, *69*, 1–60. [CrossRef]
2. Bahamon, D.; Vega, L.F. Systematic evaluation of materials for post-combustion CO_2 capture in a temperature swing adsorption process. *Chem. Eng. J.* **2016**, *284*, 438–447. [CrossRef]
3. Shu, S.; Guo, J.X.; Liu, X.L.; Wang, X.J.; Yin, H.Q.; Luo, D.M. Effects of pore sizes and oxygen-containing functional groups on desulfurization activity of Fe/NAC prepared by ultrasonic-assisted impregnation. *Appl. Surf. Sci.* **2016**, *360*, 684–692. [CrossRef]
4. Hong, K.L.; Long, Q.; Zeng, R. Biomass derived hard carbon used as a high performance anode material for sodium ion batteries. *J. Mater. Chem. A* **2014**, *2*, 12733–12738. [CrossRef]
5. Nishihara, H.; Kyotani, T. Templated nanocarbons for energy storage. *Adv. Mater.* **2012**, *24*, 4473–4498. [CrossRef]
6. He, X.J.; Li, R.C.; Qiu, J.S.; Xie, K.; Ling, P.H.; Yu, M.X.; Zhang, X.Y.; Zheng, M.D. Synthesis of mesoporous carbons for supercapacitors from coal tar pitch by coupling microwave-assisted KOH activation with a MgO template. *Carbon* **2012**, *50*, 4911–4921. [CrossRef]
7. Hu, B.; Wang, K.; Wu, L.H.; Yu, S.H.; Antonietti, M.; Titirici, M.M. Engineering carbon materials from the hydrothermal carbonization process of biomass. *Adv. Mater.* **2010**, *22*, 813–828. [CrossRef]
8. Sun, F.; Wu, H.B.; Liu, X.; Liu, F.; Zhou, H.H.; Gao, J.H.; Lu, Y.F. Nitrogen-rich carbon spheres made by a continuous spraying process for high-performance supercapacitors. *Nano Res.* **2016**, *9*, 3209–3221. [CrossRef]
9. Alabad, A.; Razzaque, S.; Yang, Y.W.; Chen, S.; Tan, B. Highly porousactivated carbon materials from carbonized biomass with high CO_2 capturing capacity. *Chem. Eng. J.* **2015**, *281*, 606–612. [CrossRef]
10. Srinivas, G.; Yue, L.; Neal, S.; Taner, Y.; Zheng, X.G. Design of hyperporous graphene networks and their application in solid-amine based carbon capture systems. *J. Mater. Chem. A* **2017**, *5*, 17833–17840. [CrossRef]
11. Srinivas, G.; Hasmukh, A.P.; Zheng, X.G. Carbon capture: An ultrahigh pore volume drives up the amine stability and cyclic CO_2 capacity of a solid-amine@carbon sorbent. *Adv. Mater.* **2015**, *27*, 4903–4909. [CrossRef]
12. Teng, H.; Ho, J.A.; Hsu, Y.F. Preparation of activated carbons from bituminous coals with CO_2 activation-Influence of coal oxidation. *Carbon* **1997**, *35*, 275–283. [CrossRef]
13. Teng, H.; Ho, J.A.; Hsu, Y.F.; Hsieh, C.T. Preparation of activated carbons from bituminous coals with CO_2 activation. 1. effects of oxygen content in raw coals. *Fuel Energy Abstr.* **1996**, *38*, 139–146. [CrossRef]
14. Akash, B.A.; OBrien, W.S. The production of activated carbon from a bituminous coal. *Int. J. Energy Res.* **1996**, *20*, 913–922. [CrossRef]
15. San, M.G.; Fowler, G.D.; Sollars, C.J. A study of the characteristics of activated carbons produced by steam and carbon dioxide activation of waste tyre rubber. *Carbon* **2003**, *41*, 1009–1061. [CrossRef]

16. Zhu, Y.W.; Gao, J.H.; Li, Y.; Sun, F. Preparation of activated carbons for SO_2 adsorption by CO_2 and steam activation. *J. Taiwan Inst. Chem. Eng.* **2012**, *43*, 112–119. [CrossRef]

17. Zhu, Y.W.; Gao, J.H.; Li, Y.; Sun, F.; Qin, Y.K. Preparation and characterization of activated carbons for SO_2 adsorption from taixi anthracite by physical activation with steam. *Korean J. Chem. Eng.* **2011**, *28*, 2344–2350. [CrossRef]

18. Liu, D.D.; Gao, J.H.; Wu, S.H.; Qin, Y.K. Effect of char structures caused by varying the amount of $FeCl_3$ on the pore development during activation. *RSC Adv.* **2016**, *6*, 87478–87485. [CrossRef]

19. Coetzee, S.; Neomagus, H.; Bunt, J.R. Improved reactivity of large coal particles by K_2CO_3 addition during steam gasification. *Fuel Process. Technol.* **2013**, *114*, 75–80. [CrossRef]

20. Fan, S.M.; Yuan, X.Z.; Park, J.C.; Xu, L.H.; Kang, T.J.; Kim, H.T. Gasification of Indonesian sub-bituminous coal with different gasifying agents using Ca and K catalysts. *Energy Fuels* **2016**, *30*, 9372–9378. [CrossRef]

21. Kopyscinski, J.; Habibi, R.; Mims, C.A. K_2CO_3-catalyzed CO_2 gasification of ash-free coal: Kinetic study. *Energy Fuels* **2013**, *27*, 4875–4883. [CrossRef]

22. Kim, H.S.; Kudo, S.; Tahara, K.; Hachiyama, Y.; Yang, H.; Norinaga, K.; Hayashi, J.I. Detailed kinetic analysis and modeling of steam gasification of char from Ca-loaded lignite. *Energy Fuels* **2013**, *27*, 6617–6631. [CrossRef]

23. Zhang, L.; Kudo, S.; Tsubouchi, N.; Hayashi, J.I.; Ohtsuka, Y.; Norinaga, K. Catalytic effects of Na and Ca from inexpensive materials on in-situ steam gasification of char from rapid pyrolysis of low rank coal in a drop-tube reactor. *Fuel Process. Technol.* **2013**, *113*, 1–7. [CrossRef]

24. Olsson, J.G.; Jaglid, U.; Petterson, J.B.C.; Hald, P. Alkali metal emission during pyrolysis of biomass. *Energy Fuels* **1997**, *11*, 779–784. [CrossRef]

25. Kowalski, T.; Ludwig, C.; Wokaun, A. Qualitative evaluation of alkali release during the pyrolysis of biomass. *Energy Fuels* **2007**, *21*, 3017–3022. [CrossRef]

26. Li, C.Z.; Sathe, C.; Kershaw, J.R.; Pang, Y. Fates and roles of alkali and alkaline earth metals during the pyrolysis of a victorian brown coal. *Fuel* **2000**, *79*, 427–438. [CrossRef]

27. Keown, D.M.; Favas, G.; Hayashi, J.; Li, C.Z. Volatilisation of alkali and alkaline earth metallic species during the pyrolysis of biomass: Differences between sugar cane bagasse and cane trash. *Bioresour. Technol.* **2005**, *96*, 1570–1577. [CrossRef]

28. Matsuoka, K.; Yamashita, T.; Kuramoto, K.; Suzuki, Y.; Takaya, A.; Tomita, A. Transformation of alkali and alkaline earth metals in low rank coal during gasification. *Fuel* **2008**, *87*, 885–893. [CrossRef]

29. Sun, X.; Li, Y. Ga_2O_3 and GaN semiconductor hollow spheres. *Angew. Chem.* **2004**, *116*, 3915–3919. [CrossRef]

30. Sugano, M.; Mashimo, K.; Wainai, T. Structural changes of lower rank coals by cation exchange. *Fuel* **1999**, *78*, 945–951. [CrossRef]

31. Gorrini, B.C.; Radovic, R.C.; Gordonb, A.L. On the potassium-catalysed gasification of a chilean bituminous coal. *Fuel* **1990**, *69*, 789–791. [CrossRef]

32. Liu, D.D.; Gao, J.H.; Cao, Q.X.; Wu, S.H.; Qin, Y.K. Improvement of activated carbon from Jixi bituminous coal by air preoxidation. *Energy Fuels* **2017**, *31*, 1406–1415. [CrossRef]

33. Heras, F.; Alonso-Morales, N.; Jimenez-Cordero, D.; Gilarranz, M.A.; Rodriguez, J.J. Granular mesoporous activated carbons from waste tires by cyclic oxygen chemisorption-desorption. *Ind. Eng. Chem. Res.* **2012**, *51*, 2609–2614. [CrossRef]

34. Gong, X.Z.; Guo, Z.C.; Wang, Z. Variation of char structure during anthracite pyrolysis catalyzed by Fe_2O_3 and its influence on char combustion reactivity. *Energy Fuels* **2009**, *23*, 4547–4552. [CrossRef]

35. Yang, K.B.; Peng, J.H.; Xia, H.Y.; Zhang, L.B.; Srinivasakannan, C.; Guo, S.H. Textural characteristics of activated carbon by single step CO_2 activation from coconut shells. *J. Taiwan Inst. Chem. Eng.* **2010**, *41*, 367–372. [CrossRef]

36. Karatepe, N.; Orbak, İ.; Yavuz, R.; Özyuǧuran, A. Sulfur dioxide adsorption by activated carbons having different textural and chemical properties. *Fuel* **2008**, *87*, 3207–3215. [CrossRef]

37. Sun, F.; Gao, J.H.; Liu, X.; Yang, Y.Q.; Wu, S.H. Controllable nitrogen introduction into porous carbon with porosity retaining for investigating nitrogen doping effect on SO_2 adsorption. *Chem. Eng. J.* **2015**, *290*, 116–124. [CrossRef]

38. Sathe, C.; Pang, Y.; Li, C.Z. Effects of heating rate and ion-exchangeable cations on the pyrolysis yields from a Victorian brown coal. *Energy Fuels* **1999**, *13*, 748–755. [CrossRef]

39. Li, T.; Zhang, L.; Li, D. Effects of gasification atmosphere and temperature on char structural evolution during the gasification of collie sub-bituminous coal. *Fuel* **2014**, *117*, 1190–1195. [CrossRef]

40. Sasezky, A.; Muckenhuber, H.; Grothe, H. Raman microspectroscopy of soot and related carbonaceous materials: Spectral analysis and structural information. *Carbon* **2005**, *43*, 1731–1742. [CrossRef]

41. He, X.F.; Jin, L.J.; Wang, D.; Zhao, Y.P.; Zhu, S.W.; Hu, H.Q. Integrated process of coal pyrolysis with CO_2 reforming of methane by dielectric barrier discharge plasma. *Energy Fuels* **2011**, *25*, 4036–4042. [CrossRef]

42. Li, W.; Zhu, Y.M. Structural characteristics of coal vitrinite during pyrolysis. *Energy Fuels* **2014**, *28*, 3645–3654. [CrossRef]

43. Liu, H.; Xu, L.F.; Zhao, D.; Cao, Q.X.; Gao, J.H.; Wu, S.H. Effects of alkali and alkaline-earth metals and retention time on the generation of tar during coal pyrolysis in a horizontal fixed-bed reactor. *Fuel Process. Technol.* **2018**, *179*, 399–406. [CrossRef]

44. Lizzio, A.A.; Debarr, J.A. Effect of surface area and chemisorbed oxygen on the SO_2 adsorption capacity of activated char. *Fuel* **1996**, *75*, 1515–1522. [CrossRef]

Article

The Effect of Various Nanofluids on Absorption Intensification of CO$_2$/SO$_2$ in a Single-Bubble Column

Soroush Karamian [1], Dariush Mowla [2] and Feridun Esmaeilzadeh [2,*]

[1] Department of Chemical Engineering, Shiraz University, Shiraz 7134851154, Iran
[2] Environmental Research Center in Petroleum and Petrochemical Industries,
    School of Chemical and Petroleum Engineering, Shiraz University, Shiraz 7134851154, Iran
* Correspondence: esmaeil@shirazu.ac.ir

Received: 23 May 2019; Accepted: 21 June 2019; Published: 26 June 2019

Abstract: Application of nanoparticles in aqueous base-fluids for intensification of absorption rate is an efficient method for absorption progress within the system incorporating bubble-liquid process. In this research, SO$_2$ and CO$_2$ were separately injected as single raising bubbles containing nanofluids to study the impact of nanoparticle effects on acidic gases absorption. In order to do this, comprehensive experimental studies were done. These works also tried to investigate the effect of different nanofluids such as water/Al$_2$O$_3$ or water/Fe$_2$O$_3$ or water/SiO$_2$ on the absorption rate. The results showed that the absorption of CO$_2$ and SO$_2$ in nanofluids significantly increases up to 77 percent in comparison with base fluid. It was also observed that the type of gas molecules and nanoparticles determine the mechanism of mass transfer enhancement by nanofluids. Additionally, our findings indicated that the values of mass transfer coefficient of SO$_2$ in water/Al$_2$O$_3$, water/Fe$_2$O$_3$ and water/SiO$_2$ nanofluids are, respectively, 50%, 42% and 71% more than those of SO$_2$ in pure water ($k_{LSO_2-water} = 1.45 \times 10^{-4}$ m/s). Moreover, the values for CO$_2$ in above nanofluids were, respectively, 117%, 103% and 88% more than those of CO$_2$ in water alone ($k_{LCO_2-water} = 1.03 \times 10^{-4}$ m/s). Finally, this study tries to offer a new comprehensive correlation for mass transfer coefficient and absorption rate prediction.

Keywords: nanofluids; absorption intensification; mass transfer coefficient; bubble column

1. Introduction

Combustion of fossil fuels led to deforestation and global warming by the emission of acidic gases such as SO$_2$ and CO$_2$ into the environment [1]. Hence, in 1992, the United Nation Conference on Environment and Development offered a new strategy for reducing the emission of acidic and other greenhouse gases to below the standard level until 2000 [1]. Consequently, the governments should finance researchers and scholars to apply new methods and techniques to reduce the amount of CO$_2$ as well as the SO$_2$ produced from large-scale industries and sources [2–6].

In order to remove acidic gases from the natural gas, the scrubbing with the amine solution is the main process in the gas refineries. In addition, various techniques including physical and chemical absorption, membrane technology and adsorption methods are applied for the high CO$_2$/SO$_2$ production industries such as metal forming plants and petrochemical companies [7–9]. One of new approaches for enhance the absorption process, is addition of nanomaterials to basefluids for obtaining novel solvent with ability to absorb gases efficiently [2,3,10,11]. This method were elucidated by several researchers due to its high efficiency, and it has received much more attention in recent years [12,13].

Krishnamurthy et al. fulfilled a comprehensive research on the application of nanoparticles for increasing of mass transfer rate within a basefluid environment. They revealed that Brownian motion of nanoparticles, leading to induce the micro-convections in nanofluids, has the most impact on mass transfer rate [14]. Ashrafmansouri et al. comprehensively studied previous research and reported an review to highlight the impacts of nanomaterials in heat and mass transfer processes [11]. They reported that much higher thorough studies are needed to disclose the impacts of main parameters including nanoparticles mean size and morphology on absorption rate by using nanofluids. They also exhibited that nanofluid reusing as well as absorption process modeling are the most important subjects for advancement of this technique. In addition, Kim et al. showed that mass transfer rate of ammonia is enhanced when a few nanoparticles are added to the basefluid. They exhibited that bubbles breaking by nanoparticles considerably enhances mass transfer through increasing interfacial area. They also reported that smaller bubbles were produced in nanofluid than in a base fluid, leading to intensification in mass transfer surface area [15].

Ma et al. declared that by adding CNTs to a basefluid, the localized micro-convection occurs due to the Brownian motion of nanotubes [16]. They reported that induced convection can intensify the ammonia molecular diffusion within the nanofluid. Moreover, they concluded that the grazing effect can be considered another mechanism enhancing the efficiency of NH_3 by means of the bubble absorption process [16]. Absorption of gas molecules by means of the nanoparticle surfaces at the bubble interface and then removing the adsorbed gas components from the nanoparticles surface into the fluid is known as grazing effect [17]. Kang et al. also assessed the impact of Carbon nanotubes on gas absorption in a nanofluid [18]. They also revealed that the mass transfer rate of gaseous ammonia in 0.001 wt. % CNTs loaded in nanofluid was 20% higher than that of pure deionized water [18,19].

Numerous researchers have focused on the application of nanofluids as a potential absorbent for the removal of acidic gases [6,11,12,20–23]. Esmaeili-Faraj et al. exhibited that the removal rate of H_2S enhanced up to 40% when 0.02 wt. % of EGO (Exfoliated Graphene Oxide) is added to deionized water as a basefluid. They showed that the main mechanism for enhanced absorption rate is the grazing effect [4].

Jung et al. performed an extensive research in which Al_2O_3 nanoparticles were scattered in methanol as with nanoparticles volume fractions range of 0.005–0.1 vol. % [24]. They observed that the maximum CO_2 removal was 8.3% at 0.01 vol. % nanoparticles compared to the conditions that pure methanol was used as an absorbent. They concluded that the enhanced CO_2 uptake is due to the mixing effect of Al_2O_3 nanoparticles, which were caused by the particle-laden flows induced by Brownian motion [24]. In addition, they observed that for the concentration above a critical value, insignificant Brownian motion can be seen since the inter-particle interactions declines this motion [24].

Darvanjooghi et al. studied the absorption of CO_2 by means of Fe_3O_4/water nanofluid during the applied alternating and constant magnetic fields [3]. Their results declared that both CO_2 solubility and mass transfer rate are increased when the strength of magnetic field is high. In addition, they found that the solubility of CO_2 and its average molar flux into the nanofluid possess a maximum value by applying an AC magnetic field. Finally, they showed that with the increment of magnetic field strength, the mass diffusivity of carbon dioxide in the nanofluid and renewal surface factor increase, whereas the diffusion layer thickness diminishes.

Although, the impacts of different parameters on gas absorption, by means of nanofluids, are studied in previous works, there are no fully agreement and comprehensive results regarding the influence of nanoparticles types on mass transfer parameters in oxides nanoparticles loaded in nanofluids.

Thus, the aim of this study is to reveal the effect of different metal oxide nanoparticles on SO_2 and CO_2 mass transfer parameters in a single-bubble absorber. Hence, comprehensive experimental studies are done to investigate the molar flux, absorption rate, mass transfer coefficient and diffusivity coefficient. In addition, a new correlation encompassing nanofluid properties was developed in order to estimate mass transfer coefficients of the mentioned gases in nanofluids.

2. Materials and Methods

2.1. Materials

In this research, SiO_2 nanoparticles with the purity of 99.99 wt. %, Al_2O_3 nanoparticles with the purity of 99.98 wt. % and Fe_2O_3 nanoparticles with the purity of 99.92 wt. % were purchased from U.S. Nano Company, United State (see Table 1) to prepare water based nanofluids. In order to perform reverse titration for measuring the quantity of CO_2 and SO_2 dissolved in nanofluids, pure NaOH pellets (99.99 wt. %) and HCl with the purity of 37 vol. % were purchased from Merck Company, Germany. Moreover, phenolphthalein and methyl orange obtained from Merck Company, Germany were used as indicators for determination of the equivalent points. Deionized water was used for the preparation and dilution of nanofluids as well as washing the laboratory glassware. All chemical materials are used as received without further purification.

Table 1. Physical properties of the nanoparticles (NPs) used in this study.

Properties	SiO_2 NPs	Al_2O_3 NPs	Fe_2O_3 NPs
Molecular weight (g/mol)	60.08	101.96	159.69
Density (g/cm^3)	2.196	3.980	5.242
Melting point (°C)	1713	2054	1539
Appearance	White solid powder	White solid powder	Red-brown solid powder

2.2. Apparatus

2.2.1. Nanofluid Preparation Instruments

In this study, the transmission electron microcopy (TEM) and dynamic light scattering (DLS) were used to estimate the size distribution of dry and dispersed metal oxides nanoparticles in deionized water, respectively. The TEM images of SiO_2, Al_2O_3 and Fe_2O_3 nanoparticles were obtained by using Hitachi, 9000 NA, Japan to characterize the size of nanoparticles and their agglomeration [25]. For preparing the sample of nanoparticles used in TEM analysis, a suspension of the nanoparticles dispersed in ethanol (0.001 wt. %) was sonicated by using an ultrasonic bath, Parsonic 30S-400W, 28 kHz, for 20 min and then was placed on the graphite surface. The samples were then put in a vacuum oven to remove the ethanol before being introduced into the TEM test device. DLS, Malvern, Zeta Sizer Nano ZS, United Kingdom, was applied to estimate the sizes of nanoparticles and the size distribution of the obtained metal oxides nanoparticles in deionized water [5,25,26]. The stability and surficial electrostatic charges of the metal oxides nanoparticles in deionized water were estimated by using Zeta Potential test (ELSZ-2000, Otsuka Electronics Co., Osaka, Japan). This analysis is a key indicator of the stability of metal oxides nanoparticles within deionized water [12]. Zeta potential accounts for the electrostatic charges on the surface of nanoparticles causing repulsive forces between dispersed particles. The negatively and positively larger magnitude of zeta potential exhibits a significant stability of nanoparticles in the basefluid, whereas a lower magnitude of maximum Zeta-potential declares the tendency of nanoparticles for agglomeration [27]. A Mass Flow Controller (MFC) model Brooks Instrument 1-888-554-flow, USA, was implemented for the injection of CO_2 and SO_2 gases into the nanofluids through the absorption apparatus. Furthermore, water based nanofluids were prepared by measuring and adding the required weight amounts of metal oxide nanoparticles. To do so, a precise electric balance (TR 120 SNOWREX, Taiwan) was implemented. A pH meter (PCE-PHD 1, PCE-Instruments holding, Southampton, UK) was used for recording the pH of solutions during the titration. Finally, an ultrasonic processor (QSONICA-Q700, NY, USA) was used in order to stop forming the agglomeration of SiO_2, Al_2O_3 and Fe_2O_3 nanoparticles, after they were under a mechanical ball-mill (YKM-2L, Changsha Yonglekang Equipment Co., Changsha, China) for grinding the clustered nanoparticles. A syringe-pump (Viltechmeda Plus SEP21S, manufactured in Vilnius, Lithuania) was also employed for injection of the titrant to the flask. Lastly, a magnetic stirrer (Model IKA-10038, Staufen, Germany) was used for stirring the solutions.

2.2.2. Experimental Set-Up

The experimental set-up contained a bubble column absorber filled with metal oxide nanoparticles loaded in nanofluids. A certain volume of CO_2 and SO_2 was injected into the nanofluid within the absorption column. Figure 1 exhibits the schematic diagram of a bubble column absorber that consists of a 1 m high and 16.2 mm diameter poly-methyl-meta-acrylate (PMMA) tube used as a semi-batch instrument to examine the absorption of acidic gases by means of nanofluids. In addition, in order to control the rate of gas absorption in nanofluids, a syringe-pump was used for the injection of the aforementioned gases through the absorber column. The gases were continuously injected into nanofluids in the absorber column with the constant flow rate of 500 mL/h in each experiment. The average bubbles diameter ranged from 6.9 to 7 mm, and the time for the rising of bubbles was found to be 2.3 s. Finally, in order to measure the concentration of gases in nanofluids in the reverse titration method, the injection of HCl solution into the discharged nanofluid was performed by means of the syringe-pump.

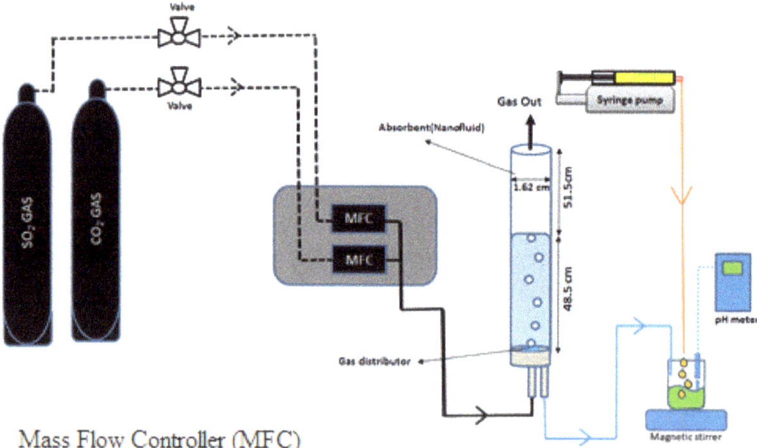

Figure 1. Schematic diagram of experimental set-up.

2.3. Methods

2.3.1. Nanofluid Preparation Procedure

At first, the nanoparticles were introduced to a ball-mill device for about 4 h to separate the agglomeration of nanoparticles. Then, water based nanofluids were prepared with the dispersing of 50 g SiO_2, Al_2O_3 and Fe_2O_3 nanoparticles in 1000 mL deionized water, separately, to produce the main suspension with the nanoparticles concentration of 5.0 wt. %, (equal to 50,000 mg/L). After adding the nanoparticles to deionized water, the suspensions were kept under stirring condition of 800 rpm for 5 h. Finally, the nanoparticles were dispersed in the basefluid by using the sonication process under three sequences of 20 min. The amplitude and cycle time of sonication were set on 70% and 0.5 s, respectively. Also for the preparation of other suspensions with different nanoparticle concentrations of 0.005, 0.01, 0.1, 1.0, and 5.0 wt. %, the stock solutions were diluted with further deionized water.

2.3.2. Experimental Procedure

Sample Analysis Procedure

The analysis for measuring the amounts of absorbed CO_2 and SO_2 in the nanofluids was carried out by using the reverse titration wherein the standard HCl and NaOH solutions were used as the titrant and reactant for producing Na_2CO_3 and Na_2SO_3, respectively [28]. Consequently, in order to determine CO_2 and SO_2 content by using the reverse titration, it is needed to convert H_2SO_3 and H_2CO_3 to Na_2SO_3 and Na_2CO_3, respectively, by the addition of a strong standard base. To do so, the nanofluids were discharged to the flask containing 15 mL of 0.1 M NaOH solution. The carbon dioxide and sulfur dioxide in the solution reacted with the sodium hydroxide and formed sodium bicarbonate or bisulfate as Equation (1) [5]:

$$RO_2 + 2NaOH \rightarrow Na_2RO_3 + H_2O, \; R = C(Carbon) \quad or \quad S(Sulfur) \tag{1}$$

The titration was then accomplished to neutralize the amount of remained NaOH, and then excess HCl (as a titrant) in the flask reacted with Na_2SO_3 and Na_2CO_3 during the titration according to the following reactions:

$$Na_2RO_3 + HCl \rightarrow NaCl + NaHRO_3, \; R = C(Carbon) \quad or \quad S(Sulfur) \tag{2}$$

$$NaHRO_3 + HCl \rightarrow NaCl + H_2O + RO_2, \; R = C(Carbon) \quad or \quad S(Sulfur) \tag{3}$$

According to Equation (2), the discharged samples were titrated with the standard acid solution, (0.1 M HCl), at first equivalent point. The titration with HCl then converted all the remained bicarbonate and bisulfate to SO_2 and CO_2 according to Equation (3). In this method, the difference of consumed HCl between two equivalent points represents the amount of CO_2 or SO_2 absorbed in the solution. Equation (4) was used for determining the value of absorbed gases by means of nanofluids [2,3,28]:

$$C_{RO_2} = \frac{(V_2 - V_1) \times M}{V} \times 10^3 \tag{4}$$

where C_{RO_2} is the absorbed CO_2 or SO_2 concentration in the nanofluids or deionized water (mol/m^3), M is HCl molarity (mol/lit), and V is the volume of absorbent used in the column, (equal to 100 mL), in all experiments. V_1 and V_2 are the volumes (mL) of standard acid solution consumed for neutralizing bicarbonate and bisulfate to SO_2 and CO_2 at two equivalent points, respectively (Figures 2 and 3).

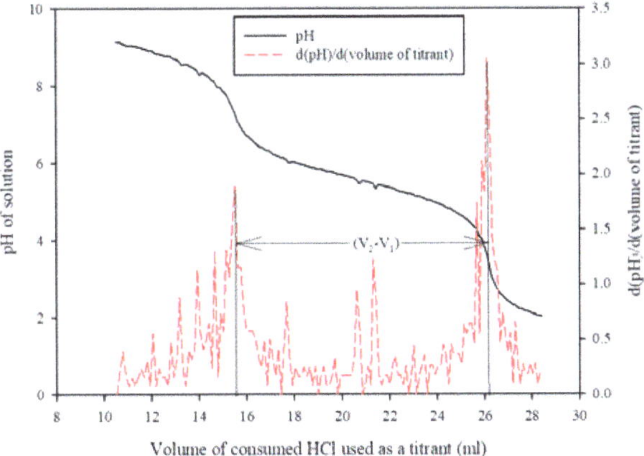

Figure 2. Plot of pH and its differentiation versus volume of consumed titrant, (HCl), for the injection of 50 mL SO_2 through deionized water.

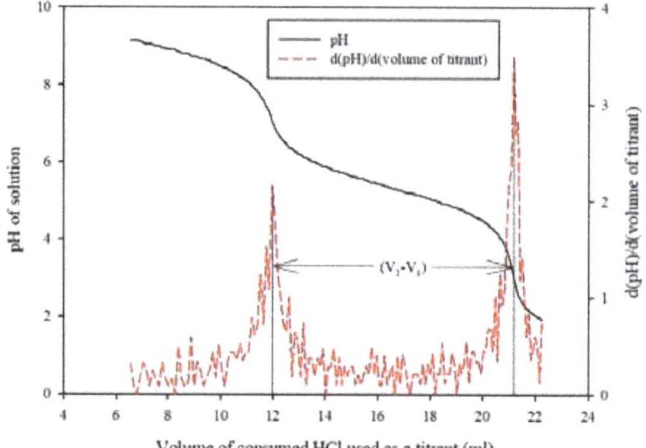

Figure 3. Plot of pH and its differentiation versus volume of consumed titrant, (HCl), for the injection of 50 mL CO_2 through deionized water.

In this work, the molar flux of absorbed CO_2 and SO_2 was calculated by means of the CO_2 and SO_2 concentration in the nanofluid according to the following equation (Equation (5)) [2,3,28]:

$$N_{ave,\,RO_2} = \frac{C_{RO_2} \times V}{(4\pi r_0^2 n) \times (\tau)} \times 10^{-6} \tag{5}$$

Here, N_{ave,RO_2} is the average molar flux transferred from gas, (pure CO_2 or SO_2), to liquid phase (mol/m^2 s), τ is the total gas-liquid contact time of bubbles passing through the nanofluids (s), which is equal to multiply of the bubbles number by raising time of one single bubble (2.3 S), n is the number of bubbles passes through nanofluids within the absorber column and r_0 is the average bubbles radius (3.5 mm) that assumed to be constant at all experiments.

Measurement of Mass Transfer Parameters

In order to obtain the mass transfer coefficient and diffusivity of CO_2 or SO_2 in a water based nanofluid, a set of experiments were performed in which the aforementioned gases were separately injected at the bottom of the column within the volumes of 20, 25, 30, 35, 40, 45 and 50 mL. The mass transfer parameters were then calculated by obtaining the absorption of CO_2 and SO_2 as well as the implementation of the model suggested by Zhao et al. [29].

Uncertainty Analysis

In this research, the uncertainty of the experimentations was calculated by the errors of measurements for parameters, incorporating time of raising bubbles, volume of liquid for the titration method and pH of solutions. The time of raising bubbles was measured by using a digital chronometer with the maximum accuracy of ±0.01 s, the pH of discharged nanofluids was measured during the titration by a pH meter with the maximum accuracy of ±0.1, and the volumes of liquids were measured by laboratories glassware with the maximum accuracy of ±0.1. According to the literature [2,3], the relative uncertainty of final experimental results was calculated as follows [30,31]:

$$U = \pm \sqrt{\left(\frac{\Delta V}{V}\right)^2 + \left(\frac{\Delta t}{t}\right)^2 + \left(\frac{\Delta pH}{PH}\right)^2} \tag{6}$$

Consequently, by substituting the values in Equation (6) the relative uncertainty of the experimental results was found to be less than 5.2 %.

3. Results and Discussion

3.1. Nanofluid Characterization

Figure 4 exhibits the TEM images of SiO_2, Al_2O_3 and Fe_2O_3 nanoparticles that used for the preparation of water based nanofluids. These images show that the diameter of SiO_2 nanoparticles ranged from 20 to 60 nm (Figure 4a), the diameter of Al_2O_3 nanoparticles ranged from 30 to 80 nm (Figure 4b) and the diameter of Fe_2O_3 nanoparticles ranged from 20 to 60 nm (Figure 4c). In addition, the results presented in Figure 4 exhibit that all metal oxides nanoparticles have a semi-spherical morphology that no considerable agglomeration was observed [32].

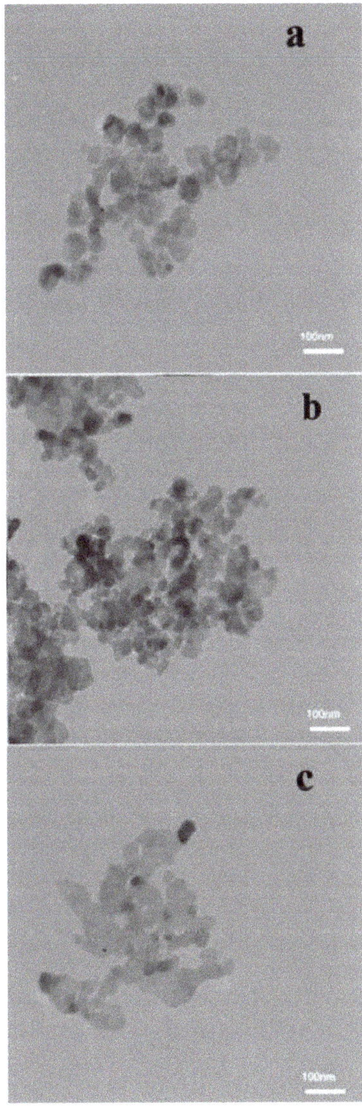

Figure 4. Transmission electron microcopy (TEM) images of (**a**) SiO_2, (**b**) Al_2O_3 and (**c**) Fe_2O_3 nanoparticles.

The results of DLS analysis for SiO_2, Al_2O_3 and Fe_2O_3 nanoparticles dispersed in deionized water exhibited that the mean diameter of nanoparticles for SiO_2 is 48.3 nm with Poly Dispersity Index, (P.D.I.), of 0.105 and the mean diameter of nanoparticles for Al_2O_3 and Fe_2O_3 is found to be 54.7 nm and 55.1 nm, respectively, with P.D.I.s of 0.145 and 0.138, respectively. These results confirm that the dispersion technique, which was used in this research, led to the well-dispersed nanoparticles diameter, with a narrow range of 48.3 to 55.1 nm. The results of this test indicate that the average size of nanoparticles is equal to that estimated by using TEM test declaring no significant agglomeration during the dispersion of nanoparticles in the basefluid.

Zeta-potential analysis can be implemented in order to quantify the stability of nanoparticles in the basefluid [33]. These results represent that nanofluids have high stability due to the fact that their zeta potential is lower than −45 mV [34]. In other words, the magnitude of the zeta potential determines the degree of electrostatic repulsion between similarly charged particles in colloidal dispersions. The large magnitude of the zeta potential for SiO_2/water, Al_2O_3/water and Fe_2O_3/water nanofluids (−97.8 mV for Al_2O_3/water, 100.2 mV for SiO_2/water and 79.5 mV for Fe_2O_3/water nanofluids) indicated high stability of nanoparticles representing high repulsive electrostatic forces [35].

3.2. Absorption

3.2.1. Maximum Absorption

Figure 5 shows the average molar flux of CO_2 into each of these three nanofluids: SiO_2/water, Al_2O_3/water or Fe_2O_3/water. The mass fraction of each metal oxides nanoparticle varies from 0.005 to 5 wt. %. The experimentations were repeated four times at a fixed mass fraction of metal oxides nanoparticles and the standard deviations are shown in this figure as the error bars. According to the results presented in this figure, the average molar flux of CO_2 increases about 21% with the increase of Al_2O_3 nanoparticles from 0.005 to 0.1 wt. % while the molar flux decreases for higher Al_2O_3 nanoparticles loads (0.1 to 5 wt. %). Moreover, the value of CO_2 molar flux increases about 45% when the mass fraction of SiO_2 nanoparticles increases from 0.005 to 0.01 wt. %. Moreover, for higher mass fractions of SiO_2 nanoparticles, a remarkable declination on CO_2 molar flux resulted. In addition, the value of CO_2 molar flux enhances about 16% when mass fraction of Fe_2O_3 nanoparticles enhances from 0.005 to 1 wt. %, and a declination of CO_2 molar flux resulted in the mass fraction range of up to 5 wt. %. Table 2 represents the mass fraction of nanoparticles where by the maximum value of CO_2 molar flux obtained. It can be concluded from this table that CO_2 absorption molar flux has a maximum value at 0.1, 0.01 and 1 wt. % for Al_2O_3/water, SiO_2/water and Fe_2O_3/water nanofluids, respectively. For all nanoparticles types, the nanoparticles intensify the micro-convections, producing larger mass transfer rate in comparison to pure basefluid; thus, initial increase in CO_2 absorption would be rationalizable with the aforementioned nanoparticles mass fractions. On the other hands, increasing a number of nanoparticles leads to enhance further the viscosity of nanofluids, thereby overcoming the nanoparticles micro-convection impacts together with diminishing the absorption of CO_2 within the nanofluids [4,12]. Furthermore, Figure 5 clearly exhibits that CO_2 absorption molar flux in metal oxides-based nanofluids is larger than that in deionized water for various nanoparticles mass loads.

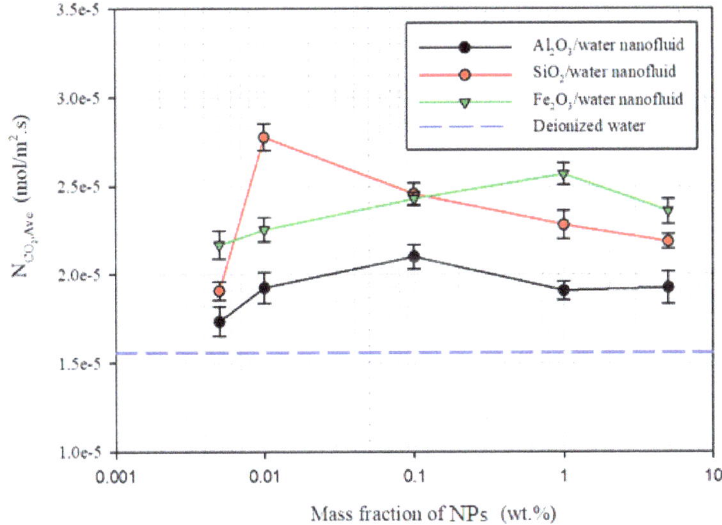

Figure 5. Average molar flux of CO_2 versus mass fraction of metal oxides nanoparticles (NPs).

Table 2. Maximum molar flux and relative absorption rate for SO_2 and CO_2.

Absorbent	SO$_2$ Absorption			CO$_2$ Absorption		
	% wt. NPs in N_{max}	N_{max}	$\frac{N_{nf}}{N_{bf}}$	% wt. NPs in N_{max}	N_{max}	$\frac{N_{nf}}{N_{bf}}$
Water (bf)		1.871×10^{-5}	1.000		1.566×10^{-5}	1.000
SiO$_2$/water	1.0	2.983×10^{-5}	1.594	0.01	2.774×10^{-5}	1.771
Al$_2$O$_3$/water	0.1	2.445×10^{-5}	1.307	0.1	2.098×10^{-5}	1.340
Fe$_2$O$_3$/water	0.1	3.312×10^{-5}	1.770	1.0	2.566×10^{-5}	1.638

Figure 6 displays the average molar flux of SO_2 into each of these three nanofluids: SiO_2/water, Al_2O_3/water or Fe_2O_3/water. The aforementioned metal oxides nanoparticles were dispersed in deionized water with different concentrations of 0.005, 0.01, 0.1, 1 and 5 wt. %. These experimentations were also repeated four times at a fixed mass fraction of each metal oxide nanoparticle, and the error bars express the standard deviation obtained from the measurements. According to the obtained results, the average molar flux of SO_2 enhances about 28% with the Al_2O_3 nanoparticles enhancement from 0.005 to 0.1 wt. %, and for higher nanoparticles loads, a substantial decrease resulted in its molar flux. In addition, the value of SO_2 absorption rate into SiO_2/water nanofluid increases about 32% when the mass fraction of SiO_2 nanoparticles in deionized water increases from 0.005 to 1 wt. %. After a further increase of mass fraction up to 5 wt. %, the absorption of CO_2 declines. Moreover, the value of CO_2 molar flux increases about 26% when mass fraction of Fe_2O_3 nanoparticles increases from 0.005 to 0.1 wt. %; and with a further increase of nanoparticles mass fraction from 0.1 to 5 wt. %, the value of CO_2 absorption declines. According to the results presented in Table 2, the maximum molar flux of SO_2 can be obtained with the nanoparticles mass fractions of 0.1, 1 and 0.1 wt. % for Al_2O_3/water, SiO_2/water and Fe_2O_3/water nanofluids, respectively. Similar to the results achieved for CO_2 absorption, the addition of nanoparticles into the deionized water enhances the micro-convections and intensifies the mass transfer rate of SO_2 while increasing the nanoparticles load increases further the viscosity of nanofluids, declining the absorption rate of SO_2 into the nanofluids [4,12]. The results presented in this figure show that SO_2 absorption in metal oxides nanofluids is more than that in deionized water.

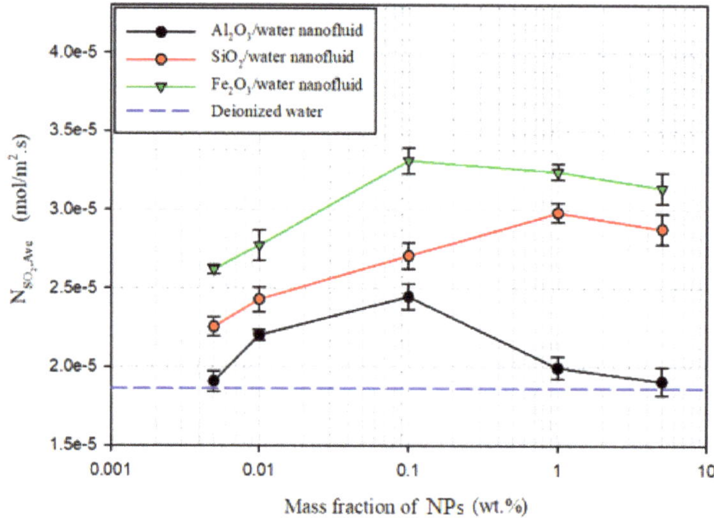

Figure 6. Average molar flux of SO$_2$ versus mass fraction of metal oxides nanoparticles.

3.2.2. Probing of Mass Transfer Rate

Volume loading rate (mL/mL s), can be attributed to the rate of gas injection divided to the total volume of gas equal to which is 50 mL. It actually represents the time which is passing during the mass transfer process and clearly shows what portion of gas is injected through the nanofluid. Therefore, this parameter can easily show the ability of nanofluid to absorb gas at the beginning of the injection or at the end of the process. Figure 7 presents the results of average CO$_2$ absorption in each of these three nanofluids: SiO$_2$/water, Al$_2$O$_3$/water or Fe$_2$O$_3$/water against the volume loading rate that was measured at the temperature of 25 °C and the optimum mass fractions of 0.1, 0.01 and 1 wt. % for SiO$_2$, Al$_2$O$_3$ and Fe$_2$O$_3$ nanoparticles in deionized water, respectively. These findings reveal that the absorption rate increases with the enhancement in volume loading rate. Additionally, it is chiefly clear when Fe$_2$O$_3$/water is used as an absorbent, the maximum value of absorption rate is obtained at any volume loading rate. Moreover, these results indicate that the minimum value of CO$_2$ absorption for the Al$_2$O$_3$/water nanofluid resulted in comparison to the other nanofluids assessed in this work. These findings indicated that type of the used nanoparticles had a major effect on mass transfer rate. In addition, it can be concluded from this figure that the mass transfer flux is low at lower volume loading rates, and it increases with the increment of loading rate due to having a higher driving force of mass transfer.

Figure 8 also shows the results of average SO$_2$ absorption in each of these three nanofluids: SiO$_2$/water, Al$_2$O$_3$/water or Fe$_2$O$_3$/water against the volume loading rate that was measured at the temperature of 25 °C and the concentrations of 0.1, 1 and 0.1 wt. % for Al$_2$O$_3$, SiO$_2$ and Fe$_2$O$_3$ nanoparticles in deionized water, respectively. These results, which are similar to those obtained for CO$_2$ absorption, show that the absorption rate increases with the growth in volume loading rate, and when SiO$_2$/water is used as an absorbent, the maximum value of absorption rate is obtained at each gas volume loading rate; while for CO$_2$ absorption by using Fe$_2$O$_3$/water nanofluid, a higher absorption rate achieved. In addition, it is chiefly evident that the minimum value of SO$_2$ absorption for the Al$_2$O$_3$/water nanofluid resulted in comparison to the other nanofluids assessed in this work, that is similar to CO$_2$ case. These findings declared that type of the used nanoparticles and their interactions with CO$_2$ and SO$_2$ had a major effect on mass transfer rate of the gas into the nanofluids. Moreover, the value of absorption rate is similar to the case of CO$_2$ absorption.

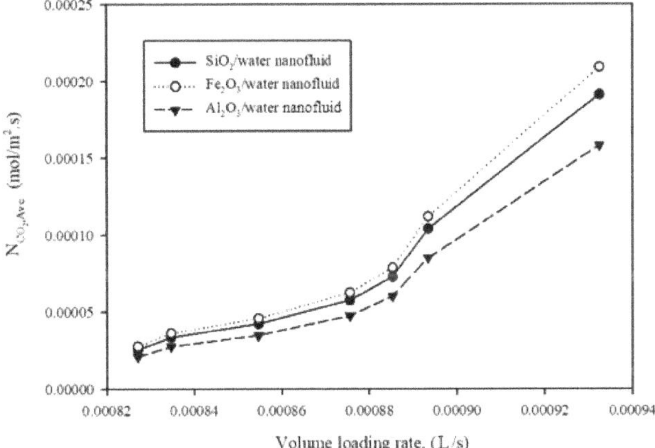

Figure 7. Average molar flux of CO_2 versus volume loading rate.

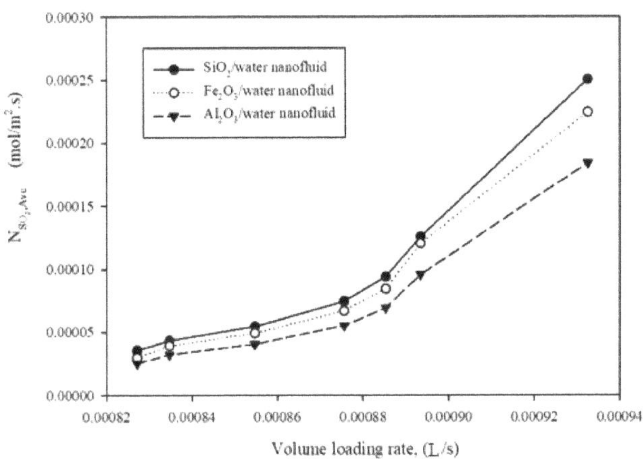

Figure 8. Average molar flux of SO_2 versus volume loading rate.

3.2.3. Mass Transfer Coefficient

For the calculation of mass transfer coefficient, in separate runs, various volumes of gases (20, 25, 30, 35, 40, 45 and 50 mL that are, respectively, equal to 7, 10, 12, 13, 15.6, 17.6 and 20 min total gas-liquid contact time) were injected into the column and then gas concentration and molar flux were measured. Figure 9 shows the average molar flux of CO_2/SO_2 against the dissolved concentration of CO_2/SO_2 in the liquid bulk. These results clearly exhibit that an increase in CO_2/SO_2 bulk concentration consecutively decreases the average value of molar flux due to the reduction of mass transfer driving force. Moreover this observation has approximately a linear behavior for all cases. In order to potpourri of this linear behavior, the principal mass transfer equation (Equation (7)) was used, and the experimental values for the absorption of CO_2/SO_2 by using different nanofluids were fitted to Equation (7):

$$N_{Avg} = k_L \left(C^*_{RO_2,Observed} - C_{RO_2} \right) \tag{7}$$

where k_L is the mass transfer coefficient at liquid phase, (m/s), C_{RO_2} is the bulk concentration of CO_2/SO_2 within the nanofluids, and $C^*_{RO_2,Observed}$ is the observed concentration of CO_2/SO_2 at gas-liquid interface, (mol/m^3). It is mentioned that observed value for gas concentration in the interface was calculated from extrapolation of line fitted on experimental data. Since linear pattern was assumed for molar flux and gas concentration. According to the results obtained for the absorption of CO_2 into each of these three nanofluids: SiO_2/water, Al_2O_3/water or Fe_2O_3/water (Figure 9a–c), the model was fitted to the experimental data with the R^2 equal to 0.9753, 0.9755 and 0.9897 declaring high accuracy of the regression analysis and low deviation of the experimental data from the fitted model.

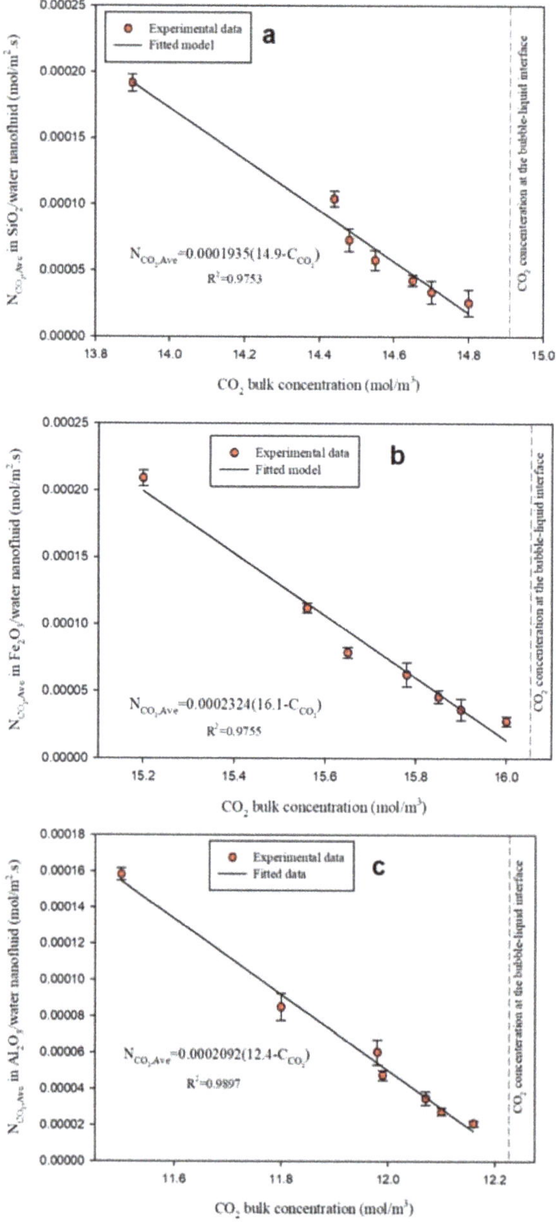

Figure 9. Average molar flux versus CO_2 bulk concentration for (**a**) SiO_2/water, (**b**) Al_2O_3/water and (**c**) Fe_2O_3/water nanofluids.

The average molar flux of SO_2 versus the bulk concentration is shown in Figure 10. These results are also similar to those obtained for CO_2 absorption declaring that an increase in SO_2 bulk concentration leads to decrease the average value of molar flux, representing a significant declination in mass transfer driving force. In order to obtain the mass transfer coefficient and SO_2 concentration at the bubbles-liquid interface, the regression analysis was also performed on Equation (7), and the equation was fitted to the values for SO_2 absorption into each of these three nanofluids: SiO_2/water, Al_2O_3/water or Fe_2O_3/water (Figure 10a–c) with the R^2 equal to 0.9711, 0.9705 and 0.9788, respectively. These values confirm the high accuracy of the regression analysis.

According to the results obtained from Figures 9 and 10, it can be concluded that for all nanofluids used in this study, the vertical diagram (dashed line) shows the observed concentration of CO_2 and SO_2 at the bubble-liquid interface. Furthermore, the diagonal plot of average molar flux versus the bulk concentration of CO_2 and SO_2 represents the operating line for gas absorption into the nanofluids. It is clearly evident that by approaching the operating line to the equilibrium concentration of CO_2 and SO_2 in each of these three nanofluids, namely SiO_2/water, Al_2O_3/water or Fe_2O_3/water, a lower molar flux resulted.

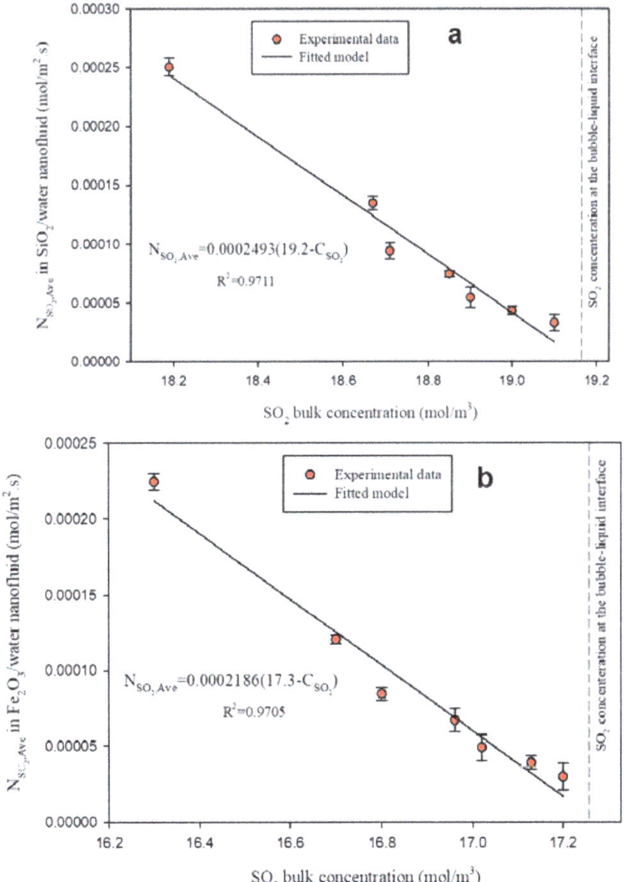

Figure 10. *Cont.*

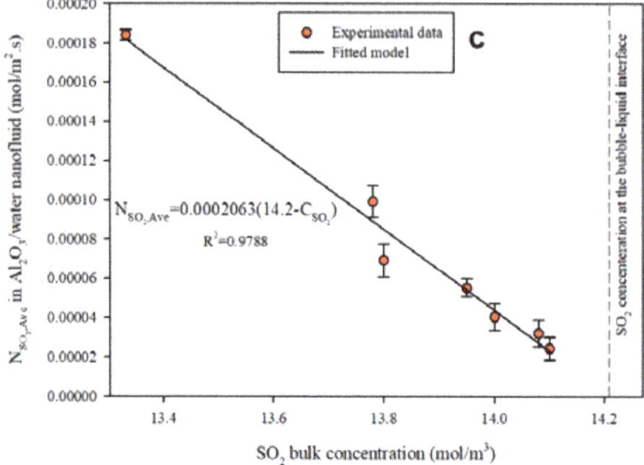

Figure 10. Average molar flux versus SO_2 bulk concentration for (**a**) SiO_2/water, (**b**) Al_2O_3/water and (**c**) Fe_2O_3/water nanofluids.

Table 3 represents the values of relative mass transfer coefficient for SO_2 and CO_2 absorption by using SiO_2/water, Al_2O_3/water or Fe_2O_3/water nanofluids with respect to water alone. These values are the slope of operating line in Figures 9 and 10. According to these results, the maximum value of relative mass transfer coefficient for CO_2 absorption was achieved by Al_2O_3/water nanofluid while the value of relative mass transfer coefficient for SiO_2/water was observed to possess a minimum value in comparison to the other nanofluids assessed in this work. Additionally, these findings exhibit that the maximum value of mass transfer coefficient for SO_2 absorption was achieved for SiO_2/water, and this parameter for Fe_2O_3/water was found to be less than the others. According to the results presented in this table, relative mass transfer coefficient intensively depend on type of the nanofluid. In fact, the absorption of SO_2 by SiO_2/water nanofluid and the absorption of CO_2 by Fe_2O_3/water nanofluid demonstrate higher values for the relative mass transfer coefficient and relative gas concentration at the bubble-liquid interface.

Table 3. Relative mass transfer coefficient for CO_2 and SO_2 in the base fluid (BF) and various nanofluids (NF).

Gas	Absorbent	$k_L \times 10^4$, (m/s)	$\dfrac{k_{L_{nf}}}{k_{L_{bf}}}$
	Water (BF)	1.030	1.00
CO_2	Water/SiO$_2$ NF	1.935	1.88
	Water/Fe$_2$O$_3$ NF	2.324	2.03
	Water/Al$_2$O$_3$ NF	2.092	2.17
	Water (BF)	1.450	1.00
SO_2	Water/SiO$_2$ NF	2.493	1.71
	Water/Fe$_2$O$_3$ NF	2.186	1.42
	Water/Al$_2$O$_3$ NF	2.063	1.50

3.3. Diffusivity Coefficient

In general, diffusivity of gases into a fluid has a higher impact on mass transfer coefficient as well as rate of gas absorption. In this study, Equation (8) was used to obtain the diffusivity of SO_2 and CO_2 into each of these three nanofluids, namely SiO_2/water, Al_2O_3/water or Fe_2O_3/water. This equation indicates a bubble-liquid mass transfer model for raising a single bubble through a liquid based on Dankwert's theory [5,29].

$$N_{Ave} = \frac{D \sinh\left(\delta \sqrt{\frac{s}{D}}\right) + D\, r_0\ \sqrt{\frac{s}{D}}\ \cosh\left(\delta \sqrt{\frac{s}{D}}\right)}{r_0\ \sinh\left(\delta \sqrt{\frac{s}{D}}\right)} \left(C_{RO_2,i} - C_{RO_2}\right) \tag{8}$$

In this model, the main factors affecting on mass transfer rate are the surface renewal rate (s), bubbles radius (r_0), diffusion layer thickness (δ) and the diffusivity of gases through a liquid (D). N_{Ave} is the molar flux $\left(mol/m^2\ s\right)$, C_{RO_2} and $C_{RO_2,i}$ are the concentration of dioxide gases within the liquid bulk and at the bubble-liquid interface (mol/m^3), respectively.

By comparing Equations (7) and (8), the mass transfer coefficient of a gas into the liquid by using a single bubble can be obtained from the following relation:

$$k_L = \frac{D \sinh\left(\delta \sqrt{\frac{s}{D}}\right) + D\, r_0 \sqrt{\frac{s}{D}}\ \cosh\left(\delta \sqrt{\frac{s}{D}}\right)}{r_0\ \sinh\left(\delta \sqrt{\frac{s}{D}}\right)} \tag{9}$$

This equation was used for estimating the diffusivity of SO_2 and CO_2 within the nanofluids. It has been reported by Darvanjooghi et al. that the effective parameters in Equation (9) (s, δ and D) intensively depend on the size of nanoparticles in the basefluid. They reported that the size of nanoparticles was about 40 to 50 nm, and the values of surface renewal rate, s, and the diffusion layer thickness, δ, were 6.85 and 0.201 mm, respectively [2]. In this research, the average mean diameter of nanoparticles ranges from 40 to 60 nm. Additionally, it can be assumed that the values of s and δ would be constant during the absorption of SO_2 and CO_2 and depend on just nanoparticles mean diameter. Additionally, the mass transfer coefficients for both SO_2 and CO_2 gases within the nanofluids studied here have been already calculated in Table 3. Therefore, Equation (9) can be simplified to the following relation:

$$F(D,s,\delta) = \exp\left(2\delta \sqrt{\frac{s}{D}}\right) \mp \frac{D - r_0 \sqrt{s.D} - r_0 k_L}{r_0 \sqrt{s.D} - r_0 k_L} = 0,\ s = 6.85\ \text{and}\ \delta = 0.201 \tag{10}$$

where $F(D,s,\delta)$ must be equal to zero for certain values of mass transfer coefficient and gas diffusivity within the different nanofluids. By using the Newton-Raphson method, Equation 10 can be solved according to the following equation in which $\partial F(D_n,s,\delta)/\partial D_n$ can be obtained by obtaining partial derivative of Equation (10). The initial value of diffusivity, D_0, was set on 10^{-10}.

$$D_{n+1} = D_n - \frac{F(D_n,s,\delta)}{\partial F(D_n,s,\delta)/\partial D_n},\ n = 0,\ 1,\ 2,\ 3,\dots \tag{11}$$

Table 4 presents the values of SO_2 and CO_2 diffusivities into SiO_2/water, Al_2O_3/water or Fe_2O_3/water nanofluids. According to the results obtained from Table 4, it is evident that the maximum value of diffusivity for the absorption of CO_2 is obtained when water/Fe_2O_3 is used as an absorbent, and the maximum diffusivity for the absorption of SO_2 is achieved when being used water/SiO_2 nanofluid. As can be seen in this table, for nanoparticles with the higher density ($\rho_{SiO2} = 2.196$ g/cm^3, $\rho_{Al2O3} = 3.980$ g/cm^3, $\rho_{Fe2O3} = 5.242$ g/cm^3) more diffusivity of CO_2 within the nanofluid is observed which is attributed to the nanoparticles Brownian motion inducing more diffusion of CO_2 molecules at the bubble-liquid interface. It has been previously reported by Attari et al. that the momentum caused by Brownian velocity of nanoparticles leading to produce micro-convections, depending on nanoparticles density according to the following relation [20]:

$$Mo_{Brownian} = \lambda \sqrt{\rho_p} \tag{12}$$

Table 4. Diffusion coefficient as well as *Re*, *Sh* and *Sc* for CO_2 and SO_2 absorption by using of nanofluids.

Gas	Absorbent	D, (m²/s)	ν (m/s)	Sc	Re_b	Sh.
CO_2	Water/SiO₂	5.38×10^{-9}	8.899×10^{-7}	165	1298	234
	Water/Fe₂O₃	7.76×10^{-9}	8.864×10^{-7}	114	1303	195
	Water/Al₂O₃	6.28×10^{-9}	8.451×10^{-7}	135	1367	217
	Deionized water	2.12×10^{-9}	8.900×10^{-7}	420	1298	316
SO_2	Water/SiO₂	8.89×10^{-9}	8.706×10^{-7}	98	1327	182
	Water/Fe₂O₃	6.85×10^{-9}	8.864×10^{-7}	129	1303	207
	Water/Al₂O₃	6.12×10^{-9}	8.852×10^{-7}	145	1305	219
	Deionized water	5.27×10^{-9}	8.900×10^{-7}	169	1298	179

According to this equation by having an increase in nanoparticles density, more momentum can be transferred through the liquid phase; and consequently, a higher magnitude of micro-convections produces. Previous efforts declared that only two significant mechanisms including Brownian micro-convections and grazing effect (absorption of gas molecules by nanoparticles at the bubble-liquid interface and desorption of them into the liquid) can be involved during the gas absorption when a nanofluid is used as an absorbent [2–5,10,11,36]. For the absorption of CO_2, Brownian mechanism has a major impact on gas molecules transfer due to the fact that CO_2 molecules have not a very polar structure and asymmetric molecular configuration to produce high molecular charges (O=C=O) for being absorbed by nanoparticles surface charge; therefore, the Brownian mechanism indicates that water/Fe₂O₃ leads to a higher diffusivity of CO_2 because of the larger micro-convections. Consequently, the minimum value of CO_2 diffusivity in water/SiO₂ nanofluid could be observed due to the lower density and lower magnitude of micro-convections produced by SiO₂ nanoparticles.

On the other hands, due to the high polarity of SO_2 molecules and formation of its Lewis structure during the absorption process [37] (Figure 11), it can be easily absorbed by means of nanoparticles surficial charge, which they are at the vicinity of the bubble-liquid interface. In addition, it is reported from the previous researches that SiO₂ nanoparticles have a high value of surface charge due to the formation of silanol bonds (Si-O-H) at the nanoparticles surface [12], which has been confirmed by Zeta Potential test presented in this study. Therefore, the main mechanism for the absorption of SO_2 is attributed to grazing effect by means of nanoparticles at the bubble-liquid interface resulting a high diffusivity of SO_2 gas when water/SiO₂ nanofluid is used (Figure 11).

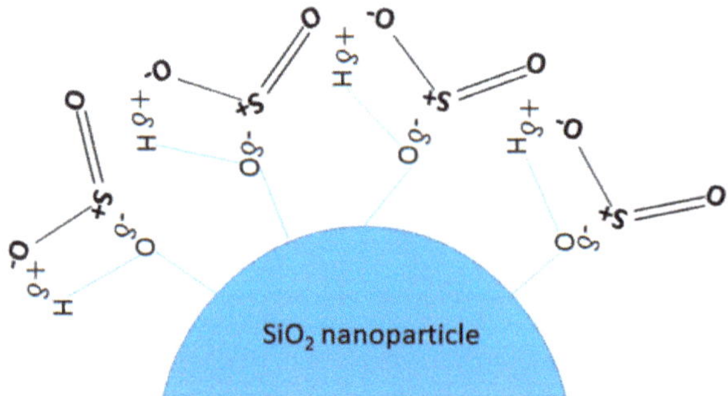

Figure 11. Schematic diagram of grazing effect of SiO₂ nanoparticles during the absorption of SO_2.

3.4. Correlation

Froessling [38] estimated the mas transfer of a raising bubble in a liquid by using Equation (13):

$$Sh = 0.6(Re)^{1/2}(Sc)^{1/3} \tag{13}$$

Equation (13) was found to be a suitable correlation for prediction of the absorption of different gases into wide ranges of liquids by means of single bubble absorber system [39]. In order to estimate *Sh* number for the gas absorption by nanofluids, other physical properties including dynamic viscosity, kinematic viscosity, and density of nanofluids were needed to obtain according to the following relations [40]:

$$\mu_{nf} = \mu_{bf}(1-\varphi)^{2.5} \tag{14}$$

$$\rho_{nf} = \varphi\rho_p + (1-\varphi)\,\rho_{bf} \tag{15}$$

$$\nu_{nf} = \mu_{nf}/\rho_{nf} \tag{16}$$

where φ is the volume fraction of oxides nanoparticles within the deionized water (can be obtained by using Equation (17) μ_{bf} is the dynamic viscosity of the deionized water, ρ_p is the bulk density of nanoparticles (presented in Table 1) and ρ_{bf} is the density of the deionized water (1000 kg/m^3).

$$\varphi(\%vol) = \frac{w(\%wt)}{w(\%wt) + \frac{\rho_p}{\rho_{bf}}(100 - w(\%wt))} \tag{17}$$

The values of *Re*, *Sc* and *Sh* can be calculated using the following equations:

$$Re_b = U_b d_b/\nu_{nf} \tag{18}$$

$$Sc_{nf} = \nu_{nf}/D_{nf} \tag{19}$$

$$Sh_{nf} = k_{L,nf}.d_b/D_{nf} \tag{20}$$

In these equations, U_b means the bubble rising velocity in the column that was approximately found to be 0.21 m/s for all the experiments. Additionally, d_b is the bubble diameter that was measured as 7 mm for all cases. Table 4 also presents the values of Re_b, *Sh* and *Sc* for the absorption of CO_2 and SO_2 by using the mentioned nanofluids.

According to Table 4 and Equation (18), the value of Reynolds number does not change significantly when either nanofluid or pure basefluid is applied during the absorption process by means of raising a single bubble absorber i.e., $\nu_{nf} \approx \nu_{bf}$. Therefore, it can be assumed that the Reynolds number has no significant effect on relative Sherwood number and this parameter is found to be just as a function of relative Schmidt number according to below:

$$\frac{Sh_{nf}}{Sh_{bf}} = K\left(\frac{Sc_{nf}}{Sc_{bf}}\right)^m \tag{21}$$

m and K were calculated by using a two-dimensional regression analysis over the experimental data shown in Figure 12. According to this figure, the following equation was obtained for the mentioned parameters with the $R^2 = 0.9919$. Equation (22) can predict the Sherwood number for various gas-nanofluid absorption systems at $Re_b \sim 1300$, accurately:

$$\frac{Sh_{nf}}{Sh_{bf}} = 1.3643\left(\frac{Sc_{nf}}{Sc_{bf}}\right)^{0.6125} \qquad for\ Re_b \cong 1300 \tag{22}$$

It is mentioned that Sh_{bf} can be calculated by the Froessling equation (Equation (13)).

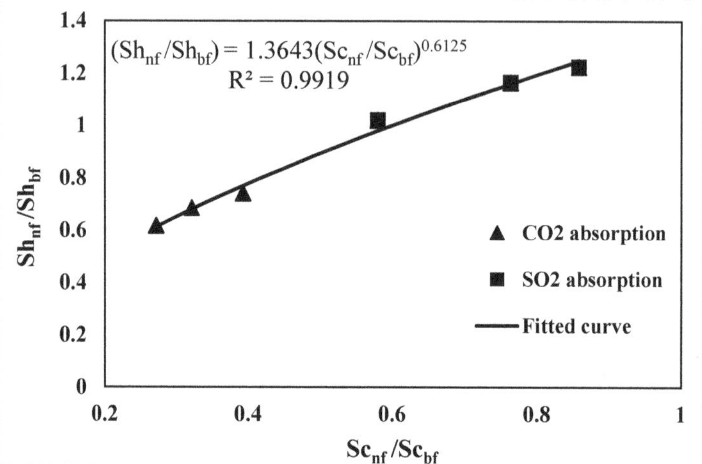

Figure 12. Effect of relative Schmidt number on relative experimental Sherwood number.

4. Conclusions

In this research, the absorption of SO_2 and CO_2 was elucidated by using a single-bubble column absorption setup into water based nanofluids containing SiO_2, Fe_2O_3 or Al_2O_3 nanoparticles. The results of this study clearly show that the aforementioned nanofluids have high stability since the zeta potential is lower than -45 mV. The results of TEM and DLS analysis also display that the average size of nanoparticles is within limit of 40–60 nm.

These results also declared that the maximum absorption of CO_2 and SO_2 could be obtained when water/SiO_2 or water/Fe_2O_3 nanofluid is utilized as an absorbent. Moreover, our findings also showed that the maximum relative absorption for SO_2 and CO_2 in the studied nanofluids in comparison to base fluid occurs when a water/Fe_2O_3 or water/SiO_2 nanofluid was used as the absorbent. Indeed, our results show that the type of gas molecules and nanoparticles determines the mechanism of mass transfer intensification of nanofluids. Therefore, both Brownian motion and grazing effect play crucial role for the increment of mass transfer in gas absorption by nanofluids. According to the type of gas and nanoparticles, the major mechanism can be distinguished.

In addition, mass transfer parameters incorporating diffusivity of gases into the oxides nanoparticles loaded in nanofluids, Sherwood number and Schmidt number were obtained. The results exhibit that the addition of nanoparticles (due to increment of Brownian momentum) increases diffusivity coefficient, and the maximum diffusivity for CO_2 and SO_2 absorption was obtained for water/Fe_2O_3 and water/SiO_2 nanofluids, respectively.

Finally, a new correlation is offered for the prediction of Sherwood number versus Schmidt number in gas-nanofluid systems (for Re_b about 1300) in which the experimental values are predicted with high accuracy.

Author Contributions: S.K.: Conceived and designed the analysis, Collected the data, Contributed data or analysis tools, Performed the analysis, Wrote the paper; F.E.: Conceived and designed the analysis, Contributed data or analysis tools, Performed the analysis, Wrote the paper; D.M.: Conceived and designed the analysis, Performed the analysis.

Funding: This research received no external funding.

Acknowledgments: The authors are grateful to the Shiraz University for supporting this research.

Conflicts of Interest: The authors declare no conflict of interest.

Nomenclature

N	Molar flux (mol/m^2 s)
C	Gas concentration at liquid bulk (mol/m^3)
C^*_{Obs}	The observed gas concentration at gas-liquid interface (mol/m^3)
V	Volume of nanofluid in the single bubble absorber (m^3)
n	Number of bubbles
τ	Average rising time for one bubble through the column (s)
r_0	Average radius of bubbles (m)
k_L	Mass transfer coefficient in liquid phase (m/s)
D	Diffusion coefficient (m^2/s)
δ	Diffusion layer thickness (mm)
s	Renewal surface factor (1/s)
Re_b	Reynolds number ($U_b d_b/\nu_{nf}$)
Sc	Schmidt number (ν_{nf}/D_{nf})
Sh	Sherwood number ($k_L d_b/D_{nf}$)
d_b	Diameter of bubbles raising through nanofluid (m)
φ	Volume fraction (%)
w	Mass fraction (%)
ρ	Density (kg/m^3)
v	Kinematic viscosity (m^2/s)
λ	Constant value for calculation of Brownian momentum transfer
R_{eff}	Relative absorption rate (N_{nf}/N_{bf})
M	HCl molarity (mol/lit)
λ	Constant value as a function of nanoparticles density, temperature, volume fraction, mean diameter, heat capacity, and Boltzmann constant.
Mo	Momentum that can be transferred by means of nanoparticle random motion
Subscript	
nf	Nanofluid
bf	Basefluid
p	Nanoparticles
B	Bubble

References

1. Nii, S.; Takeuchi, H. Removal of CO_2 and/or SO_2 from gas streams by a membrane absorption method. *Gas Sep. Purif.* **1994**, *8*, 107–114. [CrossRef]
2. Darvanjooghi, M.H.K.; Esfahany, M.N.; Esmaeili-Faraj, S.H. Investigation of the Effects of Nanoparticle Size on CO_2 Absorption by Silica-Water Nanofluid. *Sep. Purif. Technol.* **2017**. [CrossRef]
3. Darvanjooghi, M.H.K.; Pahlevaninezhad, M.; Abdollahi, A.; Davoodi, S.M. Investigation of the effect of magnetic field on mass transfer parameters of CO_2 absorption using Fe3O4-water nanofluid. *AIChE J.* **2017**, *63*, 2176–2186. [CrossRef]
4. Esmaeili-Faraj, S.H.; Nasr Esfahany, M.; Jafari-Asl, M.; Etesami, N. Hydrogen sulfide bubble absorption enhancement in water-based nanofluids. *Ind. Eng. Chem. Res.* **2014**, *53*, 16851–16858. [CrossRef]
5. Esmaeili-Faraj, S.H.; Nasr Esfahany, M. Absorption of hydrogen sulfide and carbon dioxide in water based nanofluids. *Ind. Eng. Chem. Res.* **2016**, *55*, 4682–4690. [CrossRef]
6. Kim, J.-K.; Jung, J.Y.; Kang, Y.T. Absorption performance enhancement by nano-particles and chemical surfactants in binary nanofluids. *Int. J. Refrig.* **2007**, *30*, 50–57. [CrossRef]
7. Baker, R.W. Future directions of membrane gas separation technology. *Ind. Eng. Chem. Res.* **2002**, *41*, 1393–1411. [CrossRef]
8. Richard, M.A.; Bénard, P.; Chahine, R. Gas adsorption process in activated carbon over a wide temperature range above the critical point. Part 1: Modified Dubinin-Astakhov model. *Adsorption* **2009**, *15*, 43–51. [CrossRef]
9. Whitman, W.G. The two film theory of gas absorption. *Int. J. Heat Mass Transf.* **1962**, *5*, 429–433. [CrossRef]

10. Ashrafmansouri, S.S.; Nasr Esfahany, M. Mass transfer into/from nanofluid drops in a spray liquid-liquid extraction column. *AIChE J.* **2016**, *62*, 852–860. [CrossRef]

11. Ashrafmansouri, S.S.; Esfahany, M.N. Mass transfer in nanofluids: A review. *Int. J. Therm. Sci.* **2014**, *82*, 84–99. [CrossRef]

12. Darvanjooghi, M.H.K.; Esfahany, M.N. Experimental investigation of the effect of nanoparticle size on thermal conductivity of in-situ prepared silica–ethanol nanofluid. *Int. Commun. Heat Mass Transf.* **2016**, *77*, 148–154. [CrossRef]

13. Chol, S. Enhancing thermal conductivity of fluids with nanoparticles. *ASME-Publ.-Fed* **1995**, *231*, 99–106.

14. Krishnamurthy, S.; Bhattacharya, P.; Phelan, P.; Prasher, R. Enhanced mass transport in nanofluids. *Nano Lett.* **2006**, *6*, 419–423. [CrossRef] [PubMed]

15. Kim, J.-K.; Jung, J.Y.; Kim, J.H.; Kim, M.-G.; Kashiwagi, T.; Kang, Y.T. The effect of chemical surfactants on the absorption performance during NH_3/H_2O bubble absorption process. *Int. J. Refrig.* **2006**, *29*, 170–177. [CrossRef]

16. Ma, X.; Su, F.; Chen, J.; Zhang, Y. Heat and mass transfer enhancement of the bubble absorption for a binary nanofluid. *J. Mech. Sci. Technol.* **2007**, *21*, 1813. [CrossRef]

17. Zhou, M.; Cai, W.F.; Xu, C.J. A new way of enhancing transport process–The hybrid process accompanied by ultrafine particles. *Korean J. Chem. Eng.* **2003**, *20*, 347–353. [CrossRef]

18. Kang, Y.T.; Lee, J.-K.; Kim, B.-C. Absorption heat transfer enhancement in binary nanofluids. In *International Congress of Refrigeration*; Curran Associates, Inc.: Beijing, China, 2007.

19. Yang, L.; Du, K.; Niu, X.F.; Cheng, B.; Jiang, Y.F. Experimental study on enhancement of ammonia–water falling film absorption by adding nano-particles. *Int. J. Refrig.* **2011**, *34*, 640–647. [CrossRef]

20. Attari, H.; Derakhshanfard, F.; Darvanjooghi, M.H.K. Effect of temperature and mass fraction on viscosity of crude oil-based nanofluids containing oxide nanoparticles. *Int. Commun. Heat Mass Transf.* **2017**, *82*, 103–113. [CrossRef]

21. Heris, S.Z.; Etemad, S.G.; Esfahany, M.N. Experimental investigation of oxide nanofluids laminar flow convective heat transfer. *Int. Commun. Heat Mass Transf.* **2006**, *33*, 529–535. [CrossRef]

22. Kang, Y.T.; Kim, H.J.; Lee, K.I. Heat and mass transfer enhancement of binary nanofluids for $H_2O/LiBr$ falling film absorption process. *Int. J. Refrig.* **2008**, *31*, 850–856. [CrossRef]

23. Wang, X.-Q.; Mujumdar, A.S. Heat transfer characteristics of nanofluids: A review. *Int. J. Therm. Sci.* **2007**, *46*, 1–19. [CrossRef]

24. Jung, J.-Y.; Lee, J.W.; Kang, Y.T. CO_2 absorption characteristics of nanoparticle suspensions in methanol. *J. Mech. Sci. Technol.* **2012**, *26*, 2285–2290. [CrossRef]

25. Andrade, Â.L.; Fabris, J.D.; Ardisson, J.D.; Valente, M.A.; Ferreira, J.M. Effect of tetramethylammonium hydroxide on nucleation, surface modification and growth of magnetic nanoparticles. *J. Nanomater.* **2012**, *2012*, 15. [CrossRef]

26. Andrade, Â.L.; Souza, D.M.; Pereira, M.C.; Fabris, J.D.; Domingues, R.Z. pH effect on the synthesis of magnetite nanoparticles by the chemical reduction-precipitation method. *Quim. Nova* **2010**, *33*, 524–527. [CrossRef]

27. Davoodi, S.M.; Sadeghi, M.; Naghsh, M.; Moheb, A. Olefin–paraffin separation performance of polyimide Matrimid®/silica nanocomposite membranes. *RSC Adv.* **2016**, *6*, 23746–23759. [CrossRef]

28. Terraglio, F.P.; Manganelli, R.M. The absorption of atmospheric sulfur dioxide by water solutions. *J. Air Pollut. Control Assoc.* **1967**, *17*, 403–406. [CrossRef]

29. Zhao, B.; Wang, J.; Yang, W.; Jin, Y. Gas–liquid mass transfer in slurry bubble systems: I. Mathematical modeling based on a single bubble mechanism. *Chem. Eng. J.* **2003**, *96*, 23–27. [CrossRef]

30. Skoog, D.A.; West, D.M. *Fundamentals of Analytical Chemistry*; Thomson Brooks/Cole: Pacific Grove, CA, USA, 2004.

31. Teng, T.-P.; Hung, Y.-H.; Teng, T.-C.; Mo, H.-E.; Hsu, H.-G. The effect of alumina/water nanofluid particle size on thermal conductivity. *Appl. Therm. Eng.* **2010**, *30*, 2213–2218. [CrossRef]

32. Vijayakumar, R.; Koltypin, Y.; Felner, I.; Gedanken, A. Sonochemical synthesis and characterization of pure nanometer-sized Fe_3O_4 particles. *Mater. Sci. Eng. A* **2000**, *286*, 101–105. [CrossRef]

33. áO'Brien, R.W. Electroacoustic studies of moderately concentrated colloidal suspensions. *Faraday Discuss. Chem. Soc.* **1990**, *90*, 301–312. [CrossRef]

34. Faraji, M.; Yamini, Y.; Rezaee, M. Magnetic nanoparticles: Synthesis, stabilization, functionalization, characterization, and applications. *J. Iran. Chem. Soc.* **2010**, *7*, 1–37. [CrossRef]
35. Kim, W.-G.; Kang, H.U.; Jung, K.-M.; Kim, S.H. Synthesis of silica nanofluid and application to CO_2 absorption. *Sep. Sci. Technol.* **2008**, *43*, 3036–3055. [CrossRef]
36. Koo, J.; Kleinstreuer, C. Impact analysis of nanoparticle motion mechanisms on the thermal conductivity of nanofluids. *Int. Commun. Heat Mass Transf.* **2005**, *32*, 1111–1118. [CrossRef]
37. Purser, G.H. Lewis structures are models for predicting molecular structure, not electronic structure. *J. Chem. Educ.* **1999**, *76*, 1013. [CrossRef]
38. Vasconcelos, J.M.; Orvalho, S.P.; Alves, S.S. Gas–liquid mass transfer to single bubbles: Effect of surface contamination. *AIChE J.* **2002**, *48*, 1145–1154. [CrossRef]
39. Calderbank, P.; Lochiel, A. Mass transfer coefficients, velocities and shapes of carbon dioxide bubbles in free rise through distilled water. *Chem. Eng. Sci.* **1964**, *19*, 485–503. [CrossRef]
40. Mishra, P.C.; Mukherjee, S.; Nayak, S.K.; Panda, A. A brief review on viscosity of nanofluids. *Int. Nano Lett.* **2014**, *4*, 109–120. [CrossRef]

Article

Optimization of Post Combustion CO$_2$ Capture from a Combined-Cycle Gas Turbine Power Plant via Taguchi Design of Experiment

Ben Alexanda Petrovic [1] and **Salman Masoudi Soltani** [2,*]

[1]   Department of Mechanical and Aerospace Engineering, Brunel University London, Uxbridge UB8 3PH, UK; benpetrovic@hotmail.com

[2]   Department of Chemical Engineering, Brunel University London, Uxbridge UB8 3PH, UK

*   Correspondence: Salman.MasoudiSoltani@brunel.ac.uk; Tel.: +44-(0)1895265884

Received: 20 May 2019; Accepted: 6 June 2019; Published: 12 June 2019

Abstract: The potential of carbon capture and storage to provide a low carbon fossil-fueled power generation sector that complements the continuously growing renewable sector is becoming ever more apparent. An optimization of a post combustion capture unit employing the solvent monoethanolamine (MEA) was carried out using a Taguchi design of experiment to mitigate the parasitic energy demands of the system. An equilibrium-based approach was employed in Aspen Plus to simulate 90% capture of the CO$_2$ emitted from a 600 MW natural gas combined-cycle gas turbine power plant. The effects of varying the inlet flue gas temperature, absorber column operating pressure, amount of exhaust gas recycle, and amine concentration were evaluated using signal to noise ratios and analysis of variance. The optimum levels that minimized the specific energy requirements were a: flue gas temperature = 50 °C; absorber pressure = 1 bar; exhaust gas recirculation = 20% and; amine concentration = 35 wt%, with a relative importance of: amine concentration > absorber column pressure > exhaust gas recirculation > flue gas temperature. This configuration gave a total capture unit energy requirement of 5.05 GJ/tonneCO$_2$, with an energy requirement in the reboiler of 3.94 GJ/tonneCO$_2$. All the studied factors except the flue gas temperature, demonstrated a statistically significant association to the response.

Keywords: CO$_2$ capture; Aspen Plus; CCGT; Taguchi; Minitab; optimization

---

1. Introduction

Anthropogenic greenhouse gas (GHG) emissions in 2010 reached 49 ± 4.5 GtCO$_2$-eq/year, emissions of CO$_2$ from fossil fuel combustion and industrial processes contributed approximately 80% of the total GHG emissions increase from 1970–2010 [1]. The mitigation of climate change and increasing global temperatures requires a combination of new, renewable technology and an improvement of the existing infrastructure to move towards a low and ideally zero-carbon society; in line with the Climate Change Act requirements of an 80% reduction in total emissions by 2050 [2]. The use of fossil-fueled power stations continues to grow due to their ability to respond to changes in demand [3] and offset the intermittency of current renewable technology. Coal and gas are the predominant fuels used in power generation; however, since the UK's 2016 consultation to end the use of unabated coal, its usage in power generation has declined from 22% in 2015 to 1.6% in the second quarter of 2018 [4]. Natural gas sees its share of generation at 42% and although often perceived as a much cleaner fuel at the point of use than coal [5], producing around 350 kgCO$_2$/MWh [6], reducing the carbon intensity of this growing sector is vital for stabilizing global temperature increase to below 2 °C. Amine-based carbon capture and storage (CCS) is seen as one of the best CO$_2$ abatement approaches [7]; the solvent monoethanolamine (MEA) is most commonly used due to its low material costs [8]; however, due to the energy requirements of solvent regeneration, there is a large energy penalty incurred on the power plant. To make post combustion CCS a viable option in mitigating the GHG emissions from power generation, optimization of such a plant is paramount.

The techno-economic analysis of a post combustion capture (PCC) and compression plant using MEA coupled to a 400 MW NG-CCGT conducted by Alhaja et al. [9] found that by studying the effect of the PCC unit's key operating parameters on the power plant's key performance indicators, an optimum lean loading of 0.31 molCO$_2$/molMEA, which minimized the specific reboiler duty (SRD), could be found. This represents a balance between the sensible heat required to raise the temperature of the solvent to that of the reboiler and the latent heat to vaporize water and provide the stripping steam. By increasing the pressure within the stripping column within the limits of solvent degradation, a reduction in SRD was also seen. The inclusion of packing volume as a studied parameter illustrates the importance of studying the capture process as a whole system, especially due to the capital costs associated with such plants. An optimum lean loading of the solvent that minimized the SRD was also found by Masoudi Soltani et al. [10]; the MEA-based unit demonstrated that the SRD was dependent on the concentration of MEA within the solution. The 3.98 GJ/tonneCO$_2$ SRD occurred with a 30 wt% MEA solution and a lean loading in the range of 0.19–0.21 molCO$_2$/molMEA. The SRD was also seen to vary secondly as a function of EGR, owing to the change in CO$_2$ partial pressure within the flue gas stream; employing a greater percentage EGR reduced SRD further. Another optimization of an MEA-based PCC system [11] determined strong links between the L/G ratio, lean loading, and reboiler duty. The lean solvent flow rate was determined by varying the lean CO$_2$ loading to achieve 90% capture; with a lower L/G ratio, the requirement in the reboiler is primarily for stripping steam, whereas with a higher L/G ratio, there is a larger requirement for heat to increase the temperature of the rich stream; once more corroborating the balance of sensible and latent heats in the reboiler [9,10]. A 30 wt% MEA-employing model validated against the UK CCS research centre pilot plant was used to evaluate and optimize the performance of a PCC unit [12]; a lean loading of 0.23 gave a 15% reduction in SRD from 7.1 Mj/kgCO$_2$ to 5.13 MJ/kgCO$_2$. An increase in stripper pressure from 1.25–2.50 bar added a further 17% reduction in SRD, but in order to avoid thermal degradation of the solvent, a pressure of 1.80 bar was found to be most suitable, a similar phenomenon to that found by Lindqvist et al. [13]. With an optimum lean loading of 0.21 in line with the findings of Masoudi Soltani et al. [10], the SRD was 4.4 MJ/kgCO$_2$. Packing material and heat exchanger logarithmic mean temperature difference (LMTD) were also studied, optimization of which could give SRD reductions of 40% and 5%, respectively. Xiaobo Luo [14] investigated the optimal operation of an MEA-based capture unit and found that for a 90% capture rate, a 9.58% net power efficiency decrease was seen in the NGCC using an optimum lean loading of 0.26–0.28, slightly higher than in the previous studies. The reason for this is that in the other studies, column sizing is minimized to reduce capital expenditure by reducing the L/G ratio, thus requiring a lower lean loading, whereas with this study [14], column sizing is fixed and the optimal operation is to reduce the operating expenditure, hence a higher lean loading can be exploited. The comparison between MEA and CESAR-1 (an aqueous solution of 2-amino-2-methyl-propanal and piperazine) [15] found that using MEA reduced the NGCC plant efficiency by 8.4%, with an energy requirement in the PCC unit of 3.36 GJ/tonneCO$_2$. A parametric evaluation carried out by Kothandaraman et al. [16] on a 30 wt% MEA PCC system identified that for a DOC above 95%, there was a disproportionate increase in SRD. The temperature of the solvent was shown to have little effect on the system's performance; decreasing absorption temperature increases the driving force for reaction but the rate of reaction and diffusivity decreased, effectively cancelling each other out. For the lean loading used (0.22), the SRD was 4.5 GJ/tCO$_2$ with a flue gas CO$_2$ content of 4 vol%, i.e., without the use of EGR. An MEA-based PCC unit modelled by Arachchige et al. [17] concluded that the removal efficiency was proportional to the solvent concentration and temperature whilst being inversely effected by lean loading thanks to the reduction in MEA capacity; a similar phenomenon to that found by Kothandaraman et al. [16] was observed with respect to solvent temperature. Variation of absorber pressure saw a decrease in SRD due to the increased partial pressure of the CO$_2$; 4.56 GJ/tCO$_2$ to 4.38 GJ/tCO$_2$ with an increase in absorber pressure of 0.9 to 1.2 bar. The effect of implementing EGR on the integrated MEA-based CO_2 capture plant when coupled to an 800 MW NGCC was studied exclusively by Ali et al. [18], something not considered comprehensively in other studies [12,13,15]. The use of EGR resulted in a 57% increase in CO_2 molar composition in the flue gas stream (4.16–6.53 mol%), resulting in a 2.3% reduction in SRD; they also identified that the NGCC case with EGR is the most attractive for use with CCS due to its lowest reduction in plant net efficiency. Lars Erik Øi [19] simulated the operation of a simplified MEA-based PCC coupled to a 400 MW CCGT; he found that for a removal of 85% of the emitted CO_2, the heat consumption was 3.7 GJ/tCO_2. He identified that an increased solvent circulation rate would increase the removal grade of the CO_2; an increase in the temperatures of the inlet streams to the absorber would improve CO_2 absorption due to an increase in reaction rates and; operating the stripper close to the degradation limits of the solvent would give better removal efficiency and thus lower CO_2 loading in the lean stream. Afkhamipour and Mofarahi [20] employed the methods of Taguchi to maximize the CO_2 removal efficiency using a sophisticated multilayer-perceptual-neural-network model. Focusing on the controllable inlet conditions to the absorber, they identified that CO_2 loading, amine flow rate, and amine concentration were the major factors in increasing the capture efficiency. The degree of capture (DOC), however, was not kept constant; this is dependent on amine flow rate. Instead, the response value used in the Taguchi analysis was the CO_2 removal efficiency; the optimization of removal efficiency does not inherently mean a less intensive energy requirement for the system.

It is clear that there are myriad KOPs that impact the energy intensity of a PCC unit; the importance of solvent concentration, solvent type, and CO_2 concentration were shown to be extremely influential on the energy requirements in a PCC unit; the sizing and packing of the columns were also important, but played a greater role in the economic analysis. Few studies have looked at the effect of varying absorber column pressure and its effect on the reaction kinetics within the column. Although solvent type can be seen as an optimization variable, to assess the influence of the other KOPs, this would need to be kept constant; the industry baseline solvent is MEA which has a number of advantages over other commercial solvents [21], its availability and relatively low cost also continue to make it one of the more viable options. The L/G ratio was shown to be dependent on the DOC; therefore, maintaining a DOC would require a variation of L/G ratio. It is also clear that the stripping column operates optimally near the degradation limits of the solvent, leaving little requirement for optimization; the inlet streams and conditions within the absorber were shown to be more influential in the energy demands of a PCC unit. The importance of operating the systems at near optimal configuration to mitigate the cost of capturing CO_2 and improve the viability of CCS is obvious.

A description of the chemical absorption that takes place in a PCC system can be achieved by modelling, using either an equilibrium or rate-based mass-transfer, with a number of studies being validated against pilot plant data [9,12,22–24]. Rate-based simulations can provide a greater accuracy and allow a more informed evaluation of the process [12,14,17,23–26] but equilibrium approaches can still be employed for process assessment [11,19,27,28]. The relative simplicity of an equilibrium approach and the ability to improve the accuracy of such a model with stage efficiencies [16,19,29–31] is the justification for employing such a strategy here.

In this work, the optimization of an MEA-based PCC system is carried out using a Taguchi design of experiment. The PCC system is modelled in Aspen Plus to process the flue gas from a 600 MW CCGT power plant and capture 90% of the emitted CO_2. The optimization parameters were: the inlet temperature of the flue gas to the absorber (FGT); the operating pressure of the absorber column (ACP); the amount of exhaust gas recirculation (EGR) so as to model the capture process using different molar CO_2 concentrations in the flue gas; and the concentration of the amine (CONC) in the lean stream inlet to the absorber.

2. Model Development

The modelling framework for this study employs a combination of the electrolyte-nonrandom-two-liquid (e-NRTL) [32] description of activity coefficients for the equilibrium ionic species in solution with the Soave–Redlich–Kwong (SRK) [33] cubic equation of state for the fugacities of the species in the vapor phase, a well-demonstrated combination [10,34,35]. The following sections describe briefly the thermodynamic framework that Aspen Plus uses for the calculations of chemical equilibrium, vapor-liquid equilibrium, liquid phase constitution, and regeneration energy [36].

The absorption and reaction mechanisms that occur in the MEA-CO_2-H_2O system are detailed in Equations (1)–(7) [10,21].

$$CO_{2(g)} \leftrightarrow CO_{2(aq)} \tag{1}$$

$$2H_2O \overset{K_2}{\leftrightarrow} H_3O^+ + OH^- \tag{2}$$

$$CO_2 + 2H_20 \overset{K_3}{\leftrightarrow} H_3O^+ + HCO_3^- \tag{3}$$

$$HCO_3^- + H_2O \overset{K_4}{\leftrightarrow} H_3O^+ + CO_3^{2-} \tag{4}$$

$$MEAH^+ + H_2O \overset{K_5}{\leftrightarrow} H_3O^+ + MEA \tag{5}$$

$$MEACOO^- + H_20 \overset{K_6}{\leftrightarrow} MEA + HCO_3^- \tag{6}$$

$$XO_2 + YMEA \rightarrow oxidationproducts \tag{7}$$

The absorption process begins with the dissolution of the gaseous carbon dioxide molecules into the liquid MEA-H_2O-CO_2 where the dissolved CO_2 undergoes a series of reactions described in Equations (2)–(6), resulting in the formation of a number of ionic species [10]. Reaction (2) describes the water hydrolysis resulting in the production of two ions, reaction (3) shows the formation of a bicarbonate in water, and reaction (4) demonstrates the dissociation of the bicarbonate salt into carbonate ions in the presence of liquid water. Reactions (5) and (6) describe the reactions between molecular MEA with CO_2 in the aqueous solution, specifically the dissociation of MEAH + (pronated MEA) in (5) and the carbamate reversion to bicarbonate of MEACOO-. The equilibrium constants ($K(T)$) for reactions (2)–(6) are defined on a molar basis as [10,35]:

$$K(T) = \prod \alpha_i^{v_i} \tag{8}$$

where α_i is the activity of species i; the above equation can be rewritten in terms of mole fractions, x_i and activity coefficients γ_i, to give:

$$K(T) = \prod (x_i \cdot \gamma_i)^{v_i} \tag{9}$$

where v_i is the stoichiometric constant of species i. It is worth noting that in the presence of oxygen, solvents such as MEA will react irreversibly to produce a number of oxidation products, and prediction of the accumulation of such products is limited by the incomplete knowledge of the interactions between oxygen and MEA [37]. The implication of these reactions on the energy requirements of the capture unit are insignificant but they do tend to influence the economics of the system. An economic analysis is not executed here and so the inclusion of these reactions in the model has been dismissed. The e-NRTL model is itself an excess Gibbs energy (g^{ex*}) expression comprised of three contributions [21]: (1) The long-range interactions due to the electrostatic forces between ions represented by the Pitzer–Debye–Hückel expression; (2) the ion-reference-state-transfer contribution represented by the Born expression; and (3) the short range forces between all species. The equation is given below [10,21,35]:

$$g^{ex*} = g^{ex*,PDH} + g^{ex*,Born} + g^{ex*,NRTL} \tag{10}$$

The activity coefficient, γ_i for an ionic or molecular species, solute, or solvent is derived from the partial derivative of the excess Gibbs free energy with respect to the species mole number, n_i [10,21]:

$$\ln \gamma_i = \frac{1}{RT} \left[\frac{\partial \left(n_t g^{ex^*} \right)}{\partial n_i} \right]_{T,P,n_{j \neq i}} \tag{11}$$

where i, j = molecule, cation, anionspecies. Finally, Equation (11) leads to:

$$\ln \gamma_i^* = \ln \gamma_i^{*PDH} + \ln \gamma_i^{*BORN} + \ln \gamma_i^{*NRTL} \tag{12}$$

The e-NRTL property method includes the temperature-dependent reaction equilibrium constants (K_i) listed in Equations (2)–(6) and are calculated using the Aspen Plus built-in equation [10,17,21,35]:

$$\ln(K_i) = a_i + \frac{b_i}{T} + c_i \ln(T) + d_i T \tag{13}$$

The vapor–liquid thermodynamic system can be described using an extended Henry's law to represent the behavior of solutes such as CO_2 [10,21,35,38]:

$$y_i \cdot \varphi_i \cdot P = x_i \cdot \gamma_i \cdot H_i^{p0} \cdot \exp\left[\frac{V_i^\infty \left(P - P_s^0 \right)}{RT} \right] \tag{14}$$

Similarly for the solvent species, an extended Raoult's law is employed [10,21,35,38]:

$$y_s \cdot \varphi_s \cdot P = x_s \cdot \gamma_s \cdot P_s^0 \cdot \varphi_s^0 \cdot \exp\left[\frac{V_s \left(P - P_s^0 \right)}{RT} \right] \tag{15}$$

where, y_i and y_s are the vapor phase mole fractions of species i and s; φ_i and φ_s are the fugacity coefficients for species i and s as estimated by the SRK equation of state; P is the total pressure; x_i and x_s are the liquid phase mole fractions of i and s; y_i and y_s are the liquid phase activity coefficients for i an s; H_i^{p0} is the Henry's Law constant of i in the solution at saturation pressure and P_s^0 is the saturation pressure of s; φ_s^0 is the fugacity coefficient of s under saturation pressure condition; V_i^∞ is the partial molar volume of solute at infinite dilution; and V_s is the molar volume of solvent s. In Equations (14) and (15), the exponential terms are the Poynting factors of corrections for moderate pressure and are derived from integration forms by assuming V_i^∞ and V_s to be constant over the pressure range [10]. Figure 1 depicts the process flow sheet developed for this work and is detailed in the following section; block names are given in capitals and the Aspen Plus model names are given in parentheses. The model includes two columns: the ABSORBER and the STRIPPER (RadFrac); a water wash section, WATERWAS (SEP) where any residual solvent is removed from the clean flue gas; a make-up section, MAKE-UP (MIXER) which allows the addition of both H_2O and MEA to ensure the lean amine stream inlet to the absorber is of the correct composition. The five-stage intercooled compression train [9,15,24] used to process the captured CO_2 for storage is comprised of a series of four compressors, COM1-4 (COMPR), knock out drums, KO1-4 (FLASH) to remove residual water and intercoolers, IN-COOL1-4 (HEATER) to reduce the temperature of the CO_2; this puts the CO_2 into the supercritical fluid state where the 5th stage, a pump, CO2PUMP (PUMP) increases the pressure of the CO_2 to 140 bar. The cross-heat exchanger, HEATX (HEATX) is used to heat the rich amine stream using the waste heat in the lean stream and the BLOWER (COMPR) is used to increase the flue gases pressure to overcome the pressure drop in the column. Table 1 Outlines the operating conditions for the individual units in the flow sheet.

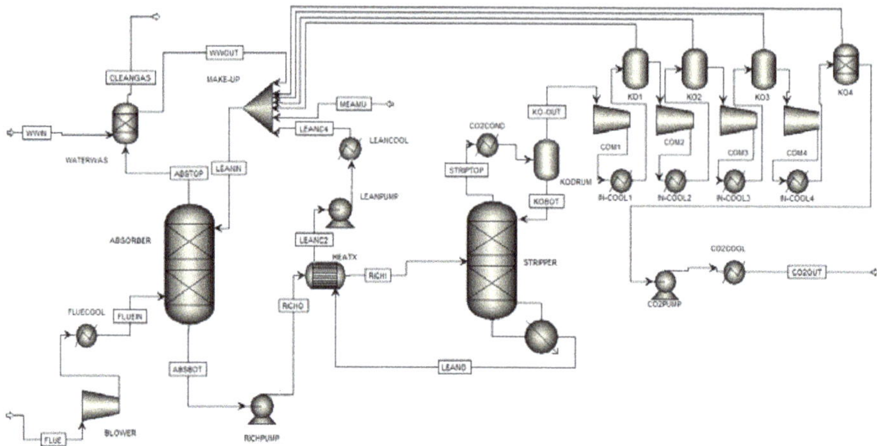

Figure 1. Post combustion CO_2 capture process flow sheet.

Table 1. The operating conditions of the main units in the simulated flowsheet.

Design Specification	Specified Value
CO_2 capture (%)	90
CO_2 loading in the inlet lean solvent (molCO_2/molMEA)	0.2
Inlet lean solvent temperature (°C)	40
Absorber no. of ideal stages	15
Absorber Murphree efficiency [19]	0.25
Absorber column pressure drop (bara)	0.1
Stripper's operating pressure (bara)	2 (isobaric)
Stripper's no. of ideal stages	10
Condenser outlet temperature (°C)	31
Knock-out drum pressure (bara)	2
Knock-out drum temperature (°C)	30
Blower ΔP (bara)	0.1
Assumption for pressure drop in pipes and equipment (bara)	0
CO_2 compression train no. of stages	5
CO_2 compression train intercooler temperature (°C)	31
CO_2 compression train outlet pressure (bara)	140

The flue gas stream flow rate inlet to the absorber is maintained at 825.31 kg/s, representing a NG-fired CCGT operating in a 1 × 1 configuration with a rated capacity of 592 MW and an efficiency of 56% on an LHV basis, modelled by Dutta et al. [27] using GT-PRO® operating at an ambient condition of 15 °C.

The Taguchi method is a statistical technique in the design of experiments (DOE), sensitivity analysis, and optimization [20]. Developed to overcome the limitations in a full factorial experiment design the method necessitates that all parameters be split into either control or noise factors which can take on a variety of preset levels; through the use of orthogonal arrays (OA) [39] a fractional-factorial sequence of experiments can be devised, upon which statistical analysis of the results can be carried out to find the optimum configuration of factors and their respective levels in order to maximize or minimize the objective function. An OA is abbreviated as L_N; the subscript N refers to the number of required trials for a given experiment; the number of levels of factors are included in parentheses next to the abbreviation. For example, L_4 (2^3) refers to a four-trial experiment to investigate the influence of three factors at two levels each on a given process [40]. The Taguchi methods serve as an offline tool for designing quality into products in a three-stage process [40,41]. The first being system design, where the factors and their appropriate levels are determined, which requires a thorough understanding of the system, the second is parametric design, where the optimum condition is determined at specific factor levels, and the third is tolerance design, where fine tuning of the optimum factor levels found in the second stage takes place. The method employed here is similar to that employed by Yusoff et al. [40] and is shown in Figure 2.

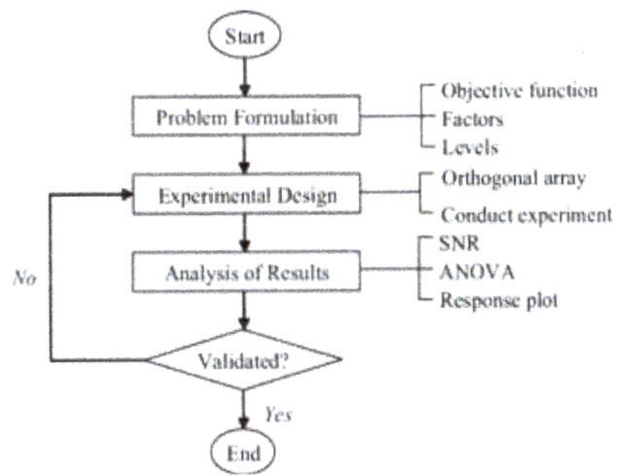

Figure 2. Flow diagram of the Taguchi method.

The problem formulation step involves defining the objective function, factors, and levels. The four factors assessed in this work are: the temperature of the flue gas stream inlet to the absorber column (FGT); the operating pressure of the absorber column (ACP); the percentage exhaust gas recirculation (EGR); and the concentration of the lean amine stream inlet to the absorber (CONC). The values of which take one of five levels detailed in Table 2.

Table 2. Description of factors and levels for the post combustion capture (PCC) plant.

Level	FGT (°C)	ACP (bar)	EGR (%)	CONC (wt%)
1	40	1	0	15
2	50	1.5	10	20
3	60	2	20	25
4	70	2.5	30	30
5	80	3	40	35

The composition of the flue gas stream is modelled so as to represent the various EGR ratios [42] and is illustrated in Figure 3.

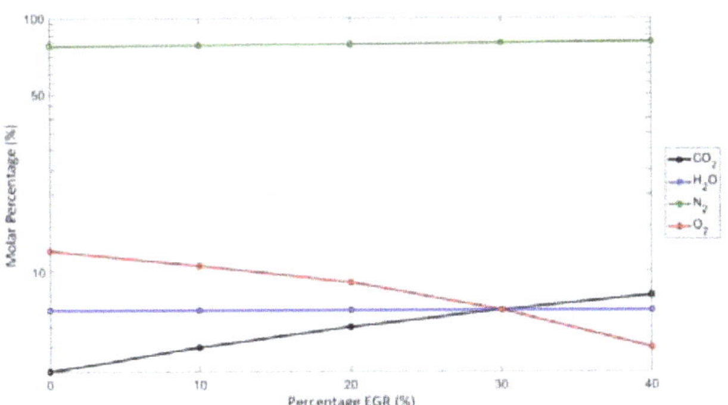

Figure 3. Molar composition of the flue gas stream at various exhaust gas recirculation (EGR) ratios.

The objective function for optimization is the minimization of the parasitic energy demand of the capture unit, given as a function of captured CO_2, A; calculated using Equation (16).

$$A\left(\frac{GJ}{t_{CO_2}}\right) = \frac{R(GW) + C(GW) + P(GW)}{F(ts^{-1})} \tag{16}$$

where R is the reboiler duty, C is the condenser duty, P is the total pumping duty for the unit including the compression train, and F is the mass flow rate of captured CO_2 from the stripper column.

The second step involves designing and conducting the experiment employing Minitab, a complete statistical software package that provides a comprehensive set of methods for data analysis. For four factors at five levels, an appropriate L_{25} OA was selected and is given Table 3.

Table 3. Orthogonal array illustrating the configuration of the simulations.

Run	FGT (°C)	ACP (bar)	EGR (%)	CONC (wt%)
1	1	1	1	1
2	1	2	2	2
3	1	3	3	3
4	1	4	4	4
5	1	5	5	5
6	2	1	2	3
7	2	2	3	4
8	2	3	4	5
9	2	4	5	1
10	2	5	1	2
11	3	1	3	5
12	3	2	4	1
13	3	3	5	2
14	3	4	1	3
15	3	5	2	4
16	4	1	4	2
17	4	2	5	3
18	4	3	1	4
19	4	4	2	5
20	4	5	3	1
21	5	1	5	4
22	5	2	1	5
23	5	3	2	1
24	5	4	3	2
25	5	5	4	3

The third step looks at the analysis of the results; two statistical tools are used in this work both of which are commonly applied in the Taguchi method: signal-to-noise ratios (SNR) and analysis of variance (ANOVA). The SNR can take three forms that are characteristic of the objective function i.e., smaller the better, nominal is best or larger the better; in this case the smaller the better SNR is used and is defined for each run, n as [20]:

$$SNR = -10\log_{10}\left[\frac{1}{n}\sum_{i=1}^{N}\frac{1}{Y_i^2}\right] \tag{17}$$

where N is the number of runs and Y_i is the response value in the i_{th} experiment. The SNR is a single response which makes a trade-off between setting the mean to a desirable level while minimizing variance; the intention is to maximize the SNR regardless of its characteristic. The SNR values can be used to determine the relative importance of each factor on the objective function and can be plotted to identify the optimum levels for each factor. The mathematical technique of variance analysis (ANOVA) dissects the total variation into accounted sources and delivers a way to interpret the results from the simulations [43]. The ANOVA is conducted using the SNR values to assess the percentage contribution of each factor in minimizing the variation of the capture unit's energy demand. Both analyses are conducted using Minitab.

3. Results

The duties of each unit taken as outputs from Aspen Plus were manipulated to give specific energy requirements within the capture unit as a function of captured CO_2 and are presented in Table 4.

Table 4. Energy duty per tonne of captured CO_2 for each individual unit of the capture plant.

Run	Energy Duties (GJ/tCO$_2$)			
	Reboiler	Condenser	Pumping	Total
1	5.06	0.92	0.28	6.26
2	4.53	1.07	0.28	5.88
3	4.34	1.24	0.28	5.85
4	4.21	1.33	0.28	5.82
5	4.16	1.41	0.28	5.85
6	4.32	1.07	0.28	5.67
7	4.19	1.21	0.28	5.67
8	4.08	1.25	0.28	5.60
9	4.79	0.96	0.28	6.02
10	4.72	1.47	0.24	6.44
11	3.97	0.90	0.28	5.14
12	5.00	0.98	0.28	6.26
13	4.55	1.16	0.28	5.99
14	4.58	1.56	0.28	6.42
15	4.37	1.56	0.28	6.21
16	4.60	1.01	0.28	5.89
17	4.41	1.26	0.28	5.95
18	4.42	1.52	0.27	6.22
19	4.28	1.51	0.28	6.07
20	4.81	1.11	0.28	6.21
21	4.18	1.13	0.28	5.60
22	4.24	1.27	0.28	5.79
23	5.02	1.18	0.28	6.49
24	4.66	1.39	0.28	6.32
25	4.53	1.58	0.28	6.39

The primary contributor to the energy requirements is the reboiler duty which demonstrated the greatest variation throughout the simulations, with a standard deviation of 0.301 when compared to 0.216 and 0.008 for the condenser and pumping duties, respectively. From the 25 simulations, the minimum total energy requirement for the unit as a whole was 5.14 GJ/tCO$_2$ with a reboiler requirement of 3.97 GJ/tCO$_2$. These requirements were found in run 11 using the following factor configuration: FGT = 60°C; ACP = 1 bar; EGR = 30% and; CONC = 35 wt%.

3.1. Signal-to-Noise Ratio (SNR) Analysis

The SNR values can be used to identify the factor levels that minimize the variability in the capture unit's energy requirements. Minitab was used to calculate the SNR for each configuration of the factors, the average of which is shown in the response Table 5. With the smaller-the-better SNR the target value is 0; as such, the values given in red/bold are the levels that minimized the energy requirement. The optimum values being: FGT = 50 °C; ACP = 1 bar; EGR = 20% and; CONC = 35 wt%.

Table 5. Response table of the signal-to-noise ratio (SNR) taken from Minitab.

Level	FGT	ACP	EGR	CONC
1	−15.46	−15.12	−15.87	−15.91
2	−15.38	−15.43	−15.65	−15.71
3	−15.54	−15.59	−15.30	−15.63
4	−15.66	−15.75	−15.54	−15.41
5	−15.71	−15.87	−15.39	−15.09
Delta	0.34	0.75	0.57	0.82
Rank	4	2	3	1

The delta values represent the variation in the mean SNR values and permit a ranking of the factor's relative importance on the energy requirements when varied in the specified range. The sequence follows CONC > ACP > EGR > FGT. The importance of each factor can be seen graphically by plotting the SNR values for each factor as shown in Figure 4. The line connecting each SNR value exemplifies whether a main effect exists for that factor; a line that demonstrates a larger difference in vertical position such as in CONC shows that the magnitude of the main effect for that factor is greater. From Figure 4 it is clear that ACP and CONC had the greatest influence with FGT and EGR showing a smaller but still notable influence.

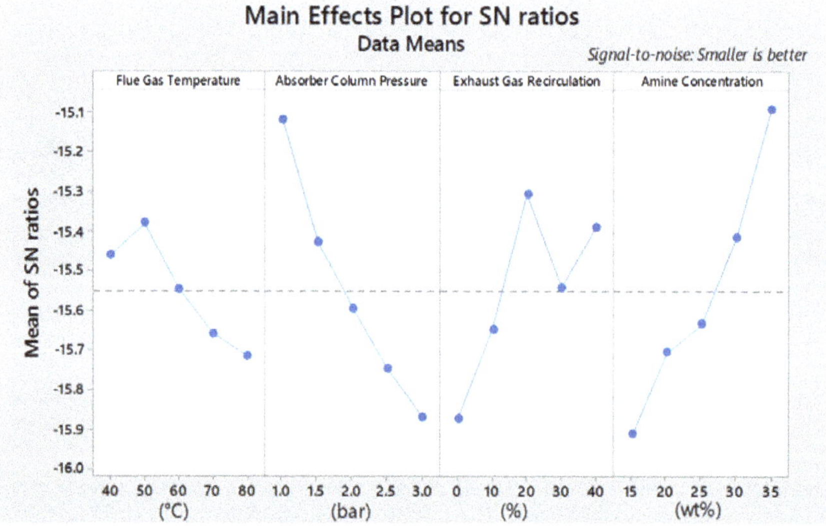

Figure 4. The main effects plot for the SNR.

3.2. Confirmation Experiment

After determining the optimum configuration of the factors, a confirmation experiment is required to validate the employed method for experimental design. The optimum factor levels were:

- Flue Gas Temperature = 50 °C
- Absorber Column = 1 bar
- Exhaust Gas Recirculation = 20%
- Amine Concentration = 35 wt%

The energy requirements for the individual units within the capture unit are given in Table 6. A total requirement of 5.05 GJ/tCO$_2$ was seen, 1.8% less than the value seen from simulation 11 and thus validating the results of the Taguchi analysis.

Table 6. Energy requirements within the capture unit during the confirmation experiment.

Energy Duties (GJ/tCO$_2$)				Percentage Contribution (%)		
Reboiler	Condenser	Pumping	Total	Reboiler	Condenser	Pumping
3.93722	0.83507	0.27812	5.05042	77.959	16.535	5.507

4. Discussion

4.1. Analysis of Variance (ANOVA)

Using a significance level of 5%, Table 7 exhibits the results of ANOVA using the SNR values from Table 5 The p-values can be used to evaluate the statistical significance of the factors influence on the capture unit's energy requirement; a p-value of less than 0.05 demonstrates a significant association of the factor to the mean of the quality characteristic value [44]. ANOVA also permits the identification of the percentage contribution of each factor. As in Figure 4 the greatest contributors to the variation in the capture unit's energy requirement are ACP and CONC, 31.07% and 35.24%, respectively. FGT and EGR demonstrated 6.92% and 18.14% contributions, respectively; the p-value for FGT, however, is 0.263, thus indicating that the flue gas temperature inlet to the absorber demonstrated no statistically significant influence on the energy requirements of the capture system. The other three factors did however show a statistically significant relationship, highlighting their importance in minimizing the parasitic energy demand of a capture unit.

Table 7. ANOVA table for the mean SNR.

Factor	Degree of Freedom (DF)	Sum of Squares (SS)	Percentage Contribution (%)	Mean of Squares (MS)	F-Value	p-Value
FGT	4	0.3845	6.92	0.09612	1.61	0.263
ACP	4	1.7254	31.07	0.43135	7.21	0.009
EGR	4	1.0076	18.15	0.25189	4.21	0.04
CONC	4	1.9569	35.24	0.48923	8.17	0.006
Error	8	0.4789	8.62	0.05986		
Total	24	5.5532	100			

4.2. Main Effect of the Factors on the Mean Energy Requirement in the Capture Unit

The influence of each factor on the energy requirement can be seen in Figure 5. An increase in ACP was seen to increase the energy requirements, potentially due to the increased energy demand to raise the pressure of the inlet streams to that of the absorber. Contradictory to this, however, the solubility of CO_2 increases in aqueous solutions at higher pressures [45] which should act to reduce the reboiler duty as found in [17] owing to an increase in CO_2 partial pressure. The equilibrium-based strategy used here may not have been able to quantify the effects of a greatly increased ACP. It is worth noting that in the aforementioned study, the variation in pressure was between 0.9 and 1.2 bar. When increasing EGR, the general trend is for SRD to decrease, owing to the increase in molar CO_2 concentration in the gas phase in the absorber column; the higher the mole fraction the feed has, the easier the MEA reaches the required rich loading for 90% capture [46]. In the literature, the optimal EGR is often around 40%; the discrepancy in this project is likely due to the fact that the mass flow rate of the flue gas was kept constant. If employing EGR in a NG-CCGT, the recycled flue gas would act to reduce the total mass flow processed by the PCC unit and as such, a smaller solvent flow rate and energy requirement would be seen [47].

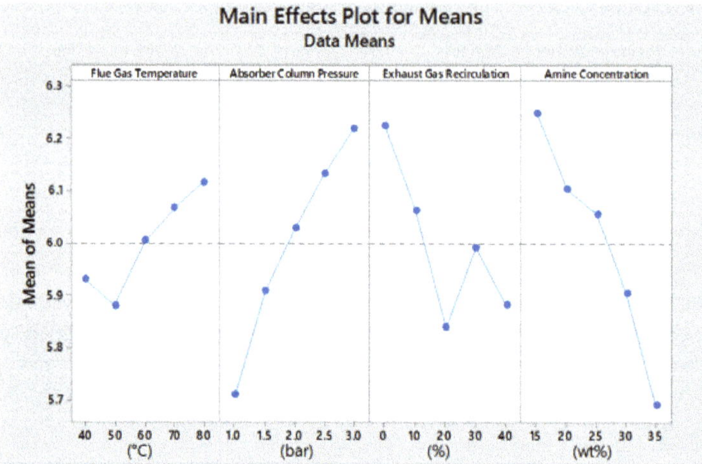

Figure 5. Main effects plot for the mean total energy requirement.

4.3. Interaction Analysis

Figure 4 illustrates the effect of each factor on the response; but, given that four variables were evaluated concurrently, an assessment of the interactions between the variables must be made so that the interpretation of the main effects is accurate. Figure 6 is the interaction plot for the four factors on the mean energy requirement in the capture unit; the optimum process parameters are illustrated by red circles. When the connecting lines are parallel, the interaction is small; the more nonparallel the lines are, the greater the level of interaction. All factors can be seen to show some level of influence on one another. It is worth noting that Figure 6 illustrates a minimum energy requirement at a FGT of 60 °C and not the 50 °C defined by the SNR analysis; but, due to the statistical insignificance of FGT, this can be disregarded and the initial interpretation can be upheld. When considering the other three factors (ACP, EGR, and CONC), the interaction plot corroborates the findings from the SNR analysis in that the minimum energy requirement was found with an ACP = 1 bar, EGR = 20%, and CONC = 35 wt%. The plot also makes evident a significant interaction between all four factors on the total energy requirement, highlighting the importance of conducting an optimization with all factors in concert.

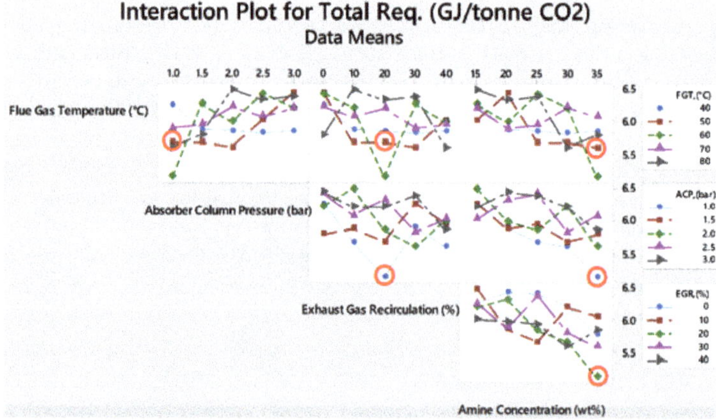

Figure 6. Interaction plot for all factors on the specific energy requirement in the capture unit.

5. Conclusions

The use of the Taguchi method allowed an accurate assessment of the effect of four control variables in the operation of an MEA-based post combustion CO_2 capture plant. The analysis of the results from the 25 simulations outlined by the Taguchi DOE demonstrated the importance of considering the capture unit as an entirety; the variation in energy demand when operating suboptimally clarifies the importance of deducing the optimal configuration to minimize the parasitic energy penalty incurred. Using the signal to noise ratios and analysis of variance for the four evaluated factors, the concentration of the amine was shown to be the greatest impetus in minimizing the energy demands; the operating pressure in the absorber and amount of exhaust gas recirculation also exhibit a significant influence on the total energy requirement, whereas the temperature of the flue gas was shown to have an insignificant effect on the specific energy requirements. The accuracy of designing experiments in this way allowed a more efficient assessment of the four factors and permitted the determination of a minimum energy requirement in the capture unit. The confirmation experiment, as outlined by the statistical analysis, further strengthens the rationale of employing a robust design of experiment. The minimum specific energy requirement for the PCC unit found with the defined optimum factor levels was 5.05 GJ/tonneCO$_2$, corresponding to a 3.94 GJ/tonneCO$_2$ requirement in the reboiler.

Author Contributions: Dr Masoudi Soltani was the project lead and the principal investigator on this research project. Mr Ben Petrovic was the active researcher in this work, responsible for the successful implementation of the research project, interpretation and analysis of the results from the start to the end.

Funding: This research received no external funding.

Acknowledgments: The authors would like to acknowledge the Chemical Engineering as well as the Mechanical and Aerospace Engineering Departments at Brunel University London, UK, to support conducting this research.

Nomenclature

Acronyms			
Acp	Absorber column pressure, Bar	GHG	Greenhouse gas
ANOVA	Analysis of variance	L/G	Liquid/gas
CCS	Carbon capture and storage	MEA	Monoethanolamine
CONC	Amine concentration, wt%	NGCC	Natural gas combined-cycle
DOC	Degree of capture	OA	Orthogonal array
DOE	Design of experiment	PCC	Post combustion capture
EGR	Exhaust gas recirculation, %	SNR	Signal-to-noise ratio
FGT	Flue gas temperature, °C	SRD	Specific reboiler duty
Symbols			
γ_i	Liquid phase activity coefficient of species i	$y_{i,s}$	Vapor phase mole fractions of species i,s
H_i^{P0}	Henry's constant of i in the amine solution at saturation pressure	α_i	Activity of species i
P_s^0	Saturation pressure of component s	v_i	stoichiometric constant of species i
V_i^∞	Partial molar volume of solute at infinite dilution	$\varphi_{i,s}$	Fugacity coefficient of species i,s
V_s	Molar volume of solvent s	φ_s^0	Fugacity coefficient of s under saturation condition
g^{ex^*}	Molar excess Gibbs free energy	P	Pressure, Bar
n_i	Mole number of species i	R	Gas constant
x_i	Liquid phase mole fraction of species i	T	Temperature, °C
		$K(T)$	Equilibrium constant

References

1. Intergovernmental Panel on Climate Change, (IPCC). *Climate Change 2014: Synthesis Report. Contribution of Working Groups I, II and III to the Fifth Assessment Report of the Intergovernmental Panel on Climate Change*; IPCC: Geneva, Switzerland, 2014.

2. The UK Government. *Climate Change Act 2008*; The UK Government: London, UK, 2008.

3. Boot-Handford, M.E.; Abanades, J.C.; Anthony, E.J.; Blunt, M.J.; Brandani, S.; Mac Dowell, N.; Fernández, J.R.; Ferrari, M.-C.; Gross, R.; Hallett, J.P.; et al. Carbon capture and storage update. *Energy Environ. Sci.* **2014**, *7*, 130–189. [CrossRef]

4. Department for Business Energy & Industrial Strategy. *Energy Trends September 2018*; Department for Business Energy & Industrial Strategy: London, UK, 2018.

5. National Energy Technology Laboratory (NETL). *Bituminous Coal (PC) and Natural Gas to Electricity. Cost and Performance Baseline for Fossil Energy Plants 2015*; National Energy Technology Laboratory (NETL): Pittsburgh, PA, USA, 2015.

6. Popa, A.; Edwards, R.; Aandi, I. Carbon capture considerations for combined cycle gas turbine. *Energy Procedia* **2011**, *4*, 2315–2323. [CrossRef]

7. Wang, M.; Lawal, A.; Stephenson, P.; Sidders, J.; Ramshaw, C. Post-combustion CO_2 capture with chemical absorption: A state-of-the-art review. *Chem. Eng. Res. Des.* **2011**, *89*, 1609–1624. [CrossRef]

8. Chen, P.; Luo, Y.X.; Cai, P.W. CO_2 Capture Using Monoethanolamine in a Bubble-Column Scrubber. *Chem. Eng. Technol.* **2015**, *38*, 274–282. [CrossRef]

9. Alhajaj, A.; Mac Dowell, N.; Shah, N. A techno-economic analysis of post-combustion CO_2 capture and compression applied to a combined cycle gas turbine: Part I. A parametric study of the key technical performance indicators. *Int. J. Greenh. Gas Control* **2016**, *44*, 26–41. [CrossRef]

10. Masoudi Soltani, S.; Fennell, P.S.; Mac Dowell, N. A parametric study of CO_2 capture from gas-fired power plants using monoethanolamine (MEA). *Int. J. Greenh. Gas Control* **2017**, *63*, 321–328. [CrossRef]

11. Oh, S.; Kim, J. Operational optimization for part-load performance of amine-based post-combustion CO_2 capture processes. *Energy* **2018**, *146*, 57–66. [CrossRef]

12. Rezazadeh, F.; Gale, W.F.; Akram, M.; Hughes, K.J.; Pourkashanian, M. Performance evaluation and optimisation of post combustion CO_2 capture processes for natural gas applications at pilot scale via a verified rate-based model. *Int. J. Greenh. Gas Control* **2016**, *53*, 243–253. [CrossRef]

13. Lindqvist, K.; Jordal, K.; Haugen, G.; Hoff, K.A.; Anantharaman, R. Integration aspects of reactive absorption for post-combustion CO_2 capture from NGCC (natural gas combined cycle) power plants. *Energy* **2014**, *78*, 758–767. [CrossRef]

14. Luo, X.; Wang, M. Optimal operation of MEA-based post-combustion carbon capture for natural gas combined cycle power plants under different market conditions. *Int. J. Greenh. Gas Control* **2016**, *48*, 312–320. [CrossRef]

15. Sanchez Fernandez, E.; Goetheer, E.L.V.; Manzolini, G.; Macchi, E.; Rezvani, S.; Vlugt, T.J.H. Thermodynamic assessment of amine based CO_2 capture technologies in power plants based on European Benchmarking Task Force methodology. *Fuel* **2014**, *129*, 318–329. [CrossRef]

16. Kothandaraman, A.; Nord, L.; Bolland, O.; Herzog, H.J.; McRae, G.J. Comparison of solvents for post-combustion capture of CO_2 by chemical absorption. *Energy Procedia* **2009**, *1*, 1373–1380. [CrossRef]

17. Arachchige, U.S.P.; Melaaen, M.C. Aspen Plus Simulation of CO_2 Removal from Coal and Gas Fired Power Plants. *Energy Procedia* **2012**, *23*, 391–399. [CrossRef]

18. Ali, U.; Font-Palma, C.; Akram, M.; Agbonghae, E.O.; Ingham, D.B.; Pourkashanian, M. Comparative potential of natural gas, coal and biomass fired power plant with post-combustion CO_2 capture and compression. *Int. J. Greenh. Gas Control* **2017**, *63*, 184–193. [CrossRef]

19. Aspen HYSYS simulation of CO_2 removal by amine absorption from a gas based power plant. In Proceedings of the 48th Scandinavian Conference on Simulation and Modeling (SIMS 2007), Göteborg, Sweden, 30–31 October 2007; Linköping University Electronic Press: Linköping, Sweden, 2007.

20. Afkhamipour, M.; Mofarahi, M. Modeling and optimization of CO_2 capture using 4-diethylamino-2-butanol (DEAB) solution. *Int. J. Greenh. Gas Control* **2016**, *49*, 24–33. [CrossRef]

21. Liu, Y.; Zhang, L.; Watanasiri, S. Representing vapor-liquid equilibrium for an aqueous MEA-CO_2 system using the electrolyte nonrandom-two-liquid model. *Ind. Eng. Chem. Res.* **1999**, *38*, 2080–2090. [CrossRef]

22. Mores, P.; Scenna, N.; Mussati, S. CO_2 capture using monoethanolamine (MEA) aqueous solution: Modeling and optimization of the solvent regeneration and CO_2 desorption process. *Energy* **2012**, *45*, 1042–1058. [CrossRef]

23. Sharifzadeh, M.; Shah, N. Carbon capture from natural gas combined cycle power plants: Solvent performance comparison at an industrial scale. *AIChE J.* **2016**, *62*, 166–179. [CrossRef]
24. Osagie, E.; Biliyok, C.; Di Lorenzo, G.; Hanak, D.P.; Manovic, V. Techno-economic evaluation of the 2-amino-2-methyl-1-propanol (AMP) process for CO_2 capture from natural gas combined cycle power plant. *Int. J. Greenh. Gas Control* **2018**, *70*, 45–56. [CrossRef]
25. Rezazadeh, F.; Gale, W.F.; Rochelle, G.T.; Sachde, D. Effectiveness of absorber intercooling for CO_2 absorption from natural gas fired flue gases using monoethanolamine solvent. *Int. J. Greenh. Gas Control* **2017**, *58*, 246–255. [CrossRef]
26. Mores, P.L.; Manassaldi, J.I.; Scenna, N.J.; Caballero, J.A.; Mussati, M.C.; Mussati, S.F. Optimization of the design, operating conditions, and coupling configuration of combined cycle power plants and CO_2 capture processes by minimizing the mitigation cost. *Chem. Eng. J.* **2018**, *331*, 870–894. [CrossRef]
27. Dutta, R.; Nord, L.O.; Bolland, O. Selection and design of post-combustion CO_2 capture process for 600 MW natural gas fueled thermal power plant based on operability. *Energy* **2017**, *121*, 643–656. [CrossRef]
28. Mores, P.; Scenna, N.; Mussati, S. Post-combustion CO_2 capture process: Equilibrium stage mathematical model of the chemical absorption of CO_2 into monoethanolamine (MEA) aqueous solution. *Chem. Eng. Res. Des.* **2011**, *89*, 1587–1599. [CrossRef]
29. Dutta, R.; Nord, L.O.; Bolland, O. Prospects of using equilibrium-based column models in dynamic process simulation of post-combustion CO_2 capture for coal-fired power plant. *Fuel* **2017**, *202*, 85–97. [CrossRef]
30. He, Z.; Ricardez-Sandoval, L.A. Dynamic modelling of a commercial-scale CO_2 capture plant integrated with a natural gas combined cycle (NGCC) power plant. *Int. J. Greenh. Gas Control* **2016**, *55*, 23–35. [CrossRef]
31. Birkelund, E.S. CO_2 Absorption and Desorption Simulation with Aspen HYSYS. Master's Thesis, Universitetet i Tromsø, Tromsø, Norway, 2013.
32. Chen, C.; Evans, L.B. A local composition model for the excess Gibbs energy of aqueous electrolyte systems. *AIChE J.* **1986**, *32*, 444–454. [CrossRef]
33. Soave, G. Equilibrium constants from a modified Redlich-Kwong equation of state. *Chem. Eng. Sci.* **1972**, *27*, 1197–1203. [CrossRef]
34. Freguia, S.; Rochelle, G.T. Modeling of CO_2 capture by aqueous monoethanolamine. *AIChE J.* **2003**, *49*, 1676–1686. [CrossRef]
35. Austgen, D.M.; Rochelle, G.T.; Peng, X.; Chen, C.C. Model of vapor-liquid equilibria for aqueous acid gas-alkanolamine systems using the electrolyte-NRTL equation. *Ind. Eng. Chem. Res.* **1989**, *28*, 1060–1073. [CrossRef]
36. Koronaki, I.P.; Prentza, L.; Papaefthimiou, V. Modeling of CO_2 capture via chemical absorption processes-An extensive literature review. *Renew. Sustain. Energy Rev.* **2015**, *50*, 547–566. [CrossRef]
37. Dickinson, J.; Percy, A.; Puxty, G.; Verheyen, T.V. Oxidative degradation of amine absorbents in carbon capture systems—A dynamic modelling approach. *Int. J. Greenh. Gas Control* **2016**, *53*, 391–400. [CrossRef]
38. Mondal, B.K.; Bandyopadhyay, S.S.; Samanta, A.N. Vapor–liquid equilibrium measurement and ENRTL modeling of CO_2 absorption in aqueous hexamethylenediamine. *Fluid Phase Equilib.* **2015**, *402*, 102–112. [CrossRef]
39. Ross, P.J. *Taguchi Techniques for Quality Engineering: Loss Function, Orthogonal Experiments, Parameter and Tolerance Design*, 2nd ed.; McGraw-Hill: New York, NY, USA, 1996.
40. Yusoff, N.; Ramasamy, M.; Yusup, S. Taguchi's parametric design approach for the selection of optimization variables in a refrigerated gas plant. *Chem. Eng. Res. Des.* **2011**, *89*, 665–675. [CrossRef]
41. Karna, S.K.; Sahai, R. An Overview on Taguchi Method. *Int. J. Eng. Math. Sci.* **2012**, *1*, 1–7.
42. Adams, T.; Mac Dowell, N. Off-design point modelling of a 420 MW CCGT power plant integrated with an amine-based post-combustion CO_2 capture and compression process. *Appl. Energy* **2016**, *178*, 681–702. [CrossRef]
43. Roy, R. *A Primer on Taguchi Method*; Van Noshtrand Reinhold Int Co. Ltd.: New York, NY, USA, 1990.
44. Liang, D.; Meng, W. The Application of the Taguchi Method in the Optimal Combination of the Parameters Design of the Spindle System. In *MATEC Web of Conference*; EDP Sciences: Julius, France, 2016; Volume 63.
45. Wilcox, J. *Carbon Capture*; Springer Science & Business Media: Berlin, Germany, 2012.

46. Hu, Y.; Xu, G.; Xu, C.; Yang, Y. Thermodynamic analysis and techno-economic evaluation of an integrated natural gas combined cycle (NGCC) power plant with post-combustion CO_2 capture. *Appl. Therm. Eng.* **2017**, *111*, 308–316. [CrossRef]
47. Canepa, R.; Wang, M.; Biliyok, C.; Satta, A. Thermodynamic analysis of combined cycle gas turbine power plant with post-combustion CO_2 capture and exhaust gas recirculation. Proc. *Inst. Mech. Eng. Part E J. Process Mech. Eng.* **2013**, *227*, 89–105. [CrossRef]

Article

Highly Selective CO$_2$ Capture on Waste Polyurethane Foam-Based Activated Carbon

Chao Ge [1], Dandan Lian [1], Shaopeng Cui [2], Jie Gao [3] and Jianjun Lu [1,4,*

[1] College of Textile Engineering, Taiyuan University of Technology, Jinzhong 030600, China
[2] College of Forestry, Shanxi Agricultural University, Taigu 030801, China
[3] State Key Laboratory of Coal Conversion, Institute of Coal Chemistry, Chinese Academy of Sciences, Taiyuan 030001, China
[4] Key Laboratory of Coal Science and Technology, Taiyuan University of Technology, Taiyuan 030024, China
* Correspondence: lujianjunktz@tyut.edu.cn; Tel.: +86-0351-3176-555

Received: 8 August 2019; Accepted: 28 August 2019; Published: 3 September 2019

Abstract: Low-cost activated carbons were prepared from waste polyurethane foam by physical activation with CO$_2$ for the first time and chemical activation with Ca(OH)$_2$, NaOH, or KOH. The activation conditions were optimized to produce microporous carbons with high CO$_2$ adsorption capacity and CO$_2$/N$_2$ selectivity. The sample prepared by physical activation showed CO$_2$/N$_2$ selectivity of up to 24, much higher than that of chemical activation. This is mainly due to the narrower microporosity and the rich N content produced during the physical activation process. However, physical activation samples showed inferior textural properties compared to chemical activation samples and led to a lower CO$_2$ uptake of 3.37 mmol·g^{-1} at 273 K. Porous carbons obtained by chemical activation showed a high CO$_2$ uptake of 5.85 mmol·g^{-1} at 273 K, comparable to the optimum activated carbon materials prepared from other wastes. This is mainly attributed to large volumes of ultra-micropores (<1 nm) up to 0.212 cm^3·g^{-1} and a high surface area of 1360 m^2·g^{-1}. Furthermore, in consideration of the presence of fewer contaminants, lower weight losses of physical activation samples, and the excellent recyclability of both physical- and chemical-activated samples, the waste polyurethane foam-based carbon materials exhibited potential application prospects in CO$_2$ capture.

Keywords: waste polyurethane foam; physical activation; high selectivity; CO$_2$ capture; ultra-micropore

---

1. Introduction

CO$_2$ emission from the combustion of coal and natural gas is mainly responsible for global warming [1,2]. The "least-cost" solution is to limit greenhouse gas emissions to meet the Paris Agreement pledges, wherein 60% of CO$_2$ emissions are hoped to be reduced in 2030 relative to 2005 [3]. To reduce the emission of CO$_2$, the selective and energy-efficient capture and storage of CO$_2$ is considered to be a satisfactory approach. Until now, three main CO$_2$ capturing strategies, namely pre-combustion, post-combustion, and oxy-combustion, have been discussed. For pre-combustion, the coal must be gasified first; the typical technology used is called an integrated gasification combined cycle. Post-combustion involves capturing CO$_2$ from the flue gases produced after fossil fuels are burned. Both of these methods have been widely accepted and used in gas-stream purification in industry. Oxy-combustion can produce a relatively pure CO$_2$ stream in emissions, but it is still in development due to the high cost of the technology involved [4]. In addition to these, several technologies have also been investigated to separate and store CO$_2$, such as membrane separation, solution absorption, cryogenic refrigeration, and adsorption approaches [5–7]. Although these CCUS (Carbon Capture, Utilization and Storage) technologies are intended to reduce CO$_2$ emission, reaching the Paris Agreement targets is still a serious challenge. Nevertheless, in these techniques, adsorption is considered to be the approach with the most potential, since it involves simple operation and has low-cost and energy-saving benefits [8].

Many adsorbents have been intensively studied for CO_2 capture, including zeolites [9], metal-organic frameworks [10], metallic oxide [11], graphene-based adsorbents [12], and porous carbon materials [13]. For future commercialization application, the selection of adsorbents is strongly dependent not only on CO_2 capture capacity but also the cost, including the availability of the raw materials, the preparation of the adsorbent, and the operating costs. Porous carbons (PCs) are considered to be the most competitive candidates due to their controlled pore structure, low cost, stable physicochemical properties, ease of chemical modification, and regeneration [14]. Micropore sizes smaller than 1 nm are beneficial to high-density CO_2 filling at ambient conditions [15,16]. In addition, a high surface area (>1000 $m^2 \cdot g^{-1}$) and a rich nitrogen environment can improve the CO_2 adsorption capacity.

The route of preparation, especially activation, will significantly affect the performance of CO_2 absorption. The porous carbons can be activated either by physical or chemical methods. Physical activation is usually achieved by carbonization in an inert atmosphere followed by oxidizing in CO_2 [17], steam [18], or air. In this way, activated carbons with a narrower pore size distribution can be obtained [19]. Generally, chemical activation takes advantage of oxidizing or dehydrating agents, such as KOH [1,15,20], NaOH, H_3PO_4, or $CaCl_2$, and being subjected to calcination under N_2 between 773 and 1223 K. Chemical activation contributes to the formation of pores and lead to carbons with higher textural development. However, it is an energy-consuming process and is also associated with the corrosion of equipment and environmental problems [21].

To achieve green and sustainable development, the production of low-cost porous carbons from original waste materials is a good, established technique [22]. Polyurethane foams (PUFs), as one of the most important thermoset polymers, are widely used in the chemical industry, electronics, textiles, and medical and other fields due to their high strength, excellent wear resistance, and wide hardness range. The total annual production of polyurethane products in the Asia Pacific region was about 11.5 million tons in 2014, and it is predicted to be over 15.5 million tons by 2019 [23]. As a result, large amounts of useless waste and spent product were generated. Since the processes used for recycling polyurethane foams, like mechanical recycling [24] or chemical depolymerization [25,26], are highly time- and energy-consuming [27], most of the wastes are discarded in landfills or directly burnt, leading to serious environmental issues [28]. However, the ease of availability of the raw materials helps to make waste PUFs a potential activated carbon adsorbent. This would not only alleviate pollution and protect the environment, but it could also convert PUF into a high-value-added product.

Herein, low-cost activated carbons were prepared from waste PUF materials using physical activation with CO_2 for the first time, and chemical activation with metal hydroxide. The synthetic procedure is simple and clear, and production costs are low. The activation conditions, like physical activation temperature and chemical activating agent, were discussed and optimized to produce porous carbons with high CO_2 adsorption capacity and CO_2/N_2 selectivity. The richness of N content and narrower microporosity produced by physical activation may be responsible for the higher CO_2/N_2 selectivity of up to 24. Furthermore, physical activation leads to less contaminant and lower weight losses but also lower CO_2 uptake. Porous carbons obtained by chemical activation with KOH possess large volumes of micropores (<1 nm) and high surface areas of up to 0.212 $cm^3 \cdot g^{-1}$ and 1360 $m^2 \cdot g^{-1}$, respectively. This material exhibits a high CO_2 adsorption capacity of 5.85 $mmol \cdot g^{-1}$ at 273 K and excellent recyclability. The easy regeneration of the PUF-based carbon adsorbent requires minimum energy input, thus reducing operational costs. Therefore, the waste PUFs have the potential to be utilized for the production of activated carbon adsorbents on an industrial scale.

2. Materials and Methods

Prior to carrying out chemical or physical activation, a batch of waste PUFs were carbonized in a porcelain boat inside a horizontal tube furnace at 673 K for 1 h under nitrogen flow (60 mL·min^{-1}). The obtained material was designated as PUF/C. Chemical activation was conducted by treating the carbonized PUFs in alkaline solutions. First, 0.8 g PUF/C was dipped into 10 mL of separate aqueous solutions containing 1.6 g of Ca(OH)$_2$, NaOH, and KOH; was stirred uniformly for 1 h at room temperature; and was subsequently dried at 383 K for 12 h to remove water. Then, the mixture was heated up to 973 K under N$_2$ flow (60 mL·min^{-1}) for 2 h. The samples were thoroughly washed in diluted HCl and distilled water until pH = 7. The products were dried at 373 K for 6 h and denoted as PUF/C-Ca(OH)$_2$-973, PUF/C-NaOH-973, and PUF/C-KOH-973, respectively.

Physical activation of PUFs was accomplished by one-step carbonization and CO$_2$ activation. Certain amounts of PUFs were carbonized in the same procedure followed above, and then N$_2$ was switched to CO$_2$; the obtained PUF/C char was heated up to 1073–1273 K for 2 h under CO$_2$ flow (15 mL·min^{-1}). After physical activation, the sample was cooled to ambient temperature under nitrogen protection and was labeled as PUF/C-CO$_2$-x, where x represents the CO$_2$ activation temperature.

Nitrogen sorption isotherms and the textural properties of the porous carbons were measured at 77 K on an ASAP 2020 apparatus. Before N$_2$ adsorption measurements, all samples were degassed at 493 K for 4 h. The specific surface area was obtained using the BET method (p/p^0 = 0–0.5). The total pore volume was estimated with the amount of N$_2$ adsorbed at p/p_0 = 0.99. The morphology and structure of the obtained samples were investigated by scanning electron microscopy (SEM, JEOL JSM-700) and transmission electron microscopy (TEM, JEOL JEM-2001F). The amount of N in this paper was determined by elemental analysis using an Elemntar Vario EL Cube microanalyzer. X-ray photoelectron spectra (XPS) were collected on a Krato AXIS Ultra DLD spectrometer. CO$_2$ temperature-programmed-desorption (CO$_2$-TPD) was measured on a Micromeritics Autochem II 2920 chemisorption analyzer.

The CO$_2$ adsorption capacity and adsorption–desorption cycles were investigated at 273 K and ordinary pressure on a BEL-SORP-max instrument. The activated carbon materials were degassed at 453 K for at least 2 h before the measurement.

3. Results and Discussion

3.1. Pyrolysis of Wasted PUF and Morphologies of Activated Carbons

Figure 1 shows the pyrolysis of the waste polyurethane foam at a heating rate of 5 K·min^{-1} in a nitrogen atmosphere. Water adsorbed on the surface is released at about 373 K. The weight loss of the raw materials mainly occurs between 523–773 K, which include three differentiated weight loss peaks. The first peak nears 593 K, resulting from the formation of amines, methylene methyl, and methane species. The second peak around 653 K is due to the decomposition of isocyanate, and the third one near 733 K results from the production of quaternary N, aromatics, CH$_4$, and other species [15]. It is clear that the carbonized precursor mainly results from the former two peaks. Therefore, the carbonization temperature was subsequently set to 673 K.

Figure 2 shows the SEM and TEM images of carbonized and chemical and physical activation samples. A slightly rough surface appeared when the carbonized precursor was activated by Ca(OH)$_2$, but the activation degree was so limited that only few macropores existed on the surface (Figure 2b). In comparison, as it was activated by NaOH and KOH, the smooth surface changes to a multi-hole morphology to a large extent (Figure 2a,c,d). In addition, NaOH activation resulted in more macropores on the surface, while KOH activation led to the plentiful generation of micropores. These observations indicate that the carbonized precursor was etched increasingly severely by the chemical activation of Ca(OH)$_2$, NaOH, and KOH (Figure 2a–d). Furthermore, in the physical activation process, the activation temperature determines the development of porosity. It can be observed that the surface of the carbonized precursor became rugged after physical activation by CO$_2$ at 1173 K and formed many micropores during the gasification reaction between CO_2 and carbon and the following diffusion processes of the generated products (Figure 2e,g). Micropore generation would possibly benefit the fast, dynamic CO_2 adsorption-desorption. Nevertheless, much higher physical activation temperatures should be avoided as they can broaden the average micropore width, producing few mesopores (Figure 2f,h), which is not good for CO_2 capture.

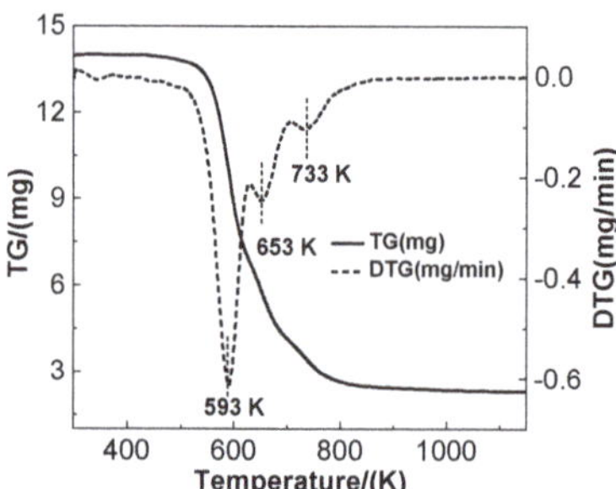

Figure 1. Pyrolysis of polyurethane foam measured at the heating rate of 5 K·min^{-1} in N_2 atmosphere.

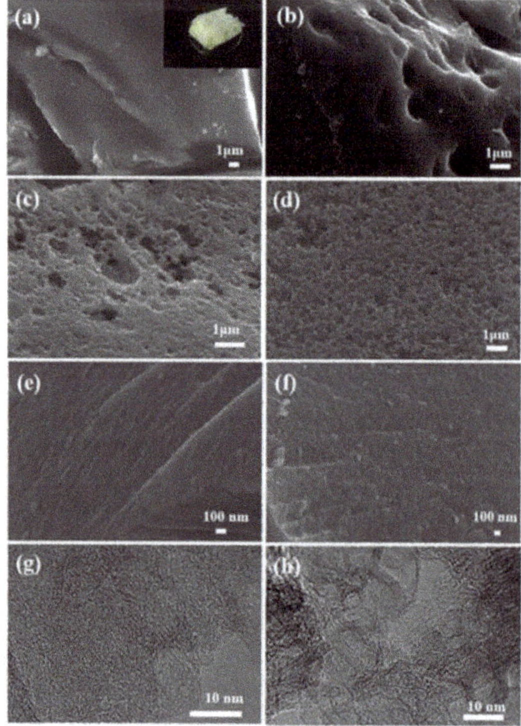

Figure 2. SEM images of the PUF/C (**a**), PUF/C-Ca(OH)$_2$-973 (**b**), PUF/C-NaOH-973 (**c**), PUF/C-KOH-973 (**d**), PUF/C-CO$_2$-1173 (**e**), PUF/C-CO$_2$-1273 (**f**), and TEM images of PUF/C-CO$_2$-1173 (**g**) and PUF/C-CO$_2$-1273 (**h**). The inset in (**a**) is a photo showing the morphology of waste polyurethane foams.

3.2. Surface Oxygen and Nitrogen Species Analyses

Figure 3 exhibits the FT-IR spectra of PUF-based activated carbons. The broad band around 3440 cm^{-1} could be ascribed to the N–H/O–H symmetric stretching vibration [29]. The intense bands at 1090 cm^{-1} and 1584 cm^{-1} are related to C–O stretching [29] and N–H in-plane deformation. The relatively weak band around 1383 cm^{-1} is attributed to the C–N stretching vibration. Bands at 2927 and 2851 cm^{-1} could be ascribed to the C–H stretching vibrations of –CH$_2$ and –CH$_3$ [15]. The presence of oxygen and nitrogen species were further confirmed by XPS. The O 1s spectra show the presence of three main peaks at 531, 532.1, and 533.5 eV, which are attributed to –C=O, C–O–C/C–OH, and O–C=O, respectively (Figure 4) [30]. The results confirm that plentiful amounts of oxygen-rich functional groups are incorporated onto the surface of the PUF-based activated carbons, which play an important role in enhancing the overall CO$_2$ adsorption.

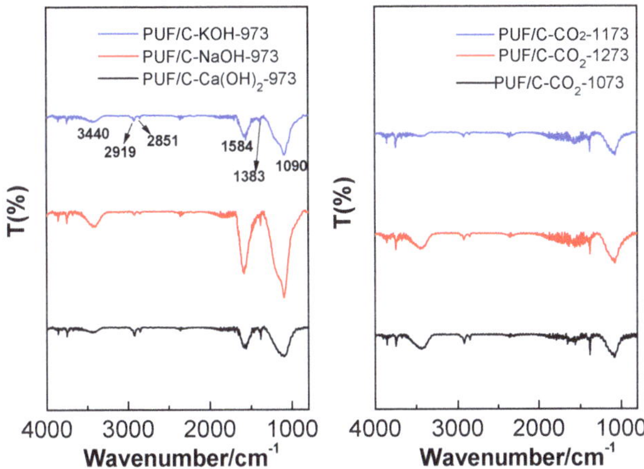

Figure 3. FT-IR spectra of the obtained samples prepared by physical and chemical activation.

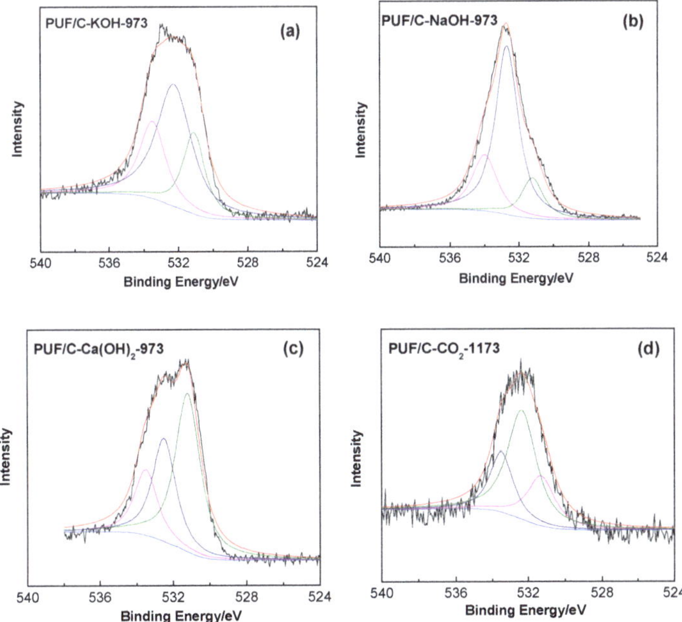

Figure 4. XPS spectra O 1s of PUF/C-KOH-973 (**a**), PUF/C-NaOH-973 (**b**), PUF/C-Ca(OH)$_2$-973 (**c**), and PUF/C-CO$_2$-1173 (**d**).

It is well-known that the type and content of N-containing species of carbon adsorbents can greatly affect CO_2 adsorption capacity. As shown in Figure 5, pyridinic (N-6), pyrrolic (N-5), quaternary N (N-Q), and pyridine N-oxide (N-X) species with binding energies of 398.5, 399.7, 400.9, and 403.0 eV, respectively, can be identified by deconvolving the N 1s signal of the activated carbons [31]. The specific components depend on the physical activation temperature or the degree of chemical activation with $R(OH)_x$ (R = Ca, Na, K; x = 1/2). The relative amount of pyridine N-oxide species increased with increasing CO_2 activation temperature. This can also apply to the quaternary N species (Figure 5a–c). Conversely, upon increasing the CO_2 activation temperature from 1073 to 1273 K, the contents of pyridinic and pyrrole N species dropped from 35% and 24% to 18% and 21%, respectively (Table 1), indicating that elevating the CO_2 activation temperature can decrease the total amount of N species and also convert the pyridinic nitrogen and pyrrole nitrogen to quaternary N and pyridine N-oxide species (Figure 5a–c). This trend was also observed upon increasing the degree of chemical activation using $Ca(OH)_2$, NaOH, and KOH as activated reagents (Figure 5d–f). The pyridinic and pyrrole-type N declined from 28% and 72% to 21% and 41%, respectively. Quaternary-N and pyridine N-oxide species are considered less effective in CO_2 capture than pyridinic and pyrrolic nitrogen [32], since basic N species are superior at capturing CO_2. Therefore, CO_2 uptake on the activated carbons would increase with decreasing the physical activation temperature or the degree of chemical activation under the conditions of similar porous structures. Actually, CO_2 uptake and physical activation temperature or the degree of chemical activation show a volcano relationship or a positive relationship, respectively, implying that the porous structure of the activated carbons may greatly affect their CO_2 capture.

Table 1. Surface nitrogen contents of polyurethane foam-based activated carbons prepared by physical and chemical activation.

Samples	Content (wt.%)				
	Total N [1]	N-6 [2]	N-5 [2]	N-Q [2]	N-X [2]
PUF/C-CO$_2$-1073	8.78	35	24	27	14
PUF/C-CO$_2$-1173	6.84	28	22	34	15
PUF/C-CO$_2$-1273	3.77	18	21	38	23
PUF/C-Ca(OH)$_2$-973	7.70	28	72	-	-
PUF/C-NaOH-973	7.51	20	60	-	20
PUF/C-KOH-973	6.92	21	41	22	16

[1] Determined by element analysis. [2] Relative content of different types of N species, as determined by XPS analysis results. pyridinic (N-6), pyrrolic (N-5), quaternary N (N-Q), and pyridine N-oxide (N-X) species.

Figure 6 shows the CO_2-TPD profiles of the porous carbons prepared by physical and chemical activation from wasted polyurethane foam. In general, physical and chemical adsorption are the two main routes for CO_2 capture. An intense peak was present between 343 and 358 K, indicating that most of the CO_2 molecules physically adsorbed on the activated carbons due to their developed pore structures. In contrast, a very weak peak was also observed around 448 K, implying that there exists a weak chemical interaction between CO_2 and N species. Thus, physical adsorption occupies the dominant position during the CO_2 capture process in this studied case.

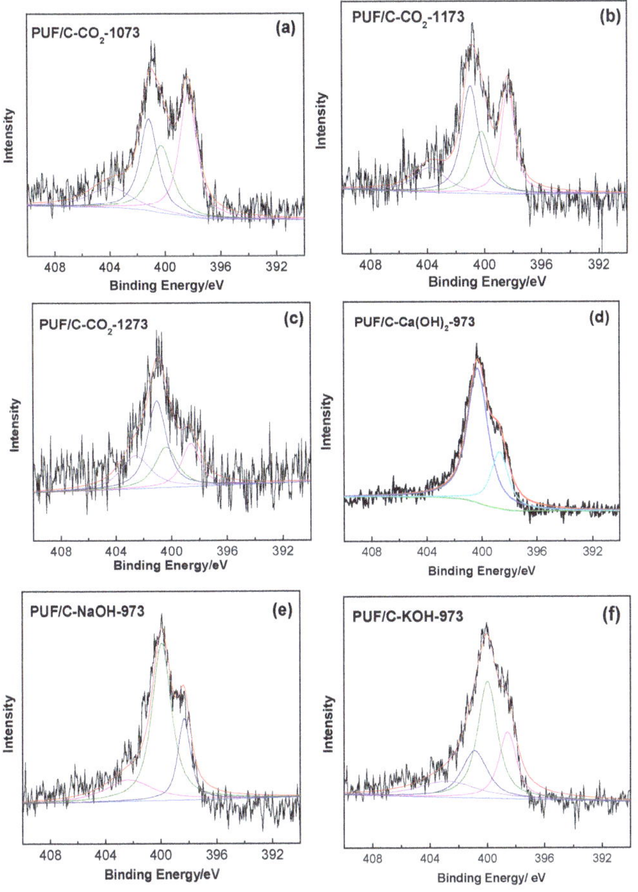

Figure 5. XPS N 1s spectra of PUF/C-CO$_2$-1073 (**a**), PUF/C-CO$_2$-1173 (**b**), and PUF/C-CO$_2$-1273 (**c**), PUF/C-Ca(OH)$_2$-973 (**d**), PUF/C-NaOH-973 (**e**), and PUF/C-KOH-973 (**f**).

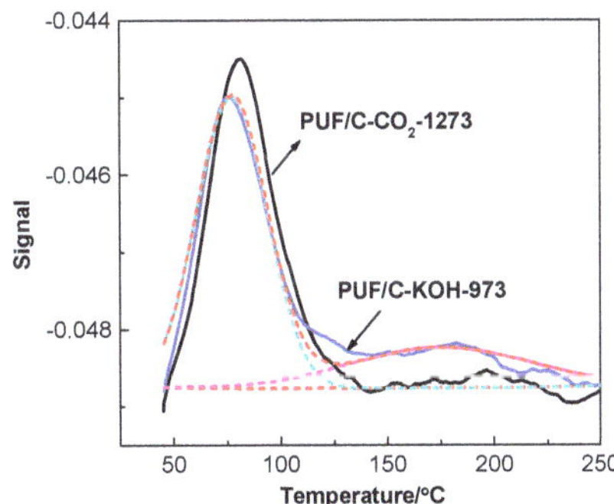

Figure 6. CO$_2$- temperature-programmed-desorption (TPD) profile of PUF/C-KOH-973 and PUF/C-CO$_2$-1273.

3.3. Effect of Physical and Chemical Activation on the Textural Properties of PUF-Based Porous Carbon Materials

Figure 7 shows the N_2 adsorption isotherms and pore size distributions (PSDs) of carbon materials after physical and chemical activation. Clearly, the prepared samples exhibit a typical micropore structure that is mainly responsible for CO_2 capture at ambient conditions [33]. Meanwhile, the hysteresis loops of the adsorption isotherm of PUF/C-KOH-973 or PUF/C-NaOH-973 are of type-IV, representing slit-shaped pores and implying the generation of mesopores (Figure 7a). It is well-known that carbon materials with narrow micropores can be obtained by physical activation [18]. It can be observed from Table 2 that, upon increasing CO_2 activation temperature from 1073 to 1273 K, the surface area and pore volume of the samples significantly rose from 15 $m^2 \cdot g^{-1}$ and 0.04 $cm^3 \cdot g^{-1}$ to 865 $m^2 \cdot g^{-1}$ and 0.42 $cm^3 \cdot g^{-1}$, respectively, and the PSDs (<1 nm) of samples widened from 0.58 to 0.87 nm (Figure 7b). In particular, the ultra-micropore (<1 nm) volume increased from 0.085 to 0.122 $cm^3 \cdot g^{-1}$ and then decreased to 0.119 $cm^3 \cdot g^{-1}$ upon elevating the activation temperature (Table 2 and Figure 8). A much higher physical activation temperature will make the reaction rate between carbon and CO_2 faster than the diffusion rate, which is not helpful for the formation of micropores. From the above, we can conclude that the pore structure of activated carbons could be controlled by adjusting the CO_2 activation temperature. In this work, the developed ultra-micropore can be obtained at 1173 K with a CO_2 flowrate of 15 $mL \cdot min^{-1}$.

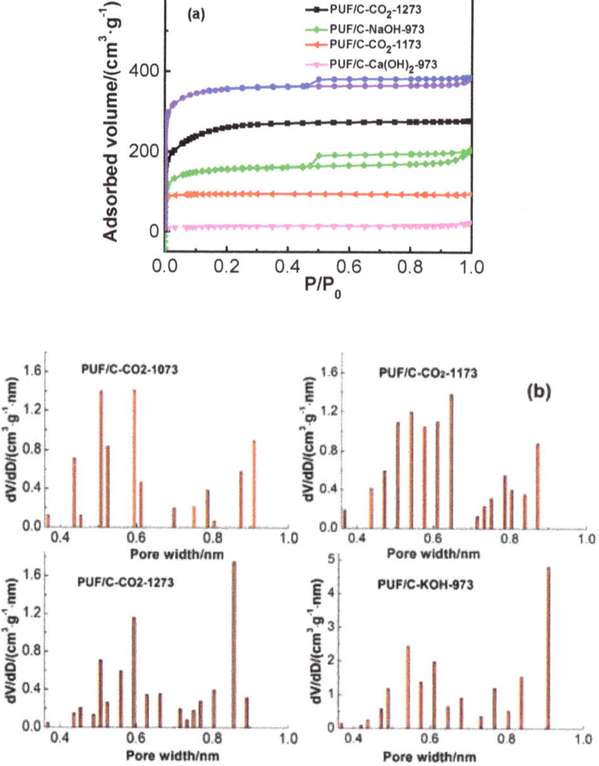

Figure 7. Adsorption isotherms of N_2 at 77 K (**a**) and pore size distributions determined by the Dubinin–Radushkevich equation applied to Table 2 adsorption data at 273 K (**b**) of the obtained samples prepared by physical and chemical activation.

Table 2. Textural parameters of prepared samples from adsorption isotherms of N_2 and CO_2.

Sample	Yield (%)	N$_2$ Adsorption at 77 K			CO$_2$ Adsorption at 273 K	
		S_{BET} (m^2·g^{-1})	V_t [1] (cm^3·g^{-1})	V_{micro} [2] (cm^3·g^{-1})	$V_{ultra-micro}$ [3] (cm^3·g^{-1}) (≤1 nm)	D [4] (nm)
PUF/C-CO$_2$-1073	16.0	15	0.04	0.03	0.085	0.58
PUF/C-CO$_2$-1173	14.0	206.7	0.10	0.08	0.122	0.64
PUF/C-CO$_2$-1273	9.3	865	0.42	0.11	0.119	0.87
PUF/C-Ca(OH)$_2$-973	25.5	39	0.04	0.01	-	-
PUF/C-NaOH-973	22.2	710	0.41	0.20	0.171	-
PUF/C-KOH-973	12.3	1360	0.59	0.52	0.212	0.91

[1] Total pore volume at $p/p_0 = 0.99$. [2] Micropore volume determined by the *t*-plot method. [3] The cumulative volume of pores smaller than 1 nm determined using CO_2 adsorption data at 273 K. [4] D is the maximum value of the PSDs. The yield of the prepared sample was calculated by the mass ratio of activated carbon and dry PUF.

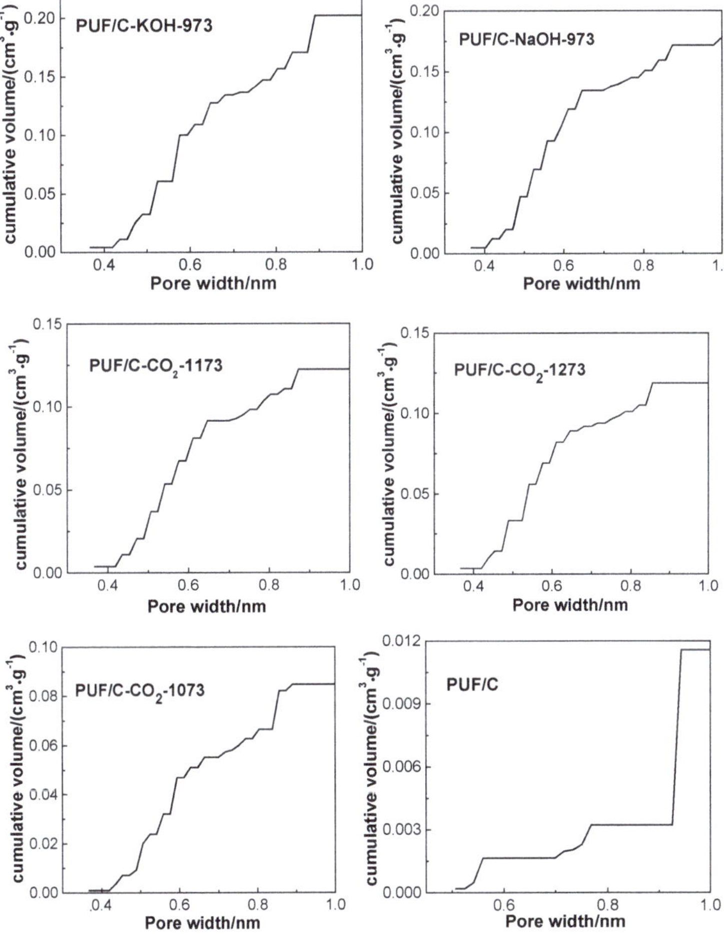

Figure 8. Cumulative volumes of ultramicropores (≤1 nm) calculated by the nonlocal density functional theory (NLDFT) method of PUF-based activated carbons. Adapted with permission from [15]. Copyright 2016 American Chemical Society.

In addition, the textural parameters of chemical activation samples are also shown in Table 2. The carbonized precursor was activated by chemical reagents $Ca(OH)_2$, $NaOH$, and KOH at 973 K. In addition to the sharp increase of surface area and pore volume from 39 $m^2 \cdot g^{-1}$ and 0.04 $cm^3 \cdot g^{-1}$ to 1360 $m^2 \cdot g^{-1}$ and 0.59 $cm^3 \cdot g^{-1}$, respectively, and the volumes of ultra-micropores (<1 nm) increased from barely detectable to 0.212 $cm^3 \cdot g^{-1}$ (Table 2 and Figure 8). It can be inferred that the pore structure of the carbon adsorbents could also be controlled by adjusting the chemical activation degree, and that increasing the degree of chemical activation is helpful for the formation of pores, especially the ultra-micropores, although a small number of mesopores were generated simultaneously.

Comparing physical activation with chemical activation, it can be concluded that physical activation mainly develops micropores and results in porous carbon materials with lower textural properties, which may affect the adsorption performance and lead to lower CO_2 uptake. However, it involves less contaminant, which avoids the emission of metal ions and the corrosion of equipment.

By increasing the activation temperature from 1073 to 1273 K (Table 2), the yield of porous carbons with physical activation decreased from 16% to 9.3%, as a result of CO_2 gas reacting with the walls of the pores (Equation (1)) [34]. In addition, the higher the burn-off temperature, the wider the pore size distribution becomes, as mentioned above. Regardless, higher carbon yield can be produced by physical activation due to lower burn-off degrees compared with effective chemical activation.

$$CO_2 + C \rightarrow 2CO \tag{1}$$

Similarly, the yield of carbon materials, as expected, decreased from 25.5% to 12.3% when enhancing the chemical activation degree using $Ca(OH)_2$, $NaOH$, or KOH as activated reagents (Table 2), which can be explained by graphitic C being oxidized by KOH beginning at 673 K, while for NaOH the temperature needs to be higher than 843 K, not even mentioning $Ca(OH)_2$ [35].

The porosity formed upon KOH, NaOH, and $Ca(OH)_2$ activation is considered to be the intercalation of metallic K, Na, and Ca in the carbon matrix, causing the deformation and expansion of the carbon lattice [36]. CO_2, H_2O, and H_2 generated in the redox reaction (take KOH and C for example) can also contribute to the development of pores. As shown in Equations (2)–(4) [35].

$$4KOH + C \rightarrow 4K + CO_2 + 2H_2O \tag{2}$$

$$4KOH + 2CO_2 \rightarrow 2K_2CO_3 + 2H_2O \tag{3}$$

$$6KOH + 2C \rightarrow 2K + 3H_2 + 2K_2CO_3 \tag{4}$$

3.4. CO$_2$ Adsorption Performances

CO_2 adsorption performances of samples with physical and chemical activation at 273 K are shown in Figure 9. For physically-activated samples, CO_2 adsorption capacities increased from 2.4 to 3.4 $mmol \cdot g^{-1}$ upon increasing activation temperature from 1073 to 1173 K. This cannot be simply attributed to surface area and pore volume (Table 2), since $PUF/C\text{-}CO_2\text{-}1273$ possesses larger surface area and pore volume but a lower CO_2 adsorption capacity than that of $PUF/C\text{-}CO_2\text{-}1173$. It was reported recently that small micropores are dominant in CO_2 capture for activated carbon materials [9]. This is attributed to the fact that the interaction between CO_2 and carbon adsorbents can be enhanced in small pores, especially at elevated temperatures and low pressures. The correlation of CO_2 uptake of PUF-based activated carbons at 273 K and 1 bar with the volumes of micropores (<1 nm) is presented in Figure 10a. Clearly, the CO_2 adsorption capacity shows a perfect linear correlation with the volumes of micropores less than 1 nm. Severe activation at higher temperature results in widening the pore size distribution, reducing the amount of ultra-micropores, which is responsible for the lower CO_2 uptake of $PUF/C\text{-}CO_2\text{-}1273$ compared to $PUF/C\text{-}CO_2\text{-}1173$.

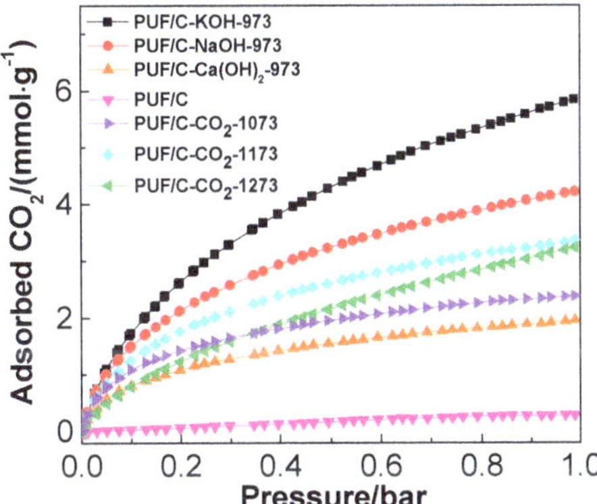

Figure 9. Adsorption isotherms of CO_2 at 273 K for the studied samples.

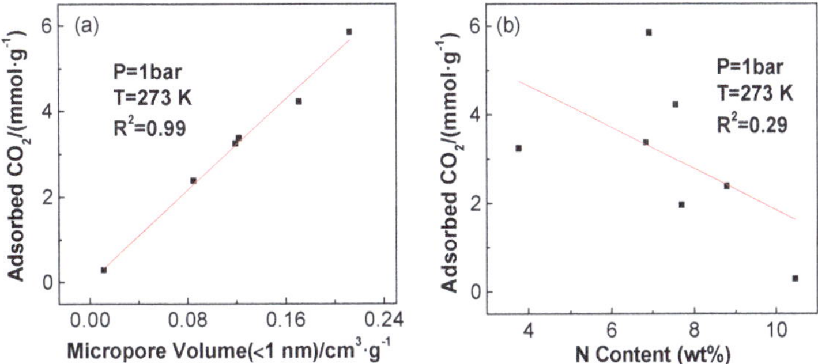

Figure 10. Correlation of CO_2 adsorption capacity with ultra-micropore volume (**a**) and with sample N content (**b**) at 273 K.

This trend can also be applied to the CO_2 adsorption performances of samples after chemical activation. Figure 8 shows that the adsorption capacity of CO_2 at 273 K and 1 bar increases as follows: PUF/C-Ca(OH)$_2$-973 < PUF/C-NaOH-973 < PUF/C-KOH-973. The PUF/C-KOH-973 sample exhibits the highest CO_2 uptake at 5.85 mmol·g^{-1}. This is mainly due to it having the largest ultra-micropore volume (<1 nm) of 0.212 cm^3·g^{-1} coupled with the largest surface area, pore volume, and abundance of basic N species. Chemical activation can be described by two steps: The formation of micropores, which is beginning with the redox reaction between activated agents and carbonized precursors; and the broadening of pores inside the opened porous channel [37]. Hence, the stage of creating micropores is primary when enhancing the activation degree from Ca(OH)$_2$ to KOH activation at 973 K.

N content in the prepared carbon materials may also contribute to the CO_2 adsorption capacity, wherein the correlation coefficient of CO_2 uptake with the N content (3.5–10.5 wt.%) of PUF-based activated carbons was only 0.29 (Figure 10b). This indicates that N content in the samples should not be the primary effect in this experiment. Thus, the superior CO_2 adsorption capabilities of the studied activated carbons predominantly result from the ultra-micropores (<1 nm).

Table 3 shows a comparation of CO_2 uptakes of carbon adsorbents prepared from PUF and other different waste materials at 273 K. It is worth mentioning that samples prepared with the waste polyurethane foams under CO_2 activation at 1173 K and KOH activation at 973 K exhibit comparable or better CO_2 adsorption capacities than those adsorbents obtained from many types of waste materials, such as ocean pollutant, sawdust, and coal fly ash (see Table 3). Whereas, some new derived activated carbons from biomass porous materials, like *Arundo donax* [29] or lotus seed [38], show a relatively higher adsorption capacity. The results mentioned above suggest that activated carbon prepared from waste PUF is a potential candidate for use in capturing CO_2.

Table 3. CO_2 adsorption capacities at 273 K and 1 bar of activated carbons prepared from different waste materials.

Sample	CO_2 Adsorption Capacity (mmol·g^{-1}) at 273 K		Reference
	Physical Activation	Chemical Activation	
Spent coffee grounds activated carbons	3.5	4.8	[39]
Sawdust-based porous carbons	-	5.8	[40]
Coal-based carbon foams activated by ZnCl$_2$	-	3.4	[41]
CO$_2$ adsorbents based on ocean pollutant	-	2.4	[42]
Coal fly ash based materials	-	5	[43]
Macroalgae based N-doped carbons	-	3.06	[44]
Arundo donax-based activated bio-carbons	-	6.3	[29]
Biomass-derived activated porous bio-carbons	-	6.8	[38]
Polyurethane foam-based activated carbons	3.37	5.85	This work

3.5. CO$_2$ Adsorption Recyclability and Selectivity

CO_2 and N_2 adsorption behaviors of PUF/C-CO$_2$-1173 and PUF/C-KOH-973 at 273 K were compared, as shown in Figure 11. The N_2 adsorption capacity is 0.59 mmol g^{-1}, which is almost one-tenth of the CO_2 uptake of chemically-activated sample PUF/C-KOH-973. While, the selectivity of CO$_2$/N$_2$ of PUF/C-CO$_2$-1173 outperforms PUF/C-KOH-973 with a lower N_2 adsorption of only 0.14 mmol g^{-1} at 273 K, which is much higher than CO_2 adsorbents from other waste materials, such as mangosteen peel waste (24 vs. 12) [45]. This can be ascribed to the rich and narrower microporosity [39], as well as the nitrogen content of 6.84 wt% and the presence of other functional groups [46].

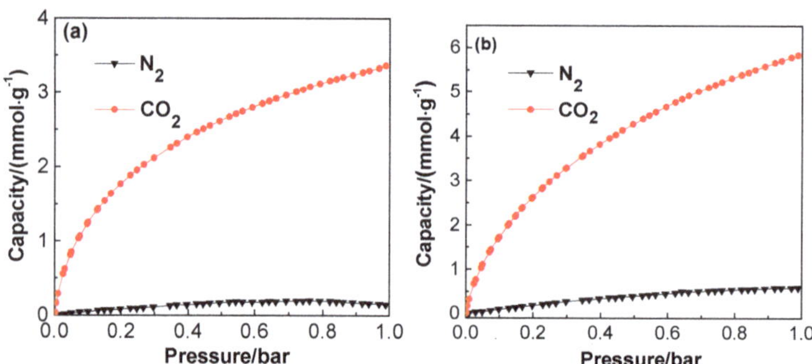

Figure 11. CO_2 and N_2 adsorption isotherms at 273 K on the PUF/C-CO$_2$-1173 (**a**) and PUF/C-KOH-973 (**b**).

The CO_2 adsorption performances of PUF/C-CO$_2$-1173 at 273 K and PUF/C-KOH-973 at 298 K are completely the same in the four repeated runs, as shown in Figure 12a,b, indicating that the physically-activated sample has as high of a recyclability as the chemically-activated sample. Hence, in comparison with the environmental pollution and cumbersome process associated with chemical activation, CO_2 sorbent from waste polyurethane foam by physical activation is encouraged, due to the higher CO$_2$/N$_2$ selectivity and the relatively higher carbon yield.

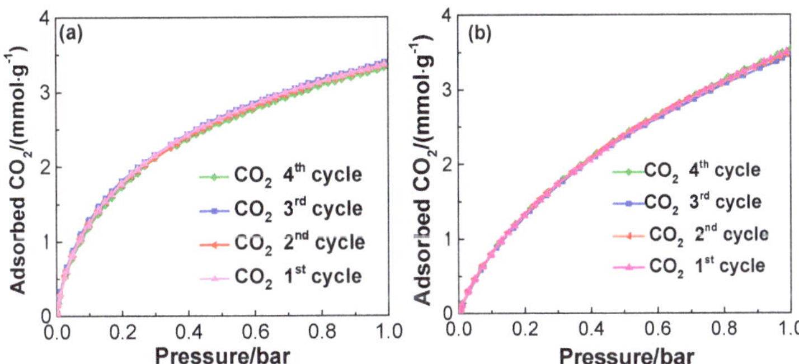

Figure 12. CO_2 adsorption performances on the PUF/C-CO_2-1173 at 273 K (**a**) and PUF/C-KOH-973 at 298 K (**b**) within four repeated cycles with regeneration.

4. Conclusions

In this work, waste polyurethane foam was physically activated by CO_2 for the first time and chemically activated with $Ca(OH)_2$, NaOH, or KOH, and was compared for the preparation of the low-cost activated carbons with a highly-developed porosity. To achieve high CO_2 adsorption capacity and CO_2/N_2 selectivity, activation conditions of the waste polyurethane foam were optimized. The low CO_2 flowrate (15 mL·min^{-1}) and temperature of 1173 K for CO_2 activation are beneficial to the formation of a narrower microporosity. This, combined with relatively high nitrogen content, resulted in the high CO_2/N_2 selectivity of up to 24, much higher than samples that underwent chemical activation. However, physical activation samples showed inferior textural properties compared to chemical activation samples and led to lower CO_2 uptake. The sample activated with KOH possessed a high volume of ultramicropores (<1 nm) and surface area up to 0.212 cm^3·g^{-1} and 1360 m^2·g^{-1}, respectively. A high CO_2 adsorption capacity of 5.85 mmol·g^{-1} was obtained at 273 K and 1 bar, which was better than the physical activation sample with 3.37 mmol·g^{-1}, and is comparable to the best reported carbon materials prepared from other waste materials. Moreover, in consideration of the decreased presence of the contaminant, the lower weight losses of physical activation samples, and the excellent recyclability of both physical and chemical activated samples, the waste polyurethane foam-based carbon materials exhibited potential application prospects in CO_2 capture.

Author Contributions: Writing—original draft, C.G.; data curation, D.L.; investigation, S.C.; formal analysis, J.G.; supervision, J.L.

Funding: This research was funded by the National Natural Science Foundation of China (No. 21802101), Scientific and Technological Innovation Programs of Higher Education Institutions in Shanxi, and Shanxi Province Science Foundation for Youths (No. 201801D221129).

Conflicts of Interest: The authors declare no conflicts of interest.

References

1. Wang, J.; Heerwig, A.; Lohe, M.R.; Oschatz, M.; Borchardt, L.; Kaskel, S. Fungi-based porous carbons for CO_2 adsorption and separation. *J. Mater. Chem.* **2012**, *22*, 13911–13913. [CrossRef]
2. Schrag, D.P. Preparing to capture carbon. *Science* **2007**, *315*, 812–813. [CrossRef] [PubMed]
3. Morris, J.; Paltsev, S.; Ku, A.Y. Impacts of China's emissions trading schemes on deployment of power generation with carbon capture and storage. *Energy Econ.* **2019**, *81*, 848–858. [CrossRef]
4. Wilberforce, T.; Baroutaji, A.; Soudan, B.; Al-Alami, A.H.; Olabi, A.G. Outlook of carbon capture technology and challenges. *Sci. Total Environ.* **2019**, *657*, 56–72. [CrossRef] [PubMed]
5. Wang, S.; Li, W.C.; Zhang, L.; Jin, Z.Y.; Lu, A.H. Polybenzoxazine-based monodisperse carbon spheres with low-thermal shrinkage and their CO_2 adsorption properties. *J. Mater. Chem. A* **2014**, *2*, 4406–4412. [CrossRef]

6. Li, P.; Wang, Z.; Li, W.; Liu, Y.; Wang, J.; Wang, S. High-performance multilayer composite membranes with mussel-inspired polydopamine as a versatile molecular bridge for CO_2 separation. *ACS Appl. Mater. Interfaces* **2015**, *7*, 15481–15493. [CrossRef] [PubMed]

7. Du, N.; Park, H.B.; Robertson, G.P.; Dal-Cin, M.M.; Visser, T.; Scoles, L.; Guiver, M.D. Polymer nanosieve membranes for CO_2-capture applications. *Nat. Mater.* **2011**, *10*, 372–375. [CrossRef]

8. Bae, Y.S.; Snurr, R.Q. Development and Evaluation of Porous Materials for Carbon Dioxide Separation and Capture. *Angew. Chem. Int. Ed.* **2011**, *50*, 11586–11596. [CrossRef]

9. Su, F.; Lu, C. CO_2 capture from gas stream by zeolite 13X using a dual-column temperature/vacuum swing adsorption. *Energy Environ. Sci.* **2012**, *5*, 9021–9027. [CrossRef]

10. Khan, I.U.; Othman, M.H.D.; Ismail, A.; Ismail, N.; Jaafar, J.; Hashim, H.; Rahman, M.A.; Jilani, A. Structural transition from two-dimensional ZIF-L to three-dimensional ZIF-8 nanoparticles in aqueous room temperature synthesis with improved CO_2 adsorption. *Mater. Charact.* **2018**, *136*, 407–416. [CrossRef]

11. Yu, H.; Wang, X.; Shu, Z.; Fujii, M.; Song, C. Al_2O_3 and CeO_2-promoted MgO sorbents for CO_2 capture at moderate temperatures. *Front. Chem. Sci. Eng.* **2018**, *12*, 83–93. [CrossRef]

12. Chandra, V.; Yu, S.U.; Kim, S.H.; Yoon, Y.; Kim, D.Y.; Kwon, A.H.; Meyyappan, M.; Kim, K.S. Highly selective CO_2 capture on N-doped carbon produced by chemical activation of polypyrrole functionalized graphene sheets. *Chem. Commun.* **2012**, *48*, 735–737. [CrossRef] [PubMed]

13. Gu, J.M.; Kim, W.S.; Hwang, Y.K.; Huh, S. Template-free synthesis of N-doped porous carbons and their gas sorption properties. *Carbon* **2013**, *56*, 208–217. [CrossRef]

14. Song, J.; Shen, W.; Wang, J.; Fan, W. Superior carbon-based CO_2 adsorbents prepared from poplar anthers. *Carbon* **2014**, *69*, 255–263. [CrossRef]

15. Ge, C.; Song, J.; Qin, Z.; Wang, J.; Fan, W. Polyurethane Foam-Based Ultramicroporous Carbons for CO_2 Capture. *ACS Appl. Mater. Interfaces* **2016**, *8*, 18849–18859. [CrossRef] [PubMed]

16. Zhang, Z.; Zhou, J.; Xing, W.; Xue, Q.; Yan, Z.; Zhuo, S.; Qiao, S.Z. Critical role of small micropores in high CO_2 uptake. *Phys. Chem. Chem. Phys.* **2013**, *15*, 2523–2529. [CrossRef]

17. Ello, A.S.; De Souza, L.K.; Trokourey, A.; Jaroniec, M. Coconut shell-based microporous carbons for CO_2 capture. *Microporous Mesoporous Mater.* **2013**, *180*, 280–283. [CrossRef]

18. Heo, Y.J.; Park, S.J. H_2O_2/steam activation as an eco-friendly and efficient top-down approach to enhancing porosity on carbonaceous materials: The effect of inevitable oxygen functionalities on CO_2 capture. *Green Chem.* **2018**, *20*, 5224–5234. [CrossRef]

19. Prauchner, M.J.; Rodríguez-Reinoso, F. Chemical versus physical activation of coconut shell: A comparative study. *Microporous Mesoporous Mater.* **2012**, *152*, 163–171. [CrossRef]

20. Sevilla, M.; Fuertes, A.B.; Solis, M.S.; Valle-Vigón, P. N-Doped Polypyrrole-Based Porous Carbons for CO_2 Capture. *Adv. Funct. Mater.* **2011**, *21*, 2781–2787. [CrossRef]

21. Nandi, M.; Okada, K.; Dutta, A.; Bhaumik, A.; Maruyama, J.; Derks, D.; Uyama, H. Unprecedented CO_2 uptake over highly porous N-doped activated carbon monoliths prepared by physical activation. *Chem. Commun.* **2012**, *48*, 10283–10285. [CrossRef] [PubMed]

22. Meng, F.Z.; Gong, Z.Q.; Wang, Z.B.; Fang, P.W.; Li, X.Y. Study on a nitrogen-doped porous carbon from oil sludge for CO_2 adsorption. *Fuel* **2019**, *251*, 562–571. [CrossRef]

23. Agrawal, A.; Kaur, R.; Walia, R. PU foam derived from renewable sources: Perspective on properties enhancement: An overview. *Eur. Polym. J.* **2017**, *95*, 255–274. [CrossRef]

24. Shinko, A. Introduction to Mechanical Recycling and Chemical Depolymerization. *Recycl. Polyurethane Foam.* **2018**, *4*, 45–55.

25. Sheel, A.; Pant, D. Chemical Depolymerization of Polyurethane Foams via Glycolysis and Hydrolysis. *Recycl. Polyurethane Foam.* **2018**, *6*, 67–75.

26. Padhan, R.K. Chemical Depolymerization of Polyurethane Foams via Combined Chemolysis Methods. *Recycl. Polyurethane Foam.* **2018**, *8*, 89–96.

27. Quadrini, F.; Bellisario, D.; Santo, L. Recycling of thermoset polyurethane foams. *Polym. Eng. Sci.* **2013**, *53*, 1357–1363. [CrossRef]

28. Ferdan, T.; Pavlas, M.; Nevrlý, V.; Šomplák, R.; Stehlík, P. Greenhouse gas emissions from thermal treatment of non-recyclable municipal waste. *Front. Chem. Sci. Eng.* **2018**, *12*, 815–831. [CrossRef]

29. Singh, G.; Kim, I.Y.; Lakhi, K.S.; Srivastava, P.; Naidu, R.; Vinu, A. Single step synthesis of activated bio-carbons with a high surface area and their excellent CO_2 adsorption capacity. *Carbon* **2017**, *116*, 448–455. [CrossRef]

30. Pevida, C.; Drage, T.; Snape, C.; Drage, T.; Snape, C. Silica-templated melamine–formaldehyde resin derived adsorbents for CO_2 capture. *Carbon* **2008**, *46*, 1464–1474. [CrossRef]

31. Su, F.; Poh, C.K.; Chen, J.S.; Xu, G.; Wang, D.; Li, Q.; Lin, J.; Lou, X.W. Nitrogen-containing microporous carbon nanospheres with improved capacitive properties. *Energy Environ. Sci.* **2011**, *4*, 717–724. [CrossRef]

32. Hao, G.P.; Li, W.C.; Qian, D.; Lu, A.H. Rapid Synthesis of Nitrogen-Doped Porous Carbon Monolith for CO_2Capture. *Adv. Mater.* **2010**, *22*, 853–857. [CrossRef] [PubMed]

33. Martín, C.; Plaza, M.; Pis, J.; Rubiera, F.; Pevida, C.; Centeno, T.A. On the limits of CO_2 capture capacity of carbons. *Sep. Purif. Technol.* **2010**, *74*, 225–229. [CrossRef]

34. Chaudhary, A.; Kumari, S.; Kumar, R.; Teotia, S.; Singh, B.P.; Singh, A.P.; Dhawan, S.K.; Dhakate, S.R. Lightweight and easily foldable MCMB-MWCNTs composite paper with exceptional electromagnetic interference shielding. *ACS Appl. Mater. Interface* **2016**, *8*, 10600–10608. [CrossRef] [PubMed]

35. Lillo-Ródenas, M.A.; Cazorla-Amorós, D.; Linares-Solano, A. Understanding chemical reactions between carbons and NaOH and KOH: An insight into the chemical activation mechanism. *Carbon* **2003**, *41*, 267–275. [CrossRef]

36. Díaz-Terán, J.; Nevskaia, D.; Fierro, J.; Lopez-Peinado, A.J.; Jerez, A. Study of chemical activation process of a lignocellulosic material with KOH by XPS and XRD. *Microporous Mesoporous Mater.* **2003**, *60*, 173–181.

37. Rodriguez-Reinoso, F.; Molina-Sabio, M. Activated carbons from lignocellulosic materials by chemical and/or physical activation: An overview. *Carbon* **1992**, *30*, 1111–1118. [CrossRef]

38. Singh, G.; Lakhi, K.S.; Ramadass, K.; Sathish, C.I.; Vinu, A. High-performance biomass-derived activated porous biocarbons for combined pre- and post-combustion CO_2 capture. *ACS Sustain. Chem. Eng.* **2019**, *7*, 7412–7420. [CrossRef]

39. Plaza, M.; Gonzalez, A.; Pevida, C.; Pis, J.; Rubiera, F. Valorisation of spent coffee grounds as CO_2 adsorbents for postcombustion capture applications. *Appl. Energy* **2012**, *99*, 272–279. [CrossRef]

40. Sevilla, M.; Fuertes, A.B. Sustainable porous carbons with a superior performance for CO_2 capture. *Energy Environ. Sci.* **2011**, *4*, 1765–1771. [CrossRef]

41. Rodríguez, E.; García, R. Low-cost hierarchical micro/macroporous carbon foams as efficient sorbents for CO_2 capture. *Fuel Process. Technol.* **2017**, *156*, 235–245. [CrossRef]

42. Zhang, Z.; Wang, K.; Atkinson, J.D.; Yan, X.; Li, X.; Rood, M.J.; Yan, Z. Sustainable and hierarchical porous Enteromorpha prolifera based carbon for CO_2 capture. *J. Hazard. Mater.* **2012**, *229*, 183–191. [CrossRef] [PubMed]

43. Liu, L.; Singh, R.; Xiao, P.; Webley, P.A.; Zhai, Y. Zeolite synthesis from waste fly ash and its application in CO_2 capture from flue gas streams. *Adsorption* **2011**, *17*, 795–800. [CrossRef]

44. Ren, M.; Zhang, T.; Wang, Y.; Jia, Z.; Cai, J. A highly pyridinic N-doped carbon from macroalgae with multifunctional use toward CO_2 capture and electrochemical applications. *J. Mater. Sci.* **2018**, *54*, 1606–1615. [CrossRef]

45. Li, Y.; Wang, X.; Cao, M. Three-dimensional porous carbon frameworks derived from mangosteen peel waste as promising materials for CO_2 capture and supercapacitors. *J. CO_2 Util.* **2018**, *27*, 204–216. [CrossRef]

46. Zhai, Q.G.; Bai, N.; Li, S.; Bu, X.; Feng, P. Design of Pore Size and Functionality in Pillar-Layered Zn-Triazolate-Dicarboxylate Frameworks and Their High CO_2/CH_4 and C_2 Hydrocarbons/CH_4 Selectivity. *Inorg. Chem.* **2015**, *54*, 9862–9868. [CrossRef] [PubMed]

Article

Modeling and Simulation of the Absorption of CO_2 and NO_2 from a Gas Mixture in a Membrane Contactor

Nayef Ghasem

Department of Chemical and Petroleum Engineering, UAE University, Al-Ain 15551, United Arab Emirates; nayef@uaeu.ac.ae; Tel.: +971-3-713-5313

Received: 9 June 2019; Accepted: 8 July 2019; Published: 11 July 2019

Abstract: The removal of undesirable compounds such as CO_2 and NO_2 from incineration and natural gas is essential because of their harmful influence on the atmosphere and on the reduction of natural gas heating value. The use of membrane contactor for the capture of the post-combustion NO_2 and CO_2 had been widely considered in the past decades. In this study, membrane contactor was used for the simultaneous absorption of CO_2 and NO_2 from a mixture of gas (5% CO_2, 300 ppm NO_2, balance N_2) with aqueous sodium hydroxide solution. For the first time, a mathematical model was established for the simultaneous removal of the two undesired gas solutes (CO_2, NO_2) from flue gas using membrane contactor. The model considers the reaction rate, and radial and axial diffusion of both compounds. The model was verified and validated with experimental data and found to be in good agreement. The model was used to examine the effect of the flow rate of liquid, gas, and inlet solute mole fraction on the percent removal and molar flux of both impurity species. The results revealed that the effect of the liquid flow rate improves the percent removal of both compounds. A high inlet gas flow rate decreases the percent removal. It was possible to obtain the complete removal of both undesired compounds. The model was confirmed to be a dependable tool for the optimization of such process, and for similar systems.

Keywords: global warming; chemical absorption; membrane contactor; removal of NO_2 and CO_2

---

1. Introduction

Harmful gases are emitted into the atmosphere from industrial plants, because of the increase in the human population and the associated economic development, energy consumption, and the requirement of burning fossil fuels for water desalination and power generation purposes. Nitrogen dioxide (NO_2) is believed to be one of the gases that contributes to smog and acid rain and which is harmful to human and animal well-being. Accordingly, there is an obligation to capture and eliminate nitrogen dioxide and other harmful gases, such NO_x, SO_2, and CO_2 from industrial emission streams, proceeding to discharge into the atmosphere [1–3]. Various methods have been established for capturing the impurity of compounds such as physical and/or chemical absorption, adsorption, membrane technology, conversion to another compound, and condensation. There are various technologies available to remove CO_2 and NO_x [3–5]. Physical absorption incorporates mass transport within the phases and mass transfer at the liquid–gas boundary. Operating conditions and gas solubility are the main factors affecting physical absorption. An example of physical absorption is the capture of CO_2 into liquid water using industrial absorption towers or gas–liquid membrane contactors. Chemical absorption is based on a chemical reaction between the absorbed substances and the liquid phase, such as the capture of CO_2 in amine solutions [4,6]. The most widely used commercial and economical method is the chemical absorption technique, used in the conventional absorption packed bed towers employed in the absorption of CO_2, H_2S, and NO_2 from chimney gas and natural gas via alkanolamine solutions, where the flue gas and the absorbent liquid are in direct contact. The conventional scrubbing method requires a huge absorption column with excess liquid absorbent and a large cross-sectional area in order to prevent foaming and channeling. The large amount of absorbent liquid utilized in the absorption process (i.e., rich solvent) increases the operating and regeneration cost as more heat is required for the regeneration and pumping of the recycled lean solvent. The main disadvantages of conventional chemical absorption processes are channeling, foaming, corrosion, and a large space area required and hence high operating and capital cost. The idea of using membrane contactor was first proposed for the absorption of carbon dioxide in sodium hydroxide as a liquid absorbent utilizing a non-dispersive microporous membrane where gas and liquid phases are not dispersed in each other [7,8]. A hollow fiber membrane contactor (HFMC) provides a greater surface contact area per unit volume, more than that of a conventional absorption column [9–13]. In most cases of membrane contactor operation, gas flows in the shell side, and liquid absorbent flows in the tube side; vice versa is also possible [14]. The performance of a HFMC declines when the micropores of the membranes are wetted with a liquid solvent [3,11,15–17]. The advantages of membrane processes are as follows: Gas and liquid flow rates are independent, high ratio of surface area per volume, easy scale up and down, and no worry about flooding and channeling [18]. The membrane acts as an obstacle between gas and liquid and delivers an exchange surface zone for the two phases, without the dispersion of the gas phase in the liquid phase. In the non-wetted mode, the membrane pores are filled with gas, and by contrast, in the wetted mode, part of the membrane pores are filled with the liquid absorbent. The pollutant gas compounds' amputation process occurs when the gas filling the membrane micropores diffuses from the gas stream and is absorbed by the liquid absorbent running in the membrane lumen ide [1,16,19]. The removal of the contaminant gas compounds from the gas stream depends on the solubility of the acid gas molecules in the absorbent liquid, and on the concentration incline among the gas stream absorbent solution. The interaction between the selected gas solute and the selective absorbent liquid defines the performance of the pollutant gas removal rate [20]. Membrane fouling and membrane wettability are the main drawbacks of the membrane contactor. The wetted portion of the membrane adds an additional mass transfer resistance. Accordingly, in order to avoid membrane wettability, researchers focused on the use of hydrophobic polymeric membranes, such those made from polypropylene (PP), polytetrafluoroethylene (PTFE), polyvinylidene fluoride (PVDF), and polyethylene (PE) [21–24]. The Gas Liquid Hollow Fiber Member Contactor (GLMC) processes were used for the removal of CO_2 from nitrogen, CO_2 from natural gas, and the instantaneous removal of CO_2 and SO_2 from combustion released gas [17,19,20,25–33]. The experimental simultaneous capture of CO_2 and NO_2 from a pretended flue gas mixture containing $CO_2/NO_2/N_2$ was first investigated using a commercial PTFE membrane by Ghobadi et al. [23]. The effects of the gas and liquid cross-flow velocity on the percent removal of these gasses were investigated. Various mathematical models were developed for the removal of carbon dioxide, sulfur dioxide, and hydrogen sulfide from simulated flue gas and natural gas streams [7,9,15,17,25,32,34–39]. To the extent of the author's knowledge, so far, there is no mathematical model published to designate the synchronized capture of CO_2/NO_2 from a mixture of gas consisting of $CO_2/NO_2/N_2$.

This study focusses on the development of a numerical model for the capture of CO_2/NO_2 from a gas containing: 5% CO_2, 300 ppm NO_2, and the balance is N_2. The principal model equations were solved using COMSOL Multiphysics (Version 5.4, Comsol Inc., Zürich, Switzerland). The developed model was used to predict the influence of various operating parameters on the percent removal and molar flux of the acid gas components. The model was verified and validated with experimental data from literature [23].

2. Model Development

The gas-liquid membrane process consists of many hollow fibers assembled in a module, where the liquid solvent flows inside the membrane lumen, and the gas flows in the shell sideways, or vice versa, in a co-current or countercurrent parallel flow. The pollutant gas compounds diffuse through the fiber walls towards the absorbent membrane–tube interface, as a result of the concentration gradient. Other gases are retained in the membrane pores because of their low diffusivity and low solubility in the liquid solvent. Figure 1 shows a schematic simplified geometry of the model domains representing the HFM module grounded on Happel's free surface [40]. The model considers the following three separate domains: the liquid phase in the tube side, the non-wetted membrane, and the gas phase in the shell side. The system is steady state, described by cylindrical coordinates, angular concentration gradients are neglected, and an asymmetrical approach is considered.

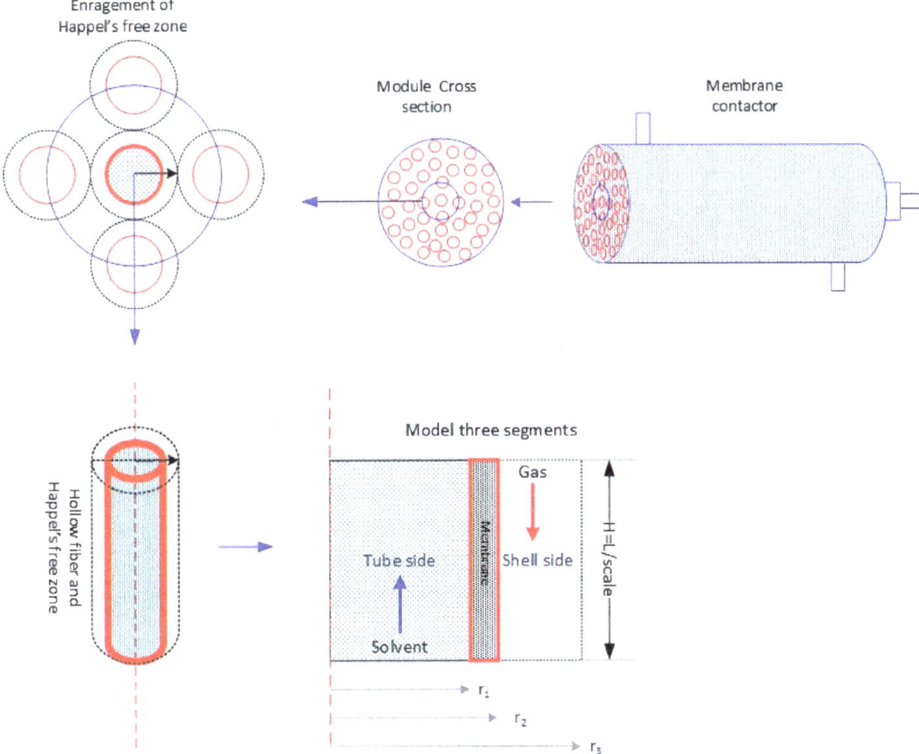

Figure 1. Simplified geometry of membrane module based on Happel's free surface method.

The sizes of the membrane used in the simulation are presented in Table 1.

Table 1. Membrane module dimensions [23].

Property	Value
Inner fiber radius (mm)	0.34
Outer fiber radius (mm)	0.60
Number of fibers	590
Module outer radius (mm)	25.4
Module effective length (mm)	200

As seen from Table 1, the fiber length is 588 times longer than the fiber radius (effective module length is 200 mm and radius are 0.34 mm). Accordingly, the membrane length is scaled up so as to avoid an excessive number of elements and nodes and for a better appearance of the module in the simulation;

therefore, a new scaled length is introduced by dividing the length by 100. The following assumptions were considered:

- The process is at isothermal and steady state conditions;
- Gas phase is an ideal gas, and the liquid phase is incompressible and Newtonian;
- Solubility is based on Henry's law at the liquid-gas interface.

The blended gas (CO_2, NO_2, and N_2) is transported in the shell side by convection and diffusion, whereas, in the membrane section, the only transport mechanism is diffusion. The liquid phase ($NaOH + H_2O$) is transported in the lumen by diffusion and convection. The following mass transport equations are formulated to describe the chemical absorption system in the model domains (tube, membrane, and shell). The developed mass transport equations are presented as follows.

2.1. Tube Side

The mass balance equations for the gas components of CO_2, NO_2, and N_2 in the tube side can be stated, as per Equation (1), as follows:

$$D_{i,t}\frac{1}{r}\left(\frac{\partial}{\partial r}r\left(\frac{\partial C_{i,t}}{\partial r}\right)\right) + D_{i,t}\frac{\partial^2 C_{i,t}}{\partial z^2} = v_{z,t}\left(\frac{\partial C_{i,t}}{\partial z}\right) + R_i \tag{1}$$

The subscripts in the material balance the following equations: t refers to tube side, m refers to membrane, and s refers to shell side, where $C_{i,t}$ refers to the concentration of component i in liquid moving in the tube side, i refers to the pollutant gas components: CO_2, NO_2, and R_i is the rate of reaction of the species i. The length of the fiber is scaled to avoid excessive computation and to make the simulation result in a better profile. The scaling is performed as follows: let $\xi = z/scale$. The scaling factor is substituted in Equation (1). Consequently, the equation becomes Equation (2), as follows:

$$D_{i,t}\frac{1}{r}\left(\frac{\partial}{\partial r}r\left(\frac{\partial C_{i,t}}{\partial r}\right)\right) + \frac{D_{i,t}}{scale^2}\frac{\partial^2 C_{i,t}}{\partial \xi^2} = \frac{v_{z,t}}{scale}\left(\frac{\partial C_{i,t}}{\partial \xi}\right) + R_i \tag{2}$$

where the velocity of liquid inside the hollow fiber ($v_{z,t}$) is described by the following parabolic equation:

$$v_{z,t} = \frac{2Q_t}{n\pi r_1^2}\left(1 - \left(\frac{r}{r_1}\right)^2\right) \tag{3}$$

where Q_t is the volumetric liquid flow rate in the tube side, and n is the number of hollow fibers. The appropriate set of boundary conditions are specified as follows:

$$\begin{aligned}
&\text{at } z = 0, &&C_{i,t} = 0 &&\text{(a)}\\
&\text{at } z = H, &&\frac{\partial^2 C_{i,t}}{\partial z^2} = 0 &&\text{(b)}\\
&\text{at } r = 0, &&\frac{\partial C_{i,t}}{\partial r} = 0 &&\text{(c)}\\
&\text{at } r = r_1, &&C_{i,t} = m_i\,C_{i,m} &&\text{(d)}
\end{aligned} \tag{4}$$

where m_i is the dimensionless physical solubility of CO_2 and NO_2 in solvent, 0.82, 0.17, respectively. The values of the dimensionless physical solubility of CO_2 and NO_2 were calculated from Henry's constant (H): 0.034 kmol/m^3 atm, 0.007 kmol/m^3 atm [23], respectively, using the relation m_i = RTxH. The liquid phase reactions between NO_2 and $NaOH$ took several steps. First, NO_2 dissolved in the aqueous $NaOH$ was reacted with H_2O, then neutralized with sodium hydroxide [41]. The controlling liquid phase reactions are as follows:

$$2NO_2 + H_2O \rightarrow HNO_2 + HNO_3 \tag{5}$$

$$HNO_2 + NaOH \rightarrow NaNO_2 + H_2O \tag{6}$$

$$HNO_3 + NaOH \rightarrow NaNO_3 + H_2O \tag{7}$$

The overall reaction is designated, as per Equation (8), as follows:

$$2NO_2 + 2NaOH \rightarrow NaNO_2 + NaNO_3 + H_2O \tag{8}$$

The general reaction rate is expressed in Equation (9), as follows:

$$r_{NO_2-NaOH} = k_{r,1}[NO_2][NaOH] \tag{9}$$

The reaction is the second order with a rate constant, $k_{r,1}\left(\text{m}^3\ \text{mol}^{-1}\text{s}^{-1}\right) = 1.0 \times 10^5$, [1]
The overall reaction of CO_2 and NaOH is represented by the following reaction [42].

$$CO_2 + 2NaOH \rightarrow Na_2CO_3 + H_2O \tag{10}$$

The rate of the reaction is determined, as per Equation (11), as follows:

$$r_{CO_2-NaOH} = k_{r,2}[CO_2][NaOH] \tag{11}$$

The reaction rate constant (in Equation (11)) is $k_{r,2} = 8.37\ (\text{m}^3\ \text{mol}^{-1}\text{s}^{-1})$ [23].

2.2. Membrane Side

The transport of the solute gas (CO_2 and NO_2) components in the membrane section confined between r_1 and r_2 can be described by the steady state material balance equation (Equation (12)), where diffusion is the only transport mechanism in the membrane phase [34], as follows:

$$D_{i,m}\frac{1}{r}\left(\frac{\partial}{\partial r}\left(r\frac{\partial C_{i,m}}{\partial r}\right)\right) + \frac{D_{i,m}}{scale^2}\frac{\partial^2 C_{i,m}}{\partial \xi^2} = 0 \tag{12}$$

The proper boundary settings are specified, as per Equation (13), as follows:

$$
\begin{aligned}
&\text{at } z = 0, & \frac{\partial C_{i,m}}{\partial z} &= 0 & \text{(a)} \\
&\text{at } z = H, & \frac{\partial C_{i,m}}{\partial z} &= 0 & \text{(b)} \\
&\text{at } r = r_1, & D_{i,m}\frac{\partial C_{i,m}}{\partial r} &= D_{i,t}\frac{\partial C_{i,t}}{\partial r} & \text{(c)} \\
&\text{at } r = r_2 & C_{i,m} &= C_{i,s} & \text{(d)}
\end{aligned}
\tag{13}
$$

2.3. Shell Side

The steady state mass transport of solute gas (CO_2 and NO_2) components in the shell side (no chemical reaction occurs in the module shell zone) is expressed in Equation (14), as follows:

$$D_{i,s}\frac{1}{r}\left(\frac{\partial}{\partial r}\left(r\frac{\partial C_{i,s}}{\partial r}\right)\right) + \frac{D_{i,s}}{scale^2}\frac{\partial^2 C_{i,s}}{\partial \xi^2} = \frac{v_{z,s}}{scale}\left(\frac{\partial C_{i,s}}{\partial \xi}\right) \tag{14}$$

The velocity profile in the shell side is described by Happel's free surface [40] and can be calculated as per Equation (15), as follows:

$$v_{z,s} = v_{z,max}\left\{1-\left(\frac{r_2}{r_3}\right)^2\right\}\left\{\frac{\left(\frac{r}{r_3}\right)^2 - \left(\frac{r_2}{r_3}\right)^2 - 2ln\left(\frac{r}{r_2}\right)}{3+\left(\frac{r_2}{r_3}\right)^4 - 4\left(\frac{r_2}{r_3}\right)^2 + 4\ln\left(\frac{r_2}{r_3}\right)}\right\} \tag{15}$$

The applicable boundary conditions are specified as follows:

$$
\begin{array}{llll}
\text{at } z = H, & C_{i,s} = C_{i,0} & \text{(a)} \\
\text{at } z = 0, & \frac{\partial^2 C_{i,s}}{\partial z^2} = 0 & \text{(b)} \\
\text{at } r = r_2, & D_{i,s}\frac{\partial C_{i,s}}{\partial r} = D_{i,m}\frac{\partial C_{i,m}}{\partial r} & \text{(c)} \\
\text{at } r = r_3, & \frac{\partial C_{i,s}}{\partial r} = 0 & \text{(d)}
\end{array}
\tag{16}
$$

The radius of the free surface (r_3) can be determined as per Equation (17), as follows:

$$
r_3 = r_2\left(\frac{1}{1-\varphi}\right)^{0.5}
\tag{17}
$$

The void fraction of the membrane module (φ) is calculated as per Equation (18):

$$
\varphi = \frac{R^2 - n\,r_2^2}{R^2}
\tag{18}
$$

where R is the module inner radius, n is the number of fibers r_1, and r_2 is the fiber outside radius. The parameters used in the model are shown in Table 2.

Table 2. Physical properties used in the model.

Physical Property	Value	Reference
Diffusivity of CO_2 in shell, $D_{CO_2,s}$	1.855×10^{-5} m^2/s	[9]
Diffusivity of CO_2 in tube, $D_{CO_2,t}$	1.92×10^{-9} m^2/s	[19]
Diffusivity of CO_2 in membrane, $D_{CO_2,m}$	$D_{CO_2,s} \times \varepsilon/\tau$	[37]
Diffusivity of NO_2 in shell, $D_{NO_2,s}$	1.54×10^{-5} m^2/s	[43]
Diffusivity of NO_2 in tube, $D_{NO_2,t}$	1.4×10^{-9} m^2/s	[43]
Diffusivity of solvent in tube, $D_{s,t}$	$0.5 \times D_{CO_2,t}$	Estimated
Diffusivity of NO_2 in membrane, $D_{NO_2,m}$	$D_{NO_2,s} \times \varepsilon/\tau$	[37]
Porosity, ε	0.52	[23]
Tortuosity, τ	$(2-\varepsilon)^2/\varepsilon$	[9]

3. Numerical Solution

The model governing the partial differential and algebraic equations was solved simultaneously using COMSOL software version 5.4. The software uses a finite element method to solve the model equations.

4. Results and Discussion

The accuracy of the mathematical model was checked prior to using the model for studying the effect of the various operating parameters on the percent deletion of CO_2 and NO_2 from the imitated flue gas. The model was authenticated with experimental data for the simultaneous absorption of an NO_2 and CO_2 from gas mixture in a PTFE polymeric gas–liquid hollow fiber membrane [23]. The percent removal of CO_2 and NO_2 was calculated as per Equation (19), as follows:

$$
\%\text{Removal} = \frac{F_{g,in}\,C_{i,in} - F_{g,out}\,C_{i,out}}{F_{g,in}\,C_{i,in}} \times 100
\tag{19}
$$

where $F_{g,in}$, $F_{g,out}$, $C_{i,in}$, and $C_{i,out}$ are the inlet gas flow rate, outlet gas flow rate, inlet gas concentration of component i, and outlet gas concentration of component i, respectively. The 2D surface plot of the CO_2 and NO_2 concentration profile throughout the model domains are shown in Figure 2. The figure reveals that, even though the solubility of CO_2 (0.75) is higher than NO_2 (0.17) in the aqueous NaOH solution, the removal rate of nitrogen dioxide is much higher than that of carbon dioxide, which is attributed to the high reaction rate of NO_2-NaOH.

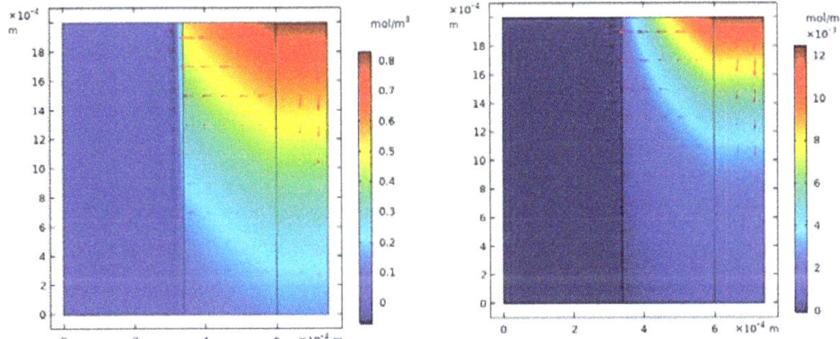

Figure 2. The 2D surface plot for the concentration profile of CO_2 (left) and NO_2 (right) at other fixed parameters (velocity of gas: 1.05 m/s; liquid rate: 0.02 m/s; 2% CO_2; 300 ppm NO_2; the balance is N_2, initial concentration of CO_2 and NO_2, 0.832 mol/m^3, 0.0125 mol/m^3, respectively).

Figure 3. illustrates the association of this model's predictions relative to the experimental results for the effect cross-flow velocity of the feed gas on the simultaneous percent removal of CO_2 and NO_2 with fixed other parameters. A comparison of the percent removal of NO_2 and the experimental data obtained from the literature was in good agreement; by contrast, there is a slight variance in the case of CO_2. The removal flux decreased with the increased inlet gas velocity, attributable to the low residence time.

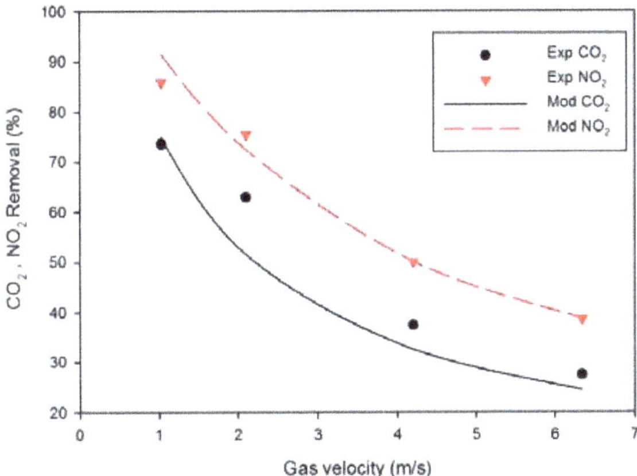

Figure 3. A comparison of this model's predictions and experimental data [23] for the influence of the inlet gas velocity on the simultaneous percent removal of CO_2 and NO_2 (velocity of liquid: 0.05 m/s; solvent concentration: 0.5 M NaOH; inlet gas composition: 2% CO_2; 300 ppm NO_2; the balance is N_2).

The predicted results were also authenticated with the experimental investigations for the case of the effects of the variable liquid velocities on the percent removal of NO_2 and CO_2 (Figure 4) at a fixed gas cross-flow velocity of 2.11 m/s, 0.5 M NaOH, 2% CO_2, 300 ppm NO_2, and the balance was nitrogen. The simulation results matched the experimental data, to a certain extent [23]. The results revealed that the increase in solvent velocity increased the percent removal of CO_2 and NO_2 sharply at low liquid velocities (below 0.05 m/s); as the liquid velocity increased further, there was a slight increase in the percent simultaneous removal of CO_2 and NO_2 from the gas mixture. The insignificant increase in liquid velocity higher than 0.05 m/s was attributed to a decrease in the residence time. The removal flux was calculated as per Equation (20):

$$J_i = \frac{\left(y_{i,in}Q_{in} - y_{i,out}Q_{out}\right) \times 273.15 \times 1000}{22.4 \times T_g \times A} \tag{20}$$

where J_i is the removal molar flux of component i (mol/m^2·s), $y_{i,in}$, $y_{i,out}$ are the inlet and outlet molar fraction of component i in the gas phase, Q_{in}, Q_{out} are the inlet and outlet gas volumetric flow rate (m^3/s), respectively, in gas molar volume (liter/mol) at standard conditions (1 atm and 0 °C) is 22.4; A is the membrane surface area (m^2); 1000 is the conversion factor (liter/m^3); T_g is the gas temperature in K. The 273.15 is the temperature at 0 °C (273.15 K).

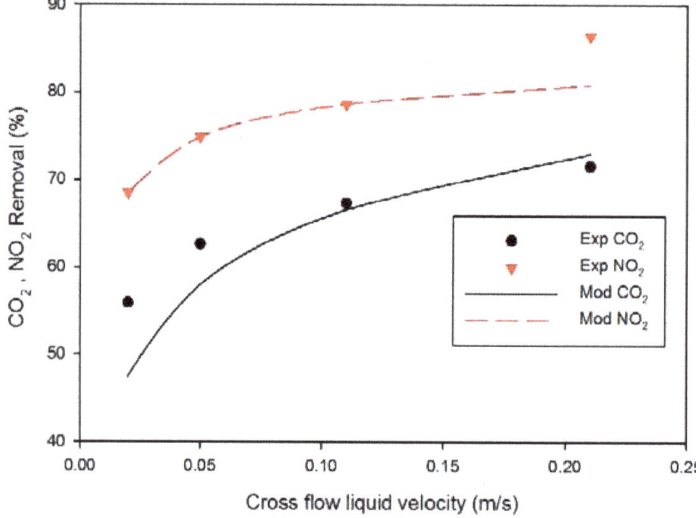

Figure 4. Effect of flow rate of absorbent on the percent removal of undesired gas (gas flow rate = 2.11 m/s, solvent concentration: 0.5 M NaOH; inlet gas composition: 2% CO$_2$; 300 ppm NO$_2$; the balance is N$_2$).

The influence of the speed of the gas on the molar flux of CO$_2$ and NO$_2$ is illustrated in Figure 5. The figure reveals that there was noticeable increase in the removal flux of CO$_2$ with the gas velocity; by contrast, the removal flux of NO$_2$ was insignificant because of its lower inlet concentration in the gas stream (300 ppm), compared with the CO$_2$ inlet concentration (2%). When the velocity of gas was increased from 1.05 m/s to 2.11 m/s, the removal flux increased from 0.003 to 0.0038 mol/m^2·s; at a high gas velocity, the increase was insignificant, from example, with the increase in gas velocity from 4.21 to 6.32 m/s, the increase in molar flux was very small. This was attributed to a decrease in residence time, as well as the insufficient amount of solvent available for the excess amount of CO$_2$ and NO$_2$ components associated with the increase in gas stream volumetric feed rate.

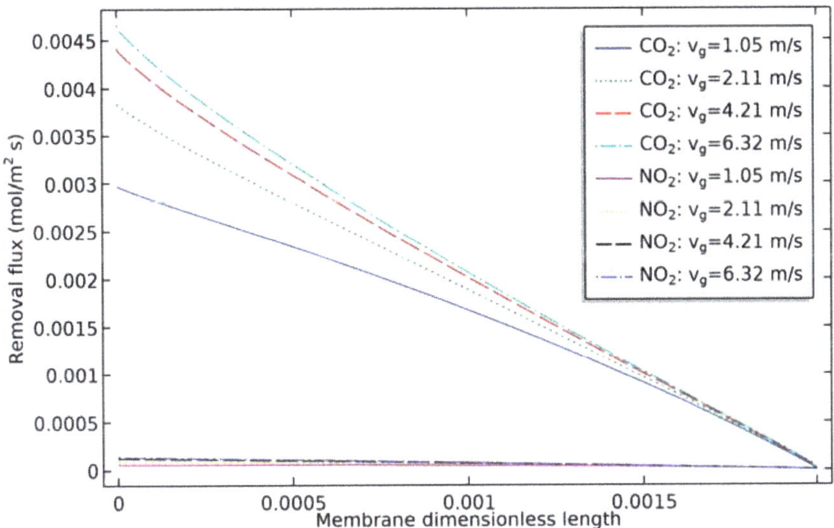

Figure 5. Effect of the variable gas velocities on the NO_2 and CO_2 removal flux along the membrane length with fixed other parameters (concentration of NaOH: 0.5 M; velocity of solvent: 0.05 m/s; gas composition: 2% CO_2; 300 ppm NO_2; the balance is N_2).

Figure 6 demonstrates the effect of the change in the inlet CO_2 mole fraction at a fixed inlet concentration of NO_2 (300 ppm) on the component's molar flux. The CO_2 molar flux increased significantly when its concentration increased, which was expected, because as the amount of CO_2 increased in the inlet gas stream, more CO_2 was being absorbed, and hence the CO_2 removal molar flux increased (molar flux: moles gas removed per area per time). By contrast, because of the fixed low concentration of NO_2 in the feed stream, its removal flux was insignificant compared to that of CO_2.

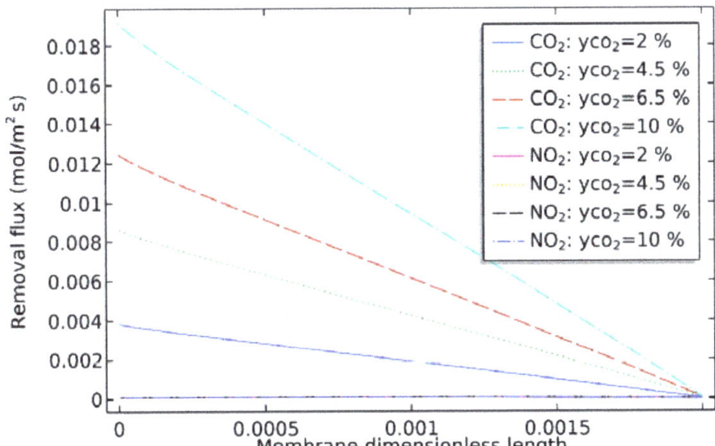

Figure 6. Impact of the inlet CO_2 feed concentration on the NO_2 and CO_2 capture flux along the membrane length at other fixed parameters (aqueous NaOH: 0.5 M; velocity of solvent: 0.05 m/s; gas velocity: 2.11 m/s; gas composition: (2% to 10%) CO_2; 300 ppm NO_2; the balance is N_2).

Figure 7 explains the effect of change in the inlet NO_2 mole fraction in the feed gas stream at a fixed concentration of CO_2 (2%) on the removal flux of CO_2 and NO_2. The predicted results are in the range of the experimental data [23] under the same conditions. The effect of change in the inlet mole fraction of NO_2 on the CO_2 removal flux was insignificant, the CO_2 removal flux was kept around 0.004 mol/m²·s and was not influenced by the change of the NO_2 inlet mole fraction. By contrast, there was a slight increase in the removal flux of NO_2 which caused an increase form 3×10^{-5} to 15×10^{-5} mol/m²·s. This was attributed to the low inlet concentration of NO_2 (in ppm) compared with the CO_2 inlet concentration (2%), and consequently, the amount absorbed from CO_2 and NO_2 did not change significantly.

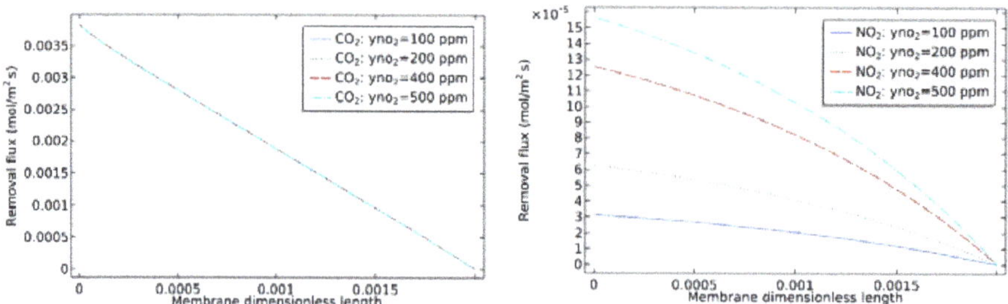

Figure 7. Effect of NO_2 mole fraction in the feed gas stream on the removal molar flux of CO_2 (left) and NO_2 (right) at other fixed parameters (liquid velocity: 0.05 m/s; gas velocity: 2.11 m/s; 0.5 M NaOH; 100 to 500 ppm NO_2; 2% CO_2; the balance is N_2).

5. Conclusions

Model equations based on material balance were utilized to describe and study the simultaneous detention of NO_2 and CO_2 with aqueous NaOH solution in a membrane module. The hollow fiber membranes were fabricated from PTFE polymer. The model equations were solved, and the model predicted results were compared with data from experimental investigation available in literature. The model was found to be in good agreement with the experimental findings. The mathematical model was then employed to study the influence of the inlet flow rate of gas and liquid, concentration of CO_2 and NO_2 in the feed stream on their percent removal and molar flux. The results revealed that the increase in CO_2 inlet mole fraction and gas cross-flow velocity shows a strong impact on the molar flux. By contrast, the change in the NO_2 inlet concentration showed insignificant influence on the CO_2 removal flux.

Conflicts of Interest: The author declare no conflict of interest.

References

1. Sun, B.; Sheng, M.; Gao, W.; Zhang, L.; Arowo, M.; Liang, Y.; Shao, L.; Chu, G.W.; Zou, H.; Chen, J.F. Absorption of nitrogen oxides into sodium hydroxide solution in a rotating packed bed with preoxidation by ozone. *Energy Fuels* **2017**, *31*, 11019–11025. [CrossRef]
2. Lee, Y.H.; Jung, W.S.; Choi, Y.R.; Oh, J.S.; Jang, S.D.; Son, Y.G.; Cho, M.H.; Namkung, W.; Koh, D.J.; Mok, Y.S.; et al. Application of pulsed corona induced plasma chemical process to an industrial incinerator. *Env. Sci. Technol.* **2003**, *37*, 2563–2567. [CrossRef]
3. Golkhar, A.; Keshavarz, P.; Mowla, D. Investigation of CO_2 removal by silica and CNT nanofluids in microporous hollow fiber membrane contactors. *J. Memb. Sci.* **2013**, *433*, 17–24. [CrossRef]
4. Kohl, A.L.; Nielsen, R.B. *Gas. Purification*, 5th ed.; Gulf Publishing Company: Houston, TX, USA, 1997.
5. Holstvoogd, R.D.; Van Swaaij, W.P.M. The influence of adsorption capacity on enhanced gas absorption in activated carbon slurries. *Chem. Eng. Sci.* **1990**, *45*, 151–162. [CrossRef]

6.   Yildirim, Ö.; Kiss, A.A.; Hüser, N.; Leßmann, K.; Kenig, E.Y. Reactive absorption in chemical process industry: A review on current activities. *Chem. Eng. J.* **2012**, *213*, 371–391. [CrossRef]

7.   Qi, Z.; Cussler, E.L. Microporous hollow fibers for gas absorption: I. Mass transfer in the liquid. *J. Memb. Sci.* **1985**, *23*, 321–332. [CrossRef]

8.   Qi, Z.; Cussler, E.L. Microporous hollow fibers for gas absorption: II. Mass transfer across the membrane. *J. Memb. Sci.* **1985**, *23*, 333–345. [CrossRef]

9.   Ghasem, N.; Al-Marzouqi, M. Modeling and experimental study of carbon dioxide absorption in a flat sheet membrane contactor. *J. Membr. Sci. Res.* **2017**, *3*, 57–63.

10.   Ghasem, N.; Al-Marzouqi, M.; Al-Marzouqi, R.; Dowaidar, A.; Vialatte, M. Removal of CO_2 from gas mixture using hollow fiber membrane contactors fabricated from PVDF/Triacetin/Glycerol cast solution. In Proceedings of the European Conference of Chemical Engineering, ECCE'10, European Conference of Civil Engineering, ECCIE'10, European Conference of Mechanical Engineering, ECME'10, European Conference of Control, ECC'10, Tenerife, Spain, 30 November–2 December 2010.

11.   Darabi, M.; Rahimi, M.; Molaei Dehkordi, A. Gas absorption enhancement in hollow fiber membrane contactors using nanofluids: Modeling and simulation. *Chem. Eng. Process. Process. Intensif.* **2017**, *119*, 7–15. [CrossRef]

12.   Kenarsari, S.D.; Yang, D.; Jiang, G.; Zhang, S.; Wang, J.; Russell, A.G.; Wei, Q.; Fan, M. Review of recent advances in carbon dioxide separation and capture. *Rsc Adv.* **2013**, *3*, 22739–22773. [CrossRef]

13.   Lv, Y.; Yu, X.; Tu, S.T.; Yan, J.; Dahlquist, E. Experimental studies on simultaneous removal of CO_2 and SO_2 in a polypropylene hollow fiber membrane contactor. *Appl. Energy* **2012**, *97*, 283–288. [CrossRef]

14.   Gabelman, A.; Hwang, S.-T. Hollow fiber membrane contactors. *J. Memb. Sci.* **1999**, *159*, 61–106. [CrossRef]

15.   Ansaripour, M.; Haghshenasfard, M.; Moheb, A. Experimental and Numerical Investigation of CO_2 Absorption Using Nanofluids in a Hollow-Fiber Membrane Contactor. *Chem. Eng. Technol.* **2018**, *41*, 367–378. [CrossRef]

16.   Zhang, Z.E.; Yan, Y.F.; Zhang, L.; Ju, S.X. Hollow fiber membrane contactor absorption of CO_2 from the flue gas: Review and perspective. *Glob. Nest J.* **2014**, *16*, 354–373.

17.   Sohrabi, M.R.; Marjani, A.; Moradi, S.; Davallo, M.; Shirazian, S. Mathematical modeling and numerical simulation of CO_2 transport through hollow-fiber membranes. *Appl. Math. Model.* **2011**, *35*, 174–188. [CrossRef]

18.   Mavroudi, M.; Kaldis, S.P.; Sakellaropoulos, G.P. A study of mass transfer resistance in membrane gas-liquid contacting processes. *J. Memb. Sci.* **2006**, *272*, 103–115. [CrossRef]

19.   Rezakazemi, M.; Darabi, M.; Soroush, E.; Mesbah, M. CO_2 absorption enhancement by water-based nanofluids of CNT and SiO_2 using hollow-fiber membrane contactor. *Sep. Purif. Technol.* **2019**, *210*, 920–926. [CrossRef]

20.   Li, J.L.; Chen, B.H. Review of CO_2 absorption using chemical solvents in hollow fiber membrane contactors. *Sep. Purif. Technol.* **2005**, *41*, 109–122. [CrossRef]

21.   Park, H.H.; Deshwal, B.R.; Jo, H.D.; Choi, W.K.; Kim, I.W.; Lee, H.K. Absorption of nitrogen dioxide by PVDF hollow fiber membranes in a G-L contactor. *Desalination* **2009**, *243*, 52–64. [CrossRef]

22.   Ghasem, N.; Al-Marzouqi, M.; Duidar, A. Effect of PVDF concentration on the morphology and performance of hollow fiber membrane employed as gas-liquid membrane contactor for CO_2 absorption. *Sep. Purif. Technol.* **2012**, *98*, 174–185. [CrossRef]

23.   Ghobadi, J.; Ramirez, D.; Khoramfar, S.; Jerman, R.; Crane, M.; Hobbs, K. Simultaneous absorption of carbon dioxide and nitrogen dioxide from simulated flue gas stream using gas-liquid membrane contacting system. *Int. J. Greenh. Gas. Control.* **2018**, *77*, 37–45. [CrossRef]

24.   Ghobadi, J.; Ramirez, D.; Jerman, R.; Crane, M.; Khoramfar, S. CO_2 separation performance of different diameter polytetrafluoroethylene hollow fiber membranes using gas-liquid membrane contacting system. *J. Memb. Sci.* **2018**, *549*, 75–83. [CrossRef]

25.   Zhang, Z.; Yan, Y.; Zhang, L.; Ju, S. Numerical simulation and analysis of CO_2 removal in a polypropylene hollow fiber membrane contactor. *Int. J. Chem. Eng.* **2014**, *2014*. [CrossRef]

26.   Rahmatmand, B.; Keshavarz, P.; Ayatollahi, S. Study of Absorption Enhancement of CO_2 by SiO_2, Al_2O_3, CNT, and Fe_3O_4 Nanoparticles in Water and Amine Solutions. *J. Chem. Eng. Data* **2016**, *61*, 1378–1387. [CrossRef]

27. Hosseinzadeh, A.; Hosseinzadeh, M.; Vatani, A.; Mohammadi, T. Mathematical modeling for the simultaneous absorption of CO_2 and SO_2 using MEA in hollow fiber membrane contactors. *Chem. Eng. Process. Process. Intensif.* **2017**, *111*, 35–45. [CrossRef]

28. Nii, S.; Takeuchi, H. Removal of CO_2 and/or SO_2 from gas streams by a membrane absorption method. *Gas. Sep. Purif.* **1994**, *8*, 107–114. [CrossRef]

29. Mavroudi, M.; Kaldis, S.P.; Sakellaropoulos, G.P. Reduction of CO_2 emissions by a membrane contacting process. *Fuel* **2003**, *82*, 2153–2159. [CrossRef]

30. Hedayat, M.; Soltanieh, M.; Mousavi, S.A. Simultaneous separation of H_2S and CO_2 from natural gas by hollow fiber membrane contactor using mixture of alkanolamines. *J. Memb. Sci.* **2011**, *377*, 191–197. [CrossRef]

31. Mohammaddoost, H.; Azari, A.; Ansarpour, M.; Osfouri, S. Experimental investigation of CO_2 removal from N_2 by metal oxide nanofluids in a hollow fiber membrane contactor. *Int. J. Greenh. Gas. Control.* **2018**, *69*, 60–71. [CrossRef]

32. Dai, Z.; Usman, M.; Hillestad, M.; Deng, L. Modelling of a tubular membrane contactor for pre-combustion CO_2 capture using ionic liquids: Influence of the membrane configuration, absorbent properties and operation parameters. *Green Energy Env.* **2016**, *1*, 266–275. [CrossRef]

33. Zhang, Z.; Cai, J.; Chen, F.; Li, H.; Zhang, W.; Qi, W. Progress in enhancement of CO_2 absorption by nanofluids: A mini review of mechanisms and current status. *Renew. Energy* **2018**, *118*, 527–535. [CrossRef]

34. Hajilary, N.; Rezakazemi, M. CFD modeling of CO_2 capture by water-based nanofluids using hollow fiber membrane contactor. *Int. J. Greenh. Gas. Control.* **2018**, *77*, 88–95. [CrossRef]

35. Wang, J.; Gao, X.; Ji, G.; Gu, X. CFD simulation of hollow fiber supported NaA zeolite membrane modules. *Sep. Purif. Technol.* **2019**, *213*, 1–10. [CrossRef]

36. Khan, M.J.H.; Hussain, M.A.; Mansourpour, Z.; Mostoufi, N.; Ghasem, N.M.; Abdullah, E.C. CFD simulation of fluidized bed reactors for polyolefin production—A review. *J. Ind. Eng. Chem.* **2014**, *20*, 3919–3946. [CrossRef]

37. Ghasem, N.; Al-Marzouqi, M.; Abdul Rahim, N. Modeling of CO_2 absorption in a membrane contactor considering solvent evaporation. *Sep. Purif. Technol.* **2013**, *110*, 1–10. [CrossRef]

38. Rosli, A.; Shoparwe, N.F.F.; Ahmad, A.L.L.; Low, S.C.; Lim, J.K.K. Dynamic modelling and experimental validation of CO_2 removal using hydrophobic membrane contactor with different types of absorbent. *Sep. Purif. Technol.* **2019**, *219*, 230–240. [CrossRef]

39. Günther, J.; Schmitz, P.; Albasi, C.; Lafforgue, C. A numerical approach to study the impact of packing density on fluid flow distribution in hollow fiber module. *J. Memb. Sci.* **2010**, *348*, 277–286. [CrossRef]

40. Happel, J. Viscous flow relative to arrays of cylinders. *Aiche J.* **1959**, *5*, 174–177. [CrossRef]

41. Villeneuve, K.; Albarracin Zaidiza, D.; Roizard, D.; Rode, S.; Nagy, E.; Feczkó, T.; Koroknai, B. Enhancement of oxygen mass transfer rate in the presence of nanosized particles. *Chem. Eng. Sci.* **2007**, *62*, 7391–7398.

42. Zanfir, M.; Gavriilidis, A.; Wille, C.; Hessel, V. Carbon dioxide absorption in a falling film microstructured reactor: Experiments and modeling. *Ind. Eng. Chem. Res.* **2005**, *44*, 1742–1751. [CrossRef]

43. Versteeg, G.F.; van Swaal, W.P.M. Solubility and diffusivity of acid gases (CO_2, N_2O) in aqueous alkanolamine solutions. *J. Chem. Eng. Data* **1988**, *33*, 29–34. [CrossRef]

Article

Adsorption of NO Gas Molecules on Monolayer Arsenene Doped with Al, B, S and Si: A First-Principles Study

Keliang Wang [1,*], **Jing Li** [1,*], **Yu Huang** [2], **Minglei Lian** [1] and **Dingmei Chen** [1]

[1] College of Chemistry and Materials Engineering, Liupanshui Normal University, Liupanshui 553004, China
[2] College of Chemistry and Chemical Engineering, Guizhou University, Guiyang 550025, China
* Correspondence: wangkeliang84@163.com (K.W.); woxinfeiyang1986@163.com (J.L.);
 Tel.: +86-858-860-0172 (K.W. & J.L.)

Received: 20 July 2019; Accepted: 9 August 2019; Published: 15 August 2019

Abstract: The structures and electronic properties of monolayer arsenene doped with Al, B, S and Si have been investigated based on first-principles calculation. The dopants have great influences on the properties of the monolayer arsenene. The electronic properties of the substrate are effectively tuned by substitutional doping. After doping, NO adsorbed on four kinds of substrates were investigated. The results demonstrate that NO exhibits a chemisorption character on Al-, B- and Si-doped arsenene while a physisorption character on S-doped arsenene with moderate adsorption energy. Due to the adsorption of NO, the band structures of the four systems have great changes. It reduces the energy gap of Al- and B-doped arsenene and opens the energy gap of S- and Si-doped arsenene. The large charge depletion between the NO molecule and the dopant demonstrates that there is a strong hybridization of orbitals at the surface of the doped substrate because of the formation of a covalent bond, except for S-doped arsenene. It indicates that Al-, B- and Si-doped arsenene might be good candidates as gas sensors to detect NO gas molecules owning to their high sensitivity.

Keywords: arsenene; doping; first principles study; gas adsorption; two-dimensional

---

1. Introduction

Owing to the adequate preparation of single-layer graphene, the research on two-dimensional (2D) materials has been increasing. Graphene, silicene, germanene, hexagonal boron nitride (h-BN), phosphorene, transition metal dichalcogenide and stanine [1–3] have attracted more and more attention in recent years. Especially, because of their ultrathin thickness and high surface-to-volume ratio, atomically thin 2D nanomaterials have been proven to be prospected nanoscale in catalyst, gas sensors and energy storage [4–6].

Good gas sensors for the detection of toxic gas plays a crucial role in industries, chemical detection and environment protection [7–9]. Due to the excellent structural properties of 2D materials, it has attracted wide attention in the field of gas sensors. Ma et al. [10] studied the small molecules (CO, H_2O, NII_3, N_2, NO_2, NO and O_2) adsorbed on the InSe monolayer and found that 2D InSe nanomaterials is a potential candidate for fabricating gas sensors. The study of Liang et al. [11] showed that the affinity between Ga-doped graphene and gas molecules is stronger than that of pristine graphene. For the adsorption of NO_x gases, there have been some experimental studies. MoS_2 nanosheets, prepared by the micromechanical exfoliation method, show a high selectivity for NO_x and NH_3 gas molecules at the ppb level [12,13]. Schedin et al. [14] have researched the adsorption of NO_2, NH_3, H_2O and CO gas molecules on graphene-based gas sensors and found that graphene was electronically quiet enough to be used as a single electronic detector at room temperature.

Recently, monolayer arsenene was predicted to be indirect semiconductors based on its excellent properties of high stability and wide band gaps [15,16]. Similar to silicene, arsenene possesses buckled honeycomb structures [17]. Liu et al. [18] have researched the adsorption of six kinds of small gas molecules on the pristine arsenene monolayer and found that in gas sensing, arsenene can be a potential candidate applied for NO and NO_2 molecules with electrical and magnetic methods. Khan and his coworkers [19] have reported NH_3 and NO_2 molecules show a significant affinity for arsenene leading to strong physisorption.

Generally, doping can improve the adsorption ability of 2D materials to gas molecules [20]. For arsenene doping, the adsorption energy of NH_3 can be enhanced by Ge- and Se-doping [21]. Bai et al. [22] have investigated the structures and properties of monolayer arsenene doped with Ge, Ga, Sb and P, and the results demonstrated that monolayer arsenene doped with Ga changes into the direct band gap. Chen et al. [23] have calculated the adsorption properties of NO_2 and SO_2 on different types of pristine, boron- and nitrogen-doped arsenene, and found that N-doped arsenene is more suitable for SO_2 gas sensors as well as that P-doped arsenene has more potential application in NO_2 gas sensors.

In this paper, we investigated the structures and electronic properties of monolayer arsenene doped with Al, B, S and Si theoretically based on the first-principles calculation. After doping, NO adsorbed on four kinds of substrates were investigated. The adsorption energy, adsorption distance and Mulliken charge transfer were calculated. These results can support a theoretical foundation to design gas sensors using 2D arsenic materials.

2. Computational Methods

The first-principles study, based on density functional theory (DFT), was calculated through the DMol^3 code. The generalized gradient approximation (GGA) with the Perdew–Burke–Ernzerhof (PBE) functional [18,19,22] was adopted in the simulation. However, GGA ignores the long-range electron effects, which led to the overestimation of van der Waals forces [24–27]. Therefore, the Grimme custom method was used to describe tiny van der Waals interaction [23,28]. Double numerical atomic orbital plus polarization (DNP) was selected as the basis set to expand electronic wave function. A $4 \times 4 \times 1$ supercell containing 32 atoms was adopted with a vacuum space of 20 Å to guarantee there was no interaction between adjacent layers. An $8 \times 8 \times 1$ and $16 \times 16 \times 1$ k-points in the Brillouin zone were adopted to optimize the configurations and calculate the electronic properties, respectively [19,21]. The convergence tolerance for energy minimizations was 1.0×10^{-6} Ha, for maximum force it was 0.002 Ha/Å and for geometry optimizations it was 0.005 Å, respectively. The flow chart of the computational process is shown in Figure 1.

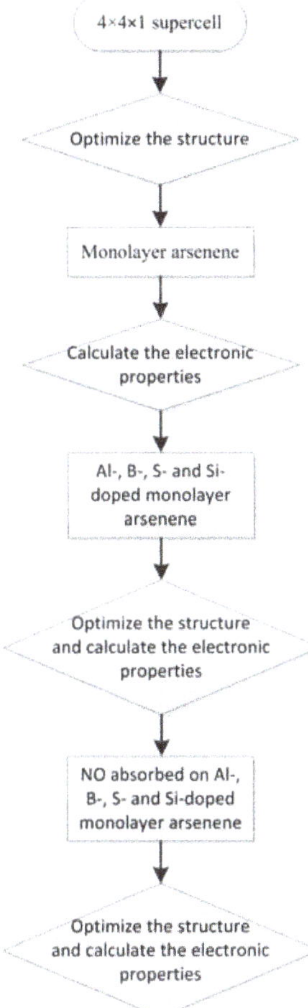

Figure 1. The flow chart of the computational process.

3. Result and Discussion

The structure and electronic properties of pristine monolayer arsenene is firstly tested to check the accuracy of the calculation procedure. The full relaxed lattice constant of monolayer arsenene is 3.624 Å, which is very close to the experimental data [29]. It can be seen from Figure 2a that the bond length of As–As and the bond angle are 2.524 Å and 91.751°, respectively. The thickness of monolayer arsenene is 1.412 Å. The monolayer arsenene with an indirect band gap of 1.718 eV shows a semiconductor property, in which the valence band maximum (VBM) displays at Γ point and the conduction band minimum (CBM) displays between M and Γ point. The results agree well with the previous results [18,22,30,31]. These results verify the accuracy of the calculation procedure and the characterization of the material. Simultaneously, it can be seen from Figure 2b that the partial density of states (PDOS) of the supercell of arsenene, which is mainly dominated by s and p orbitals of As atoms, but the influence on the total density of states (DOS) of the p orbital is greater than that of the s orbital. Similar to the feature of blue phosphorene and silicene [22,32], the energy region near the Fermi level is mainly due to the p orbital of As atoms.

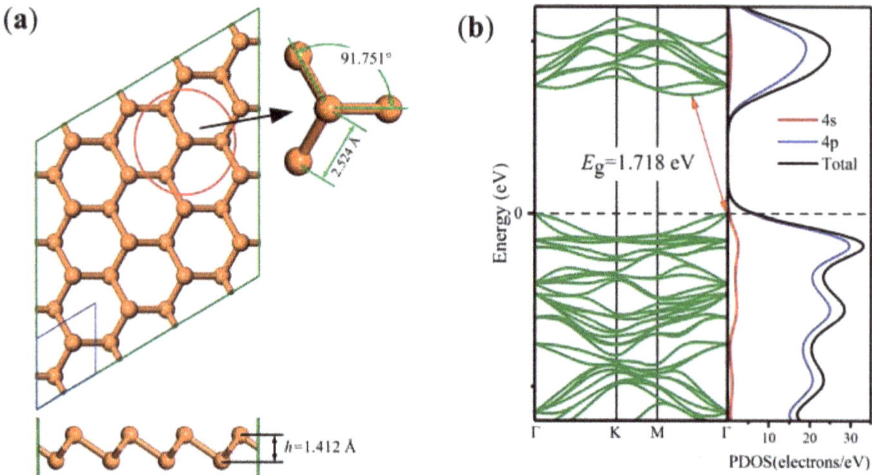

Figure 2. (**a**) Optimized structure of monolayer arsenene. The blue and green boxes in the top view (top panel) show the primitive cell and the 4 × 4 supercell, respectively. The buckling height (*h*) is indicated in the side view (bottom panel). (**b**) Electronic band structure (left panel) and partial density of states (PDOS) (right panel) of the 4 × 4 supercell of monolayer arsenene. The Fermi energy was set to zero.

Substitutional doping can effectively improve the adsorption ability of 2D materials to gas molecules. So, the structures and electronic properties of X-doped monolayer arsenene (X = Al, B, S, Si) have been firstly optimized and calculated. The binding energy (E_b) is calculated and defined as $E_b = [E_{X\text{-}As} - (n - 1)E_{As} - E_X]/n$, where $E_{X\text{-}As}$ is the total energy of substitutional systems, E_{As} and E_X are the energies of the isolated atom As and the dopant atom X, respectively, and n is the number of As atoms in arsenene supercell. It can be seen from the above equation that the greater the binding force between doping elements and the substrate, the more negative the value of E_b. As shown in Table 1, the bond length $l_{X\text{-}As}$ between the dopant X and the nearest As element is in the range of 2.064 to 2.459 Å, and $l_{B\text{-}As}$ is the shortest with 2.064 Å. The binding strength increases in the order of S < Al < Si < B with high binding energy of −4.065 eV, −4.085 eV, −4.108 eV and −4.137 eV, respectively. It indicates that S, Al, Si and B can interact strongly with its neighboring As atoms. The above results show that the dopants can effectively affect the binding energy for the doped monolayer arsenene.

Table 1. The binding energy (E_b), energy gap (E_g), Mulliken population (Q) and the bond length of X-As ($l_{X\text{-}As}$). A positive Q value means the electrons move from the dopant to the substrate. X denotes the dopant atom.

System	E_b (eV)	E_g (eV)	Q (e)	$l_{X\text{-}As}$ (Å)
Al-doped	−4.085	1.538	0.807	2.429
B-doped	−4.137	1.370	0.175	2.068
S-doped	−4.065	0	−0.120	2.447
Si-doped	−4.108	0	0.665	2.408

Meanwhile, Mulliken analysis is used to calculate the atomic charge near the dopant X and the calculated results are listed in Table 1. The Mulliken population of Al, B, S and Si atoms are 0.807, 0.175, −0.120 and 0.665 e, respectively. The results show that a large amount of electrons transfer occurs between the dopant and the substrate, which implies that there are strong interactions between the dopant and the substrate. The electrons transfer from the dopant to the substrate except for the S atom, because the S atom has a higher electronegativity than that of the As atom. Among these structures, the values of the Mulliken population of Al, B and Si atoms are positive. This leads to the formation of a huge electron depletion layer on the substrate, which is conducive to improving the adsorption properties for NO gas molecules.

The optimized adsorption configurations of NO adsorbed on four doped systems are demonstrated in Figure 3. The corresponding parameters are listed in Table 2, including the adsorption energy (E_{ad}), adsorption distance (d) and Mulliken charge (Q). It can be seen from Figure 3 that the N atom is toward the substrate and the O atom is away from the substrate in four doped systems. E_{ad} is defined as $E_{ad} = E_{NO/X\text{-}As} - (E_{X\text{-}As} + E_{NO})$, where $E_{NO/X\text{-}As}$, $E_{X\text{-}As}$ and E_{NO} denote the total energies of the NO molecule adsorbed on the doped system, the isolated doped substrate and the NO molecule with the same lattice parameters, respectively. The more negative E_{ad} is, the more stable the structure is. B-doped arsenene has the largest adsorption energy of −1.884 eV and the shortest adsorption distance of 1.428 Å with the NO molecule. For Al-, S- and Si-doped arsenene, the adsorption energy values are −1.157 eV, −0.469 eV and −1.586 eV, and the adsorption distance values are 1.942 Å, 2.548 Å and 1.864 Å, respectively. It is clear that the adsorption of NO on S-doped arsenene is physical adsorption and NO on Al-, B- and Si-doped arsenene is chemical adsorption. Mulliken population analysis is performed and the negative Q value indicates electron transfer from the substrates to the NO molecule. It shows that the NO molecule is an electron acceptor with four substrates.

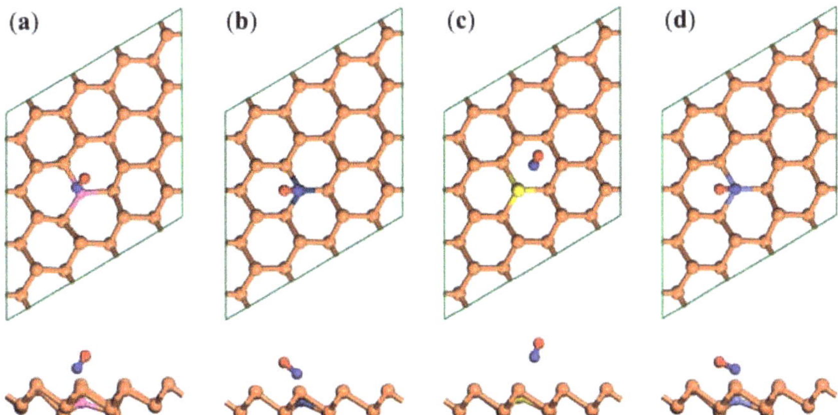

Figure 3. Top view and side view of the most energetically favorable adsorption configurations for NO adsorbed on (**a**) Al-, (**b**) B-, (**c**) S- and (**d**) Si-doped monolayer arsenene. The O and N atoms are labeled red and blue, respectively.

Table 2. Adsorption energy (E_{ad}), the shortest distance of the As-dopant atom (d) and Mulliken charge (Q) for the optimized stable configurations of gas molecules on Al-, B-, S- and Si-doped arsenene. A negative Q value indicates electron move from the doped substrates to NO molecule.

System	E_{ad} (eV)	d (Å)	Q (e)
Al-doped	−1.157	1.942	−0.184
B-doped	−1.884	1.428	−0.063
S-doped	−0.469	2.548	−0.002
Si-doped	−1.586	1.864	−0.327

Furthermore, the band structures of Al-, B-, S- and Si- doped monolayer arsenene are also calculated to investigate the effects introduced by the dopant. As shown in Figure 4a,b, the dopants of Al and B change the 2D material to the direct band gap from the indirect band gap. Both CBM and VBM are displayed on the Γ point in Brillouin Zone with the values of the band gap 1.538 and 1.370 eV, respectively. The Fermi level of S- and Si-doped arsenene systems enter into the conduction band (see Figure 4c,d) and a semiconductor-metal transition is realized. Through the band structures comparison of four doped systems, it can be concluded that the electronic properties of 2D materials are effectively tuned by substitutional doping.

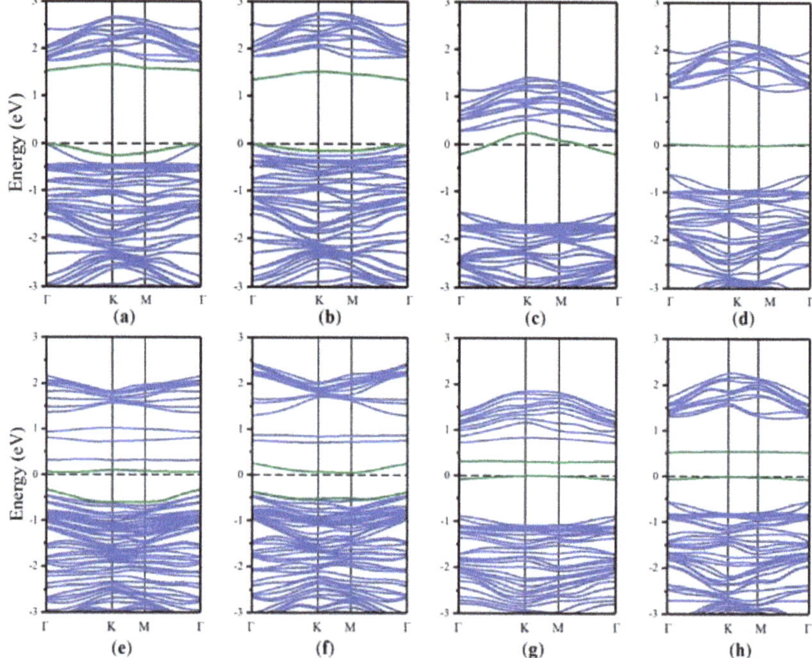

Figure 4. Band structures of (**a**) Al-, (**b**) B-, (**c**) S- and (**d**) Si-doped arsenene $As_{31}X$ systems, as well as NO adsorbed on (**e**) $As_{31}Al$, (**f**) $As_{31}B$, (**g**) $As_{31}S$ and (**h**) $As_{31}Si$ monolayers, respectively. The Fermi level is set to zero with the black dashed line.

As can be seen from Figure 4e–h, the band structures of four systems have great changes after the adsorption of the NO molecule. Interestingly, the energy gap values of NO/Al-doped arsenene and NO/B-doped arsenene both change to 0.372 and 0.407 eV, and the energy gap values of NO/S-doped arsenene and NO/Si-doped arsenene change to 0.294 and 0.521 eV. The adsorption of the NO molecule reduces the energy gap of Al- and B-doped arsenene and opens the energy gap of S- and Si-doped arsenene.

To explore the electronic properties of the four systems, the total density of states (DOS) of four systems before and after absorbing the NO molecule were analyzed and shown in Figure 5. Because of the adsorption of NO, the DOS of Al- and B-doped arsenene are shifted to the lower energy level, but the DOS of S- and Si-doped arsenene are shifted slightly to the higher energy level, which are in accordance with the changes of band structures.

To further investigate the NO adsorption on X-doped monolayer arsenene, the charge density differences of NO adsorbed on four kinds of substrates were calculated and shown in Figure 6. The charge density difference can be expressed as $\Delta\rho = \rho_{NO/X\text{-}As} - (\rho_{X\text{-}As} + \rho_{NO})$, where $\rho_{NO/X\text{-}As}$, $\rho_{X\text{-}As}$ and ρ_{NO} denote the total charge densities of the optimized NO adsorption system, isolated substrate and NO molecule, respectively. The purple and green parts correspond to the charge accumulation and the charge depletion, respectively. It can be clearly seen that charge redistribution is generated between the NO molecule and the dopant atoms due to the adsorption. The large charge depletion between the NO molecule and the dopant demonstrates that there is a strong hybridization of orbitals at the surface of the doped substrate because of the formation of a covalent bond except for S-doped arsenene. The results are consistent with the Mulliken population analysis.

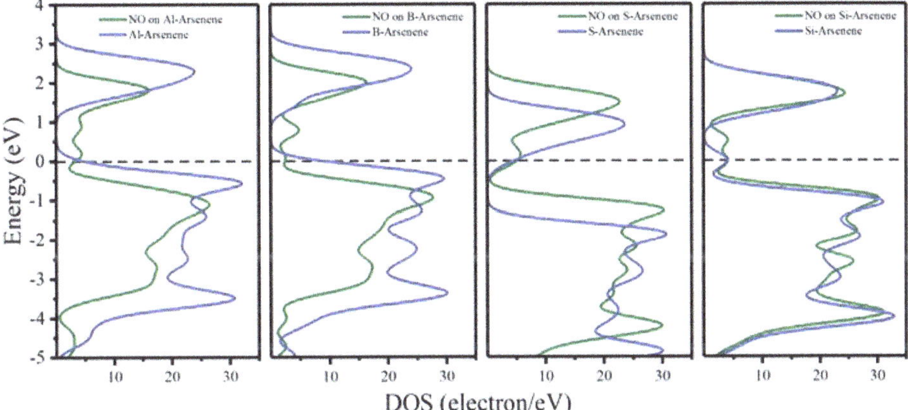

Figure 5. The density of states (DOS) of four systems. Blue and green lines present the DOS of substrates before and after NO adsorption, respectively.

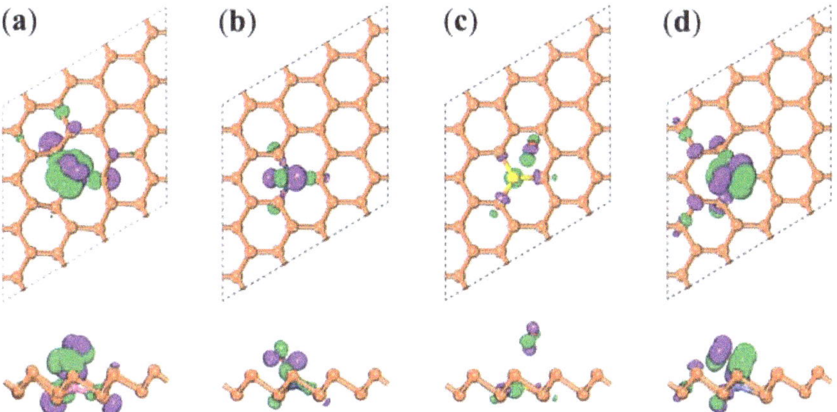

Figure 6. Charge density difference plots for NO adsorbed on (**a**) Al-, (**b**) B-, (**c**) S- and (**d**) Si-doped monolayer arsenene, respectively. The purple (green) distribution denotes the charge accumulation (depletion) with the isosurface of 0.03 e/Å^3 for (**a**) and (**d**); 0.044 e/Å^3 for (**b**) and 0.05 e/Å^3 for (**c**).

4. Conclusions

On the basis of first-principles calculation, the structures and electronic properties of monolayer arsenene doped with Al, B, S and Si were investigated. The dopants have great influences on the properties of the monolayer arsenene. The electronic properties of the substrate are effectively tuned by substitutional doping.

After doping, NO adsorbed on four kinds of substrates have been investigated. The NO molecule is an electron acceptor with four substrates. The adsorption energy, adsorption distance and Mulliken charge transfer have been calculated. The results demonstrate that NO exhibits a chemisorption character on Al-, B- and Si-doped arsenene, while a physisorption character on S-doped arsenene with moderate adsorption energy.

Due to the adsorption of NO, the band structures of the four systems have great changes. It reduces the energy gap of Al- and B-doped arsenene and opens the energy gap of S- and Si-doped arsenene. The large charge depletion between the NO molecule and the dopant demonstrates that there is a strong hybridization of orbitals at the surface of the doped substrate because of the formation of a covalent bond except for S-doped arsenene. It indicates that Al-, B- and Si-doped arsenene might be good candidates as gas sensors to detect NO gas molecules owing to their high sensitivity.

Author Contributions: K.W. and J.L. conceived and designed this case-study as well as wrote the paper; M.L. and D.C. reviewed the paper; all authors interpreted the data; and Y.H. substantively revised the work and contributed the process simulation.

Funding: This work is financially supported by Guizhou Province United Fund (Qiankehe J zi LKLS[2013]27), Excellent engineers education training plan (LPSSY zyjypyjh201702), Guizhou Solid Waste Recycling Laboratory of Coal Utilization ([2011]278), the Scientific and Technological Innovation Platform of Liupanshui (52020-2018-03-02) and Academician Workstation of Liupanshui Normal University (qiankehepingtairencai [2019]5604).

Conflicts of Interest: The authors declare no conflict of interest.

References

1. Nagarajan, V.; Bhattacharyya, A.; Chandiramouli, R. Adsorption of ammonia molecules and humidity on germanane nanosheet-a density functional study. *J. Mol. Graph. Model.* **2018**, *79*, 149–156. [CrossRef] [PubMed]

2. Komsa, H.; Krasheninnikov, A.V. Electronic structures and optical properties of realistic transition metal dichalcogenide heterostructures from first principles. *Phys. Rev. B* **2013**, *88*, 85318. [CrossRef]

3. Zhang, S.; Guo, S.; Huang, Y.; Zhu, Z.; Cai, B.; Xie, M.; Zhou, W.; Zeng, H. Two-dimensional SiP: An unexplored direct band-gap semiconductor. *2D Mater.* **2017**, *4*, 15030. [CrossRef]

4. Yang, W.; Gan, L.; Li, H.; Zhai, T. Two-dimensional layered nanomaterials for gas-sensing applications. *Inorg. Chem. Front.* **2016**, *3*, 433–451. [CrossRef]

5. Lightcap, I.V.; Kamat, P.V. Graphitic design: Prospects of graphene-based nanocomposites for solar energy conversion, storage, and sensing. *Acc. Chem. Res.* **2012**, *46*, 2235–2243. [CrossRef] [PubMed]

6. Zhang, S.; Zhou, W.; Ma, Y.; Ji, J.; Cai, B.; Yang, S.A.; Zhu, Z.; Chen, Z.; Zeng, H. Antimonene oxides: Emerging tunable direct bandgap semiconductor and novel topological insulator. *Nano Lett.* **2017**, *17*, 3434–3440. [CrossRef]

7. Nagarajan, V.; Chandiramouli, R. Adsorption behavior of NH_3 and NO_2 molecules on stanene and stanane nanosheets—A density functional theory study. *Chem. Phys. Lett.* **2018**, *695*, 162–169. [CrossRef]

8. Yamazoe, N. Toward innovations of gas sensor technology. *Sens. Actuators B Chem.* **2005**, *108*, 2–14. [CrossRef]

9. Lagód, G.; Duda, S.M.; Majerek, D.; Szutt, A.; Dołhańczuk-Śródka, A. Application of electronic nose for evaluation of wastewater treatment process effects at full-scale WWTP. *Processes* **2019**, *7*, 251. [CrossRef]

10. Ma, D.; Ju, W.; Tang, Y.; Chen, Y. First-principles study of the small molecule adsorption on the InSe monolayer. *Appl. Surf. Sci.* **2017**, *426*, 244–252. [CrossRef]

11. Liang, X.; Ding, N.; Ng, S.; Wu, C.L. Adsorption of gas molecules on Ga-doped graphene and effect of applied electric field: A DFT study. *Appl. Surf. Sci.* **2017**, *411*, 11–17. [CrossRef]

12. Late, D.J.; Huang, Y.K.; Liu, B.; Acharya, J.; Shirodkar, S.N.; Luo, J.; Rao, C.N. Sensing behavior of atomically thin-layered MoS_2 transistors. *ACS Nano* **2013**, *7*, 4879–4891. [CrossRef] [PubMed]

13. Donarelli, M.; Prezioso, S.; Perrozzi, F.; Bisti, F.; Nardone, M.; Giancaterini, L.; Ottaviano, L. Response to NO_2 and other gases of resistive chemically exfoliated MoS_2-based gas sensors. *Sens. Actuators B Chem.* **2015**, *207*, 602–613. [CrossRef]

14. Schedin, F.; Geim, A.K.; Morozov, S.V.; Hill, E.W.; Blake, P.; Katsnelson, M.I.; Novoselov, K.S. Detection of individual gas molecules adsorbed on graphene. *Nat. Mater.* **2007**, *6*, 652. [CrossRef] [PubMed]

15. Zhang, S.; Guo, S.; Chen, Z.; Wang, Y.; Gao, H.; Gómez-Herrero, J.; Ares, P.; Zamora, F.; Zhu, Z.; Zeng, H. Recent progress in 2D group-VA semiconductors: From theory to experiment. *Chem. Soc. Rev.* **2018**, *47*, 982–1021. [CrossRef] [PubMed]

16. Zhang, S.; Yan, Z.; Li, Y.; Chen, Z.; Zeng, H. Atomically thin arsenene and antimonene: Semimetal–semiconductor and indirect–direct band-gap transitions. *Angew. Chem.* **2015**, *54*, 3112–3115. [CrossRef] [PubMed]

17. Zhang, S.; Xie, M.; Li, F.; Yan, Z.; Li, Y.; Kan, E.; Liu, W.; Chen, Z.; Zeng, H. Semiconducting group 15 monolayers: A broad range of band gaps and high carrier mobilities. *Angew. Chem.* **2016**, *55*, 1666–1669. [CrossRef]

18. Liu, C.; Liu, C.; Yan, X. Arsenene as a promising candidate for NO and NO_2 sensor: A first-principles study. *Phys. Lett. A* **2017**, *381*, 1092–1096. [CrossRef]

19. Khan, M.S.; Srivastava, A.; Pandey, R. Electronic properties of a pristine and NH_3/NO_2 adsorbed buckled arsenene monolayer. *RSC Adv.* **2016**, *6*, 72634–72642. [CrossRef]

20. Xie, M.; Zhang, S.; Cai, B.; Zou, Y.; Zeng, H. N-and p-type doping of antimonene. *RSC Adv.* **2016**, *6*, 14620–14625. [CrossRef]

21. Khan, M.S.; Ranjan, V.; Srivastava, A. NH_3 adsorption on arsenene: A first principle study. In Proceedings of the 2015 IEEE International Symposium on Nanoelectronic and Information Systems, Indore, India, 21–23 December 2015; IEEE: Piscataway, NJ, USA, 2015; pp. 248–251.

22. Bai, M.; Zhang, W.X.; He, C. Electronic and magnetic properties of Ga, Ge, P and Sb doped monolayer arsenene. *J. Solid State Chem.* **2017**, *251*, 1–6. [CrossRef]

23. Chen, X.; Wang, L.; Sun, X.; Meng, R.; Xiao, J.; Ye, H.; Zhang, G. Sulfur dioxide and nitrogen dioxide gas sensor based on arsenene: A first-principle study. *IEEE Electron Device Lett.* **2017**, *38*, 661–664. [CrossRef]

24. Berland, K.; Cooper, V.R.; Lee, K.; Schröder, E.; Thonhauser, T.; Hyldgaard, P.; Lundqvist, B.I. Van der Waals forces in density functional theory: A review of the vdW-DF method. *Rep. Prog. Phys.* **2015**, *78*, 066501. [CrossRef] [PubMed]

25. Tamijani, A.A.; Salam, A.; de Lara-Castells, M.P. Adsorption of noble-gas atoms on the TiO_2 (110) surface: An Ab initio-assisted study with van der Waals-corrected DFT. *J. Phys. Chem. C* **2016**, *120*, 18126–18139. [CrossRef]

26. Grimme, S. Semiempirical GGA-type density functional constructed with a long-range dispersion correction. *J. Comput. Chem.* **2006**, *27*, 1787–1799. [CrossRef] [PubMed]

27. Lee, K.; Murray, É.D.; Kong, L.; Lundqvist, B.I.; Langreth, D.C. Higher-accuracy van der Waals density functional. *Phys. Rev. B* **2010**, *82*, 081101. [CrossRef]

28. Guo, S.; Yuan, L.; Liu, X.; Zhou, W.; Song, X.; Zhang, S. First-principles study of SO_2 sensors based on phosphorene and its isoelectronic counterparts: GeS, GeSe, SnS, SnSe. *Chem. Phys. Lett.* **2017**, *686*, 83–87. [CrossRef]

29. Schiferl, D.; Barrett, C.S. The crystal structure of arsenic at 4.2, 78 and 299 K. *J. Appl. Crystallogr.* **1969**, *2*, 30–36. [CrossRef]

30. Kamal, C.; Ezawa, M. Arsenene: Two-dimensional buckled and puckered honeycomb arsenic systems. *Phys. Rev. B* **2015**, *91*, 85423. [CrossRef]

31. Kecik, D.; Durgun, E.; Ciraci, S. Optical properties of single-layer and bilayer arsenene phases. *Phys. Rev. B* **2016**, *94*, 205410. [CrossRef]

32. Segall, M.D.; Shah, R.; Pickard, C.J.; Payne, M.C. Population analysis of plane-wave electronic structure calculations of bulk materials. *Phys. Rev. B* **1996**, *54*, 16317–16320. [CrossRef] [PubMed]

Article

Study of Various Aqueous and Non-Aqueous Amine Blends for Hydrogen Sulfide Removal from Natural Gas

Usman Shoukat, Diego D. D. Pinto and Hanna K. Knuutila *

Department of Chemical Engineering, Norwegian University of Science and Technology (NTNU),
7491 Trondheim, Norway; usman.shoukat@ntnu.no (U.S.); diego.pinto@hovyu.com (D.D.D.P.)
* Correspondence: hanna.knuutila@ntnu.no

Received: 8 February 2019; Accepted: 8 March 2019; Published: 15 March 2019

Abstract: Various novel amine solutions both in aqueous and non-aqueous [monoethylene glycol (MEG)/triethylene glycol(TEG)] forms have been studied for hydrogen sulfide (H_2S) absorption. The study was conducted in a custom build experimental setup at temperatures relevant to subsea operation conditions and atmospheric pressure. Liquid phase absorbed H_2S, and amine concentrations were measured analytically to calculate H_2S loading (mole of H_2S/mole of amine). Maximum achieved H_2S loadings as the function of pKa, gas partial pressure, temperature and amine concentration are presented. Effects of solvent type on absorbed H_2S have also been discussed. Several new solvents showed higher H_2S loading as compared to aqueous N-Methyldiethanolamine (MDEA) solution which is the current industrial benchmark compound for selective H_2S removal in natural gas sweetening process.

Keywords: H_2S absorption; amine solutions; glycols; desulfurization; aqueous and non-aqueous solutions

---

1. Introduction

Natural gas is considered one of the cleanest forms of fossil fuel. Its usage in industrial processes and human activities is increasing worldwide, providing 23.4% of total world energy requirement in 2017 [1]. Natural gas is half of the price of crude oil and produces 29% less carbon dioxide than oil per unit of energy output [2]. Methane is a major energy providing component in natural gas. However, it also contains other hydrocarbons and a variety of impurities like acid gasses (CO_2 and H_2S) and water. Besides reducing the gas energy value, the impurities can cause operational problems such as corrosion in the pipeline and other equipment [3]. Mercury, mercaptans and other sulfur components are also often found in natural gas and must be removed. Sulfur components can produce SO_2 during combustion which ultimately leads to acid rain. Therefore, it is necessary to remove acid gases, water vapors, and other impurities before the usage of natural gas.

H_2S is an extremely poisonous component, and it can cause instant death when concentrations are over 500 parts per million volume (ppmv) [4,5]. H_2S exposure limits by the Norwegian Labour Inspection Authority are 5 ppmv for an eight-hour time-weighted average (TWA) and 10 ppmv for 15 min short-term exposure limit (STEL) [6]. The most commonly used method for H_2S removal is liquid scavenging. These processes usually employ non-regenerative chemicals such as triazine or aldehydes, and because of costs and operational issues (e.g., chemicals disposal), scavengers are not used for gases with high H_2S concentrations. Alkanolamines, in particularly N-Methyldiethanolamine (MDEA), are generally used for regenerative H_2S removal processes [7].

Natural gas is commonly saturated with water increasing the chances of solid gas hydrates formation with methane at high pressure and low temperatures potentially causing plugging in gas transport pipelines. One common way to avoid hydrate formation and to achieve problem-free continuous gas transportation operations is to add hydrates inhibitors like monoethylene glycol (MEG) or triethylene glycol(TEG) in gas pipelines [8].

A system which could selectively remove H_2S and control hydrate simultaneously would not only reduce equipment footprint but also help to reduce the installation and operational costs for both subsea and platform operations. This type of system was initially proposed by Hutchinson [9]. The idea of combined H_2S and water removal was presented in 1939 by using amine glycol solution as a solvent. 2-ethanolamine (MEA) and diethylene glycol (DEG) solution in water solution was the tested solvent for the concept. McCartney [10,11] and Chapin [12] built upon Hutchinson concept and presented the idea of both absorption and regeneration process in two-stages. They discussed various arrangements to get higher efficiency and lower energy requirement. Later on, this process development discontinued due to lower selectivity of H_2S compared to CO_2, higher amine degradation and corrosion rate of MEA [7]. However, tertiary amine systems could be very interesting for this type of operations as they are known for their high selectivity to H_2S. Tertiary amine systems, like a blend of methyl diethanolamine (MDEA) with glycols (MEG/TEG), have, additionally, higher amine stability and reduce corrosion rates [7,13,14].

In the literature, there is limited data available for the tertiary amine-glycols blends and most of the data is available for aqueous Triethanolamine (TEA), diisopropanolamine (DIPA), and MDEA. TEA was the first commercially used alkanolamine for gas treating process [7]. It is now being replaced with other amines like monoethanolamine (MEA), diethanolamine (DEA), diisopropanolamine (DIPA), methyl diethanolamine (MDEA), 2-amino-2-methyl-l-L-propanol (AMP), ethyl diethanolamine (EDEA) and 2-(2-aminoethoxy) ethanol (DGA) due to its low capacity and high circulation rate [15]. MDEA based system offered advantages like selective hydrogen sulfide removal over carbon dioxide, low vapor pressure, higher thermal stability, less corrosion, lower heat of reactions and specific heat [7,13]. Equations (1)–(6) show the mechanism and overall reactions of H_2S with aqueous secondary and tertiary amines. These reactions are instantaneous and involve a proton transfer.

$$\text{Ionization of water:} \quad H_2O \leftrightarrow H^+ + OH^- \tag{1}$$

$$\text{Ionization of dissolved } H_2S: \quad H_2S \leftrightarrow H^+ + HS^- \tag{2}$$

$$\text{Protonation of secondary amine:} \quad RR'NH + H^+ \leftrightarrow RR'\,R''NH_2^+ \tag{3}$$

$$\text{Protonation of tertiary amine:} \quad RR'R''N + H^+ \leftrightarrow RR'\,R''NH^+ \tag{4}$$

$$\text{Overall reaction for secondary amine:} \quad RR'NH + H_2S \leftrightarrow RR'\,R''NH_2^+ + HS^- \tag{5}$$

$$\text{Overall reaction for tertiary amine:} \quad RR'R''N + H_2S \leftrightarrow RR'\,R''NH^+ + HS^- \tag{6}$$

The solubility of H_2S in aqueous solutions of MDEA from 11.9 wt.% to 51 wt.% in the temperature range from 10 °C to 120 °C and H_2S partial pressure from 0.141 kPa–6900 kPa were studied by various authors [16–27]. All the previous studies of aqueous MDEA showed similar trends like increasing the partial pressure of H_2S (pH_2S) increases H_2S loading at given concentration and temperature, while the increase in amine concentration at a given temperature and pH_2S decreases H_2S loading. Surplus to MDEA data, TEA from 0.09–6.32 kPa H_2S and DIPA at a pressure range of 19–1554 kPa has been reported [28,29].

Xu et al. [24] also studied H_2S absorption in 30 wt.% MDEA in MEG and MEG-H_2O solutions over a range of partial pressures of H_2S from 0.34 to 38.8 kPa and found that increasing the water content in solution increases the H_2S loading at a given temperature (40 °C). Also, the increase in temperature decreases the H_2S loading for a given concentration (30 wt.% MDEA—65 wt.% MEG— 5 wt.% H_2O). Most of the previous studies were conducted using static cell apparatus and higher liquid phase H_2S loading can be obtained by using total gas pressure (>101.3 kPa) with higher amine concentration. Therefore, very few H_2S absorption studies are available for low amine concentrations at low temperatures and low acid gas inlet partial pressure range in literature.

The objective of this study is to identify blends where the solute (amine) can give higher H_2S removal capacity as compared to MDEA in the presence of glycol. The overall goal for this process is to absorb H_2S and water simultaneously at the subsea level in two-steps. In the first step, absorption can take place at the subsea level, potentially using a co-current contactor for absorption and flash drum to separate the natural gas from solvent at subsea levels. In the second step, loaded solution can be sent to a platform for regeneration and natural gas will be transported directly from subsea allowing a system where the natural gas will not enter the platform at all. The current work focuses on the identification of amine-glycol blends with high H_2S absorption capacity. The amines for this work were chosen systematically so that insight into the influence of its structure, like amine alkanol groups, alkyl chain length, and a hydroxyl group, can be obtained. In total twelve amines were studied, one secondary sterically hindered amines (diisopropylamine), one tertiary sterically hindered amine (N-tert-butyldiethanolamine), and ten tertiary amines. The list of amines along their chemical structure used in the study is given in Figure 1.

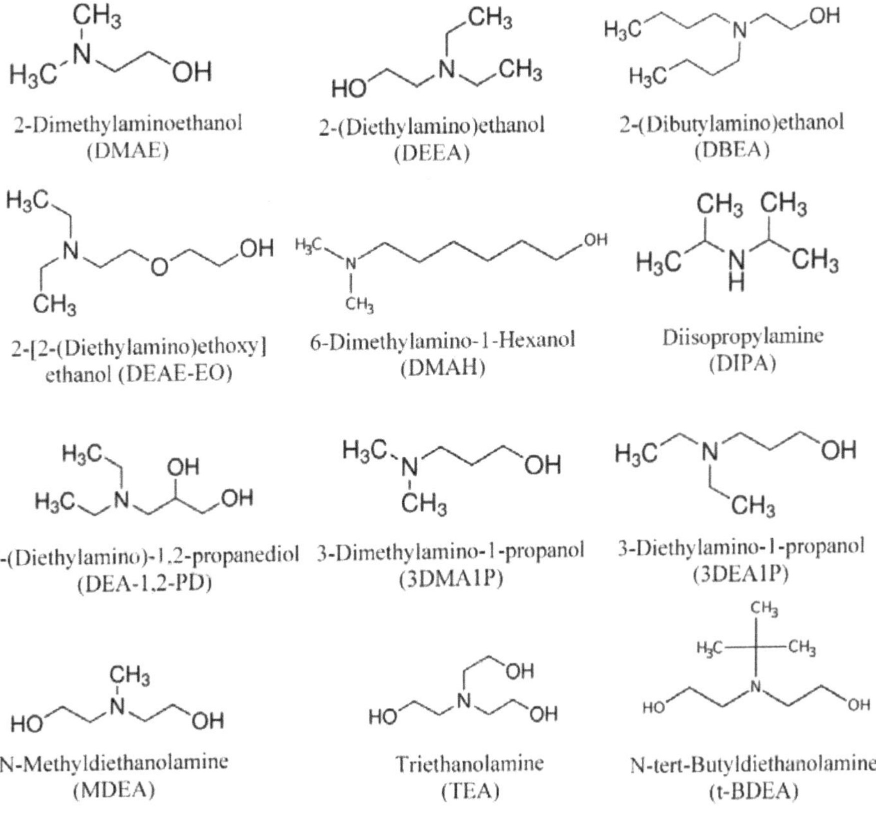

Figure 1. List of amines with chemical structures.

2. Material and Methodology

2.1. Materials

2-Dimethylaminoethanol (DMAE), 2-(Diethylamino) ethanol (DEEA), 2-(Dibutylamino) ethanol (DBAE), Diisopropylamine (DIPA), 3-Dimethylamino-1-propanol (3DEA-1P), N-Methyldiethanolamine (MDEA), Triethanolamine (TEA), Ethylene glycol (MEG), and Triethylene glycol (TEG) were bought from Sigma-Aldrich (Oslo, Norway), while 3-(Diethylamino)-1,2-propanediol (DEA-1,2-PD), 2-[2-(Diethylamino) ethoxy] ethanol (DEAE-EO), 6-Dimethylamino-1-Hexanol (DMAH), N-tert-Butyldiethanolamine (t-BDEA), and 3-Diethylamino-1-propanol (3DEA-1P) were bought from TCI Europe (Zwijndrecht, Belgium) in available maximum commercial purity. Additionally, premixed 1500 ppmv (0.15 vol.%) Hydrogen Sulphide (H_2S) in Nitrogen (N_2), 10,000 ppmv (1 vol.%) Hydrogen Sulphide (H_2S) in Nitrogen (N_2) and pure Nitrogen (N_2) (99.998 vol.%) were purchased from AGA Norway, Oslo. All chemicals were used without further purifications. Chemicals with their abbreviation, CAS number, purity, molecular weight, and pKa are given in Table 1 except hydrogen sulfide and deionized water.

Table 1. Name, abbreviation, CAS, purity (wt.%), and pKa of chemicals.

Chemical	CAS	Purity (wt.%)	Molecular Weight (g/mol)	pKa
2-Dimethylaminoethanol (DMAE)	108-01-0	≥99.5	89.14	9.49 [30]
2-(Diethylamino)ethanol (DEEA)	100-37-8	≥99.5	117.19	9.75 [31]
2-(Dibutylamino)ethanol (DBAE)	102-81-8	≥99.0	173.30	9.04 [32]
2-[2-(Diethylamino)ethoxy] ethanol (DEAE-EO)	140-82-9	>98.0	161.25	10.15 [31]
6-Dimethylamino-1-Hexanol (DMAH)	1862-07-3	>97.0	145.24	10.01 [31]
Diisopropylamine (DIPA)	108-18-9	≥99.0	101.19	8.84 [33]
3-(Diethylamino)-1,2-propanediol (DEA-1,2-PD)	621-56-7	>98.0	147.22	9.68 [31]
3-Dimethylamino-1-propanol (3DMA-1P)	3179-63-3	≥99.0	103.16	9.54 [30]
3-Diethylamino-1-propanol (3DEA-1P)	622-93-5	≥95.0	131.22	10.29 [30]
N-Methyldiethanolamine (MDEA)	105-59-9	≥99.0	119.16	8.65 [30]
Triethanolamine (TEA)	102-71-6	≥99.0	146.19	7.85 [30]
N-tert-Butyldiethanolamine (t-BDEA)	2160-93-2	≥97.0	161.24	9.06 [30]
Ethylene glycol (MEG)	107-21-1	≥99.5	62.07	14.44 [34]
Triethylene glycol (TEG)	112-27-6	≥99.8	150.17	14.50 [35]

All amine solutions were prepared by weighing the required amount of the amines using the Mettler Toledo MS6002S Scale, with an uncertainty of $\pm 10^{-5}$ kg. Aqueous solutions were made with deionized water produced by ICW-3000 Millipore water purification system, while for non-aqueous solutions MEG/TEG was used as a solvent. All amines were miscible in DI water, MEG and TEG except DBAE which made visible two phases with DI water but less visible two phases with MEG and TEG. DBAE solutions appeared homogeneous while stirring.

2.2. Methodology and Equipment

A custom-built apparatus, as shown in Figure 2, was used to screen amine solutions for hydrogen sulfide absorption study. The apparatus is designed to operate at atmospheric pressure and temperatures up to 80 °C and is similar to apparatus previously used for CO_2 absorption and desorption studies by Ma'mun et al. and Hartono et al. [31]. The apparatus consisted of the water-jacketed reactor with volume of ~200 cm^3 (NTNU, Trondheim, Norway) with a magnetic stirrer, Alicat MCS series Mass flow controllers (Tucson, AZ, USA), thermocouple (Omega Engineering Limited, Nærum, Denmark), Hubor® water bath (Huber Kältemaschinenbau AG, Offenburg,

Germany), and sodium hydroxide (NaOH) vessel for caustic wash. LabVIEW (National Instruments Norway, Drammen, Norway) was used to control and record gases flowrates and both reactor and water bath temperatures. The apparatus and H_2S gas bottles were installed in a closed fume cabinet equipped with an H_2S sensor, alarm and fail-safe system; which shut down the whole apparatus automatically in case of any H_2S leakage (limit >10 ppmv) or electrical failure. Personal protective equipment and personal H_2S sensor were used during experiments.

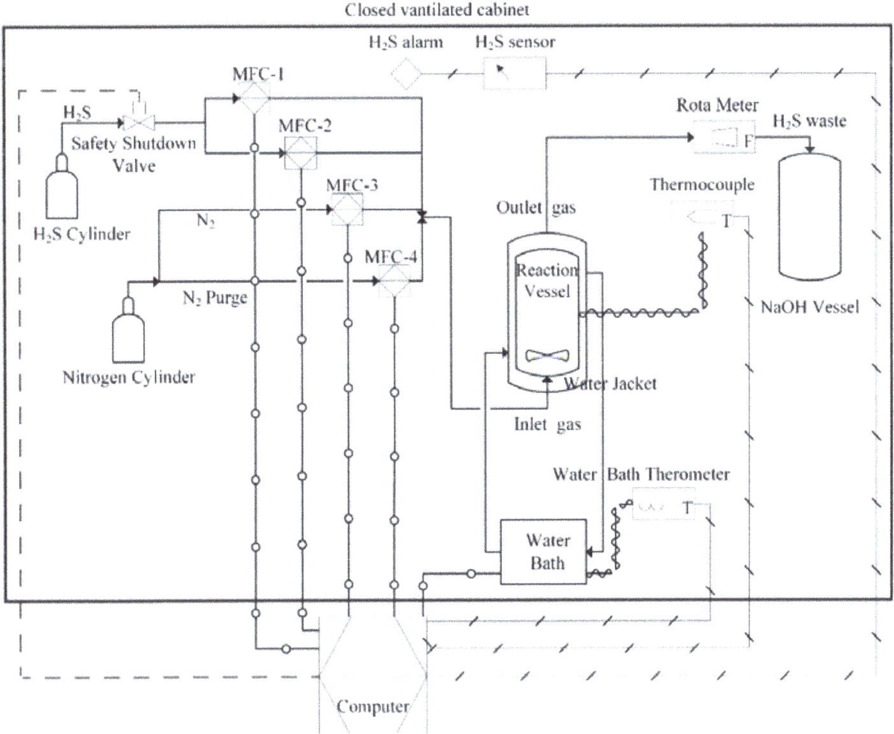

Figure 2. Schematic flow diagram of the screening apparatus.

Since the overall goal is to develop a solvent system that could be used at the subsea level, where the total gas pressure is high and H_2S content is from 50 ppm and up, higher partial pressure of H_2S up to 1 vol.% was used to achieve similar H_2S quantity during these experiments at atmospheric pressure.

At the start of each experiment 150 g of the solution was filled in the reactor and cooled/heated to the required experiment temperature after purging it with nitrogen to remove any air present. Pre-mixed H_2S and N_2 were used to achieve the required inlet hydrogen sulfide partial pressure (pH_2S) with the help of MFCs. The reactor was continuously stirred with a magnetic stirrer at isothermal and isobaric condition during the whole experiment. Huber® water bath was used to maintain the temperature constant. A thermocouple was placed in the liquid phase and used for continuous monitoring of reactor temperature. The exit gas from the reactor was sent to a series of three 10 wt.% NaOH solution vessels in order to remove residual H_2S present in it. All experiments were run for 120 min to give sample time to reach close to equilibrium between acid gas and amine solution. To ensure that 120 min is enough, experiments with 20 wt.% MDEA were performed until 240 min at 5 °C with sampling after every 15 min. The data showed that H_2S stopped absorbing after 45 min. This is in line with Lemoine et al. [20] and confirms that 120 min is enough time to reach close to equilibrium. Also, several parallel experiments for both aqueous and non-aqueous solutions of various amines were run and repeatability the data were confirmed. Different solutions were tested at different temperatures (5 °C, 25 °C and 40 °C) and inlet H_2S partial pressures (0.03 kPa, 0.5 kPa, 0.75 kPa and 1 kPa). For inlet pH_2S = 0.5 kPa to 1 kPa, 10,000 ppm H_2S gas mixture at total flow rate of 200 mL/min and for inlet pH_2S = 0.03kPa, 1500 ppm H_2S gas mixture at total flow rate of 1000 mL/min were used. The uncertainty of the inlet partial pressure of H_2S was estimated to be 2% including both the uncertainty of the ready H_2S gas mixture and the pre-calibrated mass flow controllers.

After the experiment, liquid samples were stored at <4 °C in the fridge and later on delivered to the analytical lab (St. Olav's Hospital Laboratory, Trondheim, Norway) for total sulfur analysis with inductivity coupled plasma mass spectrometry (ICP-MS). The samples were transported in ice box along with ice to keep sample temperature <5 °C. To ensure no amine loss during the experiments, amine concentration was determined by with Mettler Toledo G20 compact titrator [36] using a liquid sample of 0.2 mL that was diluted with 50 mL deionized water and titrated with 0.1 mol/L H_2SO_4. Each liquid sample was analyzed twice for total sulfur and amine concentration. The standard deviations between the duplicates of each solution were <2.5% for total sulfur and <1.5% for amine concentration. The differences in the amine concentration were less than 2% found in initial and final amine concentrations for all the solutions indicating that there was no significant amine loss during the experiments. The hydrogen sulfide loadings calculated by Equation (7), given in this work are based on the analyzed values for H_2S and amine in the liquid phase.

$$\propto_{H_2S} = \frac{mole\ of\ H_2S}{mole\ of\ amine} \tag{7}$$

3. Results

This screening apparatus was validated with a benchmarking 30 wt.% aqueous monoethanolamine (MEA) for CO_2 absorption before using it for H_2S absorption. Inlet CO_2 partial pressure was 10 kPa and absorption was done at 40 °C until 95% CO_2 absorption. Rich loading was found 0.54 mol CO_2/mol MEA after titration which was in good agreement with Hartono et al. [31] with an average deviation of 1.9%. 23.8 wt.% aqueous MDEA has been mostly used to study H_2S absorption. Therefore, the same amine concentration was used to verify the screening equipment and experimental parameters at 40 °C and pH_2S = 1 kPa. The liquid phase of H_2S loading was measure 0.14 (mol/mol) with the deviation of 4.6% from Jou et al. [16]. The experimental data are shown in with experimental uncertainties at the end is shown in Table 2.

Table 2. Experimental data.

Amine	Initial Amine (wt.%)	Solvent	H_2S Loading (α)	Inlet pH_2S (kPa)	Temperature (°C)
MDEA	20%	Water	0.015	0.03	5
DEEA	20%	Water	0.008	0.03	5
DBEA	20%	Water	0.011	0.03	5
DIPA	20%	Water	0.012	0.03	5
TEA	20%	Water	0.013	0.03	5
t-BDEA	20%	Water	0.009	0.03	5
MDEA	20%	MEG	0.010	0.03	5
MDEA	20%	TEG	0.006	0.03	5
MDEA	20%	Water	0.213	0.5	5
DEEA	20%	Water	0.237	0.5	5
3DEA-1P	20%	Water	0.258	0.5	5
DMAE	20%	Water	0.281	0.5	5
DEAE-EO	20%	Water	0.416	0.5	5
DBAE	20%	Water	0.032	0.5	5
DBAE	20%	MEG	0.089	0.5	5
3DEA-1P	20%	Water	0.271	0.75	5
DEAE-EO	20%	Water	0.389	0.75	5
DBAE	20%	Water	0.040	0.75	5
MDEA	20%	Water	0.254	1	5
MDEA	20%	Water	0.189	1	40

Table 2. *Cont.*

Amine	Initial Amine (wt.%)	Solvent	H_2S Loading (α)	Inlet pH_2S (kPa)	Temperature (°C)
DEEA	20%	Water	0.240	1	5
DEEA	20%	Water	0.249	1	40
3DEA-1P	20%	Water	0.355	1	5
3DEA-1P	20%	Water	0.272	1	25
3DEA-1P	20%	Water	0.248	1	40
3DEA-1P	30%	Water	0.120	1	5
3DEA-1P	50%	Water	0.061	1	5
DMAE	20%	Water	0.344	1	5
DMAE	20%	Water	0.260	1	40
DEAE-EO	20%	Water	0.378	1	5
DEAE-EO	20%	Water	0.413	1	25
DEAE-EO	20%	Water	0.385	1	40
DEAE-EO	50%	Water	0.094	1	5
DEAE-EO	30%	Water	0.231	1	5
DBAE	20%	Water	0.139	1	5
DBAE	20%	Water	0.043	1	25
DBAE	20%	Water	0.052	1	40
DBAE	30%	Water	0.049	1	5
DBAE	50%	Water	0.023	1	5
DIPA	20%	Water	0.185	1	5
TEA	20%	Water	0.165	1	5
t-BDEA	20%	Water	0.407	1	5
DEA-1,2-PD	20%	Water	0.306	1	5
DMAH	20%	Water	0.368	1	5
3DMA1P	20%	Water	0.299	1	5
MDEA	20%	MEG	0.189	1	5
DEEA	20%	MEG	0.155	1	5
3DEA-1P	20%	MEG	0.193	1	5
DMAE	20%	MEG	0.104	1	5
DEAE-EO	20%	MEG	0.280	1	5
DBAE	20%	MEG	0.121	1	5
DBAE	20%	MEG	0.061	1	40
MDEA	20%	TEG	0.040	1	5
DEEA	20%	TEG	0.035	1	5
3DEA-1P	20%	TEG	0.080	1	5
DMAE	20%	TEG	0.025	1	5
DEAE-EO	20%	TEG	0.073	1	5
DBAE	20%	TEG	0.049	1	5

$u(pH_2S) = \pm 2\%$; $u(T) = \pm 0.1\ °C$; $u(C_{Amine}) = \pm 1.5\%$; $u(C_{H2S}) = \pm 2.5\%$

3.1. Effect of pKa

Effect of pKa on H_2S loading in 20 wt.% aqueous amine solutions at T = 5 °C $\pm$ 0.1 °C and pH_2S = 1 kPa is shown in Figure 3. In the reaction between H_2S and aqueous amine solution, H_2S acts as weak acid whereas aqueous amine acts as a strong base, therefore, an increase in pKa increases the hydration of H_2S subsequently increasing the H_2S loading. This is also evident in tertiary amines aqueous solutions with DEEA, t-BDEA, and DBAE acting like outliers. DEEA shows lower absorption capacity than its closest pKa tertiary amine (DEA-1,2-PD), which can be due to short molecular chain of DEEA. t-BDEA. Sterically hindered amine shows the highest loading of all amines whereas DBAE shows the lowest loading, and it makes two phases with almost all the H_2S absorbed in the upper phase, i.e., amine (solute). If the amount of H_2S absorbed only in the amine phase (solute) is used to calculate H_2S loading in DBAE aqueous solutions, these solutions also start to follow the trend. The amount of H_2S absorbed in DBAE amine phase is $\approx$4.6 $\pm$ 0.2 times of absorbed H_2S in the whole solution both in aqueous and non-aqueous solutions.

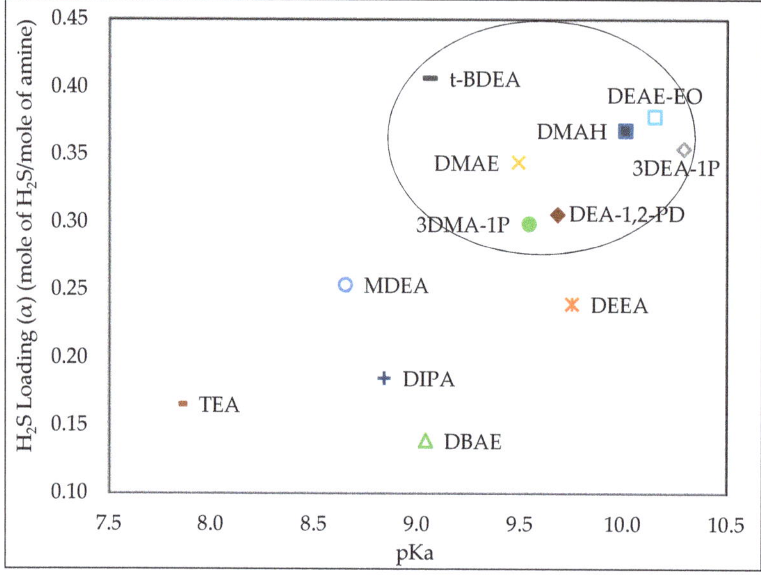

Figure 3. Effect of pKa on H$_2$S loading in aqueous amine solutions; T = 5 °C ± 0.1 °C; pH$_2$S = 1 kPa; amine concentration = 20 wt.% (unloaded); DBAE solutions make two phases.

Aqueous solutions of amines highlighted in the circle in Figure 3 presented higher H$_2$S loading as compared to MDEA and can be potential amines for further studies. t-BDEA showed the highest H$_2$S loading, but in-house data show it also degraded a lot in the presence of CO$_2$ and caused higher corrosion rates leading to damages in steel pipelines and equipment as compared to MDEA [37].

When looking into the amine structure, the results show that an increase in alkyl group decreases the H$_2$S loading in an amino-ethanol group, i.e., DMAE > DEAE (DEEA) > DBAE. It can be due to reduction in activity of nitrogen group due to increase in chain length of alkyl group in ethanol amine, a similar trend was previously observed in carbon dioxide capture studies [38,39]. Structure wise it would have been interesting to test 2-Dipropylaminoethanol (DPAE). Unfortunately, we were unable to purchase the chemical since it is commercially unavailable in Norway as it is being used in the weapon industry. A reverse trend was seen in an amino-propanol group where an increase in the alkyl group increases the H$_2$S loading, i.e., 3DMA-1P < 3DEA-1P. Hydroxyl group attracts electrons therefore, addition of more hydroxyl group reduces the activity of nitrogen atom of amine resulting in decreased H$_2$S loading in aqueous amine solutions, i.e., DMAE > MDEA > TEA and 3DEA-1P > 3DEA-1,2-PD. Also, an increase in the length of chain for hydroxyl group from -N- decreases the H$_2$S loading as seen when comparing DEAE-EO and DMAH (DEAE-EO shows higher capacity). Moreover, by adding the ethoxy group in DEEA, (DEAE-EO) increases the H$_2$S loading significantly.

Effects of pKa on H$_2$S loading in 20 wt.% amine solutions in MEG and TEG at T = 5 °C ± 0.1 °C; pH$_2$S = 1 kPa are shown in Figures 4 and 5 respectively. In each of the figures, a weak trend with respect to pKa is observed. H$_2$S loading in (DEAE-EO)-MEG solutions is higher than all other amine-MEG solutions and is even higher than aqueous MDEA solution shown in Figure 3. Increase in alkyl group in amine-TEG solutions and amine-MEG solutions are in line with each other, but not with the trend seen in aqueous solutions. However, adding an ethoxy group in DEEA has a similar effect in all three solutions.

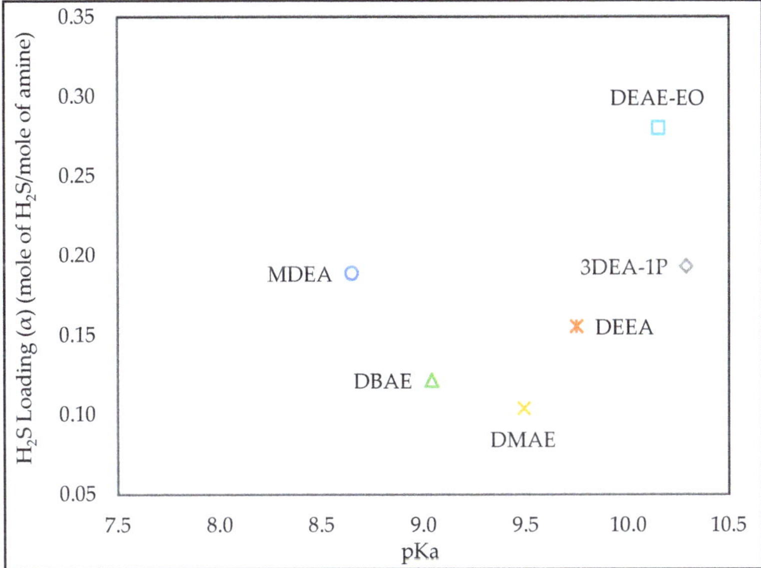

Figure 4. Effect of pKa on H$_2$S loading in amine-MEG solutions; T = 5 °C ± 0.1 °C; pH$_2$S = 1 kPa; amine concentration = 20 wt.% (unloaded); DBAE solutions makes two phases.

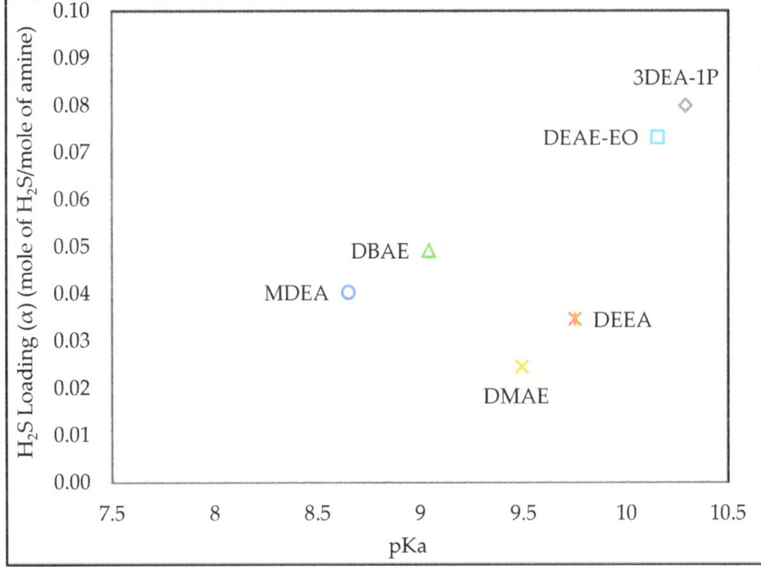

Figure 5. Effect of pKa on H$_2$S loading in amine-TEG solutions; T = 5 °C ± 0.1 °C; pH$_2$S = 1 kPa; amine concentration = 20 wt.% (unloaded); DBAE solutions makes two phases.

3.2. Effect of Solvent

Each aqueous amine solution gives more H$_2$S absorption capacity than its non-aqueous counterpart when compared on weight bases and having same system temperature, inlet partial pressure of gas and residence time of gas in the reactor as shown in Figure 6. Change of solvent from water to ethylene glycol or tri-ethylene glycol has a similar effect on all the amine solutions.

Replacing the solvent from water to monoethylene glycol decrease the H_2S loading significantly, the maximum decrease was observed in DEEA solutions while minimum has observed DBAE solutions. H_2S absorption decreased more rapidly when TEG had used as a solvent compared to MEG or water. Visual inspection also showed TEG solutions become more viscous as compared to MEG and H_2O solutions in respective amines. Furthermore, MEG shows more reactivity than TEG due to the autoprotolyses. However, the H_2S absorption capacity in TEG solutions is expected to increase significantly if water is present even at relatively low amounts [40].

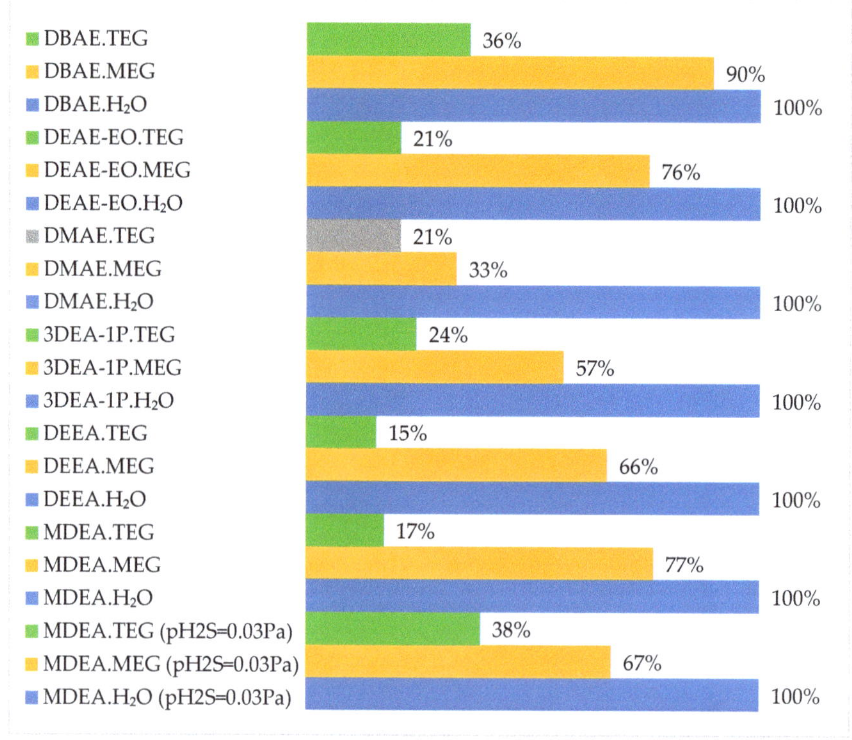

Figure 6. Effect of solvent on absorbed H_2S; T = 5 °C $\pm$ 0.1 °C; pH_2S = 1 kPa; mine concentration = 20 wt.% (unloaded); DBAE solutions makes two phases.

3.3. Effect of H_2S Partial Pressure

Hydrogen sulfide loading as the function of inlet H_2S partial pressure (pH_2S) at T = 5 °C $\pm$ 0.1 °C for 20 wt.% amine solutions is shown in Figure 7. The rise in inlet H_2S partial pressure (pH_2S) increases the H_2S loading at given temperature and amine concentration for both aqueous and non-aqueous solutions except DEAE-EO by providing more reaction sites for reaction between H_2S and amine solutions. The same trend was seen in previous studies. However, in aqueous DEAE-EO solution, H_2S loading starts to decrease with increases in pH_2S from 0.5 kPa to 1.0 kPa for an unknown reason. It is not possible to explain the behavior with the current data.

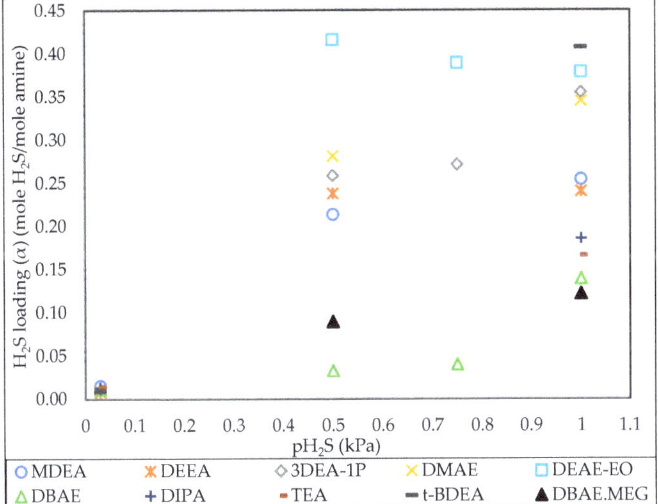

Figure 7. Effect of inlet H_2S partial pressure on H_2S loading; T = 5 °C ± 0.1 °C; amine concentration = 20 wt.% (unloaded); DBAE solutions makes two phases.

3.4. Effect of Temperature

The effect of temperature on H_2S loading on various 20 wt.% amine solutions at pH_2S = 1 kPa is shown in Figure 8. As the screening temperature increases from 5 °C to 40 °C H_2S loading decreases for all solutions except DEAE-EO and DEEA. The decrease in loading is as expected since the final loading in the experiments is almost in equilibria with the gas phase [16,20]. For DEEA, the loading difference between 5 °C and 40 °C is 0.01 mol H_2S/mol DEEA indicating that loading capacity is not as dependent on temperature as for some of the other amines. In the case of DEAE-EO, the changes are larger: The loading difference between 5 °C and 40 °C is 1.8% which is within our analytical uncertainty. However, the reason for the increase in loading seen at 25 °C, is unknown. We believe this is due to uncertainties in the analysis of H_2S and amine concentrations in the liquid samples.

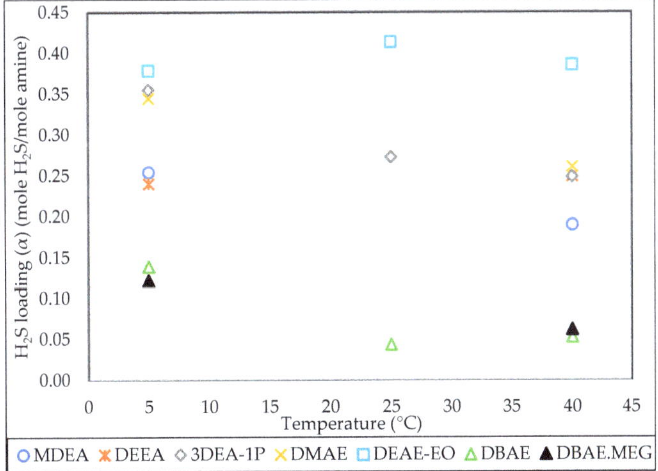

Figure 8. Effect of temperature on H_2S loading; pH_2S = 1 kPa; amine concentration = 20 wt.% (unloaded); DBAE solutions makes two phases.

3.5. Effect of Amine Concentration

Figure 9 shows the effect of amine concentration on hydrogen sulfide loading in aqueous solutions at 5 °C and inlet H_2S partial pressure of 1 kPa. The increase in amine concentration from 20 wt.% to 50 wt.% at given temperature and pressure decreases the H_2S absorption (mole/mol) subsequently decreasing H_2S loading. The trends are similar to those reported for MDEA as seen in the figure. In case of MDEA, the absorption capacity decreases by 40–50% when the MDEA concentration increases from 2.5 mol/kg to 4.2 mol/kg and it is similar to the reduction seen for DBEA. For 3DEA-1P and DEAE-EO a higher reduction in the absorption capacity is seen. Overall, the results indicate that increase in amine concentration changes the vapor-liquid equilibria behavior of the system [16,27,41,42].

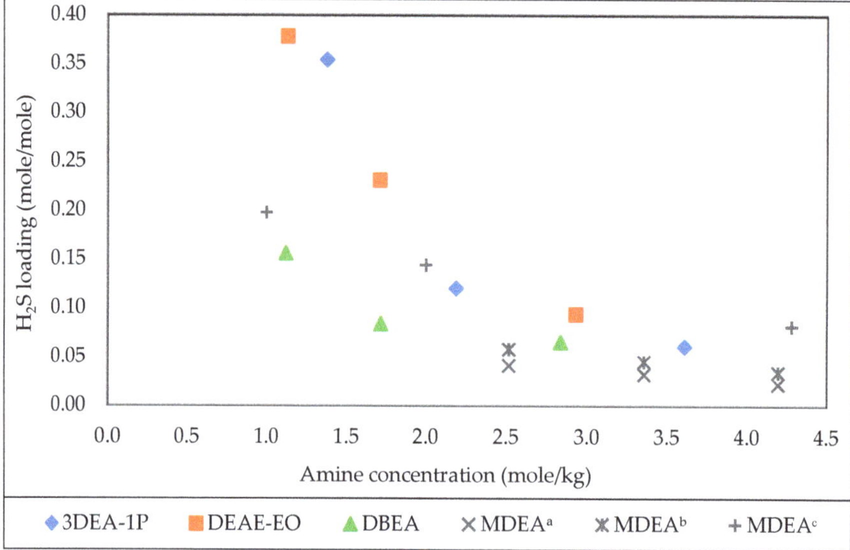

Figure 9. Effect of amine concentration on H_2S loading; T = 5 °C ± 0.1 °C; pH_2S = 1 kPa for all amines except MDEA; DBAE solutions makes two phases.; MDEA is at T = 40 °C and MDEA[a] pH_2S = 0.3 kPa [27]; MDEA[b] pH_2S = 0.5 kPa [27], MDEA[c] pH_2S = 1 kPa [16].

The data at 50 wt% allows us to compare the absorption capacity of 3DEA-1P, DEAE-EO and DBEA in aqueous and MEG solutions with similar mole fraction (mole amine/mole solution). The mole fraction of 3DEA-1P in 3DEA-1P.MEG solution (0.13) is similar to that of aqueous 50 wt% DEA-1P (0.15). Likewise, DEAE-EO and DBEA have similar mole fraction for 50 wt% aqueous solutions and 20 wt.% MEG solutions. For these three amines, the absorption capacity is 60–80% higher in the presence of MEG as compared to water (Figure 10). Further studies will be required to explain the performance differences between water and MEG based solvents.

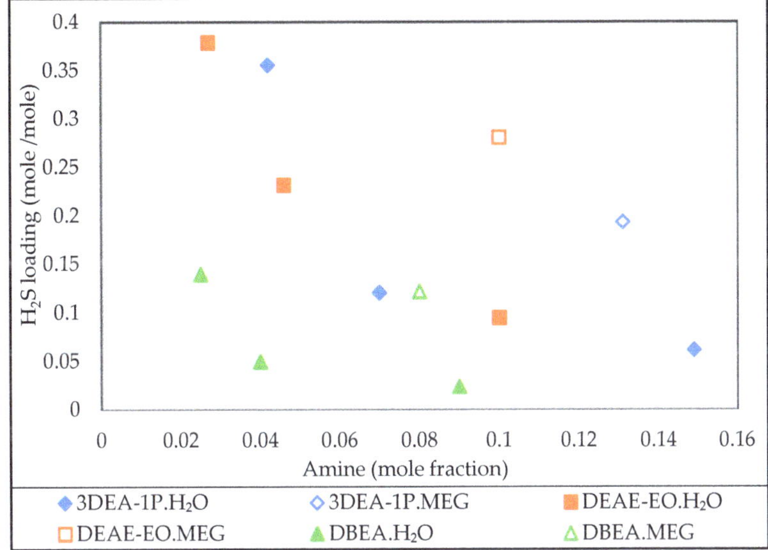

Figure 10. Effect of amine (mole fraction) on H_2S loading; T = 5 °C ± 0.1 °C; pH_2S = 1 kPa; DBAE solutions makes two phases.

4. Conclusions

In this study, various new aqueous and non-aqueous amine blends have been tested for H_2S absorption. The results show that an increase in hydroxyl group and addition of ethoxy group in amines increases the H_2S absorption in aqueous amine solutions. In general, the H_2S absorption increases also with increasing pKa. Also, increase in alkyl group enhances the H_2S loading in aqueous ethanol amines and vice versa for aqueous propanol amines. Several of the tested amines show higher H_2S absorption capacity compared to MDEA in aqueous solutions. Even though replacing water with TEG or MEG significantly decreased the H_2S loading in all tested solvents, the non-aqueous solution of (DEAE-EO)-MEG showed higher loading than aqueous MDEA at same weight concentration.

Author Contributions: Conceptualization, H.K.K., U.S. and D.D.D.P.; methodology, U.S. and H.K.K.; writing—original draft preparation, U.S.; writing—review and editing, H.K.K. and D.D.D.P.; supervision, H.K.K. and D.D.D.P.

Funding: This work was carried out under funding provided by NTNU-SINTEF Gas Technology Centre (GTS) and Faculty of Natural Sciences, Norwegian University of Science and Technology (NTNU), Trondheim, Norway.

Acknowledgments: We would like to thank Department of Chemical Engineering (IKP) at Faculty of Natural Sciences and Technology, Norwegian University of Science and Technology (NTNU), Trondheim, Norway for their support.

Conflicts of Interest: The authors declare no conflict of interest.

References

1. BP Statistical Review of World Energy. 67th Edition. 2018. Available online: https://www.bp.com/content/dam/bp/en/corporate/pdf/energy-economics/statistical-review/bp-stats-review-2018-full-report.pdf (accessed on 19 October 2018).
2. Subramaniam, R.; Yasa, S.; Bertrand, T.; Fontenot, B.; Dupuis, T.F.; Hernandez, R. Advanced simulation of H_2S scavenging process with triazine at different depths of gas well. *J. Nat. Gas Sci. Eng.* **2018**, *49*, 417–427. [CrossRef]
3. El-Gendy, N.S.; Speight, J.G. *Handbook of Refinery Desulfurization*; Taylor & Francis: Boca Raton, FL, USA, 2015.

4. Doujaiji, B.; Al-Tawfiq, J.A. Hydrogen sulfide exposure in an adult male. *Ann. Saudi Med.* **2010**, *30*, 76–80. [CrossRef] [PubMed]

5. Amosa, M.; Mohammed, I.; Yaro, S. Sulphide scavengers in oil and gas industry—A review. *Nafta* **2010**, *61*, 85–92.

6. *Grunnlag for Fastsettelse av Administrativ Norm for for Hydrogensulfid (H$_2$S)*; Arbeidstilsynet Statens Hus: Trondheim, Norway, 2011; Volume 7468.

7. Kohl, A.L.; Nielsen, R.B. Chapter 2—Alkanolamines for Hydrogen Sulfide and Carbon Dioxide Removal. In *Gas Purification*, 5th ed.; Gulf Professional Publishing: Houston, TX, USA, 1997; pp. 40–186.

8. Kohl, A.L.; Nielsen, R.B. Chapter 11—Absorption of Water Vapor by Dehydrating Solutions. In *Gas Purification*, 5th ed.; Kohl, A.L., Nielsen, R.B., Eds.; Gulf Professional Publishing: Houston, TX, USA, 1997; pp. 946–1021.

9. Hutchinson, A.J. Process for Treating Gases. U.S. Patent Application No. 2177068 A, 24 October 1939.

10. McCartney, E.R. Gas Purification and Dehydration Process. U.S. Patent Application No. 2435089, 27 January 1948.

11. McCartney, E.R. Extraction of acidic impurities and moisture from gases. U.S. Patent Application No. 2547278, 3 April 1951.

12. Chapin, W.F. Purification and Dehydration of Gases. U.S. Patent Application No. 2518752, 2 August 1950.

13. Shoukat, U.; Fytianos, G.; Knuutila, H.K. Thermal Stability and Corrosion Studies of Amines for Combined Acid Gas Removal and Hydrate Control for Subsea Gas Treatment Systems. In Proceedings of the 2016 Techno-Ocean (Techno-Ocean), Kobe, Japan, 6–8 October 2016; pp. 176–180.

14. Eimer, D. *Simultaneous Removal of Water and Hydrogen Sulphide from Natural Gas*; Department of Chemical Engineering, Norwegian University of Science and Technology: Trondheim, Norway, 1994.

15. Mathias, P.M.; Jasperson, L.V.; VonNiederhausern, D.; Bearden, M.D.; Koech, P.K.; Freeman, C.J.; Heldebrant, D.J. Assessing anhydrous tertiary alkanolamines for high-pressure gas purifications. *Ind. Eng. Chem. Res.* **2013**, *52*, 17562–17572. [CrossRef]

16. Jou, F.Y.; Mather, A.E.; Otto, F.D. Solubility of hydrogen sulfide and carbon dioxide in aqueous methyldiethanolamine solutions. *Ind. Eng. Chem. Process Des. Dev.* **1982**, *21*, 539–544. [CrossRef]

17. Macgregor, R.J.; Mather, A.E. Equilibrium solubility of H$_2$S and CO$_2$ and their mixtures in a mixed solvent. *Can. J. Chem. Eng.* **1991**, *69*, 1357–1366. [CrossRef]

18. Li, M.H.; Shen, K.P. Solubility of hydrogen sulfide in aqueous mixtures of monoethanolamine with N-methyldiethanolamine. *J. Chem. Eng. Data* **1993**, *38*, 105–108. [CrossRef]

19. Kuranov, G.; Rumpf, B.; Smirnova, N.; Maurer, G. Solubility of single gases carbon dioxide and hydrogen sulfide in aqueous solutions of N-methyldiethanolamine in the temperature range 313–413 K at pressures up to 5 MPa. *Ind. Eng. Chem. Res.* **1996**, *35*, 1959–1966. [CrossRef]

20. Lemoine, B.; Li, Y.G.; Cadours, R.; Bouallou, C.; Richon, D. Partial vapor pressure of CO$_2$ and H$_2$S over aqueous methyldiethanolamine solutions. *Fluid Phase Equilib.* **2000**, *172*, 261–277. [CrossRef]

21. Pérez-Salado Kamps, Á.; Balaban, A.; Jödecke, M.; Kuranov, G.; Smirnova, N.A.; Maurer, G. Solubility of single gases carbon dioxide and hydrogen sulfide in aqueous solutions of N-methyldiethanolamine at temperatures from 313 to 393 K and pressures up to 7.6 MPa: New experimental data and model extension. *Ind. Eng. Chem. Res.* **2001**, *40*, 696–706.

22. Sidi-Boumedine, R.; Horstmann, S.; Fischer, K.; Provost, E.; Fürst, W.; Gmehling, J. Experimental determination of hydrogen sulfide solubility data in aqueous alkanolamine solutions. *Fluid Phase Equilib.* **2004**, *218*, 149–155. [CrossRef]

23. Huttenhuis, P.J.G.; Agrawal, N.J.; Hogendoorn, J.A.; Versteeg, G.F. Gas solubility of H$_2$S and CO$_2$ in aqueous solutions of N-methyldiethanolamine. *J. Pet. Sci. Eng.* **2007**, *55*, 122–134. [CrossRef]

24. Xu, H.J.; Zhang, C.F.; Zheng, Z.S. Solubility of hydrogen sulfide and carbon dioxide in a solution of methyldiethanolamine mixed with ethylene glycol. *Ind. Eng. Chem. Res.* **2002**, *41*, 6175–6180. [CrossRef]

25. Sadegh, N.; Thomsen, K.; Solbraa, E.; Johannessen, E.; Rudolfsen, G.I.; Berg, O.J. Solubility of hydrogen sulfide in aqueous solutions of N-methyldiethanolamine at high pressures. *Fluid Phase Equilib.* **2015**, *393*, 33–39. [CrossRef]

26. Maddox, R.N.; Bhairi, A.H.; Diers, J.R.; Thomas, P.A. Equilibrium solubility of carbon dioxide or hydrogen sulfide in aqueous solutions of monoethanolamine, diglycolamine, diethanoiamine and methyldiethanolamine. In *GPA Research Report: Project 104*; GPA: Tulsa, OK, USA, 1987.

27. Tian, X.; Wang, L.; Fu, D.; Li, C. Absorption and Removal Efficiency of Low-Partial-Pressure H2S in a Monoethanolamine-Activated N-Methyldiethanolamine Aqueous Solution. *Energy Fuels* **2019**, *33*, 629–635. [CrossRef]

28. Jagushte, M.V.; Mahajani, V.V. Low pressure equilibrium between H_2S and alkanolamine revisited. *Ind. J. Chem. Technol.* **1999**, *6*, 125–133.

29. Mazloumi, S.H.; Haghtalab, A.; Jalili, A.H.; Shokouhi, M. Solubility of H_2S in aqueous diisopropanolamine + piperazine solutions: New experimental data and modeling with the electrolyte cubic square-well equation of state. *J. Chem. Eng. Data* **2012**, *57*, 2625–2631. [CrossRef]

30. Chowdhury, F.A.; Yamada, H.; Higashii, T.; Goto, K.; Onoda, M. CO_2 capture by tertiary amine absorbents: A performance comparison study. *Ind. Eng. Chem. Res.* **2013**, *52*, 8323–8331. [CrossRef]

31. Hartono, A.; Vevelstad, S.J.; Ciftja, A.; Knuutila, H.K. Screening of strong bicarbonate forming solvents for CO_2 capture. *Int. J. Greenh. Gas Control* **2017**, *58*, 201–211. [CrossRef]

32. ScolarTM, S. *Advanced Chemistry Development (ACD/Labs) Software V11.02 2016*; ACD/Labs: Toronto, ON, Canada, 2016.

33. Hamborg, E.S.; Versteeg, G.F. Dissociation Constants and Thermodynamic Properties of Amines and Alkanolamines from (293 to 353) K. *J. Chem. Eng. Data* **2009**, *54*, 1318–1328. [CrossRef]

34. Woolley, E.M.; Tomkins, J.; Hepler, L.G. Ionization constants for very weak organic acids in aqueous solution and apparent ionization constants for water in aqueous organic mixtures. *J. Solut. Chem.* **1972**, *1*, 341–351. [CrossRef]

35. Jacob, P. Potential Membrane Based Treatment of Triethylene Glycol Wastewater from Gas Separation Plant. *J. Water Sustain.* **2014**, *4*, 123–136.

36. Ma'mun, S.; Jakobsen, J.P.; Svendsen, H.F.; Juliussen, O. Experimental and Modeling Study of the Solubility of Carbon Dioxide in Aqueous 30 Mass% 2-((2-Aminoethyl)amino)ethanol Solution. *Ind. Eng. Chem. Res.* **2006**, *45*, 2505–2512. [CrossRef]

37. Shoukat, U.; Baumeister, E.; Pinto, D.D.D.; Knuutila, H.K. Thermal stability and corrosion of tertiary amines in aqueous amine and amine-glycol-water solutions for combined acid gas and water removal. *J. Nat. Gas Sci. Eng.* **2019**, *62*, 26–37. [CrossRef]

38. Bernhardsen, I.M.; Krokvik, I.R.T.; Jens, K.-J.; Knuutila, H.K. Performance of MAPA Promoted Tertiary Amine Systems for CO_2 Absorption: Influence of Alkyl Chain Length and Hydroxyl Groups. *Energy Procedia* **2017**, *114*, 1682–1688. [CrossRef]

39. El Hadri, N.; Quang, D.V.; Goetheer, E.L.V.; Zahra, M.R.M.A. Aqueous amine solution characterization for post-combustion CO_2 capture process. *Appl. Energy* **2017**, *185*, 1433–1449. [CrossRef]

40. Eimer, D. *Gas Treating: Absorption Theory and Practice*; Wiley: Hoboken, NJ, USA, 2014.

41. Fu, D.; Chen, L.; Qin, L. Experiment and model for the viscosity of carbonated MDEA–MEA aqueous solutions. *Fluid Phase Equilib.* **2012**, *319*, 42–47. [CrossRef]

42. Fu, D.; Wang, L.; Zhang, P.; Mi, C. Solubility and viscosity for CO_2 capture process using MEA promoted DEAE aqueous solution. *J. Chem. Thermodyn.* **2016**, *95*, 136–141. [CrossRef]

Article

Formation and Evolution Mechanism for Carbonaceous Deposits on the Surface of a Coking Chamber

Hao Wang [1,2], Baosheng Jin [1,*], Xiaojia Wang [1,*] and Gang Tang [3]

[1] Key Laboratory of Energy Thermal Conversion and Control of Ministry of Education, School of Energy and Environment, Southeast University, Nanjing 210096, China
[2] Huatian Engineering & Technology Corporation, MCC, Ma'anshan 243005, China
[3] School of Architecture and Civil Engineering, Anhui University of Technology, Ma'anshan 243002, China
*   Correspondence: bsjin@seu.edu.cn (B.J.); xiaojiawang@seu.edu.cn (X.W.);
    Tel.: +86-025-5209-0011 (B.J. & X.W.)

Received: 12 July 2019; Accepted: 29 July 2019; Published: 3 August 2019

Abstract: This work aimed to investigate the carbonaceous deposits on the surface of the coking chamber. Scanning electron microscopy (SEM), X-ray fluorescence spectrum (XRF), Fourier transform infrared spectrometer (FTIR), Raman spectroscopy, X-ray diffraction spectrum (XRD), and X-ray photoelectron spectroscopy (XPS) were applied to investigate the carbonaceous deposits. FTIR revealed the existence of carboxyl, hydroxyl, and carbonyl groups in the carbonaceous deposits. SEM showed that different carbonaceous deposit layers presented significant differences in morphology. XRF and XPS revealed that the carbonaceous deposits mainly contained C, O, and N elements, with smaller amounts of Al, Si, and Ca elements. It was found that in the formation of carbonaceous deposits, the C content gradually increased while the O and N elements gradually decreased. It was also found that the absorbed O_2 and H_2O took part in the oxidation process of the carbon skeleton to form the =O and –O– structure. The oxidation and elimination reaction resulted in change in the bonding state of the O element, and finally formed compact carbonaceous deposits on the surface of the coking chamber. Based on the analysis, the formation and evolution mechanisms of carbonaceous deposits were discussed.

Keywords: coke oven; carbonaceous deposits; spectral analysis; mechanism

---

1. Introduction

As an important chemical raw material, coke plays an indispensable role in the fields of metallurgy and energy. China is the largest coke supplier in the world and accounts for more than 70% of global production [1,2]. In 2016, China produced 449.1 million tons of coke [3]. Coke-making contains many processes, in which the coking chamber is the key carrier for coking. Thus, the operating status of the coking chamber significantly influences the production and quality of the coking process [4–6].

The coke-making process is a complex physical–chemical process [7–10]. The coal is pyrolyzed into many polycyclic aromatic hydrocarbon compounds: methane (CH_4), hydrogen (H_2), ammonium (NH_3), sulfur dioxide (SO_2), and so on. At the same time, the mineral composition, which contains many metal ions, also takes part in the coking process [11]. With the increase in coking operations, a compact carbonaceous deposit forms on the surface of the coking chamber, affecting its stable operation and shortening the lifetime of coke oven batteries, which not only decreases coking production, but also deteriorates the quality of the coking products [12]. Thus, it is important to investigate the formation and evolution processes of carbonaceous deposits on the surface of the coking chambers, which will benefit the enhancement of stable operations and prolong the lifetime of coke oven batteries [13].

In fact, carbonaceous deposits are significantly influenced by the temperature of the coke oven chamber, the gas phase, residence time of volatiles in the hot zone, and the surface on which the deposition takes place. A series of works have reported on the carbonaceous deposits on the surface of coke oven chambers. Furusa et al. investigated the influence of coal moisture and fine coal particles on carbonaceous deposits and clarified the formation mechanism of the carbonaceous deposits [14]. Uebo et al. researched the temperature and water presence on carbon depositions in laboratory tests, and tested brick pieces in the pilot plant oven [13]. Dumay et al. investigated the cracking conditions in the coke oven free space to better assess the parameters for the control of carbonaceous deposits. They also reported on a special device that could measure the growth of carbonaceous deposits in situ [15]. Krebs et al. investigated the influence of coal moisture content on carbonaceous deposits' yield and microstructure in detail [16]. Additionally, some strategies were applied to remove the carbonaceous deposits including manual or mechanical removal by spearing, burning-off by nature, air flow from the door or charring hole, and decomposition by blowing exhaust gas into the top space [17]. Furthermore, some methods have been proposed to prevent carbonaceous deposits. Nakagava et al. reported that injecting atomized water into the free space of the coke oven chambers could significantly decrease carbonaceous deposits [18]. Ando et al. reported the chamber wall being coated with glassy products containing 18–70 wt% of SiO_2, 10–60 wt% of Na_2O, 2–14 wt% of BaO, 0.5–25 wt% of SrO, and 0.5–20 wt% of Fe_2O_3, which could significantly inhibit carbonaceous deposits [19].

However, most of the above-mentioned studies have focused on the influencing factors of carbonaceous deposits and most of the methods used to remove the carbon depositions were based on macro experiments, with few studies focused on the composition of carbonaceous deposits. Furthermore, the formation and evolution mechanism of carbonaceous deposits need to be further investigated, which are pivotal to effectively prevent carbonaceous deposits. Thus, this paper aimed to systematically investigate the difference of the carbonaceous deposits on the surface of the coking chamber. Scanning electron microscope (SEM), X-ray fluorescence spectrum (XRF), Fourier transform infrared spectrometer (FTIR), Raman spectroscopy, X-ray diffraction spectrum (XRD), and X-ray photoelectron spectroscopy (XPS) were used to research the morphological structure, elemental composition, and bonding states of different carbonaceous deposit layers. Furthermore, the formation and evolution mechanism of carbonaceous deposits on the surface of coking chambers were discussed. The above work will provide a theoretical basis for effectively inhibiting carbonaceous deposits on the surface of coke oven chambers, which will benefit the stable operation of coke ovens.

2. Materials and Methods

2.1. Sample Preparation

The carbonaceous deposit bulk was collected from the coking chamber of the No. 3 coking plant at the Ma'anshan Iron and Steel Co. Ltd. The coals used for coking were produced by the Huainan Mining Group. The bulk sample was obtained from the inner surface of the coking chamber, which had run for 16 months. As shown in Figure 1, the bulk carbonaceous deposit sample presented a length of 25–30 cm, a width of 10–15 cm, and a thickness of 3–4 cm. Small carbonaceous deposit samples with a length of 1 cm, a width of 1 cm, and a thickness of 0.3 cm were obtained from the bulk sample for further characterization, which were marked as #1, #2, #3, and #4. Figure 2 shows the position of the four carbonaceous deposit samples in the coking chamber and bulk sample, where sample #1 was close to the coking chamber and sample #4 was connected with the coking chamber wall.

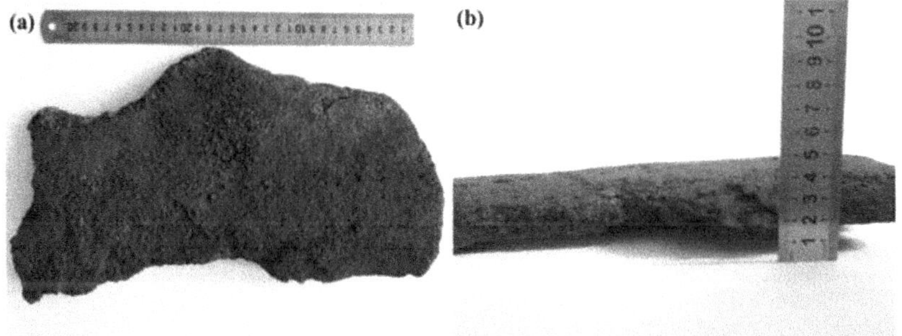

Figure 1. Dimensions of the bulk sample: (**a**) length and width; (**b**) thickness.

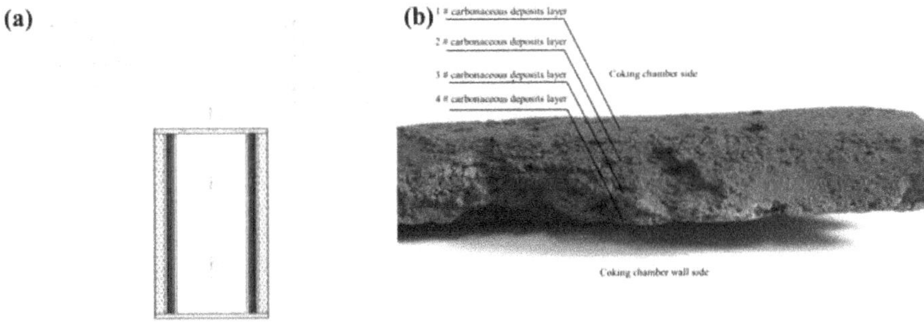

Figure 2. Distribution of each carbonaceous deposit layers: (**a**) schematic diagram; (**b**) digital photo.

2.2. Measurement and Characterization

Black carbonaceous deposit powders were obtained by the milled bulk sample in a planetary ball mill (XQM-4L, Kexi Laboratory Instrument Co Ltd., Nanjing, China) for 2 h at 300 rpm. X-ray fluorescence spectroscopy (XRF, ARL ADVANT'X Intellipower™ 3600, Thermo Scientific Nicolet, Waltham, MA, USA) was applied to investigate the elemental composition of the carbonaceous deposit powders with a working voltage of 60 kV, working current of 60 mA, and resolution of 0.01°.

Scanning electron microscopy (SEM, JSM-6490LV, JEOL Ltd., Tokyo, Japan) was applied to observe the morphology of small carbonaceous deposit samples with an accelerating voltage of 20 kV and resolution of 3 nm. Prior to observation, the sample surface was coated with a thin conductive layer.

The small carbonaceous deposit samples were ground into powders. Fourier-transformed infrared spectra spectroscopy (FTIR, Nicolet MAGNA-IR 750, Nicolet, Madison, WI, USA) was applied to characterize the powders of small carbonaceous deposit samples using a thin KBr disk. The transition mode was used and the wavenumber range was set from 4000 to 400 cm^{-1} with a resolution of 4 cm^{-1}.

The powders of small carbonaceous deposit samples were investigated by Laser Raman spectroscopy (LRS, inVia, Renishaw, London, UK). The excitation wavelength was 514 nm with a wavenumber range set from 800 to 2000 cm^{-1} with a resolution of 1 cm^{-1}.

The powders of small carbonaceous deposit samples were investigated by X-ray diffractometer (D8ADVANCE, Bruker, Karlsruhe, Germany) equipped with a Cu Kα tube and a Ni filter (λ = 0.154178 nm). The samples were scanned from 2θ = 10° to 80° with a step size of 0.02°.

X-ray photoelectron spectroscopy (XPS) with a VG Escalab Mark II spectrometer (Thermo-VG Scientific Ltd. Waltham, MA, USA) using Al Kα excitation radiation (hν = 1253.6 eV, resolution of 0.45 eV) was used to analyze the powders of small carbonaceous deposit samples.

3. Results and Discussion

3.1. Elemental Composition

Table 1 shows the XRF test results of the carbonaceous deposits on the surface of the coking chamber. It was found that the carbonaceous deposits mainly contained 34.51% of SO_2, 30.54% of SiO_2, 19.22% of Al_2O_3, and 5.6% of Fe_2O_3 (except the C element). Furthermore, small amounts of CaO, ZnO, MnO, and Cl with an abundance of 1.05–1.74% were detected in the carbonaceous deposits. All of the above data indicated that the carbonaceous deposits contained abundant S, Si, Al, Fe, and others, where S, Fe, Cr, and Al could significantly enhance the condensation reaction of polycyclic aromatic hydrocarbon compounds that result from the pyrolysis of coal, thus promoting the formation of carbonaceous deposits on the surface of the coking chamber.

Table 1. XRF data of carbonaceous deposits on the surface of the coking chamber.

Composition	Content (wt%)	Standard Error (wt%)
SO_2	34.51	0.24
SiO_2	30.54	0.23
Al_2O_3	19.22	0.20
Fe_2O_3	5.60	0.11
CaO	1.74	0.07
ZnO	1.44	0.06
Cr_2O_3	1.15	0.05
MnO	1.15	0.05
Cl	1.05	0.05
MgO	0.725	0.036
TiO_2	0.686	0.034
Na_2O	0.595	0.037
K_2O	0.591	0.029
P_2O_5	0.574	0.029

3.2. Morphology

Scanning electron microscopy (SEM) was applied to investigate the morphology of different carbonaceous deposit layers on the surface of the coking chamber. Figure 3a presents sample #1 at low magnification, which presented a loose structure with many holes. Figure 3b shows sample #1 at high magnification, from which we found combined particles of 3–5 μm, which may have come from the condensation of polycyclic aromatic hydrocarbon in the coke-making process. Figure 3c shows sample #2 at low magnification, which presented a cluster structure. Figure 3d reveals that the cluster structure was composed of carbon particles with diameters of 0.5–2 μm. It was seen that the compactness of sample #2 was significantly enhanced when compared with #1. This may be due to the primary carbon particles possessing poor stability, which can split into smaller carbon particles. These carbon particles reacted with each other to form more compact carbonaceous deposit layers. Furthermore, there were obvious gaps between the clusters, which may be due to the carbon particles having many hydroxyls, carboxyls, and carbonyls on the surface. These groups reacted and released CO_2 in the existence of metal ions (Fe, Al, Si) and high temperature. This phenomenon indicated that the primary carbon particles presented a metastable state, which can further split and combine with each other to form a compact carbonaceous deposit layer. Figure 3e,f display the morphology of sample #3, where the gap between the clusters disappeared, indicating the chemical reaction between the clusters at high temperature. Figure 3g,h displays the morphology of #4, which presented enhanced compactness compared with #3. The high magnification of sample #4 in Figure 3h showed a sponge structure with many holes, which may have resulted from the release of small molecules such as CO_2 at high temperature.

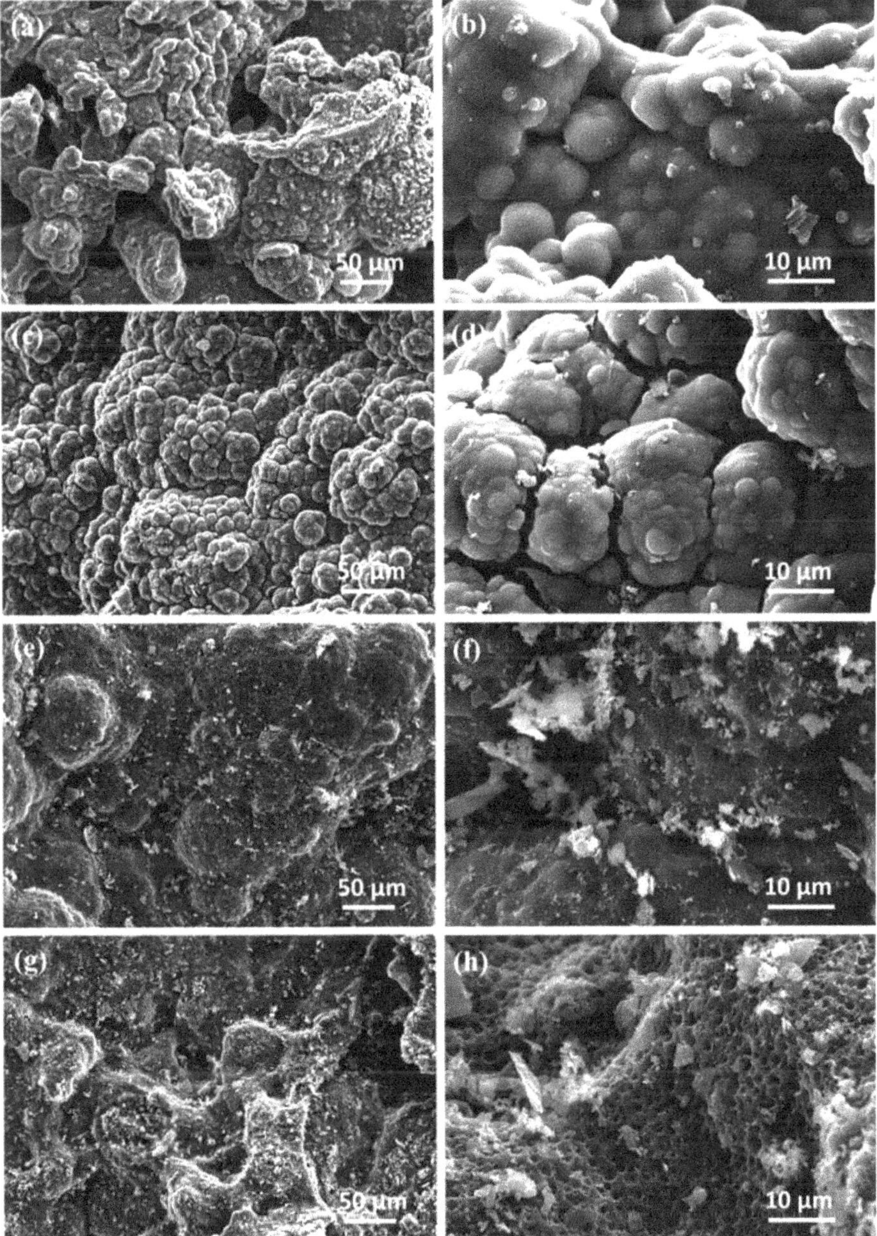

Figure 3. SEM images of each carbonaceous deposit layer in the coking chamber: sample #1 carbonaceous deposit layer, (**a**,**b**); sample #2 carbonaceous deposit layer, (**c**,**d**); sample #3 carbonaceous deposit layer, (**e**,**f**); and sample #4 carbonaceous deposit layer, (**g**,**h**).

3.3. X-ray Diffraction (XRD)

Figure 4 shows the XRD patterns of different carbonaceous deposit layers. A diffraction peak was found around 25.76° and 42.5°. The peaks at around 25.7° can be ascribed to (002), which was attributed to a hexagonal graphite structure. The peaks at around 42.5° corresponded to (100) peaks. As shown in Equations (1)–(4), the Bragg equation and Scherrer formula were introduced to calculate the structure parameter of different carbonaceous deposit layers [20–23]. In the equations, θ_{002} and θ_{100} are the diffraction angle of (002) and (100) peaks; β_{002} and β_{100} are the half-peak width of (002) and (100) peaks; d_{002} is the layer spacing; La is the diameter of the micro crystallite; Lc is the height of the layers; and N is the layer number of the aromatic structure. λ is the wavelength of the X-ray, and k_1, k_2 are the shape factors, where $k_1 = 1.84$, $k_2 = 0.94$ [24]. The calculated data are listed in Table 2.

$$d_{002} = \lambda/2\sin\theta_{002} \tag{1}$$

$$La = k_1\lambda/\beta_{100}\cos\theta_{100} \tag{2}$$

$$Lc = k_2\lambda/\beta_{002}\cos\theta_{002} \tag{3}$$

$$N = Lc/d_{002} + 1 \tag{4}$$

It can be seen from Table 2 that the #1, #2, #3, and #4 carbonaceous deposit layers presented d_{002} values from 0.3437 nm to 0.3482 nm, indicating little difference in the layer spacing between the different carbonaceous deposits. Additionally, sample #1 presented a La value of 30.81 nm, while #2 showed a decreased La value of 27.84 nm. This may be because the lamellar structure based on polycyclic aromatic compounds was not stable. Part of the lamellar was linked by chemical bonds such as ethers, esters, and aliphatics, which were destroyed at high temperature. We also found that samples #3 and #4 presented increased La values of 35.18 nm and 36.47 nm, respectively. This may have resulted from the edges of the lamellar structure reacting with each other at high temperature. Samples #1, #2, #3, and #4 presented Lc values of 9.97 nm, 9.85 nm, 8.93 nm, and 8.82 nm, respectively, indicating a decrease in the packing height of the lamellar structure from #1 to #4. This can be explained by the exfoliation of the out layered graphite lamellar by strong thermal radiation, which was consistent with the change in the N value.

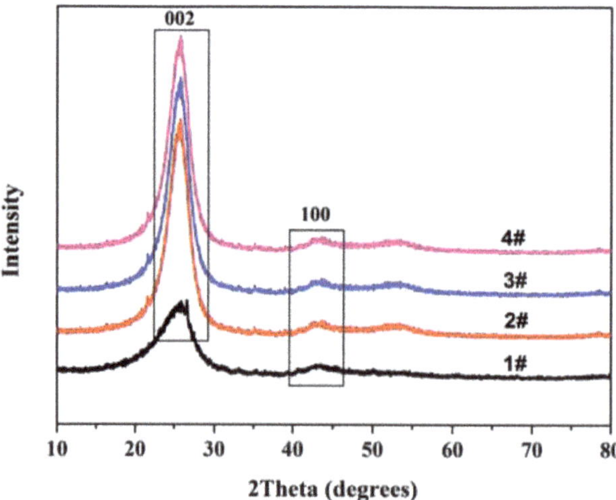

Figure 4. XRD spectra of each carbonaceous deposit layer in the coking chamber.

Table 2. XRD structure parameters of each carbonaceous deposit layer in the coking chamber.

Sample	$2\theta_{002}$	$2\theta_{100}$	d_{002}/nm	FWHM$_{002}$/nm	La/nm	FWHM$_{100}$/nm	Lc/nm	N
#1	25.92	42.98	0.3437	0.855	30.81	0.567	9.97	30
#2	25.76	42.80	0.3458	0.865	27.84	0.627	9.85	29
#3	25.68	42.64	0.3468	0.953	35.18	0.496	8.94	27
#4	25.58	42.97	0.3482	0.965	36.47	0.479	8.82	26

3.4. Fourier Transform Infrared Spectrometer (FTIR)

The FTIR spectra of samples #1, #2, #3, and #4 are shown in Figure 5. As seen in the FTIR spectrum, the peak at the range of 3200–3700 cm^{-1} can be ascribed to the stretching vibration of –OH and –NH. The peak at 1631 cm^{-1} was assigned to the C=O stretching vibration and the absorption peak at 1086 cm^{-1} was assigned to the stretching vibration of the C–O band. The peaks at 671 cm^{-1} corresponded to the bending vibration of C–H in the benzene ring structures [21]. The peak at 471 cm^{-1} confirmed the existence of the Fe–O and Al–O band [25]. The FTIR test confirmed the existence of –COOH, –NH, and –OH structure in the carbonaceous deposit layers. It was also found that the peaks at 3433 cm^{-1} in #2 and #3 were stronger than that in #1, while #4 displayed weaker absorption at 3433 cm^{-1}. This may be because there are few O_2 and H_2O entrained with polycyclic aromatic hydrocarbons to form carbonaceous deposit layers in the coke-making process. O_2 and H_2O could oxidize the carbonaceous deposits and form –COOH and –OH structures at high temperature and in the existence of metal ions (Fe, Al, etc.). However, the formed –COOH and –OH structures were not stable, and were eliminated from the carbon particle. Similar phenomena also existed at the characteristic peaks at 1631 cm^{-1} and 1086 cm^{-1}, indicating the coexistence of the formation and elimination reaction of –COOH and –OH groups during the coke-making process. In this process, the high temperature in the coking chamber promoted the oxidation reaction, and also enhanced the elimination of –COOH and –OH groups, which will consume the absorbed O_2 and H_2O, thus promoting the formation of a compact carbonaceous deposit layer.

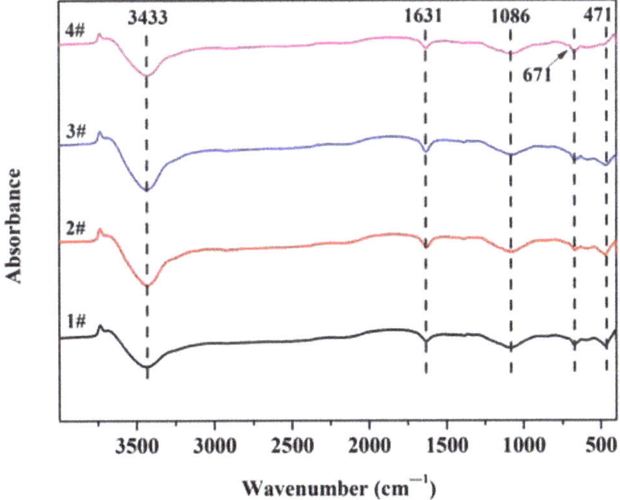

Figure 5. FTIR spectra of each carbonaceous deposit layer in the coking chamber.

3.5. Raman Spectroscopy

Figure 6 shows the Raman spectra of samples #1#, #2, #3, and #4. As shown in Figure 6, the Raman spectra of the different carbonaceous deposit layers exhibited the G band at 1591 cm^{-1} and the D band at 1347 cm^{-1} [26]. The G band corresponded to the E2g mode of graphite related to the vibration of the sp2-bonded carbon atoms in two-dimensional carbon materials, while the D band related to the defects and disorder in the hexagonal graphitic layers. The result confirmed that the carbonaceous deposits contained crystalline carbon and amorphous carbon.

Furthermore, Raman spectroscopy was used to analyze the graphitization degree of the carbonaceous material by integrating the intensity ratio of the D to G bands (I_D/I_G). A lower ratio of I_D/I_G indicates a higher graphitization degree [27]. As shown in Figure 7, the I_D/I_G values followed the sequence of #1 (2.34) > #2 (2.07) > #3 (1.79) ≈ #4 (1.84), indicating that the high temperature improved the graphitization degree of the carbonaceous deposits.

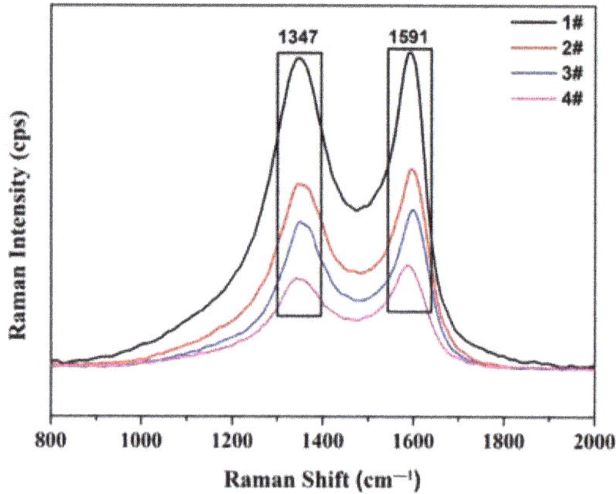

Figure 6. Raman spectra of each carbonaceous deposit layer in the coking chamber.

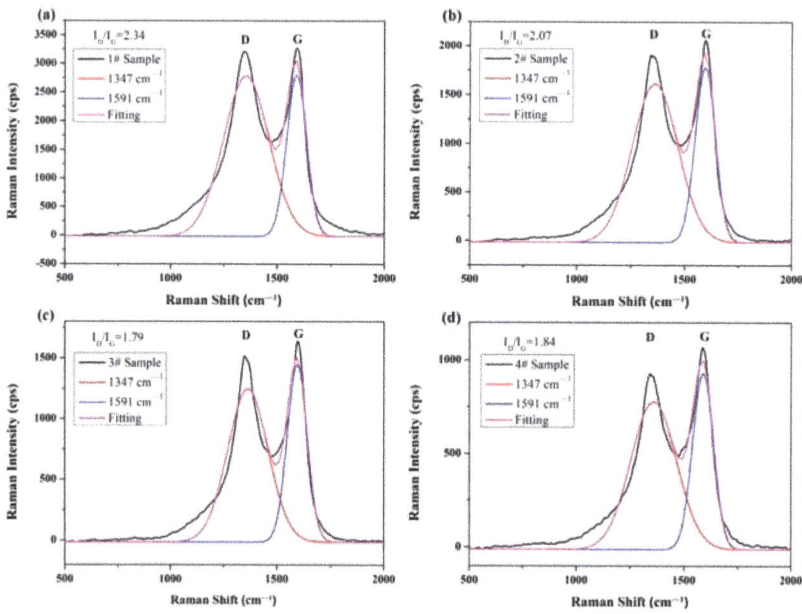

Figure 7. Raman spectra of each carbonaceous deposit layer in the coking chamber: (**a**) #1 carbonaceous deposit layer; (**b**) #2 carbonaceous deposit layer; (**c**) #3 carbonaceous deposit layer; and (**d**) #4 carbonaceous deposit layer.

3.6. X-ray Photoelectron Spectroscopy (XPS)

The chemical components of different carbonaceous deposit layers were also investigated by XPS, and the corresponding spectra are shown in Figure 8. The strong peak at around 284 eV was the characteristic peak of C1s, while the peaks at about 532 eV can be ascribed to the characteristic peak of O1s [28]. Furthermore, spots of N, Al, Si, Ca, Fe, S, and P were detected from #1 to #4. The C, O, S, and P may have come from the polycyclic aromatic hydrocarbons compounds, and Al, Si, Fe, and Ca resulted from the ore composition in coal. The above metal ions can take part in the formation of carbonaceous deposits in the form of dust particles. The test results were consistent with the XRF test.

Furthermore, the quantitative content of the above elements were conducted, which are listed in Table 3. It was found that Al, Si, Ca, Fe, S, P, etc., presented small fluctuations in samples #1, #2, #3, and #4. However, C, O, and N showed significant change from #1 to #4. The four carbonaceous deposit layers presented enhanced C elemental content of 82.41%, 89.43%, 89.69%, and 91.51% from #1 to #4, respectively. At the same time, the O element showed a decreased content of 10.91%, 6.20%, 6.32%, and 5.14%, respectively. Furthermore, the N element displayed a similar change law with a 2.20% content of #1, 1.49% content of #2, 1.17% content of #3, and a 0.76% content of #4. The C/O ratio and C/N ratio were also introduced to research change in the four carbonaceous deposit layers. It was found that samples #1–4 presented increased C/O of 7.55, 14.42, 14.19, and 17.80 with increased C/N of 37.46, 60.02, 76.66, and 120.39, respectively, indicating the O and N element release process in the formation of carbonaceous deposit layers in the coking chamber. This may be explained by many of the structures of the phenols, alcohols, ethers, and amines not being stable enough, and were eliminated at high temperature and the metal ions. The elimination of N containing and O containing groups can increase the C/O and C/N values in the carbonaceous deposits, and significantly enhance the compactness of the carbonaceous deposit layer.

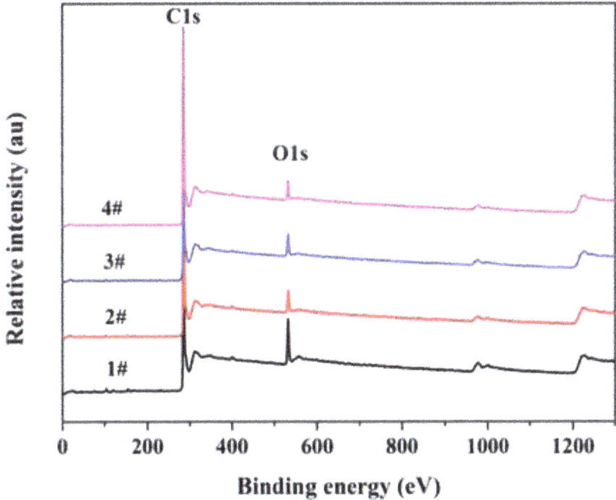

Figure 8. XPS spectra of each carbonaceous deposit layer in the coking chamber.

Table 3. XPS test data of each carbonaceous deposit layer in the coking chamber.

Sample	C (%)	O (%)	N	Al	Si	Ca	Fe	S	P	C/O	C/N
#1	82.41	10.91	2.20	1.12	1.73	0.50	0.37	0.54	0.21	7.55	37.46
#2	89.43	6.20	1.49	0.76	0.91	0.27	0.28	0.43	0.22	14.42	60.02
#3	89.69	6.32	1.17	0.77	0.72	0.30	0.31	0.46	0.26	14.19	76.66
#4	91.51	5.14	0.76	0.85	0.62	0.39	0.25	0.23	0.24	17.80	120.39

To further investigate the existing state and change law of C and O elements in the carbonaceous deposit layers, the peaks were resolved using peak analysis software (XPSPEAK4.1, kindly provided by Raymund W.M. Kwok, Department of Chemistry, the Chinese University of Hong Kong, Hong Kong, China). Figure 9 presents the peak separation fitting results of C1s for different carbonaceous deposits in the coking chamber. The peaks at 284.7 eV could be attributed to the C–C/C–H bond in aliphatic and aromatic species, which mainly resulted from a graphite carbon structure. The band at around 285.5 eV was assigned to the C–O/C–N bond in the structural formation of phenols, alcohols, ethers, and amines. The peaks at 287.4 eV were ascribed to the C=O/C=N bond in the formation of carbonyl, quinonyl, carboxylate, ester, and heterocyclic aromatic structures [28,29]. Table 4 lists the bond state content of the C element. It can be seen that #1 presented C–C/C–H, C–O/C–N, C=O/C=N contents of 69.73%, 21.68%, and 8.60%, respectively. In the carbonaceous deposit layer of #2, the content of C–C/C–H decreased to 64.59%, and the content of C–O/C–N and C=O/C=N were increased to 25.87% and 9.54%, respectively. This may be because the absorbed O_2 and H_2O in the primary carbonaceous deposits can oxidize the carbon skeletons to form C–O/C–N and C=O/C=N structures as well as decrease the C–C/C–H content. The carbonaceous deposit layer in sample #3 presented the three structure content of 66.60%, 24.51%, and 8.89%, while the carbonaceous deposit layer of sample #4 showed contents of 69.46%, 21.67%, and 8.87%, respectively. This may be due to the limited amount of absorbed O_2 and H_2O, which was consumed in the oxidation process. Furthermore, the formed C–O/C–N and C=O/C=N structures showed poor stability, which were eliminated at high temperature, resulting in the increase of C–O/C–N and C=O/C=N structures and enhanced C–C/C–H content in the carbonaceous deposits.

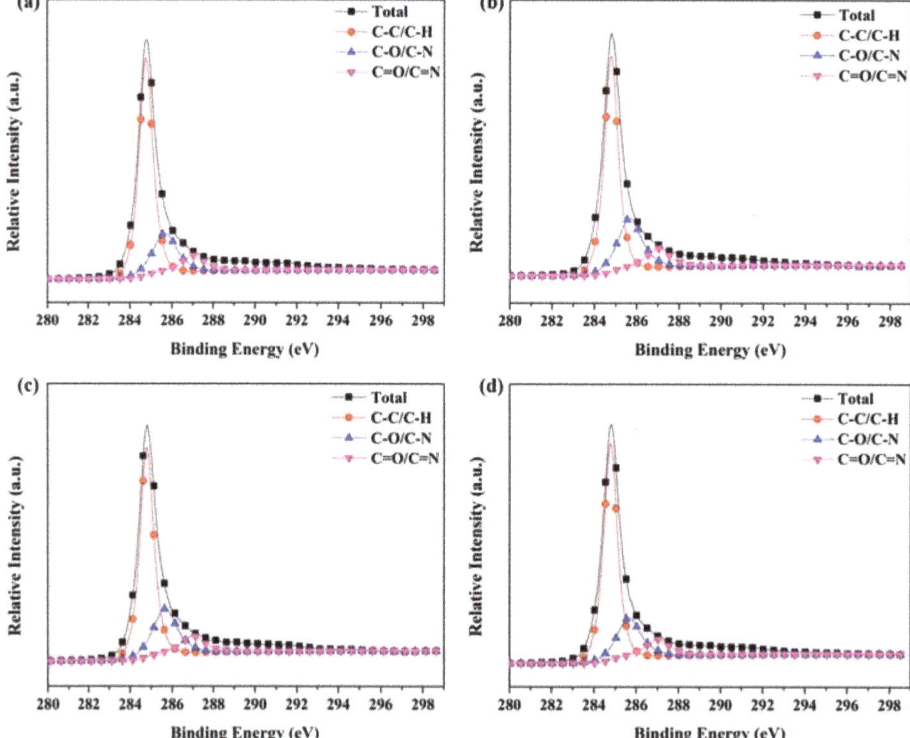

Figure 9. C1s spectra of each carbonaceous deposit layer in the coking chamber: (**a**) #1 carbonaceous deposit layer; (**b**) #2 carbonaceous deposit layer; (**c**) #3 carbonaceous deposit layer; and (**d**) #4 carbonaceous deposit layer.

Table 4. Bonding state content of the C element of each carbonaceous deposit layer.

Sample	C–C/C–H	C–O/C–N	C=O/C=N
	284.7 eV	285.5 eV	287.4 eV
#1	69.73%	21.68%	8.60%
#2	64.59%	25.87%	9.54%
#3	66.60%	24.51%	8.89%
#4	69.46%	21.67%	8.87%

Figure 10 shows the separate fitting results of O1s for different carbonaceous deposit layers in the coking chamber. The peak at 531.6 eV was assigned to the =O structure of carbonyl, quinonyl, carboxylate, and esters in the carbonaceous deposits. The bond at around 532.8 eV was ascribed to the –O– structure in the form of alcohols, phenols, and ethers in the carbonaceous deposits. The peaks at 533.8 eV can be assigned to the absorbed O_2 and H_2O in the carbonaceous deposits [30,31]. Table 5 lists the bond state content of the O element. As shown in Table 5, #1 presented =O, –O–, and O_2/H_2O contents of 30.10%, 42.10%, and 27.80%, respectively. In Sample #2, the =O and –O– contents increased to 35.16% and 43.39%, while the O_2/H_2O content decreased to 21.44%. This may be due to the consumption of the absorbed O_2 and H_2O in the primary carbonaceous deposit to form the =O and –O– structure, thus increasing the content of the =O and –O– structure in the carbonaceous deposits. Sample #3 presents the three bonding states of 33.48%, 43.19%, and 23.33%, while #4 showed the three bonding states of 32.43%, 44.89%, and 22.68%. It was found that the –O– content was almost invariant when compared with the increase in the O_2/H_2O content and decrease of the =O content in #3 and #4. This may be because the =O structure products such as carboxylate and esters presented weaker stability when compared with typical –O– structure products, which were eliminated to release H_2O, thus resulting in the decrease in the =O structure content and increase of the O_2/H_2O structure. The above results confirmed the coexistence of the oxidation and elimination process in the formation of carbonaceous deposits, which resulted in the change of contents for the O_2/H_2O, –O–, and =O structures.

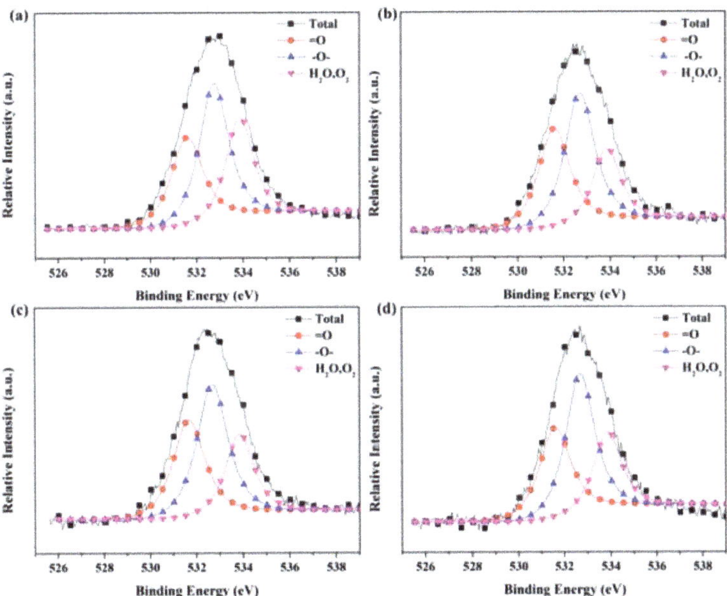

Figure 10. O1s spectra of each carbonaceous deposit layer in the coking chamber: (**a**) #1 carbonaceous deposit layer; (**b**) #2 carbonaceous deposit layer; (**c**) #3 carbonaceous deposit layer; and (**d**) #4 carbonaceous deposit layer.

Table 5. Bonding state content of the O element of each carbonaceous deposit layer.

Sample	=O	–O–	O_2/H_2O
	531.6 eV	532.8 eV	533.8 eV
#1	30.10%	42.10%	27.80%
#2	35.16%	43.39%	21.44%
#3	33.48%	43.19%	23.33%
#4	32.43%	44.89%	22.68%

3.7. Mechanism Consideration

On the basis of the above results, the possible formation and evolution mechanism of the carbonaceous deposits on the surface of the coking chamber are presented in Figure 11. In the coke-making process, many polycyclic aromatic hydrocarbon (such as anthracene and naphthalene) molecules associate with each other to form primary carbon particles with diameter of 3–5 μm. The primary carbon particles absorb O_2 and H_2O molecules combined with dust particles (containing metal ions) to form loose primary carbonaceous deposits. The primary carbon particle was not stable, and split into smaller pieces of intermediate carbon particles with diameters of 0.5–2 μm. The intermediate carbon particles reacted with each other to form a cluster structure with a diameter of 5–20 μm. The cluster structure further reacted to form compact intermediate carbonaceous deposits. There were many carboxyl, hydroxyl, and carbonyl groups on the surface of the primary and intermediate carbon particles, which can be eliminated from the particles at high temperatures and metal ions to finally form terminal carbonaceous deposits with a more compact structure at high temperature.

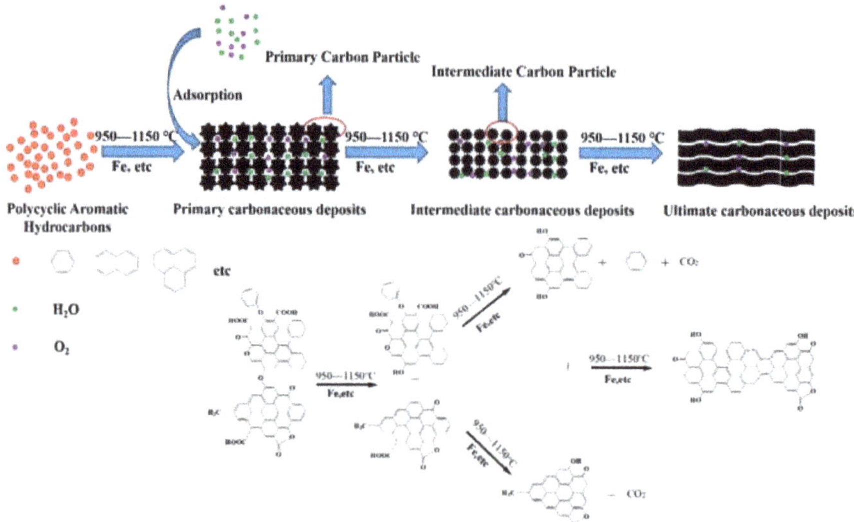

Figure 11. Schematic illustration for the formation and evolution mechanism for carbonaceous deposits on the surface of the coking chamber.

In the formation of carbonaceous deposits, the absorbed O_2 and H_2O can oxidize carbon skeletons to form –O– (such as alcohol, phenol, and ether) and =O (such as carbonyl, carboxyl, and esters) structures, resulting in the change of the bonding state of the O element. At the same time, the =O containing structure and –O– containing structure in the carbonaceous deposits conduct an elimination reaction in the condition of high temperature and metal ions. The oxidation and elimination reaction resulted in the change of the bonding state of the O element, and formed compact carbonaceous deposits on the surface of the coking chamber.

4. Conclusions

In this work, carbonaceous deposits on the surface of the coking chamber were investigated by scanning electron microscopy (SEM), X-ray fluorescence spectroscopy (XRF), Fourier transform infrared spectrometry (FTIR), Raman spectroscopy, X-ray diffraction spectroscopy (XRD), and X-ray photoelectron spectroscopy (XPS). FTIR revealed the existence of carboxyl, hydroxyl, and carbonyl groups in the carbonaceous deposits. Raman spectroscopy confirmed the decreased I_D/I_G values from the #1 carbonaceous deposit layer to the #4 carbonaceous deposit layer, indicating an enhancement in the graphitization degree of the carbonaceous deposits. SEM showed that the carbonaceous deposits resulted from polycyclic aromatic hydrocarbons, which can react to form unstable primary carbon particles with diameters of 3–5 μm. The primary carbon particles can split into intermediate carbon particles with diameters of 0.5–2 μm. The intermediate carbon particles can further react to form compact secondary carbonaceous deposits and finally form compact terminal carbonaceous deposits. XRF revealed that the carbonaceous deposits mainly contained C, O, and N elements, with a spot of Al, Si, and Ca elements. It was found from the XPS that the C content gradually increased while O and N content gradually decreased in the formation of carbonaceous deposits. The peak fitting of the XPS results revealed that absorbed O_2 and H_2O took part in the oxidation process of the carbon skeleton to form =O and –O– structures. The oxidation and elimination reaction resulted in the change in the bonding state of the O element, and formed compact carbonaceous deposits on the surface of the coking chamber. Based on the analysis, the formation and evolution mechanism of carbonaceous deposits were discussed, which provides a theoretical basis for inhibiting the formation of carbonaceous deposits on the surface of the coking oven chamber and a significantly enhanced stable operation of a coke oven.

Author Contributions: Conceptualization, H.W. and B.J.; Methodology, H.W. and B.J.; Software, G.T.; Validation, H.W., B.J., X.W., and G.T.; Formal analysis, B.J.; Investigation, H.W. and G.T.; Resources, H.W.; Data curation, H.W. and G.T.; Writing—original draft preparation, H.W. and G.T.; Writing—review and editing, H.W. and G.T.; Visualization, G.T.; Supervision, H.W.; Project administration, H.W.; Funding acquisition, H.W., B.J., and X.W.

Funding: This work was financially supported by the National Natural Science Foundation of China (51806035, 51676038), the Natural Science Foundation of Jiangsu Province (BK20170669), and the Key Research and Development Projects of Anhui Province (1804a0802195).

Conflicts of Interest: The authors declare no conflicts of interest.

References

1. Li, J.; Zhou, Y.; Simayi, M.; Deng, Y.Y.; Xie, S.D. Spatial-temporal variations and reduction potentials of volatile organic compound emissions from the coking industry in China. *J. Clean. Prod.* **2019**, *214*, 224–235. [CrossRef]
2. Bai, X.F.; Ding, H.; Lian, J.J.; Ma, D.; Yang, X.Y.; Sun, N.X.; Xue, W.L.; Chang, Y.J. Coal production in China: Past, present, and future projections. *Int. Geol. Rev.* **2017**, *60*, 535–547. [CrossRef]
3. Li, J.Y.; Ma, X.X.; Liu, H.; Zhang, X.Y. Life cycle assessment and economic analysis of methanol production from coke oven gas compared with coal and natural gas routes. *J. Clean. Prod.* **2018**, *185*, 299–308. [CrossRef]
4. Terui, K.; Matsui, T.; Fukada, K.; Dohi, Y. Influence of Temperature Distribution in Combustion Chamber on Coke Cake Discharging Behavior. *ISIJ Int.* **2018**, *58*, 633–641. [CrossRef]
5. Sobolewski, A.; Fitko, H. Individual regulation system of raw gas pressure in coke oven chambers. *Przem. Chem.* **2014**, *93*, 2115–2120.
6. Zhang, R.D.; Gao, F.R. Multivariable decoupling predictive functional control with non-zero-pole cancellation and state weighting: Application on chamber pressure in a coke furnace. *Chem. Eng. Sci.* **2013**, *94*, 30–43. [CrossRef]
7. Deng, J.; Zhao, J.Y.; Xiao, Y.; Zhang, Y.N.; Huang, A.C.; Shu, C.M. Thermal analysis of the pyrolysis and oxidation behaviour of 1/3 coking coal. *J. Therm. Anal. Calorim.* **2017**, *129*, 1779–1786. [CrossRef]

8. Wu, Y.; Li, Y.; Jin, L.; Hu, H. Integrated process of coal pyrolysis with steam reforming of ethane for improving tar yield. *Energy Fuels* **2018**, *32*, 12268–12276. [CrossRef]

9. Pretorius, G.N.; Bunt, J.R.; Gräbner, M.; Neomagus, H.; Waanders, F.B.; Everson, R.C.; Strydom, C.A. Evaluation and prediction of slow pyrolysis products derived from coals of different rank. *J. Anal. Appl. Pyrolysis* **2017**, *128*, 156–167. [CrossRef]

10. Fidalgo, B.; van Niekerk, D.; Millan, M. The effect of syngas on tar quality and quantity in pyrolysis of a typical South African inertinite-rich coal. *Fuel* **2014**, *134*, 90–96. [CrossRef]

11. Chen, X.; Zheng, D.; Guo, J.; Liu, J.; Ji, P. Energy analysis for low-rank coal based process system to co-produce semicoke, syngas and light oil. *Energy* **2013**, *52*, 279–288. [CrossRef]

12. Marcin, S.; Ludwik, K. Evaluation of the energy consumption in coal coking by using a method for reconcilation of substance and energy balances. *Przem. Chem.* **2014**, *93*, 681–685.

13. Uebo, K.; Kunimasa, H.; Suyama, S. Carbon deposition in a coke oven chamber at high productivity operation. *Tetsu Hagane* **2004**, *90*, 721–727. [CrossRef]

14. Furusawa, A.; Nakagawa, T.; Maeno, Y.; Komaki, I. Influence of coal moisture control on carbon deposition in the coke oven chamber. *ISIJ Int.* **1998**, *38*, 1320–1325. [CrossRef]

15. Dumay, D.; Gaillet, J.P.; Krebs, V.; Furdin, G.; Mareché, J.F. Formation and reduction of carbon deposits in coke ovens. *Rev. Metall. Cah. Inf. Tech.* **1994**, *91*, 1109–1116.

16. Krebs, V.; Furdin, G.; Mareché, J.; Dumay, D. Effects of coal moisture content on carbon deposition in coke ovens. *Fuel* **1996**, *75*, 979–986. [CrossRef]

17. Zymla, V.; Honnart, F. Coke oven carbon deposits growth and their burning off. *ISIJ Int.* **2007**, *47*, 1422–1431. [CrossRef]

18. Nakagava, T.; Kudo, T.; Kamada, Y.; Suzuki, T.; Komaki, I. Control of Carbon Deposition in the Free Space of Coke Oven Chamber by Injecting Atomized Water. *Tetsu Hagane* **2002**, *88*, 386–392. [CrossRef]

19. Ando, T.; Kasaoka, S.; Onozawa, T.; Nakai, S. Coating Composition for Carbonization Chamber for Coke Oven and Application Method. U.S. Patent 6,165,923, 26 December 2000.

20. Yu, J.W.; Qiao, K.; Zhu, B.; Cai, X.; Liu, D.W.; Li, M.J.; Yuan, X.M. Structural research of activated carbon fibers during a novel phosphoric acid reactivation process assisted by sonication. *Funct. Mater. Lett.* **2018**, *11*, 1850066. [CrossRef]

21. Liu, C.; Xiao, N.; Wang, Y.W.; Li, H.Q.; Wang, G.; Dong, Q.; Bai, J.P.; Xiao, J.; Qiu, J.S. Carbon clusters decorated hard carbon nanofibers as high-rate anode material for lithium-ion batteries. *Fuel Process. Technol.* **2018**, *180*, 173–179. [CrossRef]

22. Gao, R.T.; Liu, B.Y.; Xu, Z.M. Fabrication of magnetic zeolite coated with carbon fiber using pyrolysis products from waste printed circuit boards. *J. Clean. Prod.* **2019**, *231*, 1149–1157. [CrossRef]

23. Zheng, J.W.; Song, F.F.; Che, S.L.; Li, W.C.; Ying, Y.; Yu, J.; Qiao, L. One step synthesis process for fabricating NiFe$_2$O$_4$ nanoparticle loaded porous carbon spheres by ultrasonic spray pyrolysis. *Adv. Powder Technol.* **2018**, *29*, 1474–1480. [CrossRef]

24. Sonibare, O.O.; Haeger, T.; Foley, S.F. Structural characterization of Nigerian coals by X-ray diffraction, Raman and FTIR spectroscopy. *Energy* **2010**, *35*, 5347–5353. [CrossRef]

25. Chutia, S.; Narzari, R.; Bordoloi, N.; Saikia, R.; Gogoi, L.; Sut, D.; Bhuyan, N.; Kataki, R. Pyrolysis of dried black liquor solids and characterization of the bio-char and bio-oil. *Mater. Today Proc.* **2018**, *5*, 13193–23202. [CrossRef]

26. Wang, K.; Deng, J.; Zhang, Y.N.; Wang, C.P. Kinetics and mechanisms of coal oxidation mass gain phenomenon by TG–FTIR and in situ IR analysis. *J. Therm. Anal. Calorim.* **2018**, *132*, 591–598. [CrossRef]

27. Wang, S.G.; Gao, R.; Zhou, K.Q. The influence of cerium dioxide functionalized reduced graphene oxide on reducing fire hazards of thermoplastic polyurethane nanocomposites. *J. Colloid Interface Sci.* **2019**, *536*, 127–134. [CrossRef]

28. Tang, G.; Zhang, R.; Wang, X.; Wang, B.; Song, L.; Hu, Y.; Gong, X. Enhancement of flame retardant performance of bio-based polylactic acid composites with the incorporation of aluminum hypophosphite andexpanded graphite. *J. Macromol. Sci. Part A* **2013**, *50*, 255–269. [CrossRef]

29. Krawczyk, P. Effect of ozone treatment on properties of expanded graphite. *Chem. Eng. J.* **2011**, *172*, 1096–1102. [CrossRef]

30. Bourbigot, S.; Le Bras, M.; Delobel, R.; Gengembre, L. XPS study of an intumescent coating: II. Application to the ammonium polyphosphate/pentaerythritol/ethylenic terpolymer fire retardant system with and without synergistic agent. *Appl. Surf. Sci.* **1997**, *120*, 15–29. [CrossRef]
31. Zhang, P.; Hu, Y.; Song, L.; Ni, J.X.; Xing, W.Y.; Wang, J. Effect of expanded graphite on properties of high-density polyethylene/paraffin composite with intumescent flame retardant as a shape-stabilized phase change material. *Sol. Energy Mater. Sol. Cells* **2010**, *94*, 360–365. [CrossRef]

 *processes*

Article

Energy Consumption and Economic Analyses of a Supercritical Water Oxidation System with Oxygen Recovery

Fengming Zhang [1,2,*], Jiulin Chen [1], Chuangjian Su [1] and Chunyuan Ma [3]

[1] Guangzhou Institutes of Advanced Technology, Chinese Academy of Sciences, Guangzhou 511458, China; jlchen_sust@163.com (J.C.); suchuangjian501@126.com (C.S.)

[2] Shenzhen Institutes of Advanced Technology, Chinese Academy of Sciences, Shenzhen 518055, China

[3] National Engineering Laboratory for Coal-fired Pollutants Emission Reduction, Shandong University, Jinan 250061, China; pheres8@mail.sdu.edu.cn

* Correspondence: fm.zhang@giat.ac.cn; Tel.: +86-020-2291-2754

Received: 15 October 2018; Accepted: 14 November 2018; Published: 16 November 2018

Abstract: Oxygen consumption is one of the factors that contributes to the high treatment cost of a supercritical water oxidation (SCWO) system. In this work, we proposed an oxygen recovery (OR) process for an SCWO system based on the solubility difference between oxygen and CO_2 in high pressure water. A two-stage gas–liquid separation process was established using Aspen Plus software to obtain the optimized separation parameters. Accordingly, energy consumption and economic analyses were conducted for the SCWO process with and without OR. Electricity, depreciation, and oxygen costs contribute to the major cost of the SCWO system without OR, accounting for 46.18, 30.24, and 18.01 $\$ \cdot t^{-1}$, respectively. When OR was introduced, the total treatment cost decreased from 56.80 $\$ \cdot t^{-1}$ to 46.17 $\$ \cdot t^{-1}$, with a reduction of 18.82%. Operating cost can be significantly reduced at higher values of the stoichiometric oxygen excess for the SCWO system with OR. Moreover, the treatment cost for the SCWO system with OR decreases with increasing feed concentration for more reaction heat and oxygen recovery.

Keywords: supercritical water oxidation; high-pressure separation; oxygen recovery; energy recovery; economic analysis

---

1. Introduction

Supercritical water (SCW) ($P > 22.1$ MPa, $T > 374$ °C) has unique thermo-physical characteristics [1], which can dissolve organic compounds and gases to form a homogeneous mixture without mass transfer resistance [2,3]. SCW has been widely used to treat organic waste by supercritical water oxidation (SCWO) or supercritical water gasification (SCWG) for high efficiency and low residence time [3–5]. In the SCWO process, no SO_2 or NO_X by-products during organic waste degradation emit [6–8]. Although SCWO has many unique advantages in treating wastewater, some technical problems, such as corrosion and salt plugging, have hindered its development for years [9,10]. Inorganic acids (e.g., HCl and H_2SO_4), combined with high temperature and high oxygen concentration, can cause severe corrosion of the reactor and other devices [11]. Inorganic salt is hardly soluble in SCW, which leads to the plugging of the reactor, as well as the preheating and cooling sections [12]. At present, an effective solution for corrosion and salt plugging is the use of a transpiring wall reactor (TWR). A TWR typically consists of a dual shell with an outer pressure-resistant vessel and an inner porous tube. Transpiring water at subcritical temperatures passes through the porous pipe to form a protective film on its inner surface. This water film can prevent reactants from spreading to the porous wall and dissolve inorganic salt. Numerous studies have proven that TWR plays an effective role in preventing corrosion and salt plugging [13–15].

A high treatment cost, which is due to material input (such as oxidants and additives) and energy consumption during the pressurization and heating steps, is another obstacle that hinders the SCWO application. Treatment cost is determined according to the adopted equipment, treatment capacity, concentrations, and types of organic matter, operating conditions, and staff costs. At present, the treatment cost for an SCWO system with 1000 kg/h wet organic waste and an organic matter content of 10 wt% typically ranges from tens to hundreds of dollars [16]. Energy recovery is the leading method for reducing energy consumption and operating cost. An autothermal operation with a certain feed concentration (>2 wt%) can be achieved under ideal conditions [17–19]. Power generation is another potential application that uses high-pressure and high-temperature reactor effluent [20–22]. However, the reactor effluent in an SCWO system with TWR is cooled to subcritical temperature (<360 °C) for transpiring water injection at a low temperature to avoid salt plugging [13]. Accordingly, feedstock preheating and hot water/steam production may be more realistic and effective choices [23,24].

Oxygen, the most popular oxidant in SCWO systems, is another major source of cost. Results have indicated that a stoichiometric oxygen excess (R) of 1.05 may be sufficient for complete oxidation of organic wastewater [16]. However, a higher amount of oxygen is required in the pilot or industrial plant, which is mainly due to the local heterogeneous state in the reactor. Thus, twice the value of R (or even higher) is obtained, which leads to oxygen loss. Xu et al. [25] conducted an economic analysis of urban sludge via SCWO using a 2.5 t/day demonstration device. The operating cost was approximately 83.1 $\cdot t^{-1}$, with oxygen cost accounting for 25% of the total amount. Zhang et al. [26] analyzed a 100 t/day SCWO system for high-concentration printing and dyeing wastewater; the operating cost of the system was 10.3 $\cdot t^{-1}$, with oxygen cost accounting for 37% of the total amount. Shen et al. [27] conducted an economic analysis of an SCWO system with TWR. The feed was 300 m^3/day, with an initial concentration of 40,000 mg/L chemical oxygen demand; the cost was 4.99 $\cdot t^{-1}$, with oxygen cost accounting for 71.8% of the total amount. Thus, oxygen consumption control will be an important solution for reducing operating cost.

In addition, CO_2 is another primary gas in reactor effluent. However, it is low in purity due to excess oxygen consumption, which is the main obstacle that inhibits CO_2 recovery and utilization. Thus, recovering CO_2 with high purity may be another solution for reducing the operating cost of SCWO systems. The low operating cost calculated by Shen [27] is mainly attributed to the benefit of CO_2 recovery. Abeln [28] reported that the operating cost of a 100 kg/h SCWO–TWR plant is approximately 659 €/t, and by-product income, such as surplus heat energy and CO_2, must be ensured to obtain a comparably low operating cost.

Species recovery can considerably reduce operating cost for less input and additional income. However, only a few studies have focused on this issue, and a simple operation process with low energy consumption is urgently required for species recovery. In the current work, a species recovery process based on high-pressure water absorption was proposed to separate and recover oxygen and CO_2. A two-stage gas–liquid separation process was established using Aspen Plus V8.0. Reasonable thermodynamic models for high-pressure separation were evaluated to identify the optimized separation parameters. Accordingly, SCWO processes with and without oxygen recovery (OR) were simulated, and energy consumption and economic analyses were conducted.

2. Proposal of OR for SCWO Systems

Baranenko et al. [29] tested the solubility of oxygen and CO_2 in high-pressure water at temperatures ranging from 0 °C to 360 °C and pressures from 1 MPa to 20 MPa. The solubility of oxygen (Figure 1a) and CO_2 (Figure 1b) increases with increasing pressure, but the effect of temperature on solubility does not exhibit a distinct trend. At low pressures, an evident reduction in solubility is observed as temperature increases. At high pressures, solubility initially decreases, then increases, and finally decreases with increasing temperature. Thus, two solubility extremes occur in the wave curve of the high-pressure zone. Moreover, the solubility of CO_2 is nearly one order of magnitude higher than that of oxygen under the same conditions. Given that reactor effluent is mostly composed of oxygen, CO_2, and water, the ratio of oxygen to CO_2 in the gaseous phase under different conditions is calculated using the typical effluent composition in our previous pilot plant.

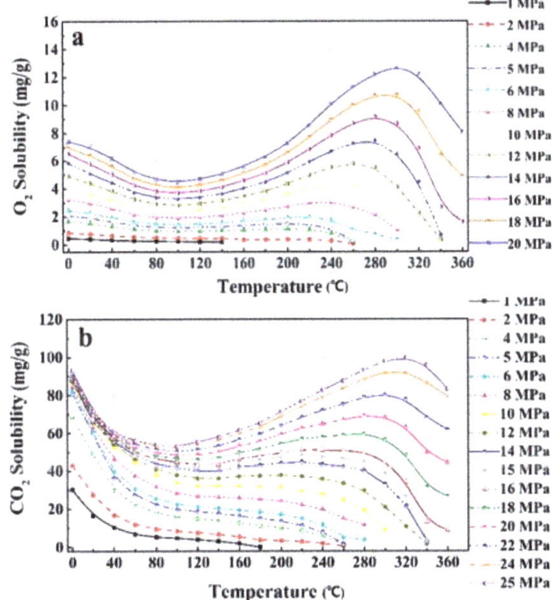

Figure 1. The solubility of oxygen (**a**) and carbon dioxide (**b**) in the high-pressure water.

Over 80% of oxygen cannot be dissolved in high-pressure water and occurs in gaseous phase at $P < 9$ MPa and 20 °C $< T <$ 360 °C, as shown in Figure 2a. In addition, Figure 2b shows that CO_2 can be dissolved completely in water under certain conditions. Moreover, low temperatures are conducive to dissolving CO_2 in water. CO_2 can dissolve completely in water at temperatures below 20 °C when $P = 2$ MPa; however, temperature can be increased to 280 °C when $P = 10$ MPa. Thus, the temperatures for completely dissolving CO_2 in water can be increased at high pressures. Figure 2c shows the releasing ratio difference in gaseous phase between oxygen and CO_2. The temperature zone gradually widens with increasing pressure to obtain a high releasing ratio, but the releasing ratio difference slowly decreases. The temperature zone between 20 °C and 60 °C can reach a releasing ratio difference of 80% at 2 MPa. However, when pressure is increased to 8 MPa, the temperature zone can be widened to a range of 20 °C to 240 °C. These results have motivated us to develop a solution for separating oxygen and CO_2 by adjusting reactor effluent parameters. Thus, a new process for improving oxygen utilization rate in SCWO systems [30] is proposed, as shown in Figure 3.

In the proposed SCWO process, excess oxygen and preheated organic waste are injected from the top of the TWR, which initiates the oxidation reaction. Simultaneously, two branches of transpiring water with different temperatures are injected from the side of the TWR to protect the reactor. The reactor effluent first enters a high-pressure gas–liquid separator after heat exchange and depressurization. Most of the oxygen is released in gaseous phase, whereas most of the CO_2 is dissolved in aqueous phase for the solubility difference between oxygen and CO_2 in water, thereby achieving the separation of oxygen and CO_2. Subsequently, oxygen is reused through the oxygen circulation pump. The aqueous fluid from the high-pressure gas–liquid separator is adjusted further and injected into a low-pressure separator, whereas CO_2 is released and collected. Therefore, oxygen and CO_2 are separated and recovered.

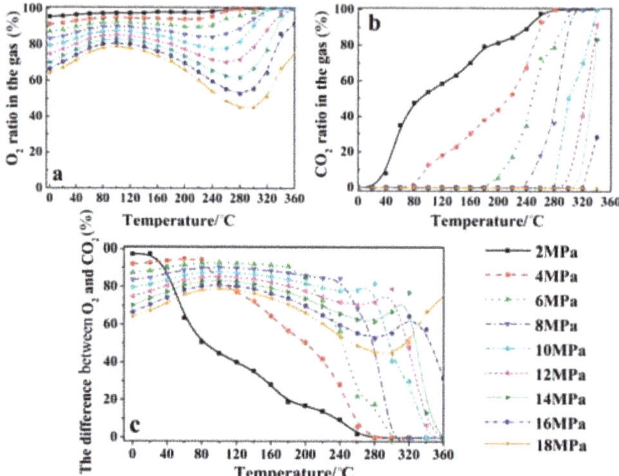

Figure 2. The releasing ratio difference between oxygen and carbon dioxide at different pressures and temperatures based on our previous experimental results, in the reactor effluent, water flow: 46.044 kg/h, oxygen flow: 0.448 kg/h, carbon dioxide flow: 0.836 kg/h, (**a**) O_2 ratio in the gas, (**b**) CO_2 ratio in the gas, (**c**) the ratio difference between O_2 and CO_2.

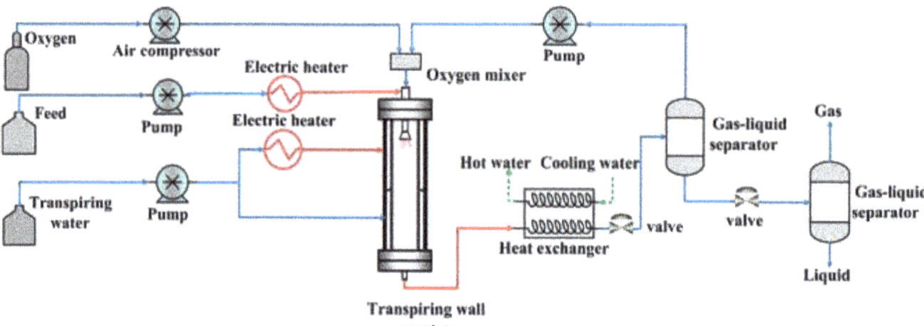

Figure 3. The simplified diagram of a SCWO system to increase the oxygen utilization rate.

3. High-Pressure Separation for Reactor Effluent

3.1. High-Pressure Separation Process

To identify optimized parameters for OR, a simulation flow of a two-step separation process for reactor effluent based on high-pressure water absorption was first established using Aspen Plus V8.0 (Figure 4). High- and low-pressure gas–liquid separators were introduced to separate and recover oxygen and CO_2.

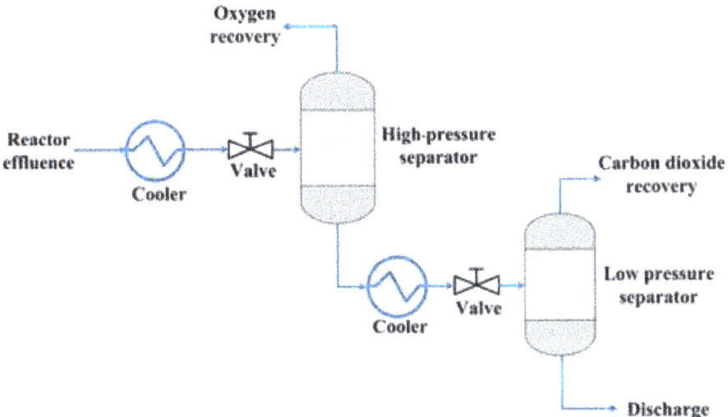

Figure 4. The simulation flow of the high-pressure water absorption for oxygen recovery.

3.2. Definition of Process Parameters

The OR ratio (γ_{O2}) is defined as the oxygen in the gaseous phase of the high-pressure separator divided by the oxygen in the reactor effluent:

$$\gamma_{O_2} = \frac{F'_{O_2,g}}{F'_{O_2,g} + F'_{O_2,l}} \tag{1}$$

where $F'_{O_2,g}$ and $F'_{O_2,l}$ are the oxygen mass flows in the gaseous and aqueous phases, respectively, of the high-pressure separator.

Oxygen purity (β_{O2}) is defined as the oxygen ratio in the gaseous phase of the high-pressure separator, which can be calculated as follows:

$$\beta_{O_2} = \frac{F'_{O_2,g}}{F'_{O_2,g} + F'_{CO_2,g} + F'_{H_2O,g}} \tag{2}$$

where $F'_{CO_2,g}$ and $F'_{H_2O,g}$ are the mass flows of CO_2 and water in the gaseous phase, respectively. Water can be typically disregarded when its content is small.

Similarly, the CO_2 recovery ratio (γ_{CO2}) is defined as the CO_2 in the gaseous phase of the low-pressure separator divided by the CO_2 in the reactor effluent:

$$\gamma_{CO_2} = \frac{F''_{CO_2,g}}{F'_{CO_2,g} + F'_{CO_2,l}} \tag{3}$$

where $F''_{CO_2,g}$ is the CO_2 mass flow in the gaseous phase of the low-pressure separator, and $F'_{CO_2,l}$ is the CO_2 mass flow in the aqueous phase of the high-pressure separator.

CO_2 purity (β_{CO2}) is defined as the CO_2 ratio in the gaseous phase of the low-pressure separator, which can be calculated as follows:

$$\beta_{CO_2} = \frac{F''_{CO_2,g}}{F''_{O_2,g} + F''_{CO_2,g} + F''_{H_2O,g}} \tag{4}$$

where $F''_{O_2,g}$ and $F''_{H_2O,g}$ are the mass flows of oxygen and water, respectively, in the gaseous phase of the low-pressure separator, and water can be typically disregarded when its content is small.

The mass flow rate of oxygen is calculated using a constant R [18], which is defined as follows:

$$R = \frac{F_{O_2}}{1.5F_f\omega} \tag{5}$$

where F_{O_2} (kg/h) and F_f (kg/h) are the mass flow rates of oxygen and the feed, respectively; and ω (wt%) is the methanol concentration in the feed.

3.3. Thermodynamic Property Models

The selection of an appropriate model for estimating the thermodynamic properties of reactor effluent is one of the most important steps that can affect the simulation results. To date, no model has been adopted for all the components and processes. Moreover, the same model cannot be used under all operating conditions, especially at wide ranges of pressure (0.1–23 MPa) and temperature (20–360 °C). Therefore, an appropriate method for estimating the separation process should be carefully selected. Aspen Plus includes a large databank of thermodynamic properties and transport models with the corresponding mixing rules for estimating mixture properties. Several potential thermodynamic models recommended by Aspen Plus were selected and tested (as listed in Table 1) based on the composition of our reactor effluent (i.e., water, CO_2, and oxygen) and the range of the operating conditions. The selected models were simulated with default interaction parameters for the preliminary assessment due to the lack of component interaction coefficients within a large range. The γ_{O2} and γ_{CO2} values at different pressure and temperature values with 10 recommended thermodynamic models were plotted in Figure 5. Additionally, the ideal results calculated from the experimental solubility data of Baranenko et al. [29] were also plotted for the comparison and verification of the thermodynamic models. In the ideal results calculation, the reactor effluent was assumed to conduct an ideal separation in the high-pressure and low-pressure separators, and the interaction between O_2 and CO_2 has been ignored.

Identifying an accurate thermodynamic model that can fulfill the standard for CO_2 and oxygen within a wide range of temperature and pressure values is difficult, as shown in Figure 5. The γ_{O2} (Figure 5a,c,e,g,i,k) calculated using the BWR-LS, PR-BM, SR-POLAR, SRK, PSRK, RKS-BM, and LK-Plock models agree well with the ideal results calculated from the experimental solubility data (red curves) of Baranenko et al. [29]. By contrast, the comparison of γ_{CO2} between the thermodynamic models and the ideal results present more complex information. At 0.1 MPa (Figure 5b), all the models can accurately predict γ_{CO2}. At 2 MPa (Figure 5d) and 4 MPa (Figure 5f), only the PSRK, RKS-BM, and RKS-MHV2 models exhibit accurate prediction performance in terms of trend and value. At higher pressures (i.e., 6, 8, and 10 MPa), only the PSRK model (magenta curves) can achieve good prediction performance, with a maximal deviation of less than 20% (Figure 5h,j,l). Thus, PSRK is selected as the thermodynamic model for the high-pressure separation process in this study under the comprehensive consideration of γ_{O2} and γ_{CO2}. A detailed model expression for PSRK is available in the literature [31].

Table 1. Potential thermodynamic property models in Aspen Plus for the process.

Aspen Plus Property Model	Model Name
BWR-LS	Benedict-Webb-Rubin-Lee-Starling
PR-BM	Peng-Robinson-Boston-Mathias
SR-POLAR	Schwarzentruber-Renon-POLAR
SRK	Soave-Redlich-Kwong
PSRK	Predictive Redlich-Kwong-Soave
RKS-BM	Redlich-Kwong-Soave-Boston-Mathias
LK-Plock	Lee-Kesler-Plock
RK-SWS	Redlich-Kwong-Soave-Wong-Sandler
PR-MHV2	Peng-Robinson-MHV2
PR-WS	Peng-Robinson-Wong-Sandler
RKS-MHV2	Redlich-Kwong-Soave-MHV2

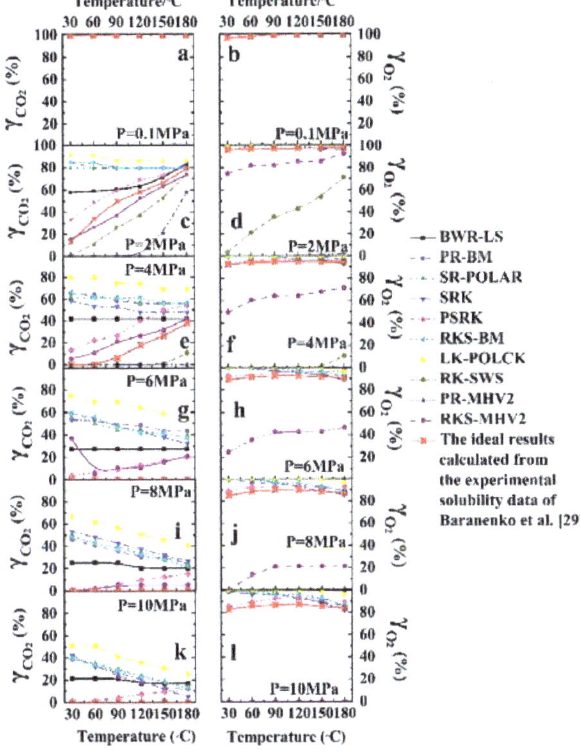

Figure 5. Comparisons of the ideal results calculated from the experimental solubility data and simulation results at different pressures and temperatures, (**a**) γ_{O2} at $P = 0.1$ MPa, (**b**) γ_{CO2} at $P = 0.1$ MPa, (**c**) γ_{O2} at $P = 2$ MPa, (**d**) γ_{CO2} at $P = 2$ MPa, (**e**) γ_{O2} at $P = 4$ MPa, (**f**) γ_{CO2} at $P = 4$ MPa, (**g**) γ_{O2} at $P = 6$ MPa, (**h**) γ_{CO2} at $P = 6$ MPa, (**i**) γ_{O2} at $P = 8$ MPa, (**j**) γ_{CO2} at $P = 8$ MPa, (**k**) γ_{O2} at $P = 10$ MPa, (**l**) γ_{CO2} at $P = 10$ MPa.

3.4. Effects of Operating Parameters

3.4.1. Stoichiometric Oxygen Excess

The interaction between the high- and low-pressure separators typically results in different recovery ratio and purity values for oxygen and CO_2. For convenience, the separation parameters of the low-pressure separator are set under ambient conditions ($P = 0.1$ MPa, $T = 20$ °C) and, thus, we focus only on the separation parameters of the high-pressure separator.

Figure 6(a1–a4,b1–b4) show that a temperature increase or a pressure decrease is favorable for increasing γ_{O2} but unfavorable for increasing β_{O2}. $R = 1.5$ is used as an example. γ_{O2} is 89.3% at $P = 8$ MPa and $T = 20$ °C, and it increased to 92.8% when pressure decreased to 5 MPa. γ_{O2} increased further to 96.4% when pressure and temperature were modified to 5 MPa and 90 °C, respectively (Figure 6(a1)). Similarly, β_{O2} is 78.5% at $P = 8$ MPa and $T = 20$ °C. It decreased to 70.1% when pressure was reduced to 5 MPa and to 56.5% when pressure and temperature were adjusted to 5 MPa and 90 °C, respectively (Figure 6(b1)).

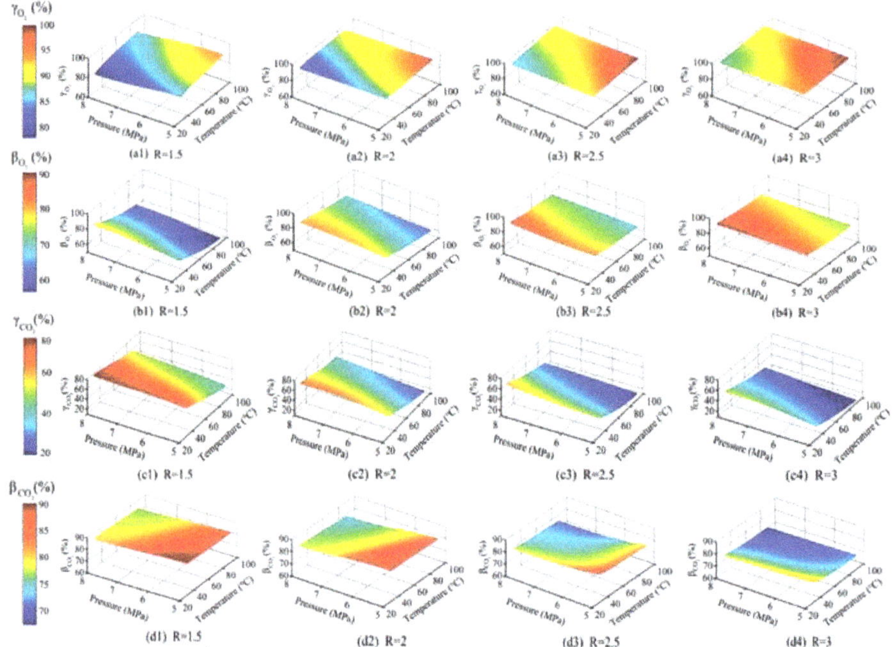

Figure 6. The effect of R on the performance of the high-pressure and low-pressure separators, (**a1**) γ_{O2} at R = 1.5, (**a2**) γ_{O2} at R = 2, (**a3**) γ_{O2} at R = 2.5, (**a4**) γ_{O2} at R = 3, (**b1**) β_{O2} at R = 1.5, (**b2**) β_{O2} at R = 2, (**b3**) β_{O2} at R = 2.5, (**b4**) β_{O2} at R = 3, (**c1**) γ_{CO2} at R = 1.5, (**c2**) γ_{CO2} at R = 2, (**c3**) γ_{CO2} at R = 2.5, (**c4**) γ_{CO2} at R = 3, (**d1**) β_{CO2} at R = 1.5, (**d2**) β_{CO2} at R = 2, (**d3**) β_{CO2} at R = 2.5, and (**d4**) β_{CO2} at R = 3.

The input of the low-pressure separator came from the aqueous mixture of the high-pressure separator. Thus, γ_{CO2} and β_{CO2} in the low-pressure separator are dependent on the separation parameters of the high-pressure separator. A standard for the high-pressure separator is first defined with high values of γ_{O2} (>80%) and β_{O2} (>70%) to narrow down the parameter range. The separating pressure and temperature values that can fulfill the standard can then be obtained. Subsequently, γ_{CO2} and β_{CO2} are analyzed based on the high-pressure separation results. Figure 6(c1–c4) show that a temperature increase or a pressure decrease in the high-pressure separator decreases γ_{CO2}, which is contrary to the effects of pressure and temperature on γ_{O2}. R = 1.5 is used as an example. γ_{CO2} is 78.9% at 8 MPa and 30 °C, and it decreased to 42.1% at 5 MPa and 90 °C (Figure 6(c1)). Moreover, Figure 6(d1–d4) show that a decrease in temperature and pressure are beneficial for β_{CO2}. β_{CO2} is 83.3% at 8 MPa and 90 °C, and it increased to 86.7% when pressure decreased to 5 MPa. Moreover, β_{CO2} increased further to 88.7% when pressure and temperature were decreased to 5 MPa and 30 °C, respectively (Figure 6(d1)).

Figure 6 shows that an increase in R contributes to an increase in γ_{O2} and β_{O2}, but decreases the values of γ_{CO2} and β_{CO2}. γ_{O2}, β_{O2}, γ_{CO2}, and β_{CO2} at P = 5 MPa and T = 90 °C are 92.8%, 56.5%, 42.1%, and 88.9%, respectively, at R = 1.5. γ_{O2} and β_{O2} increased to 98.2% and 78.5%, respectively, whereas γ_{CO2} and β_{CO2} decreased to 15.7% and 75%, respectively, when R increased to 3. An increase in R increases the amount of oxygen in reactor effluent, whereas the amount of CO_2 remains constant (constant (Table 2(A1−A4))). An increase in R is conducive to OR, but reduces CO_2 recovery and purity. The optimized parameters are provided in Table 2(A1−A4). pressure range of 6 MPa to 7 MPa and a temperature range of 30 °C to 40 °C are appropriate for the high-pressure separator.

3.4.2. Feed Concentration

The effects of pressure and temperature at different feed concentrations on species recovery and purity (Figure 7(a1−a5,b1−b5,c1−c5,d1−d5)) are similar to those discussed in the previous section. The values of γ_{CO2} and β_{O2} will be lower at higher feed concentrations, but the values of γ_{O2} and β_{CO2} will be higher. Although an increase in feed concentration will increase the amounts of oxygen and CO_2 in the reactor effluent with the same proportion, the solubility difference between oxygen and CO_2 in the water achieves the following results. The γ_{O2}, β_{O2}, γ_{CO2}, and β_{CO2} at P = 5 MPa and T = 90 °C are 88.9%, 80.0%, 50.0%, and 66.7% at ω = 2 wt%, respectively (Figure 7(a1,b1,c1,d1)). When ω increased to 10 wt%, γ_{O2} and β_{CO2} increased to 98.2% and 85.7%, respectively, but β_{O2} and γ_{CO2} decreased to 68.2% and 19.3%, respectively (Figure 7(a5,b5,c5,d5)).

Therefore, an increase in feed concentration is also conducive to OR, but oxygen purity will be lower. Moreover, an increase in feed concentration is unfavorable for CO_2 recovery, but high CO_2 purity will be obtained. The optimized parameters at different feed concentrations are provided in (B1−B4) in Table 2. A pressure range of 5 MPa to 7 MPa and a temperature range of 30 °C to 70 °C are appropriate for the high-pressure separator.

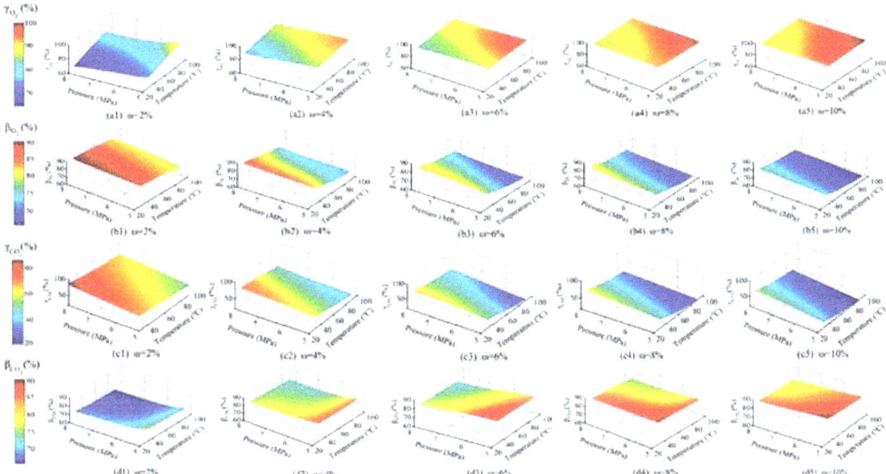

Figure 7. The effect of feed concentration on the performance of the high-pressure and low-pressure separator, (**a1**) γ_{O2} at ω = 2 wt%, (**a2**) γ_{O2} at ω = 4 wt%, (**a3**) γ_{O2} at ω = 6 wt%, (**a4**) γ_{O2} at ω = 8 wt%, (**a5**) γ_{O2} at ω = 10 wt%, (**b1**) β_{O2} at ω = 2 wt%, (**b2**) β_{O2} at ω = 4 wt%, (**b3**) β_{O2} at ω = 6 wt%, (**b4**) β_{O2} at ω = 8 wt%, (**b5**) β_{O2} at ω = 10 wt%, (**c1**) γ_{CO2} at ω = 2 wt%, (**c2**) γ_{CO2} at ω = 4 wt%, (**c3**) γ_{CO2} at ω = 6 wt%, (**c4**) γ_{CO2} at ω = 8 wt%, (**cd**) γ_{CO2} at ω = 10 wt%, (**d1**) β_{CO2} at ω = 2 wt%, (**d2**) β_{CO2} at ω = 4 wt%, (**d3**) β_{CO2} at ω = 6 wt%, (**d4**) β_{CO2} at ω = 8 wt%, and (**d5**) β_{CO2} at ω = 10 wt%.

Table 2. Detailed parameters of the high-pressure and low-pressure parameters.

	R α	ω [β] (wt%)	F_{O2} (kg·h^{-1})	F_{CO2} [γ] (kg·h^{-1})	F_{H2O} [δ] (kg·h^{-1})	P' /MPa	T' (°C)	$F'_{O2,g}$ (kg·h^{-1})	$F'_{CO2,g}$ (kg·h^{-1})	$F'_{H2O,g}$ (kg·h^{-1})	γ_{O2} (%)	β_{O2} (%)	P'' /MPa	T'' (°C)	$F'_{CO2,l}$ (kg·h^{-1})	$F'_{O2,l}$ (kg·h^{-1})	$F''_{O2,g}$ (kg·h^{-1})	$F''_{CO2,g}$ (kg·h^{-1})	γ_{CO2} (%)	β_{CO2} (%)
A1	1.5	6	0.450	0.825	37.820	6–7	30–40	0.353–0.385	0.122–0.21	<0.018	78.6–85.7	70.6–78.6	0.1	20	0.616–0.704	0.064–0.096	0.064–0.096	0.572–0.616	68.42–73.68	82.4–87.5
A2 [ε]	2.0	6	0.900	0.825	39.293	6–7	30–40	0.803–0.835	0.254–0.340	<0.018	89.3–92.86	76.5–80.6	0.1	20	0.484–0.572	0.064–0.096	0.064–0.096	0.44–0.528	52.63–63.16	80–84.62
A3	2.5	6	1.350	0.825	40.766	6–7	30–40	1.252–1.283	0.342–0.430	<0.018	92.86–95.24	80–83.1	0.1	20	0.396–0.484	0.064–0.096	0.064–0.096	0.35–0.442	42.10–53.63	76–81.1
A4	3.0	6	1.800	0.825	42.239	6–7	30–40	1.699–1.730	0.386–0.474	<0.018	94.6–96.43	83.1–85.5	0.1	20	0.352–0.44	0.064–0.096	0.064–0.096	0.308–0.396	36.84–47.36	73–80
B1	2.0	2	0.300	0.275	34.916	5	60–70	0.234–0.266	0.054–0.098	<0.018	78–89	82–86	0.1	20	0.176–0.22	0.032–0.064	0.032	0.132–0.176	50.00–66.67	75–80
B2	2.0	4	0.600	0.550	37.105	6	30–40	0.540	0.112–0.156	<0.018	89.5	81–84	0.1	20	0.396–0.44	0.064	0.064	0.352–0.396	64.58–69.23	80–82
B3	2.0	6	0.900	0.825	39.293	6–7	30–40	0.800–0.832	0.254–0.340	<0.018	89.3–92.86	76.5–80.6	0.1	20	0.484–0.572	0.064–0.096	0.064–0.096	0.44–0.528	52.63–63.16	80–84.62
B4	2.0	8	1.200	1.100	41.482	6–8	30–40	1.107–1.142	0.396–0.528	<0.018	92.1–94.7	75–80	0.1	20	0.572–0.704	0.064–0.096	0.064–0.096	0.528–0.66	48.00–60.00	81.25–86.7
B5	2.0	10	1.500	1.375	43.670	6–8	30–40	1.408–1.420	0.582–0.758	<0.018	93.62–95.7	73–77.2	0.1	20	0.616–0.792	0.064–0.096	0.064–0.096	0.572–0.748	41.93–54.84	82.4–88.2

[α] The stoichiometric oxygen excess at the reactor inlet. [β] the feed concentration at the reactor inlet, and corresponding feed flow F_f is 10 kg·h^{-1}. [γ] the oxygen (F_{O2}) and carbon dioxide (F_{CO2}) flows at the reactor outlet were calculated from the completed oxidation of methanol based on the parameters of the reactor inlet. [δ] the water flow (F_{H2O}) at the reactor outlet was originated from the injection of transpiring water with a transpiring intensity of 0.06 [13] and methanol oxidation product. [ε] A2 = B3.

4. Aspen Model for SCWO System Simulation with Energy and Species Recovery

In this section, our pilot plant was amplified similar to an SCWO industrial plant with a treatment capacity of 1000 kg/h based on the optimized parameters for energy and species recovery. The simulation process can be established without considering the complex equipment structure in Aspen Plus, which is a 1D simulation software based on mass and energy conservation.

4.1. TWR

A TWR is the most important equipment of an SCWO system, and Figure 8a shows the diagram of the TWR in our pilot plant [13]. Five streams were introduced into the reactor. The oxygen and the feed were injected into the reactor via a coaxial nozzle from the top of the reactor, with oxygen in the central tube and the feed in the outer tube. The transpiring tube is divided into three zones using two retaining rings to ensure that the transpiring streams can pass through the porous tube at different temperatures and flow rates. The transpiring water (tw) is divided into three branches, namely, the upper (tw1), middle (tw2), and lower (tw3) branches of transpiring water.

Considering the complicated flow, transpiring heat, and reaction characteristics, the reactor was separated into three sections for simplicity, namely, the mixing, adiabatic reacting, and cooling sections. A simplified model was proposed to simulate the TWR (Figure 8b) in Aspen Plus. The mixing section provides a sufficient mixing space for the reactants. Among the three branches of transpiring water, the upper branch is the only one that can directly influence the reaction [11]. For simplicity, the feed, oxygen, and upper branch of transpiring water will first flow into a mixer to fully mix the reactants. The adiabatic reacting section is simulated by a plug flow reactor (PLUG). When reaction is done, the product flows into the cooling section where the middle and lower branches of transpiring water are injected sequentially into the reactor, and the two mixers are used to simulate the mixing process. Finally, hot effluent flows out of the reactor.

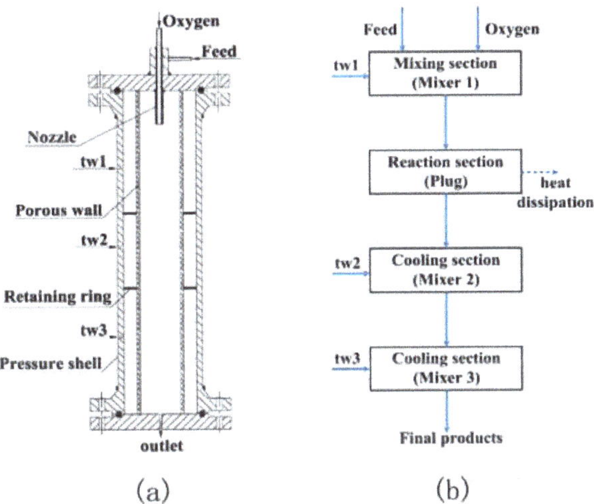

Figure 8. (**a**) The experimental diagram of the TWR and (**b**) the simplified model for the TWR in Aspen plus.

4.2. Reaction

A desalinated water–methanol mixture is also used as artificial wastewater in Aspen Plus. Previous experimental results [13,32] have proven that CO is the major intermediate during the

SCWO of methanol. Thus, a two-step mechanism based on Arrhenius law is created and implemented in the simulation, as shown in Equations (6) to (9):

$$CH_3OH + O_2 = CO + H_2O \tag{6}$$

$$CO + 0.5O_2 = CO_2 \tag{7}$$

$$r_{CH_3OH} = -\frac{d[CH_3OH]}{dt} = 2.0 \times 10^{21} \times \exp\left(\frac{-303.85 \text{ kJ/mol}}{RT}\right)[CH_3OH] \tag{8}$$

$$r_{CO} = -\frac{d[CO]}{dt} = 3.16 \times 10^6 \times \exp\left(\frac{-88 \text{kJ/mol}}{RT}\right)[CO] \tag{9}$$

The kinetic data used in the present study were based on the literature [33–36], and the reaction order of oxygen was assumed zero because of the large excess amount.

4.3. Process Flow

The simulation process, including energy recovery and OR, was developed and presented in Figure 9. After the feed is pressurized by pump 1 (P1), it first flows into heat exchanger 1 (HE1) to be heated by one branch of the final products (FINAL), and then it flows into electric heater 1 (EH1) for further heating. Simultaneously, oxygen is pressurized by the air compressor (AC), and then flows into mixer 1 (M1) to fully mix with the feed and tw1. After transpiring water is pressurized by pump 2 (P2), it splits into three branches (tw1, tw2, and tw3). Before tw1 reaches M1, it first flows into heat exchanger 2 (HE2) to be preheated, and then flows to electric heater 2 (EH2) for further heating. tw2 is preheated by heat exchanger 3 (HE3), and then it mixes with the effluent in mixer 2 (M2). tw3 mixes with the effluent in mixer 3 (M3) to form the final products (FINAL). Oxygen and tw3 are injected into the reactor at room temperature.

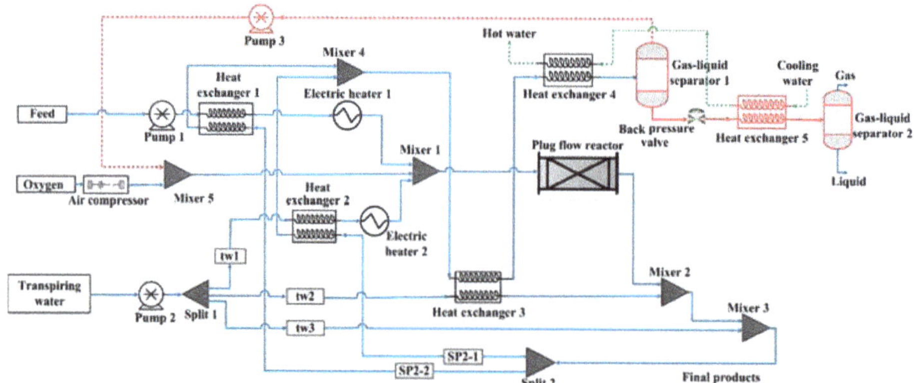

Figure 9. The Aspen Plus diagram of supercritical water oxidation system with oxygen recovery (lines and equipment with red color are specially for OR).

FINAL is split into two branches in split 2, and these branches are treated as hot streams to preheat the feed and tw1. The two branches of FINAL then reunite in mixer 4 (M4) and are cooled down in heat exchanger 3 (HE3). Moreover, the effluent was further cooled in heat exchangers 4 (HE4) and 5 (HE5) by cooling water before gas–liquid separators 1 (S1) and 2 (S2), respectively. The recovered oxygen from S1 is pressurized by pump 3 (P3) and mixed with the supplement oxygen in mixer 5 (M5).

5. Energy and Economic Analysis

5.1. Equipment Investment Calculation

Several alternatives are available to estimate the cost of a major piece of equipment, such as obtaining a quotation from a suitable vendor, using the cost data of a previously purchased equipment of the same type, or utilizing available summary graphs for various types of common equipment. Considering that no similar SCWO industrial plant exists, the last option may be more accurate for our preliminary cost estimation. This methodology allows the estimation of equipment and installation costs according to certain base conditions (e.g., low pressure and construction materials with the lowest cost) and a particular year. Deviations from the base conditions are corrected by a factor that depends on working pressure and construction materials. The obtained cost is then translated into the current time by using an index that considers the time variation of equipment cost.

On the basis of the results obtained for the pilot plant under typical conditions (Table 3, B3, and D3), economic analyses for the 1000 kg/h SCWO plant with and without OR were performed. The investment costs for the TWR, high-pressure pumps, compressors, electric heaters, and gas–liquid separators can be calculated as follows [37]:

$$C_{PM} = C(B_1 + B_2 F_M F_P) \tag{10}$$

$$\lg C = K_1 + K_2 \lg X + K_3 (\lg X)^2 \tag{11}$$

$$\lg F_P = C_1 + C_2 \lg P + C_3 (\lg P)^2 \tag{12}$$

where C is the equipment investment that uses carbon steel under environmental conditions, and X is the design parameter (e.g., pump power and reactor volume). P is the design pressure, which is set as 30 MPa. K_1, K_2, K_3, C_1, C_2, C_3, B_1, and B_2 are constant for each piece of equipment. F_P and F_M are the pressure and material correction coefficients, respectively. Detailed data are provided in Table 4.

Table 3. The operating parameters for the SCWO systems with and without oxygen recovery.

	NO	R	ω (wt%)	F_f (kg·h^{-1})	F_{tw1}[a] (kg·h^{-1})	F_{tw2} (kg·h^{-1})	F_{tw3} (kg·h^{-1})	F_{cw}[b] (kg·h^{-1})	F_{Final} (kg·h^{-1})	F_{sp2-1}[c] (kg·h^{-1})	F_{sp2-2} (kg·h^{-1})	$F_{O2,tot}$ (kg·h^{-1})	$F_{O2,re}$ (kg·h^{-1})	$F_{O2,su}$ (kg·h^{-1})	$F_{CO2,re}$ (kg·h^{-1})	T_{EH1}[d] (°C)	T_{EH2} (°C)	T_{EH3} (°C)	T_f (°C)	T_{tw1} (°C)	T_{tw2} (°C)	T_{M1} (°C)	T_{out} (°C)	T_{sp2-1} (°C)	T_{sp2-2} (°C)	T_{M4} (°C)	CO_{out} (%)	TOC_{out} /ppm	γ_{O2} (%)	γ_{CO2} (%)	β_{O2} (%)	β_{CO2} (%)
With oxygen recovery	A1	1.5	6	1000	1857	620	1236	19,700	4927	2956	1970	135	35.8	99.2	79.2	301	301	181	380	350	160	360	333	191	223	204	0	0	79.56	73.7	73.89	86.78
	A2[e]	2	6	1000	1931	644	1287	20,000	5121	3072	2048	180	82.1	97.9	79.2	300	300	179	380	350	159	357	331	188	220	201	0	0	91.22	58.4	77.89	82.13
	A3	2.5	6	1000	2005	668	1336	20,250	5308	3185	2123	225	124.8	100.2	74.8	299	299	179	381	350	158	355	330	189	221	202	0	0	92.44	47.4	81.56	78.89
	A4	3	6	1000	2078	693	1385	20,500	5500	3300	2200	270	169.95	100.05	74.8	298	298	178	382	350	161	352	328	190	221	203	0	0	94.42	42.1	84.39	74.56
	B1	2	2	1000	1735	578	1156	17,500	4551	2731	1820	60	20.61	39.39	22.1	296	296	155	390	350	150	368	316	161	183	170	0	0	79.2	68.91	85.89	80.21
	B2	2	4	1000	1832	611	1222	18,500	4833	2900	1933	120	53.64	66.36	48.4	298	298	165	385	350	155	363	324	175	204	187	0	0	89.4	62.14	82.36	81.34
	B3	2	6	1000	1931	644	1287	20,000	5121	3073	2048	180	82.1	97.9	79.2	300	300	179	380	350	159	357	331	188	220	201	0	0	91.22	58.4	77.89	82.13
	B4	2	8	1000	2029	676	1353	21,500	5399	3239	2160	240	113.64	126.36	101.2	303	302	194	376	350	167	352	335	198	236	214	0	0	94.7	52.1	76.20	84.36
Without oxygen recovery	C1	1.5	6	1000	1857	620	1236	20,200	4848	2908	1939	135	-	-	-	301	301	182	380	350	161	361	334	192	224	205	0	0	-	-	-	-
	C2	2	6	1000	1931	644	1287	21,000	5042	3025	2016	180	-	-	-	300	300	181	380	350	160	360	333	190	223	204	0	0	-	-	-	-
	C3	2.5	6	1000	2005	668	1336	21,200	5234	3140	2093	225	-	-	-	299	299	180	381	350	159	356	330	189	221	202	0	0	-	-	-	-
	C4	3	6	1000	2078	693	1385	21,500	5426	3255	2170	270	-	-	-	298	298	180	382	350	161	354	330	190	222	203	0	0	-	-	-	-
	D1	2	2	1000	1735	578	1156	18,000	4529	2717	1812	60	-	-	-	297	296	156	390	350	150	368	316	161	183	170	0	0	-	-	-	-
	D2	2	4	1000	1832	611	1222	19,500	4785	2871	1914	120	-	-	-	299	298	167	385	350	155	364	325	175	204	187	0	0	-	-	-	-
	D3	2	6	1000	1931	644	1287	21,000	5042	3025	2017	180	-	-	-	300	300	181	380	350	160	360	333	190	223	204	0	0	-	-	-	-
	D4	2	8	1000	2029	676	1353	22,500	5298	3179	2119	240	-	-	-	302	302	195	376	350	168	354	337	199	237	215	0	0	-	-	-	-

[a] F_{tw1}, F_{tw2}, and F_{tw3} are the mass flows for tw1, tw2, and tw3, respectively. F_{cw}, and F_{Final} are the mass flows for cooling water and the effluent from the reactor outlet, respectively. $F_{O2,tot}$, and $F_{O2,su}$ are the mass flows for total oxygen and supplemental oxygen, respectively. $F_{O2,re}$, and $F_{CO2,re}$ are the mass flows recovered from the S1, respectively. [b] The outlet temperature of the cooling water is set as 60 °C by adjusting flow for hot water production. [c] The F_{SP2-1}/F_{SP2-2} is kept at 1.5 for energy recovery optimization [24]. [d] T_{EH1}, T_{EH2}, and T_{EH3} are the outlet temperatures of feed, tw1, and tw2 in EH1, EH2, and EH3, respectively. T_{tw1}, T_f, and T_{tw2} are the reactor inlet temperatures for tw1, feed, and tw2, respectively. T_{out} is the temperatures of the reactor effluent. T_{SP2-1}, and T_{SP2-2} are the temperatures of the reactor effluent after cooling by HE2 and HE1, respectively. T_{M1}, and T_{M4} are the mixing temperatures after M1 and M4, respectively. [e] A2 = B3, C2 = D3. [f] The T_w for reaction initiation is usually higher (380–420 °C) than that of the steady state for reaction heat releasing.

Table 4. The coefficient for each equipment.

Equipment	K_1	K_2	K_3	C_1	C_2	C_3	B_1	B_2	F_P	F_M
Reactor	4.7116	0.4479	0.0004	-	-	-	-	-	-	4
Pump	3.8696	0.3161	0.1220	−0.3935	0.3957	−0.0023	1.89	1.35	2.2	-
Electric heater	1.1979	1.4782	−0.0958	−0.01635	0.05687	−0.00876	-	-	-	1.4
Compressor	2.2897	1.3604	−0.1027	0	0	0	-	-	2.2	-
Gas-liquid separator	3.4974	0.4485	0.1074	-	-	-	1.49	1.52	-	1.25

Directly estimating the cost of the TWR is difficult because no similar reactor is available for comparison. The cost of a plug flow reactor was first estimated with the same volume for sufficient residence time, and then the cost of the TWR was calculated based on our empirical relationship. The reactor was divided into three sections according to our previous TWR design [24]. The total required volume of the reactor is 570 L. Thus, the actual reactor volume is 695 L when a loading coefficient of 0.82 is considered [38].

Shell and tube heat exchangers were selected in the SCWO system, and the cost of the regular heat exchanger can be calculated as follows [39]:

$$C_{HE} = 3.28 \times 10^4 \left(\frac{A}{80} \right)^{0.68} \delta_M \delta_P \delta_T \tag{13}$$

where A is the heat exchanger area. Considering that the heat exchanger was used in high-pressure and high-temperature conditions, δ_M, δ_P, and δ_T are the material, pressure, and temperature correction coefficients (Table 5), respectively, which were used to modify cost estimation.

Table 5. The coefficients for heat exchanger.

Equipment	Temperature (°C)	Pressure (MPa)	Material	δ_M	δ_P	δ_T
Heat exchanger 1	500	30	Stainless steel 316L	2.9	1.9	2.1
Heat exchanger 2	500	30	Stainless steel 316L	2.9	1.9	2.1
Heat exchanger 3	500	30	Stainless steel 316L	2.9	1.9	2.1
Heat exchanger 4	300	30	Stainless steel 316L	2.9	1.9	1.6

The obtained cost is then translated into the present time by using an index that considers the time variation of equipment cost for the process industries, which was calculated using the following equation [37]:

$$Cost_{2016} = Cost_{2001} \left(\frac{CEPCI_{2016}}{CEPCI_{2001}} \right) \tag{14}$$

Given the aforementioned considerations, the total equipment cost for the SCWO pilot plant with and without OR in 2016 was calculated as \$2,592,096 and \$2,522,654, respectively. Details on equipment sizing assumptions, construction materials, and estimated cost per piece of equipment are presented in Table 6.

Table 6. Equipment investment.

NO	Equipment	Parameter	SCWO System with Oxygen Recovery				SCWO without Oxygen Recovery			
			Theoretical Value	Safety Factor [a]	Design Value	Cost ($)	Theoretical Value	Safety Factor [a]	Design Value	Cost ($)
1	Heat exchanger 1	Area/m^2	5.77	1.2	7	72,412	5.76	1.2	7	72,412
2	Heat exchanger 2	Area/m^2	12.26	1.2	15	121,586	12.24	1.2	15	121,586
3	Heat exchanger 3	Area/m^2	1.146	1.2	1.5	25,403	1.14	1.2	1.5	25,403
4	Heat exchanger 4	Area/m^2	19.73	1.2	24	127,522	22.5	1.2	27	143,166
5	Heat exchanger 5	Area/m^2	3.78	1.2	4.5	25,533	–	–	–	–
6	Waste water pump	Power/kW	11.28	1.2	14	155,452	11.28	1.2	14	155,452
7	Compressor [b]	Power/kW	36.88	1.6	60	132,840	36.88	1.6	60	132,840
8	Transpiring water pump	Power/kW	41.32	1.2	50	145,800	41.32	1.2	50	145,800
9	Oxygen circulation pump	Power/kW	4.59	2.1	9.5	21,034	–	–	–	–
10	Transpiring wall reactor	Volume/m^3	0.695	3	2.1	531,798	0.695	3	2.1	531,798
11	High-pressure separator	Volume/m^3	0.189	3	0.6	7142	–	–	–	–
12	Low-pressure separator	Volume/m^3	0.186	1.5	0.3	3570	0.189	1.5	0.3	3570
13	Electric heater 1	Power/kW	306.81	1.2	370	31,365	306.81	1.2	370	31,365
14	Electric heater 2	Power/kW	190.56	1.2	228	19,380	190.56	1.2	228	19,380
15	Total equipment cost/1996	$			1,420,837				1,382,772	
16	Total equipment cost/2016	$			2,253,997				2,193,612	
17	Installation cost [c]	$			338,099				329,042	
18	Total investment cost/2016	$			2,592,096				2,522,654	

[a] obtained from reference [38]. [b] based on the power consumption in the system startup. [c] set as 15% of the equipment cost.

5.2. Treatment Cost Calculation and Distribution

The treatment cost of an SCWO system includes investment and operating costs. The basic operating costs were determined using the procedure parameters in Table 3 (B3) and (D3), which were estimated under the assumption that the plant operates 330 days a year and 24 h a day. The operating cost includes energy consumption, raw material, labor, and capital-related costs [36]. Energy consumption cost includes the cost of electricity required to operate the process equipment and the plant. Raw material cost, which includes the costs of oxygen, cooling water, and transpiring water, was estimated from the amount of required raw materials. Labor cost includes the salaries of operation and supervisory employees. The depreciation time of the system is 10 years, and the maintenance cost is 3% of the equipment cost.

Figure 10 shows the treatment cost comparisons of the SCWO systems with and without OR. In the SCWO system without OR, electricity, depreciation, and oxygen contribute to the primary treatment cost, accounting for 46.18, 30.24, and 18.01 $\cdot t^{-1}$, respectively, of the total cost. Although the heat of the reactor effluent has been recovered, energy (electricity) consumption remains high. This phenomenon is attributed to the low-grade heat of the reaction effluent (<370 °C) due to the injection of transpiring water at a low temperature to avoid salt plugging. Hot water, which comprises the major income of the system, was calculated as a negative value in the treatment cost and accounted for -56.72 \$$\cdot t^{-1}$. Thus, the total treatment cost for the SCWO system without OR is 56.80 \$$\cdot t^{-1}$, with electricity and oxygen cost accounting for 81.30% and 31.69% of the total treatment cost, respectively.

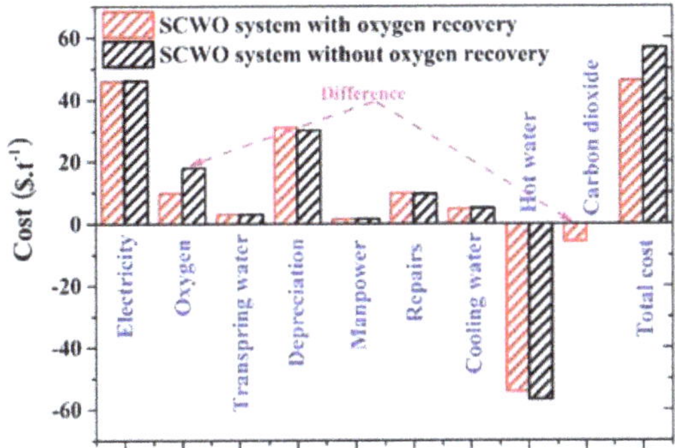

Figure 10. The treatment cost comparisons for SCWO systems with and without OR, the prices for electricity, oxygen, transpiring water and cooling water, are 0.079 \$/kW·h, 100 \$$\cdot t^{-1}$, 0.8 \$$\cdot t^{-1}$, and 0.24 \$$\cdot t^{-1}$, respectively; the manpower is 6000 \$/man·year; the income for hot water and CO_2 are 2.7 \$$\cdot t^{-1}$ and 71.4 \$$\cdot t^{-1}$, respectively.

Electricity, depreciation, and oxygen still contribute to the primary treatment cost of the SCWO system with OR. Electricity consumption slight decreases from 46.18 \$$\cdot t^{-1}$ to 45.88 \$$\cdot t^{-1}$ due to OR, but oxygen cost significantly decreased from 18.01 \$$\cdot t^{-1}$ to 9.77 \$$\cdot t^{-1}$. Additionally, the additional income of CO_2, which accounted for -5.65 \$$\cdot t^{-1}$, was obtained due to OR. Treatment cost considerably decreased from 56.80 \$$\cdot t^{-1}$ to 46.17 \$$\cdot t^{-1}$, with a reduction rate of 18.82%. Thus, OR considerably contributes to reducing the treatment cost of an SCWO system.

5.3. Effect of Stoichiometric Oxygen Excess

On the basis of the previously designed system, this section investigates the effects of the operating parameters on energy consumption and treatment cost. Similar to the previous analysis, several episodes of actual oxygen consumption may be necessary for complete feed degradation. Thus, the effect of R on the treatment cost of the SCWO systems with and without OR is analyzed in this section, and the operating parameters and detailed results are listed in Table 3(A1–A4, C1–C4) and Table 7(A1–A4, C1–C4). Electricity consumption and hot water income increase slightly with an increase in R in both SCWO systems (Figure 11a,e). Oxygen consumption increases linearly with an increase in R in the SCWO system without OR. When R increased from 1.5 to 3, oxygen consumption considerably increased from 13.5 $\$\cdot t^{-1}$ to 27 $\$\cdot t^{-1}$ (Figure 11b). Furthermore, a slight increase in cooling water consumption (Figure 11d) occurs with an increase in R. An increase in R has minimal effect on depreciation, repair (Figure 11c), transpiring water consumption, manpower (Figure 11d), and CO_2 income (Figure 11e). Thus, the treatment cost of the SCWO system without OR can increase from 53.89 $\$\cdot t^{-1}$ to 65.25 $\$\cdot t^{-1}$ (Figure 11f) when R increased from 1.5 to 3. In the SCWO system with OR, oxygen consumption in the start-up stage is equal to that of the SCWO system without OR. However, the supplemental oxygen content is gradually reduced to a value that is slightly higher than the actual oxygen consumption after attaining OR equilibrium (Table 3). Thus, an increase in R exerts minimal effect on oxygen consumption (Figure 11b). Moreover, high-purity CO_2 can be recovered as an income due to OR (Figure 11e). In addition, equipment repairs and depreciation (Figure 11c), cooling water, transpiring water, and manpower consumption (Figure 11d) also exhibit minimal differences with varying R values. Figure 11f shows that the treatment cost of the SCWO system with OR slightly increased from 46.63 $\$\cdot t^{-1}$ at $R = 1.5$ to 48.89 $\$\cdot t^{-1}$ at $R = 3$, which motivates us to operate the SCWO system with a high R value for complete feed degradation.

Table 7. Electricity consumption for the SCWO system.

		R	ω(wt%)	P1(kW)	P_{AC}(kW)	P2(kW)	P3(kW)	P_{EH1}(kW)	P_{EH2}(kW)	Total(kW)
With oxygen recovery	A1	1.5	6	11.28	25.6	41.31	2.23	306.81	190.56	577.79
	A2	2	6	11.28	24.53	42.95	4.59	306.81	190.56	580.08
	A3	2.5	6	11.28	26.12	44.59	6.67	312.89	193.38	594.93
	A4	3	6	11.28	26.05	46.22	8.99	316.87	196.47	605.88
	B1	2	2	11.18	11.89	38.05	1.56	361.4	206.32	630.4
	B2	2	4	11.23	16.71	39.70	2.82	331.45	198.15	600.06
	B3	2	6	11.28	24.53	42.95	4.59	306.81	190.56	580.08
	B4	2	8	11.34	30.66	43.97	6.45	281.64	182.82	556.88
Without Oxygen recovery	C1	1.5	6	11.28	27.66	41.31	-	302.11	190.01	572.37
	C2	2	6	11.28	36.88	42.95	-	304.24	189.26	583.97
	C3	2.5	6	11.28	46.10	44.59	-	310.19	191.98	604.14
	C4	3	6	11.28	55.32	46.22	-	315.77	195.17	623.76
	D1	2	2	11.18	12.29	38.05	-	360.40	204.82	626.74
	D2	2	4	11.23	24.59	39.70	-	329.34	197.61	602.47
	D3	2	6	11.28	36.88	42.95	-	304.24	189.26	583.97
	D4	2	8	11.34	49.18	43.97	-	277.14	180.82	562.45

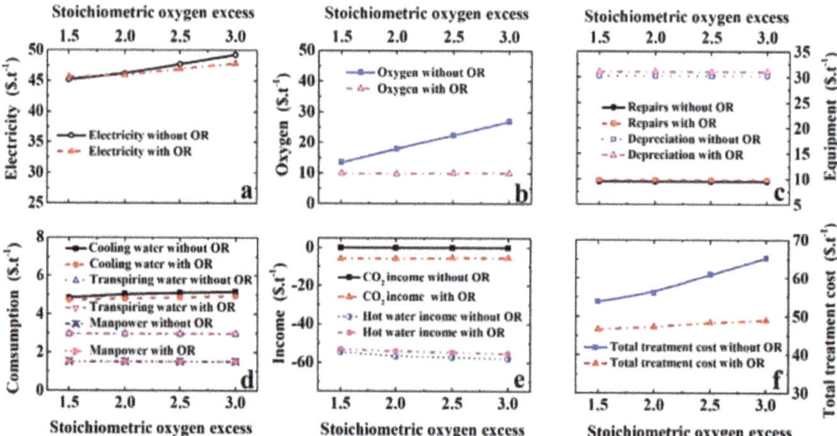

Figure 11. The effect of R on the treatment cost for the SCWO system with and without OR, (**a**) electricity consumption, (**b**) oxygen consumption, (**c**) equipment repairs and depreciation, (**d**) cooling water, transpiring water, and manpower consumption, (**e**) CO_2 and hot water income, and (**f**) total treatment cost.

5.4. Effect of the Feed Concentration

The treatment cost for feed concentration between 2 wt% and 8 wt% is tested in this section under operating conditions, and the detailed results are listed in Table 3(B1–B4, D1–D4) and Table 7(B1–B4, D1–D4). When feed concentration increases, oxygen and transpiring water flow rates will also increase for feed degradation and reactor protection, and consequently, the electricity consumption of the pumps will also increase. However, reaction heat linearly increases with increasing feed concentration, and more heat can be recovered from the reactor effluent. Moreover, the preheating temperature of the feed at the starting and steady states can be reduced at a high feed concentration [40]. Thus, the total electricity consumption of the systems with and without OR decreased from 49.51 $\text{\$·t}^{-1}$ and 49.80 $\text{\$·t}^{-1}$ to 44.35 $\text{\$·t}^{-1}$ and 43.91 $\text{\$·t}^{-1}$, respectively, when feed concentration was increased from 2 wt% to 8 wt% (Figure 12a).

Oxygen consumption significantly increased from 6.00 $\text{\$·t}^{-1}$ at ω = 2 wt% to 24.00 $\text{\$·t}^{-1}$ at ω = 8 wt% (Figure 12b), and hot water income considerably increased from 48.6 $\text{\$·t}^{-1}$ to 60.75 $\text{\$·t}^{-1}$ (Figure 12e) in the SCWO system without OR. Thus, treatment cost can increase from 54.82 $\text{\$·t}^{-1}$ at ω = 2 wt% to 57.93 $\text{\$·t}^{-1}$ at ω = 8 wt% (Figure 12f). In the SCWO system with OR, when feed concentration was increased from 2 wt% to 8 wt%, the supplemental oxygen content increased from 3.26 $\text{\$·t}^{-1}$ to 13.05 $\text{\$·t}^{-1}$, respectively (Figure 12b), and hot water and CO_2 income increased from 47.25 $\text{\$·t}^{-1}$ and 1.57 $\text{\$·t}^{-1}$ to 58.05 $\text{\$·t}^{-1}$ and 7.22 $\text{\$·t}^{-1}$, respectively (Figure 12e). Figure 12f shows that the treatment cost of the SCWO system with OR decreased from 54.27 $\text{\$·t}^{-1}$ at ω = 2 wt% to 42.06 $\text{\$·t}^{-1}$ at ω = 8 wt%. Thus, an increase in feed concentration is conducive to reducing both the energy consumption and the treatment cost of the SCWO system with OR.

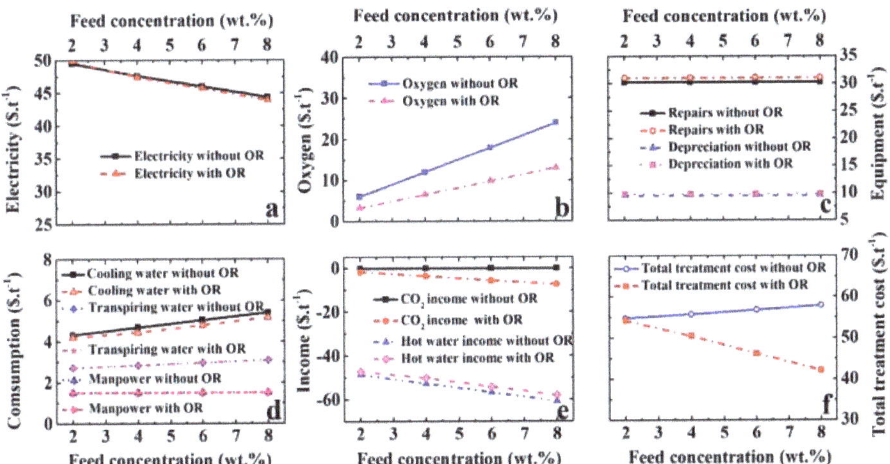

Figure 12. The effect of feed concentration on the treatment cost for a SCWO system with and without OR, (**a**) electricity consumption, (**b**) oxygen consumption, (**c**) equipment repairs and depreciation, (**d**) cooling water, transpiring water, and manpower consumption, (**e**) CO_2 and hot water income, and (**f**) total treatment cost.

6. Conclusions

In this work, a species recovery process for an SCWO system with a TWR was first proposed based on the solubility difference between oxygen and CO_2 in high-pressure water. Thus, oxygen and CO_2 can be separated and recovered from the reactor effluent to reduce operating cost.

A two-step separation process was first established using Aspen Plus software to increase species recovery rate. Then, 10 potential thermodynamic models for high-pressure separation were evaluated and selected. The detailed recovery rates of oxygen and CO_2 were compared with the ideal results calculated from the experimental solubility data. The PSRK model was proven to be an appropriate thermodynamic model for predicting the separation process of the reactor effluent under a wide range of conditions. Accordingly, the detailed optimized parameters for species separation were obtained.

The SCWO processes with and without OR were simulated and economic analyses were conducted. Electricity, depreciation, and oxygen costs contribute to the major treatment cost of the SCWO system without OR, accounting for 46.18, 30.24, and 18.01 $\cdot t^{-1}$, respectively. When OR was introduced, oxygen cost decreased from 18.01 $\cdot t^{-1}$ to 9.77 $\cdot t^{-1}$, and additional CO_2 income, which amounted to -5.65 $\cdot t^{-1}$, was gained due to OR. The total treatment cost considerably decreased from 56.80 $\cdot t^{-1}$ to 46.17 $\cdot t^{-1}$, with a reduction rate of 18.82%. Thus, OR contributes to reducing the treatment cost of an SCWO system. In addition, R and feed concentration increased and contributed to reducing the operating cost of the SCWO system with OR.

As a preliminary study of new SCWO system with OR, more experiments are needed to obtain more accurate results based on the simulation results in the future.

Author Contributions: F.Z. and C.M. put forward the idea of this work, F.Z. wrote this paper, J.C. conducted the simulation and calculation, F.Z., J.C., and C.S. contributed to the results analysis and post-processing.

Funding: This work is supported by National Natural Science Foundation (no. 51706049), Youth Innovation Promotion Association CAS (no. 2017412), and Science research project of Guangzhou City (201707010407).

Conflicts of Interest: The authors declare no conflict of interest.

Nomenclature

Abbreviations

A	area
AC	air compressor
C	capital
COD	chemical oxygen demand
EH	electric heater
F	mass flow rate, $kg \cdot h^{-1}$
FINAL	Final products
F_P	pressure correction coefficient
F_M	material correction coefficient
HE	heat exchanger
M	mixer
OR	oxygen recovery
P	pressure/pump
R	transpiring intensity, universal gas constant
S	Separator
SCW	supercritical water
SCWG	supercritical water gasification
SCWO	supercritical water oxidation
r	reaction rate
R	stoichiometric oxygen excess
t	time, s
T	temperature, °C
tw1	upper branch of transpiring water
tw2	middle branch of transpiring water
tw3	lower branch of transpiring water
TWR	transpiring wall reactor
TOC	total organic carbon, ppm
X	Design parameter

Greek letters

β purity

ρ fluid density, kg·m^{-3}

φ transpiring intensity

γ recovery ratio

δ correction coefficient

ω feed concentration, wt%

Subscripts

cw   cooling water

f   feed

g   gas

l   liquid

m   material

su   supplement

out   outlet

tot   total

References

1. Akiya, N.; Savage, P.E. Roles of Water for Chemical Reactions in High Temperature Water. *Chem. Rev.* **2002**, *33*, 2725–2750. [CrossRef]
2. Xu, D.H.; Huang, C.B.; Wang, S.Z.; Lin, G.K.; Guo, Y. Salt deposition problems in supercritical water oxidation. *Chem. Eng. J.* **2015**, *279*, 1010–1022. [CrossRef]
3. Queiroz, J.P.S.; Bermejo, M.D.; Mato, F.; Cocero, M.J. Supercritical water oxidation with hydrothermal flame as internal heat source: Efficient and clean energy production from waste. *J. Supercrit. Fluid* **2015**, *96*, 103–113. [CrossRef]
4. Molino, A.; Migliori, M.; Blasi, A.; Davoli, M.; Marino, T.; Chianese, S.; Catizzone, E.; Giordano, G. Municipal waste leachate conversion via catalytic supercritical water gasification process. *Fuel* **2017**, *206*, 155–161. [CrossRef]
5. Fedyaeva, O.N.; Vostrikov, A.A.; Shishkin, A.V.; Dubov, D.Y. Conjugated processes of black liquor mineral and organic components conversion in supercritical water. *J. Supercrit. Fluids* **2019**, *143*, 191–197. [CrossRef]
6. Vadillo, V.; Belén García-Jarana, M.B.; Sánchez-Oneto, J.; Portela, J.R.; de la Ossa, E.J.M. Simulation of Real Wastewater Supercritical Water Oxidation at High Concentration on a Pilot Plant Scale. *Ind. Eng. Chem. Res.* **2011**, *50*, 2512–2520. [CrossRef]
7. Zhang, F.M.; Chen, S.Y.; Xu, C.Y.; Chen, G.F.; Ma, C.Y. Energy consumption analysis of a transpiring-wall supercritical water oxidation pilot plant based on energy recovery. *Desalin. Water Treat.* **2013**, *51*, 7341–7352. [CrossRef]
8. Fourcault, A.; García-Jarana, B.; Sánchez-Oneto, J.; Marias, F.; Portela, J.R. Supercritical water oxidation of phenol with air. Experimental results and modelling. *Chem. Eng. J.* **2009**, *152*, 227–233. [CrossRef]
9. Marrone, P.A. Supercritical water oxidation-Current status of full-scale commercial activity for waste destruction. *J. Supercrit. Fluid* **2013**, *79*, 283–288. [CrossRef]
10. Vadillo, V.; Sánchez-Oneto, J.; Portela, J.R.; de la Ossa, E.J.M. Problems in Supercritical Water Oxidation Process and Proposed Solutions. *Ind. Eng. Chem. Res.* **2013**, *52*, 7617–7629. [CrossRef]
11. Kritzer, P. Corrosion in high-temperature and supercritical water and aqueous solutions: A review. *J. Supercrit. Fluid* **2004**, *29*, 1–29. [CrossRef]
12. Hodes, M.; Marrone, P.A.; Hong, G.T.; Smith, K.A.; Tester, J.W. Salt precipitation and scale control in supercritical water oxidation–Part A: Fundamentals and research. *J. Supercrit. Fluids* **2004**, *29*, 265–288. [CrossRef]
13. Zhang, F.M.; Chen, S.Y.; Xu, C.Y.; Chen, G.F.; Zhang, J.M.; Ma, C.Y. Experimental study on the effects of operating parameters on the performance of a transpiring-wall supercritical water oxidation reactor. *Desalination* **2012**, *294*, 60–66. [CrossRef]
14. Bermejo, M.D.; Cocero, M.J. Supercritical water oxidation: A technical review. *AIChE J.* **2006**, *52*, 3933–3951. [CrossRef]

15. Xu, D.H.; Wang, S.Z.; Huang, C.B.; Tang, X.Y.; Guo, Y. Transpiring wall reactor in supercritical water oxidation. *Chem. Eng. Res. Des.* **2014**, *92*, 2626–2639. [CrossRef]

16. Kritzer, P.; Dinjus, E. An assessment of supercritical water oxidation (SCWO)-Existing problems, possible solutions and new reactor concepts. *Chem. Eng. J.* **2001**, *83*, 207–214. [CrossRef]

17. Kodra, D.; Balakotaiah, V. Autothermal oxidation of dilute aqueous wastes under supercritical conditions. *Ind. Eng. Chem. Res.* **1994**, *33*, 575–580. [CrossRef]

18. Lavric, E.D.; Weyten, H.; Ruyck, J.D.; Plesu, V.; Lavric, V. Delocalized organic pollutant destruction through a self-sustaining supercritical water oxidation process. *Energy Convers. Manag.* **2005**, *46*, 1345–1364. [CrossRef]

19. Cocero, M.J.; Alonso, E.; Sanz, M.T.; Fdz-Polanco, F. Supercritical water oxidation process under energetically self-sufficient operation. *J. Supercrit. Fluids* **2002**, *24*, 37–46. [CrossRef]

20. Bermejo, M.D.; Cocero, M.J.; Fernández-Polanco, F. A process for generating power from the oxidation of coal in supercritical water. *Fuel* **2004**, *83*, 195–204. [CrossRef]

21. Marias, F.; Mancini, F.; Cansell, F.; Mercadier, J. Energy recovery in supercritical water oxidation process. *Environ. Eng. Sci.* **2008**, *25*, 123–130. [CrossRef]

22. Lavric, E.D.; Weyten, H.; De Ruyck, J.; Plesu, V.; Lavric, V. Supercritical water oxidation improvements through chemical reactors energy integration. *Appl. Therm. Eng.* **2006**, *26*, 1385–1392. [CrossRef]

23. Jimenez-Espadafor, F.; Portela, J.R.; Vadillo, V.; Sánchez-Oneto, J.; Becerra Villanueva, J.A.; Torres García, M.; Martínez de la Ossa, E.J. Supercritical water oxidation of oily wastes at pilot plant: Simulation for energy recovery. *Ind. Eng. Chem. Res.* **2011**, *50*, 775–784. [CrossRef]

24. Zhang, F.M.; Shen, B.Y.; Su, C.J.; Xu, C.Y.; Ma, J.N.; Xiong, Y.; Ma, C.Y. Energy consumption and exergy analyses of a supercritical water oxidation system with a transpiring wall reactor. *Energy Convers. Manag.* **2017**, *145*, 82–92. [CrossRef]

25. Xu, D.H.; Wang, S.Z.; Gong, Y.M.; Guo, Y.; Tang, X.Y.; Ma, H.H. A demonstration plant for treating sewage sludge by supercritical water oxidation and its economic analysis. *Mod. Chem. Ind.* **2009**, *29*, 55–59.

26. Zhang, J.; Wang, S.Z.; Lu, J.L.; Chen, S.L.; Li, Y.H.; Ren, M.M. A system for treating high concentration textile wastewater and sludge by supercritical water oxidation and its economic analysis. *Mod. Chem. Ind.* **2016**, *36*, 154–158.

27. Shen, X.F.; Ma, C.Y.; Wang, Z.Q.; Chen, G.F.; Chen, S.Y.; Zhang, J.M.; Yi, B.K. Economic analysis of organic waste liquid treatment through supercritical water oxidation system. *Environ. Eng.* **2010**, *28*, 47–51.

28. Abeln, J.; Kluth, M.; Pagel, M. Results and rough costestimation for SCWO of painting effluents using a transpiring wall and a pipe reactor. *J. Adv. Oxid. Technol.* **2007**, *10*, 169–176.

29. Baranenko, V.I.; Fal'kovskii, L.N.; Kirov, V.S.; Kurnyk, L.N.; Musienko, A.N.; Piontkovskii, A.I. Solubility of oxygen and carbon dioxide in water. *At. Energy* **1990**, *68*, 342–346. [CrossRef]

30. Ma, C.Y.; Zhang, F.M.; Chen, S.Y.; Chen, G.F.; Zhang, J.M. A Method of Improving Oxygen Utilization Rate in Supercritical Water Oxidation System. Chinese Patent ZL 201010174846.9, 18 May 2010.

31. Holderbaum, T.; Gmehling, J. PSRK: A group contribution equation of state based on UNIFAC. *Fluid Phase Equilib.* **1991**, *70*, 251–265. [CrossRef]

32. Zhang, F.M.; Xu, C.Y.; Zhang, Y.; Chen, S.Y.; Chen, G.; Ma, C.Y. Experimental study on the operating characteristics of an inner preheating transpiring wall reactor for supercritical water oxidation: Temperature profiles and product properties. *Energy* **2014**, *66*, 577–587. [CrossRef]

33. Li, L.; Chen, P.; Gloyna, E.F. Generalized kinetic model for wet oxidation of organic compounds. *AIChE J.* **1991**, *37*, 1687–1697. [CrossRef]

34. Vogel, F.; Blanchard, J.L.D.; Marrone, P.A.; Rice, S.F.; Webley, P.A.; Peters, W.A.; Smith, K.A.; Tester, J.W. Critical review of kinetic data for the oxidation of methanol in supercritical water. *J. Supercrit. Fluid* **2005**, *34*, 249–286. [CrossRef]

35. Dagaut, P.; Cathonnet, M.; Boettner, J. Chemical Kinetic Modeling of the Supercritical-Water Oxidation of Methanol. *J. Supercrit. Fluids* **1996**, *98*, 33–42. [CrossRef]

36. Tester, J.W.; Webley, P.A.; Holgate, H.R. Revised global kinetic measurements of methanol oxidation in supercritical water. *Ind. Eng. Chem. Res.* **1993**, *32*, 236–239. [CrossRef]

37. Turton, R.; Bailie, R.C.; Whiting, W.B.; Bhattacharyya, D. *Analysis, Synthesis and Design of Chemical Processes*; Prentice Hall: Upper Saddle River, NJ, USA, 2012.

38. Shen, X.F. Design and Technical Economic Analysis on Heat Supply System by Supercritical Water Oxidation Energy Conversion. Master's Thesis, Shandong University, Jinan, China, 2009.

39. Wildi-Tremblay, P.; Gosselin, L. Minimizing shell-and-tube heat exchanger cost with genetic algorithms and considering maintenance. *Int. J. Energy Res.* **2007**, *31*, 867–888. [CrossRef]
40. Zhang, F.M.; Zhang, Y.; Xu, C.Y.; Chen, S.Y.; Chen, G.F.; Ma, C.Y. Experimental study on the ignition and extinction characteristics of the hydrothermal flame. *Chem. Eng. Technol.* **2015**, *38*, 2054–2066. [CrossRef]

Communication

Analysis of the Excess Hydrocarbon Gases Output from Refinery Plants

Jerzy Szpalerski [1] and Adam Smoliński [2,*]

[1] PKN ORLEN S.A., Chemików Str. 7, 09-411 Płock, Poland; jerzy.szpalerski@orlen.pl
[2] Central Mining Institute, Plac Gwarków 1, 40-166 Katowice, Poland
*   Correspondence: smolin@gig.katowice.pl; Tel.: +48-32-259-2252

Received: 8 March 2019; Accepted: 25 April 2019; Published: 1 May 2019

Abstract: The article presents the ideas of maximizing recovery of flare gases in the industrial plants processing hydrocarbons. The functioning of a flare stack and depressurization systems in a typical refinery plant is described, and the architecture of the depressurization systems and construction of the flares are shown in a simplified way. The proposal to recover the flare gases together with their output outside the industrial plant, in order to minimize impact on the environment (reduction of emissions) and to limit consumption of fossil fuels is presented. Contaminants that may be found in the depressurization systems are indicated. The idea presented in the article assumes the injection of an excess stream of gases into an existing natural gas pipelines system. A method of monitoring is proposed, aiming to eliminate introduction of undesirable harmful components into the systems.

Keywords: refinery plants; industrial gas streams; petrochemical processes; waste gases

1. Introduction

There are many technologies applied in processing of so widely understood charges [1–5] in the industrial plants dealing with crude oil or individual hydrocarbons processing. Typical production processes carried out in the refineries are: fractional distillation process, catalytic cracking process, gasoline reforming process, diesel/oil/hydro desulfurization, hydrocracking, gasoline isomerization, asphalt oxidation, and storage of raw materials, semi-finished products and finished products [1–5].

As a part of the refinery, many auxiliary processes are carried out, without which a modern refinery could not function, particularly the refinery in which extended are conversion processes that lead to greater destruction of the hydrocarbon chain. Among these processes, the following should first and foremost be distinguished [5–7]:

- Hydrogen production;
- Claus's process;
- Energy media production processes;
- Wastewater treatment, production of circulating water; and
- Hydrogen recovery processes.

Refineries produce engine fuels, lubricating oils, other commercial fluids, and some refinery products are used as raw materials for petrochemical processes. This is particularly important when the refinery plant is integrated with a petrochemical plant [8]. Among the petrochemical processes, above all should be mentioned [9–12] production of olefins, butadiene, aromatic hydrocarbons, cumene, or ethylene oxide.

These processes aim at obtaining products that comply with the relevant technical specifications (technical conditions, company standards, national standards, etc.). During almost every production process various types of by-products (gaseous, liquid) are formed. Great efforts are made to achieve a situation in which the by-products are utilized within the production plants. In the case of gas products, the aim is also to direct them to a further process. However, there is always a certain amount of gas products that for various reasons cannot be utilized. Some of these streams are directed to the flare stacks. These are not only streams coming from production but also streams from the preparation of entire production units to stop (e.g., repairs), from commissioning, and finally from the preparation of individual apparatus for repair, revisions provided for by law, system switching, etc.

It happens that industrial gas streams, directed to depressurization systems, are full-value products which should be used outside the plants as a fuel for energy combustion whether they cannot be used in the industrial plant at the moment.

The aim of this paper is to analyze the possibilities of maximizing recovery of the flare gases in the hydrocarbon processing plants.

2. Flare Gases in the Industrial Plants

Some of the gas streams coming from various types of operations, carried out on production installations, are directed to the flare stacks. However, it is not possible to consider the flare stack as isolated from the entire depressurization system that is understood as a system of pipelines and collectors with assigned devices supplying the flare gases to the flare [13,14]. The flare stacks are applied and used most often in the crude oil and natural gas mining facilities, refineries, coke plants, chemical plants, and garbage dumps. The flare stack (utilization flare) is a device, usually in a shape of a stack, which burns the not utilized gas (generally), or which excess is impossible (or inexpedient) to be managed or stored at a given moment. Gas combustion in flares is mainly used in order to:

- Protect the surrounding (locally) against the effects of uncontrolled migration of gas, e.g., against explosion, poisoning, and fetor;
- Protect the environment by replacing emissions of more harmful gases by emission of less harmful fumes (e.g., acid gases—hydrogen sulfide, ammonia—occurring primarily as waste gases during the production process; directing acid gases to the flares happens extremely rarely).

In the most commonly used open flares, gas combustion takes place at their outlet [14]. In the closed flares, gas combustion takes place in the inlet (lower) part of the flare. In view of construction, it is usually a stack supported by a lattice tower supporting system with a triangle or a square base. What distinguishes it from a typical stack it is the use of a different technological solution. In the typical solutions, the stack serves for removal of waste gases that remain after the process, e.g., combustion. In case of the flare stack, it is used to transport the waste gases coming from the technological processes that will be burnt with open fire at its outlet, by the burner installed there. For this reason, the entire technological system of the flare consists of a knockout drum with liquid seal drum located at the base, constituting the beginning of the stack pipe, which can be a self-supporting element, self-supporting with guy-ropes or can be supported on the tower lattice structure. In the upper part of the stack pipe there is a head part with a molecular seal, protecting against the penetration of fire inside the stack and equipped with a main burner in which the waste gases are burnt. The discharge gases ignite from the "pilot" burner, on which a burning flame is continuously maintained. The flares are devices classified to the first class of reliability. Properly functioning discharge and flare systems guarantee safe operation of production plants, they can be described as the most important safety valve. The unavailability of the flare stack disqualifies the work of the installation connected to it. For economic reasons and space saving, in practice, a single flare is usually dedicated to many installations. Therefore, the failure of one flare generates the need to stop not only the flare itself, but also all installations associated with it. To avoid similar situations, the principle of combining discharge systems is used—several flares are connected into one system. It is possible after performing appropriate calculations of hydraulics for the discharge systems and proper design of water closures. In case of emergency, the flares with such a system are provided with the appropriate counter pressure at the outlet of the discharge gases of the individual installations. Figure 1 shows the flares most commonly applied in the hydrocarbon processing plants. The flares serving for the largest amounts of discharges are most often supported by a lattice construction.

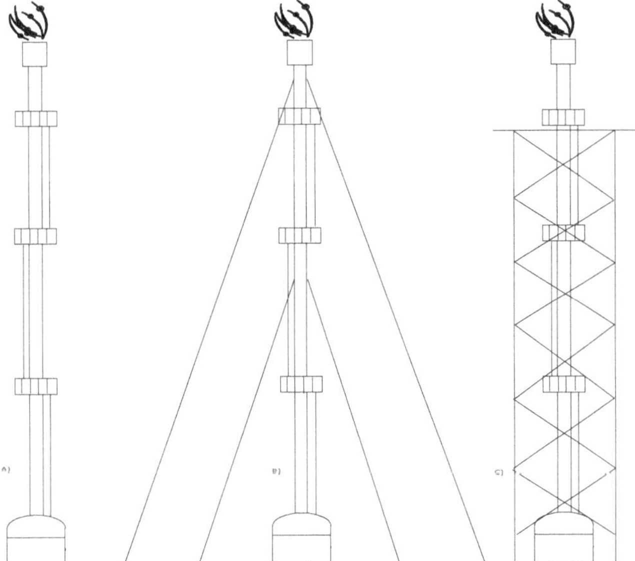

Figure 1. Division of flares in view of their construction: (**A**) self-supporting flare, (**B**) flare with guy-ropes, and (**C**) flare with a lattice construction.

The flares make the end of the entire depressurization systems. The beginning of such systems starts at the individual production installations from safety valves, built into the individual apparatus that are the equipment of a production installation in order to protect them against the uncontrolled pressure increase, flow increase, etc. The safety valves are connected by individual pipelines to gathering collectors that remove discharge gases beyond the installation. The collectors outputs the gases out of the installation usually pass through the tanks—knockout drums built into the battery limit of the installation. Most often, the liquid caught in the separators consists of valuable hydrocarbons or hydrocarbon fractions. Usually these hydrocarbons are recycled back to the process. Separation of liquid hydrocarbons takes place in the tanks, then they are returned for recycling by means of sloping systems. Collectors removing gases out of the installations without a liquid phase can send them directly to the flare from a single installation. An example of this are the acid gases, the so-called discharges with oxygen or other dedicated discharges, or they can be connected to the collective depressurization collectors with larger diameters, located in the direct vicinity of the flares. These collectors are connected to the water seal, whose role is to maintain the appropriate pressure in the installation line to the water seal and protect for back of flame. As a part of the water seal, there may be a separation chamber (KO) in which the liquid, entrained by the flare gas flowing through the closure, is separated again from the gas. Then, the flare gas leaves the separation chamber and is directed by a pipeline to the flare socket on which the flare stack is set. The flare gases are transported to the end (up) of the stack, on which the molecular seal is built, the task of which is to protect the discharge system against backing of the flame and aspiration of the ambient air into the system. The fire closure (with protect the system from fire flashback) is a part of the molecular closure. The molecular closure is made by the purging gas that can be fed into the flare gases line to the battery limit, to the discharge line behind the water closure or to the flare tube. Fuel gas or inert gas—nitrogen—may be used as a purge gas. The gas transported to the depressurization line or the flare tube is fed continuously in the appropriate projected quantity.

At the molecular closure, a main burner is installed in which the flare gases are burnt. The discharge gases ignite from the pilot burners installed on the main burner. The pilot burners burn all the time and are fed with external fuel gas or the fuel gas coming from the internal heating gas systems of a given refinery or another production plant, where waste flares are located. The fire closure as a part of the molecular closure is shown in Figure 2. The red line illustrates the flow direction of the discharge gases. The blow-through gas passes the same route. At the top of the fire closure there is a main burner of the flare with a flame atomization system, a pilot burner system and a fuel gas distribution system. In turn, a typical water closure is shown in Figure 3.

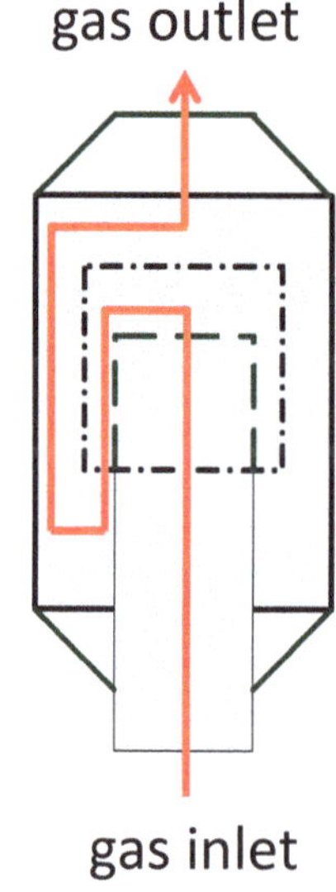

Figure 2. A fire closure as a part of the molecular closure.

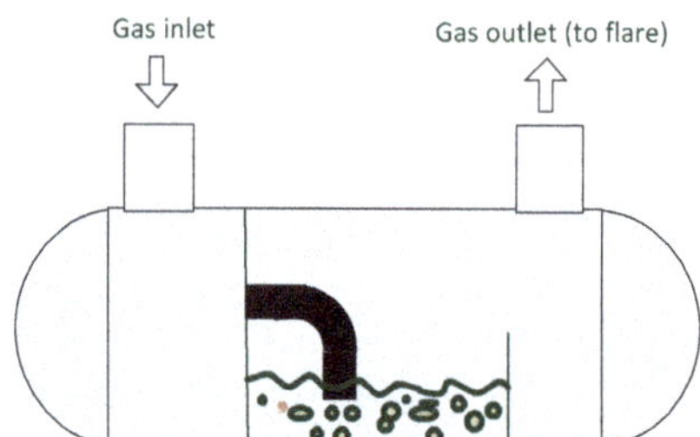

Figure 3. A typical water closure in horizontal arrangement.

The flare stacks allow for utilization of a few up to several thousand Mg of discharge gases per hour, depending on the production systems for which needs they are designed [14]. In the case of very large hydrocarbon processing plants as well as for economic reasons and savings of investment areas (to a large extent also an economic effect in the aspect of infrastructure is achieved) multi-stack flares are designed and built. The economics of these solutions is that one lattice structure holds two, three, or a maximum of four stacks.

3. Results and Discussion

3.1. Existing Solutions for Utilization of Discharge Gases

Installation of flare gases disposal systems usually has a very good rate of return on the invested funds, because on the other side of the bill is unproductive burning of hydrocarbon gas, generating a physical loss for a given plant. On the other hand, the costs of recovery systems for discharge gases consist mainly of supply in energy media (electricity, circulating water, costs of servicing by operators, costs of ongoing repairs, costs of major repairs, costs of technical supervision, etc.). The legitimacy of the construction of systems recovering the flare gases is meaningful first of all when there is a large number of production installations operating within a given production plant. This results in more frequent preparation of the installation for repairs, commissioning of these installations, larger number of equipment which should be periodically prepared for overhaul, and a larger number of safety valves (sometimes safety valves allow a certain amount of gases to pass, with such situations appear particularly at the end of the period between scheduled repairs). The use of such systems is especially justified in the situations:

- When refinery plants are dealing with a large number of conversion processes;
- During implementation of destructive/conversion processes, a larger number of gas streams is created than in the case of conserving processes;
- The amount of processed batch for all installations in total is large.

Figure 4 presents a simplified scheme of the installation recovering the discharge gases, recycled to the internal gas network system of the production plant.

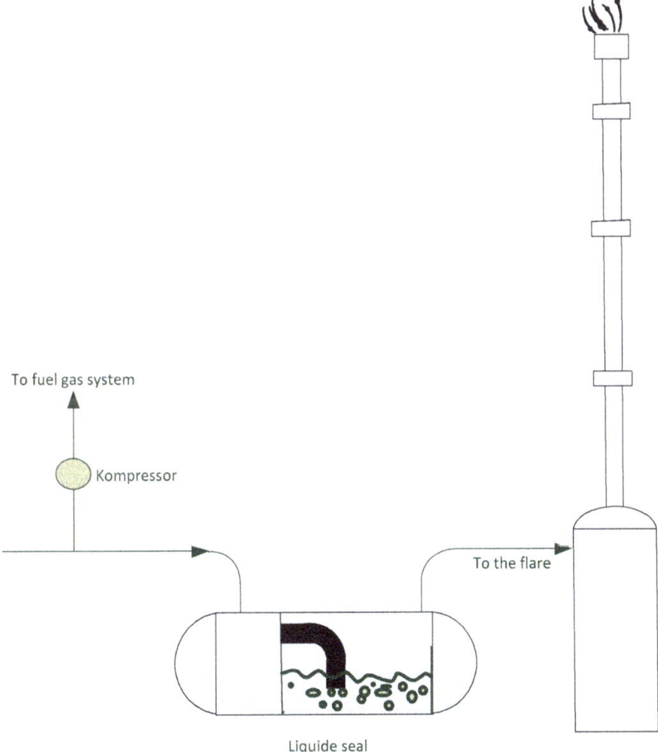

Figure 4. *Cont.*

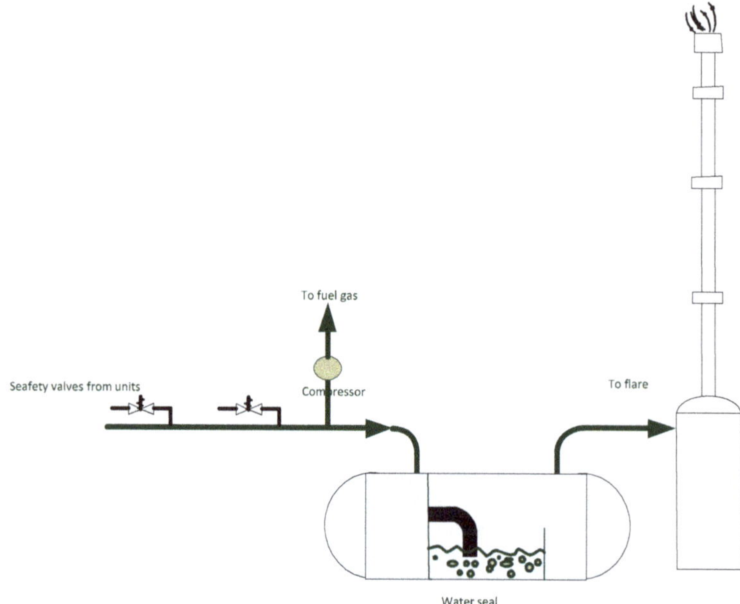

Figure 4. Diagram of the installation recovering the discharge gases, which are "recycled" to the internal heating gas network of the production plant.

In every case, the aim is to recover a large amount of flare gases. The analyzed configurations revealed wide differences in the profitability of gas disposal due to economic effects and the discussed conditions. For each process in the given installation, the most desirable condition is a durable, stable technological situation that does not generate the discharge gases. Without loss of a part of the load, or a part of the streams at various stages of the process in the installation, the profits are maximized, because the high-value products always have higher price than the discharge gases streams.

Practical design solutions in the field of recovery of the discharge gases are almost always identical. The differences are small and first of all come from structural differences. These differences are primarily visible in the construction of water closures (which can be oriented horizontally or vertically), in solutions regarding the used compressors, preparation of gas for compression, etc. Functioning of the gas recovery can be presented in several stages:

- Gases are discharged from a given installation into a pipeline system and discharge collectors,
- Within a given installation, the liquid phase is separated;
- Gases leave the knockout drum (separator) built into the production plant and are further directed to the collectors cooperating directly with the flare stacks;
- Flare gas streams pass through a knockout drums (separator), the KO may be integrated with a liquid seal that maintains the pressure of the flare gases at the required level;
- The maintained pressure of the flare gases ensures their influx into the suction of the compressor,
- The compressor the pressure of the gases is increased to the required value,
- Directing gases to the internal fuel gas system.

The problem is the quality of recovered streams, which can be very variable over time. By controlling individual streams through the use of existing pipelines or construction of dedicated pipelines, it is possible to control and maintain the stream of recovered gases with constant and more stable quality.

The discharge gases, after a few, usually two, stages of separation and compression liquid hydrocarbon, are directed to the internal fuel gas system, which is used for heating of the process furnaces [15,16].

3.2. The Use of Surplus of Discharge/Flare Gases beyond Manufacturing Plants

In production practice, there are technological situations during which it is not possible to use all the gas streams in the existing networks of fuel gas. In manufacturing plants, one type or several types of fuel gas networks, in the aspect of system-work pressure, can be used, which may differ in terms of gas pressure as well as its quality. During normal stable operation of the refinery, these systems are mostly largely balanced. The balancing can also take place in a different way, appropriate for a given refinery or a given production plant. There are refineries which do not have connection to an external gas source, e.g., natural gas. Then balancing may be made, for example, by evaporation of a portion of LPG and directing it to the main fuel gas collector. In the fuel gas system, the most common is participation of light hydrocarbons: C_1, C_2, and lower amounts of C_3 and C_4 hydrocarbons and H_2. Introduction of a larger part of C_3 and C_4 hydrocarbons into the fuel gas network causes a number of difficulties in the functioning of such a system, especially in periods of low ambient temperatures. In this case, there is a risk of condensation of heavier hydrocarbons and problems with burners built on technological furnaces. There may also be problems on pilot burners of flares. However, there are also situations in which the excess fuel gas is unproductively burnt in flares—the system is imbalanced. An example of such a case is the failure of the production installation, which is a large consumer of the fuel gas. The average composition of the flare gas and natural gas is presented in Table 1 [17,18].

Table 1. The average composition of the flare gases and natural gas [17,18].

No	Gas Components	Flare Gases, %v/v	Natural Gas, %v/v
1	Methane	43.600	96.470
2	Ethane	3.660	1.300
3	Propane	20.030	0.278
4	n-Butane	2.780	0.050
5	Isobutane	14.300	0.040
6	n-Pentane	0.266	0.010
7	Isopentane	0.530	0.020
8	neo-Pentane	0.017	0.000
9	n-Hexane	0.635	0.000
10	Ethylene	1.050	0.000
11	Propylene	2.730	0.000
12	1-Butene	0.696	0.000
13	Carbon monoxide	0.186	0.000
14	Carbon dioxide	0.713	0.000
15	Hydrogen sulfide	0.256	0.577 *
16	Hydrogen	5.540	0.000
17	Oxygen	0.357	0.000
18	Nitrogen	1.300	1.778
19	Water	1.140	0.0095 **

* mg/m^3; ** g/m^3.

Based on the results presented in Table 1 it is possible to conclude that the amount of energy contained in the flare gases is 63.14 MJ/m^3, while in the case of the natural gas stream it is 35.8 MJ/m^3, whereas the calorific values are equal 48.33 MJ/kg and 46.09 MJ/kg, respectively. In case of simple injection of 1 Mg/h of the flare gas into the natural gas pipeline (with flow equals 30 Mg/h), the methane content will decrease to 94.97%, nitrogen slightly decrease to 1.77%, while hydrogen sulphide content increase to 8 ppm/kg. Gas density increase from 0.740 kg/m^3 to 0.748 kg/m^3. Based on that simple calculation it is possible to conclude that injection of the flare gases to the natural gas system is technically feasible. Thus, it is possible to inject the non-balancing flare gasses produced in the refinery plats into existing methane pipeline system. Moreover for the presented in Table 1 composition of flare gases the amount of produced CO_2 in the combustion of 1 Mg of this gas is equal 2.91 Mg.

There are also refineries that are connected to the natural gas system. Then, during the shortage of own fuel gas caused by stopping for repair or emergency stop of the installation, which is a large producer of fuel gas, the refinery complements the shortages by supplying gas from the external natural gas network.

In periods in which the current technological system in the refinery generates surplus fuel gas, or surpluses of other gas streams—technological streams and hydrogen gas streams (as mentioned, these are situations related to the retention of individual production installations or their groups for renovation or when we are dealing with situations Emergency systems that generate such streams).

A solution that would eliminate the need for unproductive combustion of the discharge gas in utilization flares is to take them out of production plants and take them, for example, to an existing natural gas transmission pipeline or to a nearby heating plant or a combined heat and power plant burning gas for energy.

Before leaving the industrial plant, the gases have to be properly identified, catalogued and prepared. Among the most important activities that should be performed before implementing such solution, first of all it is necessary to specify the following:

- Gas streams that can be found in the flare discharge system, which are dedicated to individual processes/installations;
- Quality of these streams;
- Amount of discharge gas;
- Variability of their composition; and
- Content of impurities.

Knowing the gas streams (based on the processes carried out) that can be found in a given depressurization system, one may initially determine the quality of these gases. Based on this knowledge, one may decide whether a given stream should be sent for recovery or it should be burnt in a flare stack. Knowledge about the gas streams makes possible determination of the composition of these gases (the content of individual hydrocarbons) by applying analytical methods dedicated for this purpose. As a consequence, it is possible to select carefully gases that should undergo recovery. Determination of the number of discharge gas streams will allow determining the output capacity of the discharge gas recovery installation. In the refinery plants located in Europe, this is not a problem as a result of provisions of Commission Regulation (EU) No 601/2012 of 21 June 2012 on the monitoring and reporting of greenhouse gas emissions pursuant to Directive 2003/87/EC of the European Parliament and of the Council. The most commonly used measurement method is ultrasonic measurement. The measuring systems were installed on the flare stacks, i.e., at the end of the depressurization system. This solution indicates all the flare gases (excluding inert gases). A better way to determine accurately the amount of discharge gases is to use the plant balance. This method is much more precise. Knowing the quantity and quality of these streams, you can determine their variability. This parameter allows to determine the impact of a given industrial gas stream on the quality of natural gas, transported by a specific pipeline system in the event, when the industrial gas was pumped into the transmission gas pipeline. The necessary condition is knowing the quality and quantity of natural gas flowing through the pipeline. It is also important to identify impurities in the industrial stream. In the case of high-methane natural gas, the content of CH_4 is very stable and basically does not decrease below 95%. The other components are methane homologs—as their molecular weight increases, their share in the natural gas stream decreases. From the point of view of natural gas utility, it is very important to ensure in the stream of industrial gas, which could be directed to the transmission system, that there is as little as possible heavier hydrocarbons, nitrogen, and sulfur compounds.

A larger amount of heavier hydrocarbons is a threat of their condensation and subsequent problems during transport and use. Nitrogen is an inert gas which is unnecessary from the point of view of transport and use—it significantly reduces the calorific value of fuel. Sulfur compounds can be very toxic (e.g., hydrogen sulfide) and could pose hazards related to the safety of use, they also constitute a significant corrosion hazard. The above mentioned components are always undesirable from the point of view of the industrial use of natural gas, for example for the production of hydrogen and heating industrial furnaces.

4. Conclusions

The recovery and output of excessive flare gases (in order to use them for energetic purpose) beyond the production plants is a solution possible to introduce in case of refinery and petrochemical plants. There are situations in which there is a surplus of gases: heating, technological, or discharge ones, characterized by good quality. This solution is possible for production plants in the vicinity of which natural gas transmission pipelines exist.

The following are facts confirming the applicability of the assumptions described in the paper:

1. The quality of the discharge gases generated at individual production installations is known. It is possible to take samples for analysis and have them analyzed in the laboratory. It is also possible to install a chromatograph, which would be used for continuous monitoring and quality control of the selected gas parameters.
2. The quality of the natural gas, pumped through the individual transmission pipelines, is known. The technical parameters of transmission pipelines are also known. Also technical requirements for these objects defined in the regulations and technical specifications are known.
3. It is possible to determine the impact of the compressed and injected stream of flare gas on the quality of natural gas transported by the selected transmission pipeline, or under the selected system of transit. Knowing the quality and quantity of the discharge gas and natural gas, it is possible to develop a mathematical model by which the gas stream quality can be determined after mixing two streams and then refer the results to the required natural gas quality described in the relevant standards.
4. The solution would generate the following beneficial effects on the production plants: Less use and thus consumption of infrastructure.
5. Reduced consumption of energy media in the depressurization flare stack systems, thus reducing the consumption of steam-producing fuels, used to atomize the flame on the flare main burner.

The demonstrated effects for industrial plants would at the same time limit the negative impact on the natural environment—the use of less fuels for steam production and a significant reduction in the flare gas burnt in the flares.

Author Contributions: Conceptualization, J.S. and A.S.; Methodology, J.S.; Investigation, J.S.; Writing—Original Draft Preparation, J.S. and A.S.; Writing—Review and Editing, J.S. and A.S.; Supervision, A.S.

Funding: This research received no external funding.

Conflicts of Interest: The authors declare no conflict of interest.

References

1. Mehraban, M.; Shahraki, B.H. A mathematical model for decoking process of the catalyst in catalytic naphtha reforming radial flow reactor. *Fuel Process. Technol.* **2019**, *188*, 172–178. [CrossRef]
2. Shi, J.; Hou, Y. Practice and Analysis of FCCU Efficient Operation Based on Synergy of Refinery. *Petr. Process. Petrochem.* **2019**, *50*, 42–46.
3. Zhang, J.; Cao, Z.; Wu, Z. Investigation of the technology that can maximize chemical feedstock production by hydrocracking straight-run diesel for Tianjin Petrochemical Company. *Petr. Refin. Eng.* **2018**, *48*, 25–27.
4. Li, R.; Zhang, X.; Li, S.; Li, Z. Analysis of Guo VI gasoline upgrading processes for refineries producing ethanol gasoline blendstocks. *Petr. Refin. Eng.* **2017**, *47*, 9–13.

5. Ozren, O. *Oil Refineries in the 21st Century: Energy Efficient, Cost Effective, Environmentally Benign*; Wiley-VCH Verlag GmbH & Co. KGaA: Weinheim, Germany, 2005.

6. He, Y. Analysis and practice of operation optimization of hydrogen system. *Petr. Refin. Eng.* **2017**, *47*, 14–17.

7. Al-Subaie, A.; Maroufmashat, A.; Elkamel, A.; Fowler, M. Presenting the implementation of power-to-gas to an oil refinery as a way to reduce carbon intensity of petroleum fuels. *Int. J. Hydrog. Energy* **2017**, *42*, 19376–19388. [CrossRef]

8. Wu, C.; Li, Y.; Zhou, Y.; Li, Z.; Zhang, S.; Liu, H. Upgrading the Chinese biggest petrochemical wastewater treatment plant: Technologies research and full scale application. *Sci. Total Environ.* **2018**, *633*, 189–197. [CrossRef] [PubMed]

9. Hou, M.; Miao, X. Analysis of factors affecting light olefins' yields of catalytic cracking unit and optimization. *Petr. Refin. Eng.* **2018**, *48*, 5–9.

10. Yang, J.C.; Chang, P.E.; Chie, W.C.; Liu, J.P.; Wu, C.F. Large-scale search method for locating and identifying fugitive emission sources in petrochemical processing areas. *Process Saf. Environ. Protect.* **2016**, *104*, 382–394. [CrossRef]

11. Arendt, J.S.; Casada, M.L.; Rooney, J.J. Reliability and hazards analysis of a cumene hydroperoxide plant. *Plant/Oper. prog.* **1986**, *5*, 97–102. [CrossRef]

12. Franek, L.; Galbfach, R.; Rosciszewski, A.; Urbanski, J.; Zebrowski, M. Cumene Phenol Production—Development of the Technology in Mazovian Refining and Petrochemical Plants in Plock. *Przem. Chem.* **1985**, *64*, 522–524.

13. Kalat Jari, H.R.; Borhani Sazeh, A. Minimize Flaring with Modification to Flare Gas Recovery Unit. Available online: http://gasprocessingnews.com/features/201804/minimize-flaring-with-modifications-to-flare-gas-recovery-unit.aspx (accessed on 8 March 2019).

14. Baukal, C.E. *Combustion Handbook, Design and Operations*; CRC Press, Taylor & Francis Group: Boca Raton, FL, USA, 2013. Available online: http://vr360app.ternium.com/the_john_zink_hamworthy_combustion_handbook_second_edition_volume_2_design_and_operations_industrial_combustion.pdf (accessed on 8 March 2019).

15. Environmental Protection Agency (EPA). Installing Vapor Recovery Units on Storage Tanks. 2006. Available online: https://www.epa.gov/sites/production/files/2016-06/documents/ll_final_vap.pdf (accessed on 8 March 2019).

16. Mokhatab, S.; Mak, J.Y.; Valappil, J.V.; Wood, D.A. *Handbook of Liquefied Natural Gas*; Elsevier: Oxford, UK, 2014.

17. Peterson, J.; Cooper, H.; Baukal, C. Minimize facility flaring. *Hydrocarb. Process.* **2007**, *6*, 111–115.

18. Gas System (GS). 2019. Available online: https://www.gaz-system.pl/ (accessed on 8 March 2019).

Article

Theoretical and Experimental Insights into the Mechanism for Gas Separation through Nanochannels in 2D Laminar MXene Membranes

Yun Jin [1], Yiyi Fan [1], Xiuxia Meng [1], Weimin Zhang [1,*], Bo Meng [1], Naitao Yang [1,*] and Shaomin Liu [2,*]

1   School of Chemical Engineering, Shandong University of Technology, Zibo 255049, China; yunjin@sdut.edu.cn (Y.J.); yiyifan5@126.com (Y.F.); mengxiux@126.com (X.M.); mb1963@126.com (B.M.)
2   Department of Chemical Engineering, Curtin University, Perth 6102, Australia
*   Correspondence: wmzhang@sdut.edu.cn (W.Z.); naitaoyang@126.com (N.Y.); shaomin.liu@curtin.edu.au (S.L.)

Received: 4 September 2019; Accepted: 10 October 2019; Published: 15 October 2019

Abstract: Clarifying the mechanism for the gas transportation in the emerging 2D materials-based membranes plays an important role on the design and performance optimization. In this work, the corresponding studies were conducted experimentally and theoretically. To this end, we measured the gas permeances of hydrogen and nitrogen from their mixture through the supported MXene lamellar membrane. Knudsen diffusion and molecular sieving through straight and tortuous nanochannels were proposed to elucidate the gas transport mechanism. The average pore diameter of 5.05 Å in straight nanochannels was calculated by linear regression in the Knudsen diffusion model. The activation energy for H_2 transport in molecular sieving model was calculated to be 20.54 kJ mol^{-1}. From the model, we can predict that the gas permeance of hydrogen (with smaller kinetic diameter) is contributed from both Knudsen diffusion and molecular sieving mechanism, but the permeance of larger molecular gases like nitrogen is sourced from Knudsen diffusion. The effects of the critical conditions such as temperature, the diffusion pore diameter of structural defects, and the thickness of the prepared MXene lamellar membrane on hydrogen and nitrogen permeance were also investigated to understand the hydrogen permeation difference from Knudsen diffusion and molecular sieving. At room temperature, the total hydrogen permeance was contributed 18% by Knudsen diffusion and 82% by molecular sieving. The modeling results indicate that molecular sieving plays a dominant role in controlling gas selectivity.

Keywords: MXene; gas separation; Knudsen diffusion; molecular sieving; transport mechanism

---

1. Introduction

Gas separation techniques using membranes have many merits such as high efficiency, facile operation, and low energy cost. [1,2]. Traditionally, polymeric membranes are often used for this application while they suffer from the well-known problem of permeability–selectivity trade-off (e.g., the Robinson upper bound) [3]. Recently, two-dimensional (2D) material membranes have been developed for gas separation because it exhibits the promising potential to overcome the permeability–selectivity trade-off problem. Therefore, in the past decade, they have gained enormous attention [1,2]. Typically, the 2D nanosheets assembled lamellar microporous inorganic membranes have interlayer galleries which can provide abundant molecular pathways [4,5]. For this attractive property, a variety of the 2D materials have been exploited to assemble lamellar membranes for gas separation. From the current published studies, the 2D materials include layered double hydroxides (LDH) [6], graphene (GA) [7,8], MXene 2D materials [2,9], graphene oxide (GO) [10,11], tungsten disulfide (WS$_2$) [12], molybdenum disulphide (MoS_2) [13,14], and metal-organic frameworks (MOFs) [15,16] have been receiving particular interest.

Generally, MXene 2D materials are a large family of 2D carbides and nitrides with the general formula of $M_{n+1}X_nT_x$, where M represents a transition metal, X is carbon and/or nitrogen, and T is referred to the surface termination [17,18]. Very recently, MXene-based 2D materials have been introduced to fabricate 2D lamellar membranes because of their tunable nanochannel width, excellent mechanical strength, and easy fabrication and integration [2,19,20]. It has been reported that the assembled MXene 2D membranes exhibit a range of attractive characters in separation, e.g., precise ion sieving [21], ultrafast water permeation [22], and gas separation [19,20].

Even a promising potential for highly efficient gas separation has been proved experimentally, to fully clarify the gas transportation mechanism in 2D lamellar membranes, the insight into the theoretical and simulation clarification is scarce and the corresponding investigation is still needed to be conducted. Especially, there are only a few reports on theoretical simulation of gas separation through 2D lamellar nanochannels [11,23,24] while the mechanism of transportation and separation of gas through 2D nanochannels is far away to be fully clarified. For example, Li et al. [24] studied various gas transportations in MXene nanogalleries with molecular dynamic (MD) simulations and activated and Knudsen diffusion being observed for gas diffusion in through MD simulations. Nevertheless, Fan et al. [20] found the gas transportation in MXene nanogalleries via the molecular sieving mechanism. Therefore, in order to optimize and promote the performance and the efficiency of the 2D MXene lamellar membranes, the gas transport mechanisms are still needed to be further clarified because they can efficiently provide the guidance for tuning the channel width of 2D lamellar membranes and gas molecule–channel wall interactions in gas transportation. That is due to the gas transportation mechanism in porous media is primarily related with pore diameter, pore geometry, and interconnectivity of the interlayer distance and defects in the 2D lamellar membranes structure [25].

In this work, the related parameters were firstly determined theoretically and then the gas transport modeling to permeate through different nanochannels was developed to reveal the diffusion of different gas molecules (H_2 and N_2) in 2D MXene lamellar membranes. The simulation results are well consistent with the experimental results and their significance to the gas diffusion (e.g., permeance and selectivity) was discussed. The structural effects from MXene nanochannels formed during MXene nanosheets assembling on the transportation of different gas molecules (e.g., size and mass) were studied. Moreover, we provided the effects of temperature, pore diameter of structural defects, and the MXene thickness of the lamellar membrane on hydrogen and nitrogen permeances.

2. Experimental

The MXene lamellar membrane was prepared using the similar method illustrated previously [20]. Briefly, $Ti_3C_2T_x$ was synthesized by etching Ti_3AlC_2 powders using 50 wt% HF solution at a certain temperature followed with DMSO intercalation. The interlayer interaction became weak due to the removal of Al atom between layers, leading to a facile exfoliation to form MXene nanosheets under sonication. The suspended MXene nanosheets were deposited by filtration on the anodic aluminum oxide (AAO) support with a pore diameter of 200 nm (Whatman Co., Maidstone, UK) by a vacuum pump. The resultant MXene membrane was dried at 120 °C for 8 h in a vacuum oven to remove the water molecular between interlayers.

H_2 and N_2 gas permeances were derived from their gas mixture separation performance measurement through the MXene membrane supported on AAO [20]. The gas mixture permeance was carried out in a home-made device as reported previously [20]. The gas mixture containing 50 vol% H_2 and 50 vol% N_2 as the feed gas was supplied to the membrane feed side with a flow rate of 50 mL min^{-1}, while the argon sweep gas with a flow rate of 40 mL min^{-1} at a standard pressure was supplied to the sweep side. The exit gas from the membrane sweep side was transferred to an online

GC (6890N, Agilent Technologies, Inc, Waldbronn, Germany) with TCD to measure the permeated H_2 or N_2 concentration. The permeance and mixture selectivity or separation factor were defined by:

$$F_i = \frac{N_i}{P_{i1} - P_{i2}} \tag{1}$$

where F_i is the permeance of the derived gas (i) (mol m^{-2}s^{-1}Pa^{-1}), N_i is the molar flux of the gas (i) (mol m^{-2} s^{-1}), P_{i1} and P_{i2} are the partial pressures of gas (i) at the feed side and sweep side.

$$S = \frac{y_{i2}/y_{j2}}{y_{i1}/y_{j1}} \tag{2}$$

where S is the mixture selectivity or separation factor y_{i1}, y_{i2}, y_{j1} and y_{j2}, the volumetric fraction of the gas component i or j in the feed or permeated side gas mixtures, respectively. Experimental results are summarised in Table 1.

Table 1. Experimental results of the supported 800-nm-thick MXene lamellar membrane for temperature dependent nitrogen and hydrogen permeances measured from gas mixture separation performance tests [20].

Testing Temperature (K)	Permeance of N_2 (10^{-8} mol m^{-2} s^{-1} Pa^{-1})	Permeance of H_2 (10^{-7} mol m^{-2} s^{-1} Pa^{-1})	Separation Factor (H_2/N_2)
295	1.71	2.34	13.45
343	1.13	2.22	19.65
423	0.84	2.11	25.11
443	0.74	2.09	28.24
473	0.64	2.07	32.34
503	0.62	2.06	33.22
533	0.59	2.05	34.74
563	0.55	2.03	36.90
593	0.50	2.03	40.60

The crystalline characteristics of MXene nanosheets and composite membranes were studied by XRD (Bruker D8 Advance with Cu-Kα radiation $\lambda = 0.154$ nm at 40 kV and 40 mA, in the 2 θ range 20–80° with a sacn step of 0.01°). The surface topology, cross-section, and microstructures of the MXene lamellar membrane were investigated by using a field emission scanning electron microscopy (SEM, FEI Sirion 200, Philips, The Netherlands).

3. Theoretical Models of Transport Mechanism

We propose the presence of two kinds of gas transport nanochannels as illustrated in Figure 1. One is straight channels from structural defects and its width is assumed between 2 nm and the mean free path of transport gases (λ). Another nanochannel consists of randomly distributed nanoscale wrinkles and interlayer spacing between stacked MXene sheets and its width is between the kinetic diameter of gas molecular (Φ_k) and 2 nm. Accordingly, two mathematical models of the gas transport mechanism for the permeation through the MXene lamellar membrane can be proposed based on the different nanochannels width using the following assumptions:

(1) One-dimensional transport model is used.
(2) There are two kinds of transport channels as illustrated in Figure 1, which are the straight nanochannels (2 nm $< d_p < \lambda$) and tortuous nanochannels ($\Phi_k < d_p < 2$ nm). Their size remains unchanged with the temperature.
(3) The MXene/AAO lamellar membrane operates at a steady-state isothermal condition.
(4) The linear adsorption isotherm (or non-adsorption) for all gases on the surface of MXene lamellar membranes is neglected.
(5) The mass transfer resistance of gas within the porous AAO layer is neglected due to its much large pore size [10,26].

(6)   Gas mixture (hydrogen and nitrogen) transport through the membrane is under ideal conditions on which the actual separation factor is equal to the ideal selectivity and pure gas permeance is equal to the mixture permeance.

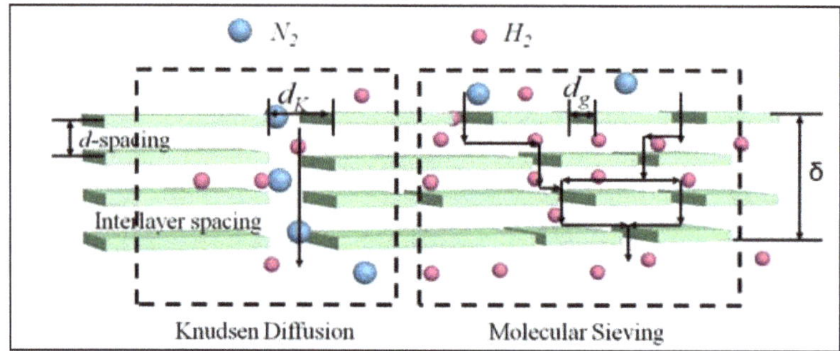

Figure 1. Schematic diagram of the two kinds of gas transport model through different nanochannels within the MXene lamellar membranes.

The relevant schematic transport models for gas permeation through the MXene lamellar membranes are shown in Figure 1. Here, H_2 (kinetic diameter of 2.89 Å) and N_2 (3.64 Å) diffuse through the MXene lamellar membrane based on our experimental results [20], so the geometrical structure of nanochannels derived from the structural defects and interlayer spacing plays a significant role in the gas transportation. For gas permeation, two transport models were proposed corresponding to the nanochannels in order to explain the experimental results. The Knudsen diffusion is assumed to occur within the straight nanochannels with the lager pore diameter ($2\ nm < d_p < \lambda$). The tortuous nanochannels are mainly composed of randomly distributed nanoscale wrinkles, inter-galleries, and interlayer spacing between stacked MXene sheets. Therefore, the gas transportation within tortuous nanochannels containing the interlayer spacing ($\Phi_k < d_p < 2\ nm$) is mainly related to the molecular sieving.

Therefore, for MXene lamellar membranes, both Knudsen diffusion and molecular sieving contribute to the total mass transportation. The total permeance can be written by Equation (3) as follows:

$$F_{total} = F_{defects} + F_{interlayer} = F_{Kn} + F_g \tag{3}$$

Here, F_{total}, $F_{defects}$, and $F_{int\ erlayer}$ are the total gas permeance, the gas permeance through straight nanochannels of structural defects, and tortuous nanochannels containing interlayer spacing, respectively. F_{Kn} and F_g are the permeances contributed by the Knudsen diffusion and the molecular sieving, respectively.

3.1. Knudsen Diffusion (KD) Model

Typically, Knudsen diffusion dominates in the mesoporous nanochannels with the size range between 2 nm and the mean free path of transport gases ($2\ nm < d_p < \lambda$), i.e., the average distance a molecule traversed by collisions, which is comparable or larger than transport channels, transport falls in Knudsen regime [27]. The mass transport of gas may be described by Fick's first law as Equation (4).

$$J_{K,i} = -\varphi D_K \frac{dc}{dz} = -\frac{\varphi D_K}{RT} \frac{dP}{dz} \tag{4}$$

where φ is the factor for structure geometrical effects by Equation (16). R is the ideal gas constant, T is the absolute temperature. Thus, the expression of Knudsen diffusion flux ($J_{K,i}$) for gas can be obtained in terms of pressure gradient. In this case, D_K is the Fick diffusion coefficient (Knudsen diffusivity), which may be expressed as the product of a geometric factor by diffusion pore diameter and the velocity of gas molecules by Equation (5):

$$D_{K,i} = \frac{d_p}{3}\left(\frac{8RT}{\pi M}\right)^{\frac{1}{2}} \tag{5}$$

where d_p is the diffusion pore diameter, the velocity of diffusing molecules is given by the kinetic theory of gases, and M is the molecular weight of the diffusing gas. The geometric factor is 1/3 since only these molecules moving in the considered direction will be taken into account [28]. The expression for Knudsen diffusion flux ($J_{K,i}$) obtained by combining Equations (4) and (5) is expressed as Equation (6).

$$J_{K,i} = -\frac{\varphi d_p}{3}\left(\frac{8}{\pi MRT}\right)^{\frac{1}{2}}\frac{dP}{dz} \tag{6}$$

The Knudsen diffusion permeance through a porous membrane can be determined after integration of Equation (6) over the membrane thickness (δ):

$$F_{K,i} = \frac{J_K}{\Delta P} = \frac{\varphi d_p}{3\delta}\left(\frac{8}{\pi MRT}\right)^{\frac{1}{2}} \tag{7}$$

In order to obtain the diffusion pore diameter (d_p), Equation (7) was transformed to Equation (8) to reveal the gas permeance dependence on the temperature. Thus, the gas permeance in Knudsen regime is pressure-independent and decreases with temperature as indicated by Equation (8).

$$F_{K,i}\left(\frac{8}{\pi MRT}\right)^{\frac{1}{2}} = \frac{8\varphi d_p}{3\delta\pi MR}\left(\frac{1}{T}\right) \tag{8}$$

3.2. Molecular Sieving (MS) Model

The molecular sieving model was firstly used for zeolite, which can be illustrated by the kinetic theory of gases [28]. In the small channels with the size range between kinetic diameter of gas molecular and 2 nm ($\Phi_k < d_p < 2$ nm), channel size changes into the molecular dimensions and molecules are no longer as free as these in Knudsen diffusion. For simplification purpose, the individual gas molecular adsorption difference is not considered. This is a reasonable assumption for these gases with less adsorption like He, H_2, and N_2 than CO_2. The molecular sieving flux can be expressed in terms of pressure gradient. As a result, the molecular sieving flux ($J_{s,i}$) can be written as:

$$J_{s,i} = -\varphi D_{s,i}\frac{dc}{dz} = -\frac{\varphi D_{s,i}}{RT}\frac{dP}{dz} \tag{9}$$

where $D_{s,i}$ is the molecular sieving coefficient in the MXene laminates, which is given by Equation (10).

$$D_{s,i} = \frac{l_s}{Z}\left(\frac{8RT}{\pi M}\right)^{\frac{1}{2}}\exp\left(-\frac{E_{a,g}}{RT}\right) \tag{10}$$

where l_s is the diffusion distance (the distance between two adjacent sites of the low energy regions), Z is the number of adjacent sites [28], and $E_{a,g}$ is the activation energy, which is required for molecules to surmount the attractive constrictions imposed by the nanochannels structure. However, the geometrical factor ($1/Z$) is the probability of a molecule moving in the direction under consideration. The expression of molecular sieving flux ($J_{s,i}$) obtained by combining Equations (9) and (10) is shown as the following.

$$J_{s,i} = -\frac{\varphi l_s}{Z}\left(\frac{8}{\pi MRT}\right)^{\frac{1}{2}}\exp\left(-\frac{E_{a,g}}{RT}\right)\frac{dP}{dz} \tag{11}$$

The gas permeance for molecular sieving through a microporous membrane is obtained after integration of Equation (12) over the membrane thickness δ:

$$F_{s,i} = \frac{J_{s,i}}{\Delta P} = \frac{\varphi l_{s,i}}{Z\delta}\left(\frac{8}{\pi MRT}\right)^{\frac{1}{2}} \exp\left(-\frac{E_{a,g}}{RT}\right) \tag{12}$$

Equation (12) reveals an exponential dependence of gas permeance on the temperature, which is different from that of the Knudsen diffusion model. Taking logs of the both sides of Equation (12) can obtain the activation energy $E_{a,g}$ and the diffusion distance.

$$\mathrm{Ln}(F_{s,i}T^{\frac{1}{2}}) = \left(-\frac{E_{a,g}}{R}\right)\frac{1}{T} + \mathrm{Ln}\left[\frac{\varphi l_{s,i}}{Z\delta}\left(\frac{8}{\pi MR}\right)^{\frac{1}{2}}\right] \tag{13}$$

The gas permeance tests at different temperatures were carried out, the activation energy $E_{a,g}$ and the diffusion distance in the logarithmic plots can be regressed based on experimental data by Equation (13).

3.3. Diffusion Contribution to Total Transport

The fractional diffusion of Knudsen diffusion and molecular sieving, respectively can be expressed as the following Equations (14) and (15).

$$f_{Kn} = \frac{F_{Kn}}{F_{Kn} + F_s} = \frac{Z d_p}{Z d_p + 3l_s \exp\left(-\frac{E_{a,g}}{RT}\right)} \tag{14}$$

$$f_s = \frac{F_s}{F_{Kn} + F_s} = \frac{3l_s \exp\left(-\frac{E_{a,g}}{RT}\right)}{Z d_p + 3l_s \exp\left(-\frac{E_{a,g}}{RT}\right)} \tag{15}$$

The relative individual diffusion contribution to total transport from Equations (3), (7), and (12) obtained using Equations (14) and (15) can be used to determine the rate-dominated diffusion process.

4. Results and Discussion

4.1. Morphology and Structure

The scanning electron microscopy (SEM) images of the cross-section of MXene ($Ti_3C_2T_x$), high magnification over the cross-section of MXene membrane, external surface of MXene membrane, and low magnification of the cross-section of MXene membrane are displayed as Figure 2a–d, respectively.

As shown in Figure 2a, the MXene membrane was assembled by the stacked 2D MXene nanosheets. Additionally, the nanosheets exhibit plicate feature on their surface. We can find there are structure defects and interlayer spacing in the bulky membrane, which can provide channels for the gas transportation. It is worth noting that the inner-sheet structural defect is assumed to be correlated with straight channels. The tortuous nanochannels consist of randomly distributed inter-galleries and nanoscale wrinkles between the stacked nanosheets. Here, the 2D MXene laminar membrane with 800-nm-thickness was assembled and supported on the AAO substrate (Figure 2b), which shows similar morphology to that of other laminar materials such as GO membranes [11]. MXene (Ti_3C_2Tx) nanosheets were deposited as an outer layer on top of a porous AAO support with a pore diameter of 200 nm using the vacuum impregnation method to form the supported MXene membrane (Figure 2c,d).

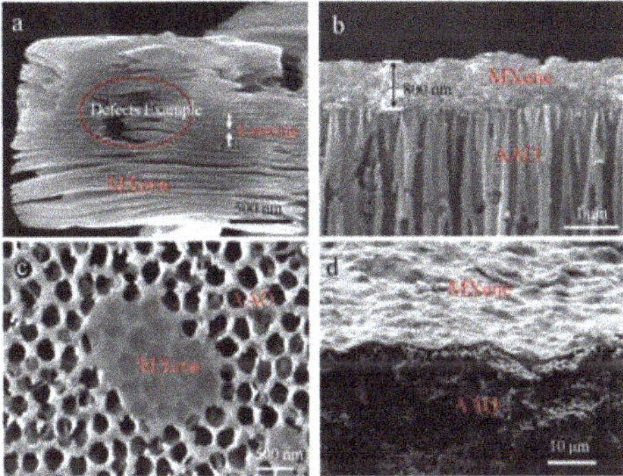

Figure 2. SEM images of the cross-section of MXene (Ti₃C₂Tₓ) (**a**), the cross-section of MXene membrane (**b**), external surface of MXene membrane (**c**), and the cross-section of MXene membrane (low magnification) (**d**).

From the XRD patterns in Figure 3, the (002) plane at 6.6° can be used to determine the d-spacing between the MXene nanosheets and the monolayer thickness, which were calculated to be ~13.4 Å and 10 Å, respectively. These data are inconsistent with the previous publications [2,20]. Moreover, the interlayer spacing of MXene membrane is estimated to be ~3.4 Å (Figure 1), which is favorable to sieve small molecules such as hydrogen (kinetic diameter of 2.89 Å) and helium (2.60 Å).

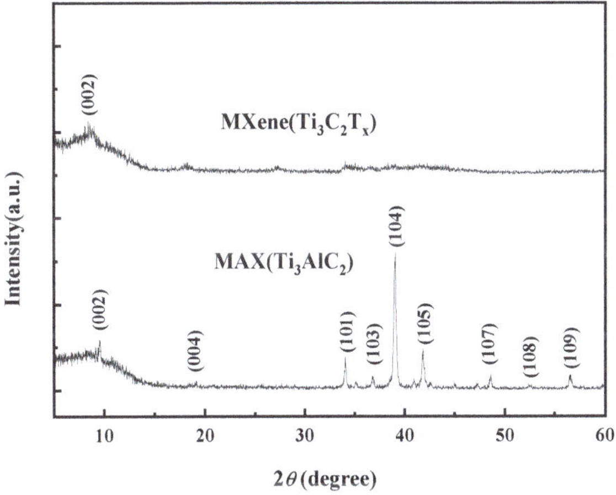

Figure 3. XRD patterns of the MAX (Ti₃AlC₂) and MXene (Ti₃C₂Tₓ) membrane.

The determination of geometrical effects in nanochannel structure is quite critical for the gas transport mechanisms model, and which can be described by [16]:

$$\varphi = \frac{\varepsilon}{\tau} = \frac{\left(1 - \frac{a}{d}\right)}{\tau} \tag{16}$$

where φ is the geometrical effect of the porous structure (i.e., the ratio of the membrane porosity ε to the tortuosity factor τ), a is the thickness of monolayer thickness MXene lamellar which is ~10 Å, d is the *d*-spacing of MXene laminates which is about ~13.4 Å, and τ is the tortuosity factor which can be approximated as the ratio of diffusion length to the MXene laminar thickness (Equation (17)).

$$\tau = \frac{l_s}{\delta} \tag{17}$$

In our work, the thickness of the MXene laminar membrane is 800 nm. The tortuosity factor τ can be estimated to be ~1 for straight nanochannels of inner-sheet structural defects. Moreover, the tortuosity factor τ ($\tau > 1$) in tortuous nanochannels must be calculated using molecular sieving model.

4.2. The Experimental Nitrogen Permeance and the Parameters Regression of Knudsen Diffusion (KD) Model

There are two transport nanochannels in the MXene lamellar membrane, which are straight nanochannels from inner-sheet structural defects and tortuous nanochannels in wrinkles and inter-galleries contained interlayer spacing between stacked MXene sheets. Additionally, the gases transport mechanisms for the diffusion in porous membrane are primarily dependent on the transport channel width, geometry, and interconnectivity [2,25,28]. Therefore, it is extremely critical to figure out the dimension of the nanochannels before we perform the modeling. To calculate the interlay spacing, the XRD measurement was conducted.

We can determine from the XRD results (Figure 3) for MXene membranes that the interlayer spacing between the MXene sheets is 3.4 Å, which is smaller than the kinetic diameter of N_2 (3.64 Å), CH_4 (3.84 Å), C_3H_6 (4.30 Å), and C_3H_8 (4.50 Å) (Figure 4). Since we assume that the width of straight nanochannels are larger than the kinetic diameters of these gases, therefore, for the transport permeance of N_2, CH_4, C_3H_6, and C_3H_8, Knudsen diffusion is the main process in the gas transportation in the straight channels. Since straight nanochannels flow dominates gas permeation, the Knudsen diffusion modeling was performed by the ordinary least squares method using MATLAB 7.0 (The Math Works Inc., Natick, MA, USA) [29] to obtain the parameters in Equations (7) and (8) and the regression results were shown in Figure 5. We can observe that the calculations using the Knudsen diffusion model with the obtained parameters can fit the experimental data well with the resultant correlation coefficient up to 0.9966.

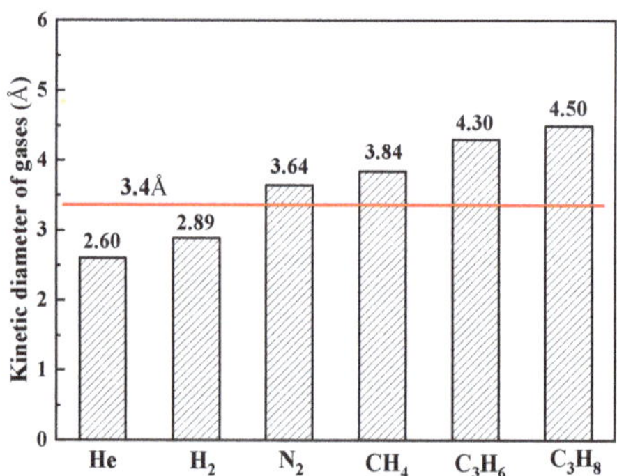

Figure 4. The kinetic diameter of gases (Φ_k) of different gases.

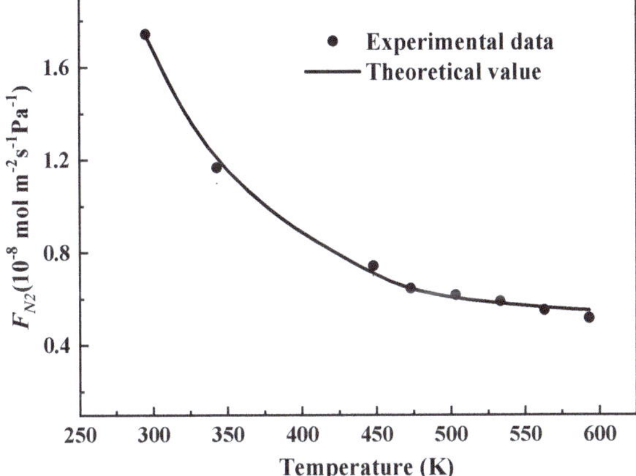

Figure 5. Comparison of the experimental data and the model predictions of the temperature dependent nitrogen permeance through the MXene lamellar membrane.

The values for the model parameters φ and d_p can be obtained from the regression. Since the tortuosity factor τ in Equation (17) can be estimated to ~1 in inner-sheet structural defects because of the straight diffusion channels. The geometrical effects (φ) of the porous structure derived from Equation (16) is 0.25, and the average diffusion pore diameter d_p is 5.05 Å which is larger than the kinetic diameter of N_2 (3.64 Å), CH_4 (3.84 Å), C_3H_6 (4.30 Å), and C_3H_8 (4.50 Å). It can be concluded that the Knudsen diffusion through straight nanochannels for nitrogen is reasonable.

4.3. The Experimental Hydrogen Permeance and the Parameters Regression of Molecular Sieving (MS) Model

The interlayer spacing width between the MXene sheets is 3.4 Å which is larger than the kinetic diameter of He (2.60 Å) and H_2 (2.89 Å) (Figure 4), thus it is favorable to separate them by molecular sieving diffusion. Therefore, hydrogen transport mechanism models in the MXene membrane are the combined Knudsen diffusion and molecular sieving, which correspond to the diffusion through straight nanochannels of structural defects, and tortuous nanochannels contained interlayer spacing between stacked MXene sheets, respectively. Hence, H_2 permeance is higher than N_2 permeance in the temperature range of 295~593 K in experimental (Table 1 and Figure 6). For example, H_2 and N_2 permeances were 2.34 and 0.174 × 10^{-7} mol m^{-2} s^{-1} Pa^{-1}, respectively, at 295 K. The calculated separation factor or selectivity of H_2/N_2 was ~13 based on the gas permeance, and it greatly exceeded the Knudsen selectivity of 3.74 for H_2/N_2 pairs. It indicates the promising potential application of the 2D MXene membrane to separate H_2 from its gas mixture. The modeling results show that molecular sieving plays a dominant role in the selectivity of gas separation.

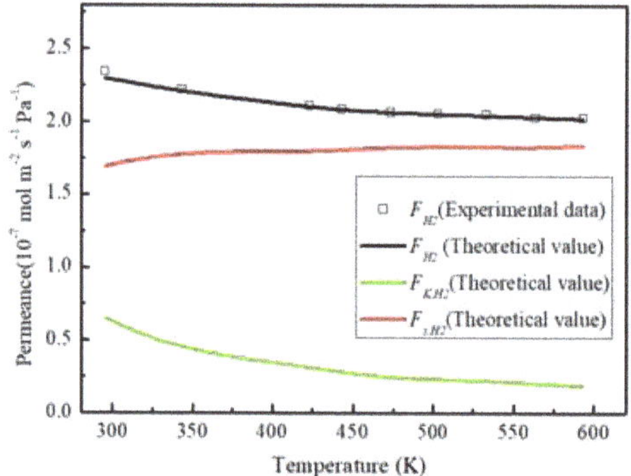

Figure 6. Comparison of the experimental data and the model predictions of temperature dependent hydrogen permeance through the MXene lamellar membrane.

Hydrogen permeances through the MXene lamellar membrane were obtained via the gas permeation test as a function of temperature between 22 °C (295.15 K) and 320 °C (593.15 K) using the mixture of hydrogen and nitrogen with a flow rate of 50 mL (STP) min^{-1} in the feed side of the membrane, and an argon sweep gas with a flow rate of 40 mL (STP) min^{-1} on the permeate side to remove the permeated hydrogen and nitrogen. The values for the model parameters φ, l_s, and $E_{a,g}$ can be obtained by Equations (12) and (13).

The calculated activation energies ($E_{a,g}$) for H$_2$ permeance displayed in Figure 6 is 20.54 kJ mol^{-1}. Although the geometrical effects of the porous structure (φ) and diffusion distance (l_s) are difficult to be determined in Equation (12), by performing the ordinary least squares method using MATLAB 7.0 (The Math Works Inc., Natick, MA, USA) [29] for regression, the value of multiple ($l_s \times \varphi$) can be calculated to be 1.73×10^{-12}. In addition, φ is the ratio of the membrane porosity ε to the tortuosity factor τ (see Equation (16)) and the tortuosity factor τ is difficult to be determined in Equation (17) as transport nanochannels from wrinkles and inter-galleries between stacked MXene sheets. The calculations using the model incorporating the obtained parameters fit the experimental data well with the resultant correlation coefficient of 0.9966.

Molecular sieving mechanism is also generally characterized by the activated diffusion [24,30]. Thus, the hydrogen permeance contributed from molecular sieving increased while that from Knudsen diffusion decreased with the temperature increment, so leading to the total permeance change with temperature variation.

4.4. Temperature Dependent Permeance and Relative Contribution to Total Gas Transport from Knudsen Diffusion and Molecular Sieving

To simulate the hydrogen or nitrogen gas permeance through the MXene lamellar membrane under the same conditions, we calculated the permeance using the models (Equations (7) and (12)) with the related regressed parameters and analyzed fractional diffusion to determine the rate dominates diffusion described by Equations (14) and (15) at different temperatures.

Figure 7 displays the gas permeance of hydrogen and nitrogen, and selectivity (H$_2$/N$_2$) between 260 and 700 K. We can see that the hydrogen permeances are higher than nitrogen. For example, the hydrogen and nitrogen permeance through the MXene lamellar membrane is 2.11 and 0.11×10^{-7} mol m^{-2}s^{-1}Pa^{-1} at 363 K, respectively. Moreover, the corresponding selectivity of H$_2$/N$_2$ is ~19 based on the gas permeance calculation. These results can be explained by the gas diffusion mechanism. The nitrogen transport is dominated by Knudsen diffusion through straight nanochannel of structural defects with the width of ~5.05 Å, while hydrogen transports based on Knudsen diffusion and molecular sieving through straight nanochannels of structural defects and tortuous nanochannels from interlayer spacing.

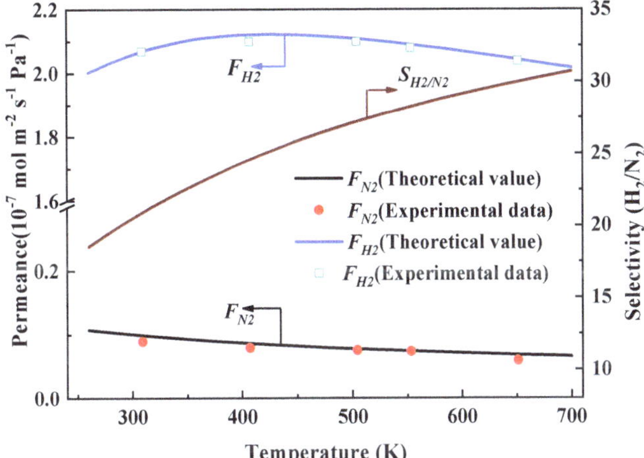

Figure 7. The gas permeance and selectivity (H_2/N_2) through the MXene lamellar membrane as a function of temperature.

Meanwhile, we can also see that in Figure 7 the nitrogen permeance decay upon the temperature increased in the range between 260 and 700 K. This trend can be ascribed to its Knudsen diffusion mechanism via which the permeance decreased with temperature as illustrated by Equation (7). Compared with nitrogen, the hydrogen permeance displays a different trend and reaches its maximum value of 2.12×10^{-7} mol m^{-2}s^{-1}Pa^{-1} at 408 K. However, the selectivity of H_2/N_2 is always enhanced. Such results can be interpreted by the joint effects of molecular sieving and Knudsen diffusion. From molecular sieving (Equation (12)), it reveals an exponential dependence of gas permeance on the temperature, which is obviously different from that of Knudsen diffusion. However, these trends are ascribed to the coverage of functional groups such as $-OH$ and $-O$ which will affect the adsorbed amount of hydrogen [31,32].

The effect of temperature on the fractional diffusion of Knudsen and molecular sieving for hydrogen permeance through the MXene membrane is depicted in Figure 8. It is clear that the molecular sieving permeances are always higher than these of Knudsen diffusion between 260 and 700 K. Meanwhile, the fraction of molecular sieving increases steadily from ~80% to ~88%, and the Knudsen diffusion fraction slightly decreases from ~20% to ~12%. On average, the molecular sieving diffusion is about four times higher than the Knudsen diffusion, which indicates tortuous nanochannels containing interlayer spacing that dominates the whole transport channels.

Figure 8 also reveals that the increase of temperature will result in the decrease of Knudsen diffusion permeance which is consistent with Equation (7). However, the molecular sieving permeance increases steadily upon the increase of temperature which reveals that temperature is more readily affected in the exponent than those in pre-exponential coefficients in Equation (12).

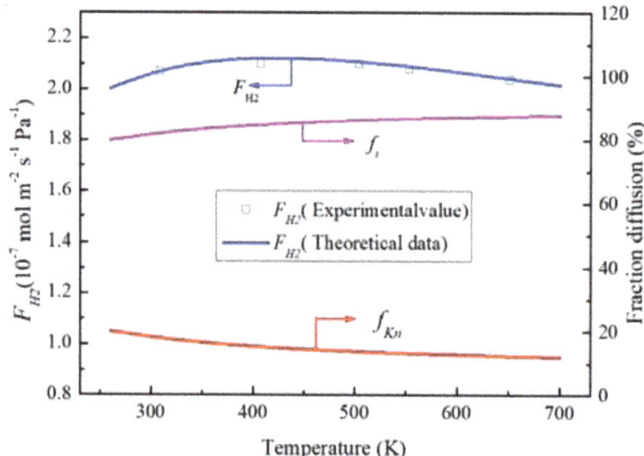

Figure 8. Temperature dependent fractional diffusion of H_2 to total transport from Knudsen diffusion and molecular sieving. Note: Data point represents experimental data while the continuous line indicates model predictions.

4.5. The Effect of the Diffusion Pore Diameter of Straight Nanochannels on the Gas Permeance and Relative Contribution from Knudsen Diffusion and Molecular Sieving to Total Gas Transport

The change of the gas permeance and the transport fractions of Knudsen diffusion and molecular sieving with the pore diameter of the straight nanochannels can be calculated from Equations (7) and (12). According to the relationship between the gas permeance and the average pore diameter of straight nanochannels at 295 K as shown in Figure 9, the hydrogen and nitrogen permeances increase with the average diffusion pore diameter of straight nanochannels in the MXene lamellar membrane. We can see the hydrogen permeance increased from 1.955 to 9.0×10^{-7} mol m^{-2}s^{-1}Pa^{-1} and nitrogen permeance 0.73 to 18.6×10^{-8} mol m^{-2}s^{-1}Pa^{-1} with the pore diameter of straight nanochannel alternation between 0.364 and 1.0 nm, respectively. The discrepancy between them becomes more pronounced with the increasing average pore diameter of straight nanochannels. This reflects that the larger nanochannel of structural defects will enhance Knudsen diffusion more significantly for permeance through the MXene lamellar membrane. Therefore, the selectivity of H_2/N_2 decreases from 26 to 4.87 (close to 3.74 of the Knudsen selectivity). This indicates that it is important to reduce the average pore diameter of straight nanochannels resulting from structural defects to ensure the good gas selectivity.

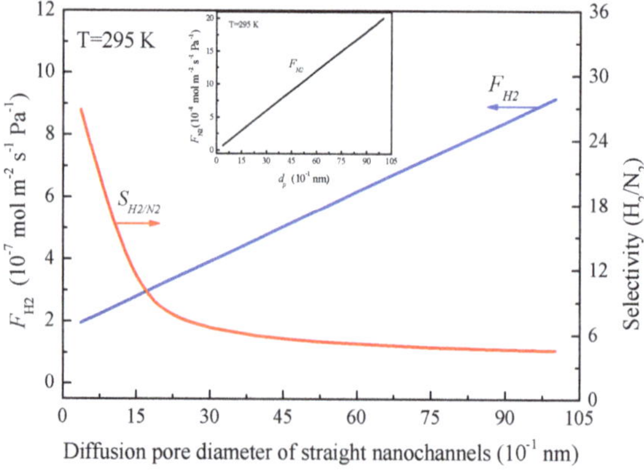

Figure 9. The effect of the average pore diameter of straight nanochannels on gas permeance at 295 K.

Figure 10 shows the fractions of Knudsen diffusion and molecular sieving contribution as a function of the average pore diameter of straight nanochannels at 295 K. As can be seen, the increase of the average pore diameter of straight nanochannels from 0.364 to 1.0 nm enlarges the fractions of Knudsen diffusion permeance from 14% to 82%, while the fractions of molecular sieving decreases from 86% to 18%. Moreover, the characteristic pore diameter (d_c) of 2.32 nm (23.2 Å) was also obtained from Figure 10, at which the Knudsen diffusion and molecular sieving equally share the transport. in addition, increasing the average diffusion pore diameter of defects more than such characteristic value (d_c) will lead to the higher proportion of Knudsen diffusion than molecular sieving in the total permeance (Figure 10).

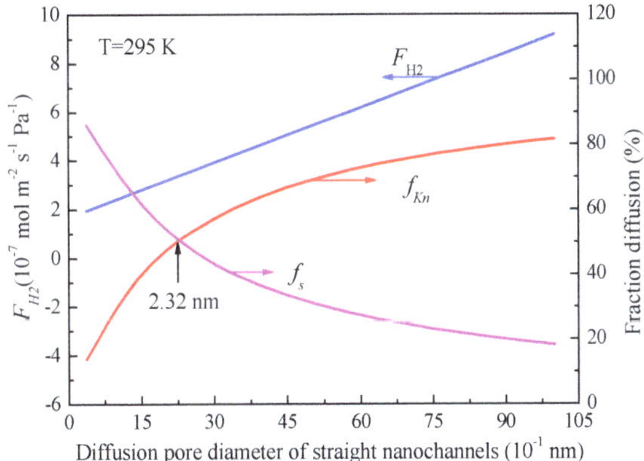

Figure 10. Average pore diameter of the straight nanochannels dependent fractional diffusion to total transport of Knudsen diffusion and molecular sieving for H_2 at 295 K.

4.6. The Effect of MXene Layer Thickness on the Gas Permeance and Fractional Diffusion of Knudsen Diffusion and Molecular Sieving

The dependence of the gas permeance and the fractions of Knudsen diffusion and molecular sieving on the effect of the thickness was determined by Equations (7) and (12), and Figure 11 reveals the effect of the MXene membrane thickness on gas permeance at 295 K. From Figure 11 we can see that the decrease of the thickness of the MXene layer leads to the increase for both hydrogen and nitrogen permeances. For example, the hydrogen and nitrogen permeances increase from 1.60 to 13.7 and 0.081 to 0.675×10^{-7} mol m^{-2}s^{-1}Pa^{-1}, respectively, when the MXene thickness is reduced from 1000 to 20 nm. On the other hand, the H_2/N_2 selectivity maintains at 13.5. It indicates that the thinner MXene layer can lead to higher gas permeance for hydrogen or nitrogen (Equations (7) and (12)) but maintain the gas selectivity unaltered.

Figure 12 presents the transport fractions of Knudsen diffusion and molecular sieving to the total permeance as a function of the MXene membrane thickness (operated at 295 K). As can be seen, the decrease of the thickness of the MXene layer results in the substantial increase in the hydrogen permeance. However, the fractions of Knudsen diffusion and molecular sieving keep constant at around 18% and 82%, respectively. This can be explained from Equations (14) and (15), where the MXene membrane thickness does not affect the fractional diffusion to total transport from Knudsen diffusion and molecular sieving.

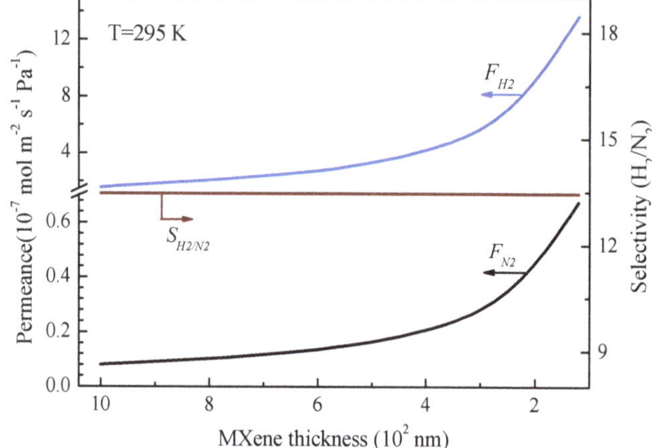

Figure 11. The effect of MXene membrane thickness on gas permeance at 295 K.

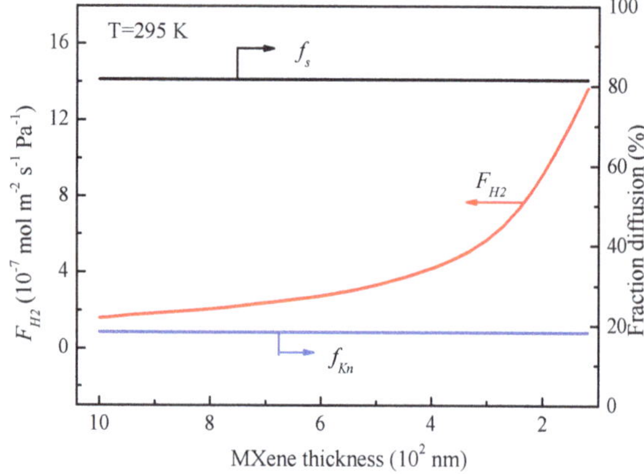

Figure 12. Effects of the MXene membrane thickness on the transport fractions from Knudsen diffusion and molecular sieving to total gas transport operated at 295 K.

5. Conclusions

In conclusion, in order to determine the diffusion mechanism of gases in the MXene lamellar membrane, we evaluated the hydrogen and nitrogen permeation properties as a function of temperature using the prepared membrane. We proposed Knudsen diffusion and molecular sieving through the respective straight nanochannels that stemmed from structural defects and tortuous nanochannels formed by interlayer spacing. Furthermore, we performed linear regression on the experimental data to obtain the model parameters values for the MXene lamellar membrane, which were further applied to explain the Knudsen diffusion and molecular sieving mechanisms for hydrogen and nitrogen transport. Based on the modeling results, we simulated the effects of temperature, the pore size of the structural defects, and the thickness of the lamellar MXene membranes on hydrogen and nitrogen permeances. The relative contribution of Knudsen diffusion and molecular sieving to the total hydrogen permeance was also investigated. The model provides insights into the dominant diffusion at different operational condition and geometry variables. The results of theoretical and experimental study show that molecular sieving through tortuous nanochannels plays a dominant role in controlling the gas selectivity.

Author Contributions: Conceptualization, X.M., B.M. and S.L.; methodology, Y.J.; investigation, Y.F.; writing—original draft preparation, Y.J.; writing—review and editing, W.Z. and S.L.; supervision, N.Y.; funding acquisition, Y.J., W.Z., X.M., B.M., N.Y. and S.L.

Funding: The authors gratefully acknowledge the research funding provided by the National Natural Science Foundation of China (No. 21878179, 21776165, and 21776175), Project of Shandong Province Higher Educational Science and Technology Program (J18KA095) and Natural Science Foundation of Shandong Province (ZR2019MB056). S. Liu acknowledges the financial support provided by the Australian Research Council through the Discovery Program (DP180103861).

Acknowledgments: The authors would thank the facilities support from Analysis & Testing Center of Shandong University of Technology.

Conflicts of Interest: The authors declare no conflict of interest.

Nomenclature

a	Thickness of monolayer MXene lamellar, m
c	Concentration of gas species, mol m^{-3}
D_K	Knudsen diffusivity coefficient, m^2 s^{-1}
$D_{g,i}$	Molecular sieving coefficient, m^2 s^{-1}
d	d-spacing of MXene laminates, m
d_p	Diffusion pore diameter, m
d_c	Characteristic diffusion pore diameter, m
$E_{a,g}$	Activation energy of gas diffusion, J mol^{-1}
F	Permeance of the gas, mol m^{-2} s^{-1} Pa^{-1}
F_{total}	Total gas permeance, mol m^{-2} s^{-1} Pa^{-1}
$F_{defects}$	Permeance through structural defects, mol m^{-2} s^{-1} Pa^{-1}
$F_{interlayer}$	Permeance through interlayer spacing, mol m^{-2} s^{-1} Pa^{-1}
f_{Kn}	Fractional of Knudsen diffusion contribution
f_s	Fractional of molecular sieving contribution
l_s	Distance between two adjacent sites, m
M	Molecular weight of the diffusing gas
N	Molar flux of gas, mol m^{-2} s^{-1}
P_1	Pressures at the feed side, Pa
P_2	Pressures at permeate side, Pa
R	Ideal gas constant, J mol^{-1} K^{-1}
S	Selectivity of the gas pair
T	Absolute temperature, K
Z	Number of adjacent sites
z	Distance coordinate, m
Greek Letters	
Φ_k	Kinetic diameter, m
λ	Mean free path, m
φ	Geometrical effects of the porous structure
τ	Tortuosity factor
δ	Membrane thickness, m
ε	Membrane porosity
Subscripts	
i, j	Gas species i and j
Kn	Knudsen diffusion
s	Molecular sieving

References

1. Sholl, D.S.; Lively, R.P. Seven chemical separations to change the world. *Nature* **2016**, *532*, 435–437. [CrossRef] [PubMed]

2. Ding, L.; Wei, Y.; Li, L.; Zhang, T.; Wang, H.; Xue, J.; Ding, L.X.; Wang, S.; Caro, J.; Gogotsi, Y. MXene molecular sieving membranes for highly efficient gas separation. *Nat. Commun.* **2018**, *9*, 155. [CrossRef] [PubMed]

3. Park, H.B.; Kamcev, J.; Robeson, L.M.; Elimelech, M.; Freeman, B.D. Fructose-driven glycolysis supports anoxia resistance in the naked mole-rat. *Science* **2017**, *356*, 1138–1148. [CrossRef] [PubMed]

4. Mi, B. Graphene oxide membranes for ionic and molecular sieving. *Science* **2014**, *343*, 740–742. [CrossRef] [PubMed]

5. Cheng, L.; Liu, G.; Jin, W. Recent progress in two-dimensional-material membranes for gas separation. *Acta Phys. Chim. Sin.* **2019**, *35*, 1090–1098.

6. Liu, Y.; Wang, N.; Cao, Z.; Caro, J. Molecular sieving through interlayer galleries. *J. Mater. Chem. A* **2014**, *2*, 1235–1238. [CrossRef]

7. Koenig, S.P.; Wang, L.; Pellegrino, J.; Bunch, J.S. Selective molecular sieving through porous grapheme. *Nat. Nanotech.* **2012**, *7*, 728–732. [CrossRef] [PubMed]

8. Celebi, K.; Buchheim, J.; Wyss, R.M.; Droudian, A.; Gasser, P.; Shorubalko, I.; Kye, C.L.J.I.; Park, H.G. Ultimate permeation across atomically thin porous graphene. *Science* **2014**, *344*, 289–292. [CrossRef]

9. Gao, X.; Li, Z.K.; Xue, J.; Qian, Y.; Zhang, L.Z.; Caro, J.; Wang, H. Titanium carbide $Ti_3C_2T_x$ (MXene) enhanced PAN nanofiber membrane for air purification. *J. Membr. Sci.* **2019**, *586*, 162–169. [CrossRef]

10. Shen, J.; Liu, G.; Huang, K.; Chu, Z.; Jin, W.; Xu, N. Subnanometer two-dimensional graphene oxide channels for ultrafast gas sieving. *ACS Nano* **2016**, *10*, 3398–3409. [CrossRef]

11. Jin, Y.; Meng, X.; Yang, N.; Meng, B.; Sunarso, J.; Liu, S. Modeling of hydrogen separation through porous YSZ hollow fiber-supported graphene oxide membrane. *AIChE J.* **2018**, *64*, 2711–2720. [CrossRef]

12. Sun, L.; Ying, Y.; Huang, H.; Song, Z.; Mao, Y.; Xu, Z.; Peng, X. Ultrafast molecule separation through layered WS_2 nanosheet membranes. *ACS Nano* **2014**, *8*, 6304–6311. [CrossRef] [PubMed]

13. Wang, D.; Wang, Z.; Wang, L.; Hua, L.; Jin, J. Ultrathin membranes of single-layered MoS_2 nanosheets for high-permeance hydrogen separation. *Nanoscale* **2015**, *7*, 17649–17652. [CrossRef] [PubMed]

14. Achari, A.S.; Eswaramoorthy, S.M. High performance MoS_2 membranes: Effects of thermally driven phase transition on CO_2 separation efficiency. *Energy Environ. Sci.* **2016**, *9*, 1224–1228. [CrossRef]

15. Wang, X.; Chi, C.; Zhang, K.; Qian, Y.; Gupta, K.M.; Kang, Z.; Jiang, J.; Zhao, D. Reversed thermo-switchable molecular sieving membranes composed of two-dimensional metal-organic nanosheets for gas separation. *Nat. Commun.* **2017**, *8*, 14460. [CrossRef]

16. Peng, Y.; Li, Y.; Ban, Y.; Jin, H.; Jiao, W.; Liu, X.; Yang, W. Metal-organic framework nanosheets as building blocks for molecular sieving membranes. *Science* **2014**, *346*, 1356–1359. [CrossRef]

17. Naguib, M.; Kurtoglu, M.; Presser, V.; Lu, J.; Niu, J.; Heon, M.; Hultman, L.; Gogotsi, Y.; Barsoum, M.W. Two-dimensional nanocrystals produced by exfoliation of Ti_3AlC_2. *Adv. Mater.* **2011**, *23*, 4248–4253. [CrossRef]

18. Ghidiu, M.; Lukatskaya, M.R.; Zhao, M.Q.; Gogotsi, Y.; Barsoum, M.W. Conductive two-dimensional titanium carbide 'clay' with high volumetric capacitance. *Nature* **2014**, *516*, 78–82. [CrossRef]

19. Ding, L.; Wei, Y.; Wang, Y.; Chen, H.; Caro, J.; Wang, H. A two-dimensional lamellar membrane: MXene nanosheet stacks. *Angew. Chem. Int. Ed.* **2017**, *56*, 1825–1829. [CrossRef]

20. Fan, Y.; Wei, L.; Meng, X.; Zhang, W.; Yang, N.; Jin, Y.; Wang, X.; Zhao, M.; Liu, S. An unprecedented high-temperature-tolerance 2D laminar MXene membrane for ultrafast hydrogen sieving. *J. Membr. Sci.* **2019**, *569*, 117–123. [CrossRef]

21. Ren, C.E.; Hatzell, K.B.; Alhabeb, M.; Ling, Z.; Mahmoud, K.A.; Gogotsi, Y. Charge-and size-selective ion sieving through $Ti_3C_2T_x$ MXene membranes. *J. Phys. Chem. Lett.* **2015**, *6*, 4026–4031. [CrossRef] [PubMed]

22. Wang, J.; Chen, P.; Shi, B.; Guo, W.; Jaroniec, M.; Qiao, S.Z. A regularly channeled lamellar membrane for unparalleled water and organics permeation. *Angew. Chem. Int. Ed.* **2018**, *57*, 6814–6818. [CrossRef] [PubMed]

23. Li, W.; Zheng, X.; Dong, Z.; Li, C.; Wang, W.; Yan, Y.; Zhang, J. Molecular dynamics simulations of CO_2/N_2 separation through two-dimensional graphene oxide membranes. *J. Phys. Chem. C* **2016**, *120*, 26061–26066. [CrossRef]

24. Li, L.; Zhang, T.; Duan, Y.; Wei, Y.; Dong, C.; Ding, L.; Qiao, Z.; Wang, H. Selective gas diffusion in two-dimensional MXene lamellar membranes: Insights from molecular dynamics simulations. *J. Mater. Chem. A* **2018**, *6*, 11734–11742. [CrossRef]

25. Lito, P.F.; Cardoso, S.P.; Rodrigues, A.E.; Silva, C.M. Kinetic modeling of pure and multicomponent gas permeation through microporous membranes: Diffusion mechanisms and influence of isotherm type. *Sep. Purif. Rev.* **2014**, *44*, 283–307. [CrossRef]

26. Eliseev, A.A.; Poyarkov, A.A.; Chernova, E.A.; Eliseev, A.A.; Chumakov, A.P.; Konovalov, O.V.; Petukhov, D.I. Operando study of water vapor transport through ultra-thin graphene oxide membranes. *2D Mater.* **2019**, *6*, 035039. [CrossRef]

27. Knudsen, M. Die Gesetze der Molekularströmung und der inneren Reibungsströmung der Gase durch Röhren. *Ann. Phys.* **1909**, *333*, 75–130. [CrossRef]

28. Xiao, J.; Wei, J. Diffusion mechanism of hydrocarbons in zeolites—I. Theory. *Chem. Eng. Sci.* **1992**, *47*, 1123–1141. [CrossRef]

29. The Math Works Inc. *MATLAB 7.0.4 [Computer Program]*; The Math Works Inc.: Natick, MA, USA, 2005.

30. Ibrahim, A.; Lin, Y.S. Gas permeation and separation properties of large-sheet stacked graphene oxide membranes. *J. Membr. Sci.* **2018**, *550*, 238–245. [CrossRef]

31. Gao, G.; O'Mullane, A.P.; Du, A. 2D MXenes: A new family of promising catalysts for the hydrogen evolution reaction. *ACS Catal.* **2016**, *7*, 494–500. [CrossRef]

32. Seh, Z.W.; Fredrickson, K.D.; Anasori, B.; Kibsgaard, J.; Strickler, A.L.; Lukatskaya, M.R.; Gogotsi, Y.; Jaramillo, T.F.; Vojvodic, A. Two-dimensional molybdenum carbide (MXene) as an efficient electrocatalyst for hydrogen evolution. *ACS Energy Lett.* **2016**, *1*, 589–594. [CrossRef]

Article

Simulation Study on the Influence of Gas Mole Fraction and Aqueous Activity under Phase Equilibrium

Weilong Zhao [1], Hao Wu [1], Jing Wen [2], Xin Guo [1], Yongsheng Zhang [1,*] and Ruirui Wang [2]

[1] Henan Province Engineering Laboratory for Eco-architecture and the Built Environment, Henan Polytechnic University, Jiaozuo 454000, China; zhaowl@hpu.edu.cn (W.Z.); 211607020050@home.hpu.edu.cn (H.W.); 211607010023@home.hpu.edu.cn (X.G.)

[2] School of Mechanical and Power Engineering, Henan Polytechnic University, Jiaozuo 454000, China; 211705010028@home.hpu.edu.cn (J.W.); jxndwrr@hpu.edu.cn (R.W.)

* Correspondence: zhangyongsheng@hpu.edu.cn; Tel.:+86-183-368-63941

Received: 28 December 2018; Accepted: 17 January 2019; Published: 22 January 2019

Abstract: This work explored the influence of gas mole fraction and activity in aqueous phase while predicting phase equilibrium conditions. In pure gas systems, such as CH_4, CO_2, N_2 and O_2, the gas mole fraction in aqueous phase as one of phase equilibrium conditions was proposed, and a simplified correlation of the gas mole fraction was established. The gas mole fraction threshold maintaining three-phase equilibrium was obtained by phase equilibrium data regression. The UNIFAC model, the predictive Soave-Redlich-Kwong equation and the Chen-Guo model were used to calculate aqueous phase activity, the fugacity of gas and hydrate phase, respectively. It showed that the predicted phase equilibrium pressures are in good agreement with published phase equilibrium experiment data, and the percentage of Absolute Average Deviation Pressures are given. The water activity, gas mole fraction in aqueous phase and the fugacity coefficient in vapor phase are discussed.

Keywords: gas mole fraction; activity; UNIFAC; phase equilibrium; threshold value

---

1. Introduction

Gas hydrate is a non-stoichiometric crystalline compound, which consists of a lattice formed by hydrogen bonds of water molecules as the host under high pressure and low temperature. Some gas molecules, such as methane, nitrogen, carbon dioxide and propane, are as the guest firmly surrounded by the crystal network formed by hydrogen bond of water molecules. The ice-like structure enables and stabilizes the existence of gas hydrates at higher temperatures and elevated pressures. The guest molecules must be of correct size to fit inside and stabilize the crystal lattice via weak van der Waals forces with the host water molecules [1,2].

Gas hydrate technology can be applied in many applications, such as gas storage and transportation [3], gas separation [4], refrigeration [5,6], etc. Meanwhile, in response to increasing carbon emissions, the hydrate-based gas separation (HBGS) has attracted the interest of researchers as an effective CO_2 capture and storage (CCS) technology [7]. In the last couple of years, vast quantities of natural gas hydrate in the permafrost and deep seabed was found, which is twice as much as the amount of the other fossil fuels combined under a conservative estimate [8]; it makes natural gas hydrate as a kind of potential energy possible. However, gas hydrates can block oil and gas pipelines with high pressure and low temperature inside subsea oil and gas flow line [9]. Furthermore, the methane trapped in gas hydrates is a potent greenhouse gas [10]. In order to solve these problems, scholars conducted a lot of studies and found that adding thermodynamic inhibitors can effectively change the conditions of hydrate formation into higher pressure and lower temperature.

On the contrary, the formation conditions of hydrate can be changed to lower pressure and higher temperature by adding thermodynamic promoters. Regardless of whether inhibiting or promoting hydrate formation, it is necessary to predict phase equilibrium conditions for the above-mentioned applications, and it is important to use reliable and accurate predictive models for predicting hydrate phase equilibria [11].

Among the gas hydrate predictive models of three-phase equilibrium, there are two common thermodynamic prediction models for calculating the phase equilibrium conditions. One is the classical statistical thermodynamic model of van der Waals and Platteeuw [12]. Some improved models designed for the phase equilibrium conditions in distilled water and aqueous solutions were proposed by Nasrifar et al. [13], Haghighi et al. [14], and Martin and Peters [15]. Another hydrate model is the Chen-Guo model [16,17], based on equality of the fugacity in the hydrate and vapor phases, which assumed the activity of water to be in unity and neglected the influence of gas solubility in aqueous phase. However, the changes in water activity caused by the solubility of gases, especially acid gases (e.g., carbon dioxide and hydrogen sulfide), and the addition of inhibitors/promoters should not be ignored [18,19]. Therefore, a better precision activity of water in aqueous phase is the key to improving the Chen-Guo model. Ma et al. [20,21] used the Patel–Teja equation of state (PT EoS), coupled with the Kurihara mixing rule and the one-fluid mixing rule to calculate the water activity in aqueous phase. Sun and Chen [18] combined the modified method introducing the Debye–Hückel electrostatic contribution term with the PT EoS to predict the nonideality of aqueous phase including ionic components. Liu et al. [22] built a simple correlation to calculate the activity of water in methanol–water solutions of sour gases ($CH_4/CO_2/H_2S/N_2$), in which parameters were determined from the phase equilibrium data. Moreover, among the models of aqueous phase activity, the UNIQUAC model [23] and the modified UNIFAC model [24] were constantly used to calculate aqueous phase. Delavar and Haghtalab [25,26] used the Chen-Guo and UNIQUAC models, referring the Soave-Redlich-Kwong-Huron-Vidal equation of state (SRK-HV EoS) conjunction with the Henry's law, to calculate the gas hydrate formation conditions. Dehaghani and Karami [27] employed the predictive Soave-Redlich-Kwong equation of state (PSRK-EoS) along with the modified Huron-Vidal (MHV1) missing rule and UNIQUAC model to calculate fugacity and activity coefficient of water in equilibrated fluid phases. Klauda and Sander [19,28] applied the modified UNIFAC model and PSRK-EoS coupled with the classical mixing rules, and the results obtained were in good agreement with the experimental data.

However, regardless of using UNIFAC or UNIQUAC model to calculate aqueous phase activity, the gas mole fraction in aqueous phase must be obtained first. Thus, in previous literature [19,25–28], Henry's law was used to describe the gas solubility in aqueous phase, and the gas mole fraction in aqueous phase relied on simultaneous Henry constants, the partial molar volume at infinite dilute (presented by Heidemann and Prausnitz [29]) and vapor phase fugacity calculated by the equation of state. It should be noted that both Henry's law and the partial molar volume at infinite dilution are defined on the basis of an imaginary ideal system. Furthermore, some parameters used in calculating the Henry's law constants and the partial molar volume at infinite dilution are also obtained by regression under the assumed ideal state. As a result, when acid gases exist in the actual system, there is a deviation in the water activity and the gas mole fraction in aqueous phase. Therefore, the gas mole fraction in aqueous phase as a function of temperature and pressure is considered one of the factors that influence the phase equilibrium conditions in this work.

In order to minimize the adverse impacts of the Henry's law constants and infinite dilution partial molar volumes on the hydrate equilibrium conditions prediction, we fitted a correlation of gas mole fraction in aqueous phase according to experimental data. Moreover, the UNIFAC model [30–32] and the correlation proposed in this work were employed to calculate the activity coefficient of aqueous phase; PSRK [33–35] was used to calculate vapor phase fugacity, and the Chen-Guo model was used to calculate the fugacity of the hydrate phase. These models not only alleviate empirically fitting the intermolecular parameters required in the van der Waals and Platteeuw model but also avoid the calculation error of water activity caused by Henry's law and the infinite dilution partial molar volume. It is noteworthy that the framework proposed in this work only applied in the pure gas (such as CH_4, CO_2, N_2 or O_2) and distilled water system; the mixed gas system and the additive system will be further studied in future work. Finally, the results calculated are compared with the experimental data in literatures, and the calculated fugacity coefficient of vapor phase and water activity are given.

2. Thermodynamic Framework

To predict the phase equilibrium conditions of gas hydrate, the iso-fugacity rule is used in the three phases (vapor, water, and hydrate) and a fugacity approach is considered for both gas and water as:

$$f_i^H(z, T, P) = f_i^L(x, T, P) = f_i^V(y, T, P) \tag{1}$$

where f_i^H, f_i^L and f_i^V are the fugacity of component i including water in the hydrate, liquid and vapor phases, respectively; z, x and y represent the mole fraction of component i in the hydrate, liquid and vapor phases, respectively. The fugacity of water in hydrate phase, f_w^H, is expressed as:

$$f_w^H(T, P) = f_w^{MT}(T) \times exp\left(\frac{-\Delta\mu_w^{MT-L}}{RT}\right) \tag{2}$$

where f_w^{MT} represents the fugacity of water in the hypothetical empty hydrate lattice and is assumed equal to the saturated vapor pressure of the empty hydrate lattice [36]; $-\Delta\mu_{L\,w}^{MT-}$ is the chemical potential difference calculated by the method of Holder et al. [37]; and R is the universal gas constant.

2.1. Thermodynamic Model of Vapor Phase

The PSRK group-contribution method is based on the SRK equation of state [38], which is used to calculate the fugacity of components in vapor phase, as:

$$P = \frac{RT}{v_m - b} - \frac{a}{v_m(v_m - b)} \tag{3}$$

where P and T are the system pressure and temperature, respectively; v_m represents the mole volume, which is obtained by solving the cubic equation derived from Equation (3), and the value is the same as the largest real root of the equation [35]; a and b are parameters of PSRK.

The parameters a_i and b_i of pure component i can be calculated from the critical properties $T_{c,i}$ and $P_{c,i}$.

$$a_i = \frac{0.42748R^2T_{c,i}^2}{P_{c,i}}f(T) \tag{4a}$$

$$b_i = \frac{0.08664RT_{c,i}}{P_{c,i}} \tag{4b}$$

$$f(T) = \left[1 + c_1\left(1 - T_r^{0.5}\right)\right]^2 \tag{5}$$

where $T_r = T/T_c$; the pure fluid parameter c_1 is taken from the study of Holderbaum and Gmehling [35]. The PSRK mixing rule is written as:

$$a = b\left[\frac{RT\sum y_i ln\gamma_i}{A_1} + \sum y_i\frac{a_i}{b_i} + \frac{RT}{A_1}\sum y_i ln\frac{b}{b_i}\right] \tag{6}$$

$$b = \sum y_i b_i \tag{7}$$

where γ_i stands for the activity coefficient of component i calculated by UNIFAC model; the recommended value of $A_1 = -0.64663$ in PSRK model. The activity coefficient γ_i is a correction factor that accounts for deviations of real systems from that of an ideal solution, which can be estimated from chemical models (such as UNIFAC). Thus, the fugacity coefficient is given by:

$$ ln\ \varphi_i = \frac{b_i}{b}(z-1) - ln\left[z\left(1 - \frac{b}{v_m}\right)\right] - \sigma\ ln(1 + \frac{b}{v_m}) \tag{8} $$

$$ \sigma = \frac{1}{A_1}\left(ln\gamma_i + ln\frac{b}{b_i} + \frac{b_i}{b} - 1\right) + \frac{a_i}{b_i RT} \tag{9} $$

where φ_i is the fugacity coefficient of component i; $z = Pv_m/RT$.

2.2. Thermodynamic Model of Hydrate Phase

Chen and Guo [16,17] proposed a two-step hydrate formation mechanism for gas hydrate formation. The following two processes are considered simultaneously in the nucleation process of hydrate: (1) A quasi-chemical reaction process to form basic hydrate, and (2) an adsorption process of smaller gas molecules in the linked cavities of basic hydrate, which leads to the non-stoichiometric property of hydrate. The model is expressed as:

$$ f_i^H = z_i f_i^0 \left(1 - \sum_j \theta_j\right)^\alpha \tag{10} $$

where z_i denotes the mole fraction of the basic hydrate formed by gas component i, and $z_i = 1$ for pure gas; θ_j represents the fraction of the linked cavities occupied by the gas component j; α is the ratio of linked cavities and basic cavities [39], which equals 1/3 for sI hydrates and 2 for sII hydrates, respectively.

$$ \sum_j \theta_j = \frac{\sum_j f_j c_j}{1 + \sum_j f_j c_j} \tag{11} $$

where f_j denotes the fugacity of component j in vapor phase calculated by PSRK method; c_j stands for the rigorous Langmuir constant, which is calculated from the Lennard-Jones potential function.

In Equation (10), f_i^0 represents the fugacity of component i in vapor phase in equilibrium with the unfilled pure basic hydrate i ($\sum\theta_j = 0$) [21]. According to the Chen-Guo model, it can be calculated as:

$$ f_i^0 = exp\left(\frac{-\sum_j A_{ij}\theta_j}{T}\right)\left[A_i'exp\left(\frac{B_i'}{T - C_i'}\right)\right]exp\left(\frac{\beta P}{T}\right)a_w^{\frac{-1}{\lambda_2}} \tag{12} $$

where A_{ij} is the binary interaction coefficient which stands for the interplays between gas molecule i in the basic hydrate and gas molecule j in the linked cavities. A_i', B_i' and C_i' are the Antoine constants, as reported by Chen and Guo [17]. β is the function of water volume difference between that in the unfilled basic hydrate phase and the water phase, and the large cavity number per water molecule [20], $\beta = 0.4242$ K/bar for sI hydrates, $\beta = 1.0224$ K/bar for sII hydrates. a_w is the activity of water in aqueous phase, which is calculated by the UNIFAC method. For sI and sII hydrates, $\lambda_2 = 3/23$ and $\lambda_2 = 1/17$, respectively.

2.3. Thermodynamic Model of Aqueous Phase

In order to calculate the activity coefficient of components in aqueous phase, the mole fraction of each component in aqueous phase should be obtained first. Therefore, gas is also considered to be a component in aqueous phase. The pressure-corrected Henry's law is employed to calculate the mole fraction of gas component in aqueous phase exclude water as:

$$ f_i^L(x_i, T, P) = x_i^L H_i exp\left(\frac{P\overline{V_i}^\infty}{RT}\right) \tag{13} $$

where subscript i represents the gas component in aqueous phase; H_i is the Henry's constant of component i, given by the Krichevsky-Kasarnovsky correlation [36,40]; $\overline{V_i}^\infty$ is the infinite partial molar volume of the component i in water, given by Heidemann and Prausnitz [29].

With the phase equilibrium, the gas mole fraction in aqueous phase can be calculated by the correlation as:

$$x_i^L = \frac{f_i^V(y_i, T, P)}{H_i exp(\frac{P\overline{V_i}^\infty}{RT})} \tag{14}$$

Considering the presence of the additive, whether it is a promoter that raises the phase equilibrium temperature (pressure) or an inhibitor that lowers the temperature (pressure), the components in the liquid phase should be recalculated. Furthermore, Delavar and Haghtalab [25,26] point out that the mole fraction of each component in aqueous phases can be calculated as:

$$n_i = \frac{x_i^L(1 - wt\%)}{M_i} \tag{15}$$

$$n_p = \frac{wt\%}{M_p} \tag{16}$$

where i represents water and gas component; p represents the promoter (inhibitor); M_i and M_p are the molecular weight of component i and the promoter (inhibitor); $wt\%$ stands for the weight percentage of the promoter (inhibitor) in aqueous phase.

$$x_i = \frac{n_i}{\sum n_i} \tag{17}$$

where i represents water, gas components and the promoter (inhibitor); n_i represents the mole fraction of water, gas component and the promoter (inhibitor) in aqueous solutions of a unit mass.

The activity coefficient of the components in aqueous phase is calculated by UNIFAC model [30–32], which consisting of the combinatorial and residual terms, as:

$$ln\gamma_i = ln\gamma_i^C + ln\gamma_i^R \tag{18}$$

where γ_i^C and γ_i^R stand for the combinatorial term and residual term of component i, respectively. The combinatorial term takes the different sizes and shapes of the molecules into account.

$$ln\gamma_i^C = ln\frac{\psi_i}{x_i} + 1 - \frac{\psi_i}{x_i} - \frac{1}{2}Zq_i(ln\frac{\varphi_i}{\theta_i} + 1 - \frac{\varphi_i}{\theta_i}) \tag{19}$$

$$\psi_i = \frac{x_i r_i^{\frac{2}{3}}}{\sum_j x_j r_j^{\frac{2}{3}}} \tag{20a}$$

$$\varphi_i = \frac{x_i r_i}{\sum_j x_j r_j} \tag{20b}$$

$$\theta_i = \frac{x_i q_i}{\sum_j x_j q_j} \tag{20c}$$

where $Z = 10$; j represents all components in aqueous phase; φ_i and θ_i are the volume and surface area fraction of component i, respectively; r_i and q_i are the volume and surface area parameters of component i, respectively. They are calculated by the van der Waals volumes R_k and surface areas Q_k of the individual group k using equations as follows:

$$r_i = \sum_k V_k^{(i)} R_k \tag{21a}$$

$$q_i = \sum_k V_k^{(i)} Q_k \tag{21b}$$

where $V_k^{(i)}$ is the number of group k in component i. The volume R_k parameters and surface areas parameters Q_k of group k are listed in Table 1.

Table 1. The UNIFAC group volume and surface-area parameters.

Main Group	Sub Group	Number	R_k	Q_k
H_2O	H_2O	4	1.506	1.732
CO_2	CO_2	56	2.592	2.522
CH_4	CH_4	57	2.244	2.312
O_2	O_2	58	1.764	1.910
N_2	N_2	60	1.868	1.970

The residual term of component i in Equation (18) is replaced by the solution-of-groups concept [32] as:

$$ln\gamma_i^R = \sum_k V_k^{(i)} \left(ln\Gamma_k - ln\Gamma_k^{(i)} \right) \tag{22}$$

where k represents all groups in aqueous phase, including water; $ln\Gamma_k$ stands for the residual activity coefficient of functional group k; $ln\Gamma_k^{(i)}$ is the residual activity coefficient of group k in the reference solution containing only component i. Both $ln\Gamma_k$ and $ln\Gamma_k^{(i)}$ are calculated as:

$$ln\Gamma_k = Q_k \left[1 - ln\left(\sum_m \theta_m \Psi_{mk} \right) - \sum_m \frac{\theta_m \Psi_{km}}{\sum_n \theta_n \Psi_{nm}} \right] \tag{23}$$

$$\theta_m = \frac{Q_m X_m}{\sum_n Q_n X_n} \tag{24}$$

$$X_m = \frac{\sum_m V_m^{(i)} x_i}{\sum_j \sum_k V_k^{(j)} x_j} \tag{25}$$

where m and n are the summations cover different groups of all components in aqueous phase; i in Equation (25) is the same as component i in Equation (22); θ_m and X_m are the surface area fraction and the mole fraction of group m in the mixture, respectively. The group interaction parameter Ψ_{nm} proposed by Sander et al. [30] is described as:

$$\Psi_{nm} = exp\left(-\frac{u_{nm} - u_{mm}}{T} \right) \tag{26}$$

where u_{nm} and u_{mm} are the adjustable group interaction parameters (energy parameters). For each group–group interaction, the two parameters have the relation of $u_{nm} = u_{mn}$. The gas–gas group interaction-energy parameters u_{nm} and temperature range are given in Table 2.

Table 2. Gas–gas group interaction-energy parameters u_{nm} and temperature range.

Gas	Temperature Range (K)	Gas (56, 57, 58, 60) [a]
CO_2	280–475	84.2
CH_4	275–375	−80
O_2	250–330	−260
N_2	210–330	−250

[a] 56, 57, 58 and 60 are the group numbers of CO_2, CH_4, O_2 and N_2 in the UNIFAC group parameter list, respectively.

In order to properly describe the temperature dependence gas solubility, the correlation proposed by Sander et al. [30] has been used as follows:

$$u_{gas-water} = u_0 + \frac{u_1}{\left(\frac{T}{K}\right)} \tag{27}$$

where u_0 and u_1 are temperature-independent parameters, shown in Table 3.

Table 3. Constants for the calculation of gas–water interaction-energy parameters in the temperature range 273–348K.

Gas	u_0	$u_1(\times 10^{-5})$
CO_2	980.1	−1.6895
CH_4	1059.8	−2.3172
O_2	1259.9	−3.0295
N_2	1260.4	−2.7416

3. Calculation Procedure

The equations described above were solved using codes generated by MATLAB 2014b (MathWork, Beijing, China). The calculation procedure to obtain the phase equilibrium conditions of gas hydrates is summarized in the schematic flow diagram shown in Figure 1.

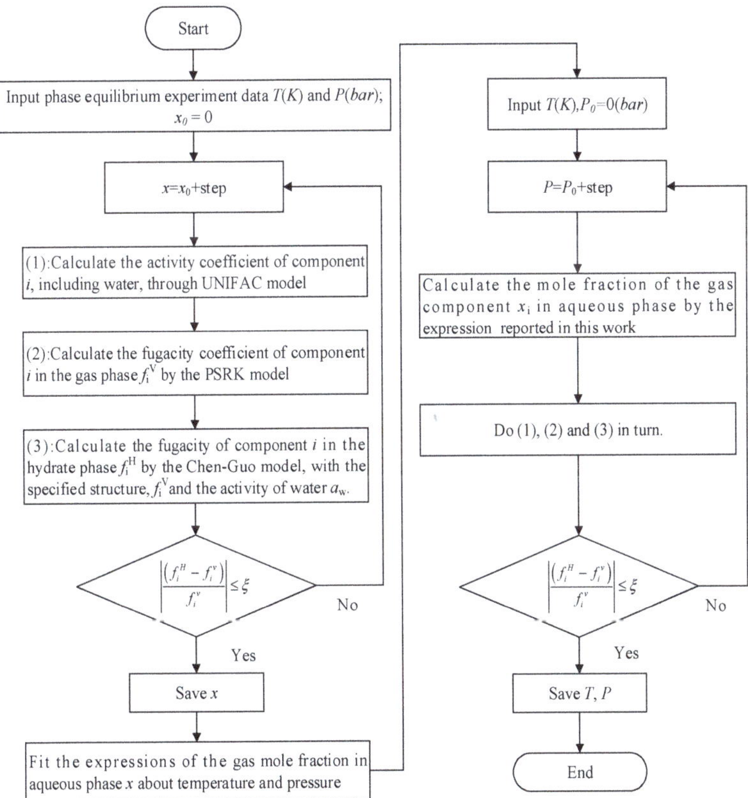

Figure 1. Calculation procedure for the prediction of phase equilibrium pressures at given temperatures.

The percentage of the Average Absolute Deviation in Pressure (AADP) is calculated as:

$$AADP(\%) = 100 \sum_{i=1}^{N} \left| \frac{P_i^{exp} - P_i^{cal}}{P_i^{exp}} \right| / N \tag{28}$$

where N is the number of experimental points; P_i^{exp} and P_i^{cal} stand for the experimental and calculated pressure, respectively.

4. Results and Discussion

When using the UNIFAC model to calculate the activity of gas components, the mole gas fraction in aqueous phase needs to be obtained first. Thus, the Henryconstant H_i and the infinitely diluted partial molar volume $\overline{V_i}^{\infty}$ are employed to calculate the gas mole fraction in aqueous phase. Heidemann and Prausnitz [29] provided a correlation for solving $\overline{V_i}^{\infty}$ as follows:

$$\frac{P_{c,i}\overline{V_i}^{\infty}}{RT_{c,i}} = 0.095 + 2.35 \left(\frac{TP_{c,i}}{c_{11}T_{c,i}} \right) \tag{29}$$

$$c_{11} = \frac{\left(h^0 - h^s - P_w^s v_w^s + RT \right)}{v_w^s} \tag{30}$$

where c_{11} represents the cohesive energy for water, which was evaluated at each temperature from thermodynamic properties tabulated; h^0 is the molar enthalpy at the given temperature but at zero pressure, and v_w^s is the molar volume of the saturated liquid [29].

However, it is not convenient to get the infinitely diluted partial molar volume of gas in the actual system by the parameters regressed from the assumed ideal state system, especially in the industrial application. Moreover, the parameters are not available for a liquid of unknown components. Therefore, as one of the factors affecting the formation of gas hydrate, the gas mole fraction in aqueous phase cannot be obtained accurately.

As seen from Equation (14), when gas type was given, the gas mole fraction in aqueous phase is only a function of temperature and pressure in phase equilibrium. Furthermore, the gas mole fraction in aqueous phase should be a fixed value in the three-phase equilibrium, which is related to the phase equilibrium temperature and pressure. Therefore, when the equilibrium temperature and pressure are given, the gas mole fraction in aqueous phase can be found by interval search using the framework mentioned in this work. Remarkably, in the process of numerical calculation, due to the unreasonable search step length and the inevitable error of experimental data, a set of phase equilibrium temperature and pressure may correspond to multiple gas mole fraction values. In this case, the average of these values can be taken as the gas mole fraction in aqueous phase under the phase equilibrium. As a result, the correlation of the gas mole fraction in aqueous phase is fitted by temperature and pressure, which is determined by the experimental data and defined as:

$$x_i = a + b \times T + c \times P \tag{31}$$

where a, b and c are constants for gas component i + water system, given in Table 4.

Table 4. Parameters for the correlation of the gas mole fraction in aqueous phase.

Guests	a	$b\,(\times 10^5)$	$c\,(\times 10^7)$	N_p	R-Square
CH_4	0.01713	−2.58066	−24.6488	500	0.99924
CO_2	0.00454	0.403588	−49.4532	200	0.89458
N_2	0.0226	−5.50785	1.49362	409	1
O_2	0.02468	−5.90016	1.33948	200	0.99999

Table 5. The phase equilibrium pressure and temperature range of experimental data and the Average Absolute Deviation in Pressure (AADP) for predicted results.

Guests	Temperature (K)	P-Range (bar)	N_p	Reference	AADP (%)
CH_4	273.27–289.9	26.33–159.52	66	[41]	1.0634
	276.81–281.3	37.79–62.02	5	This work	1.6317
CO_2	270.15–283.15	10.19–45.05	41	[42]	1.5911
N_2	273.15–291.05	160.09–958.53	34	[43]	1.0224
	274.55–283.05	190.93–453.55	3	[44]	1.4142
	279.30–284.00	303.00–500.00	3	[45]	1.5697
O_2	273.15	121.50	1	[43]	1.6786
	273.78–284.55	138.21–441.30	4	[44]	1.8001
	286.27–291.18	527.13–953.65	6	[46]	0.9462

Although the correlation of the gas mole fraction in aqueous phase is simple and is multivariate linear form, its precision and calculation accuracy are satisfactory. The number of data points used in fitting Equation (31) and the R-Square are given in Table 4.

As shown in Figure 2, there are two kinds of tendencies for the fugacity coefficient in methane system with the increase of temperature. First, when the temperature is lower than 286.5 K, the fugacity coefficient decreases with the increase of temperature, which has the same tendency with the fugacity coefficient in the carbon dioxide system, although the reduction rate is smaller. On the other hand, when the temperature is higher than 286.5 K, the fugacity coefficient increases exponentially with the temperature increment, which can also be seen in nitrogen and oxygen systems, as displayed in Figure 2. It should be pointed out that the fugacity coefficient decreases with the temperature increment in carbon dioxide system. This indicates that carbon dioxide is more likely to yield to pressure when the temperature is below the critical temperature. Moreover, the exponential growth of the fugacity coefficient in the N_2 and O_2 systems is mainly related to the temperature increment.

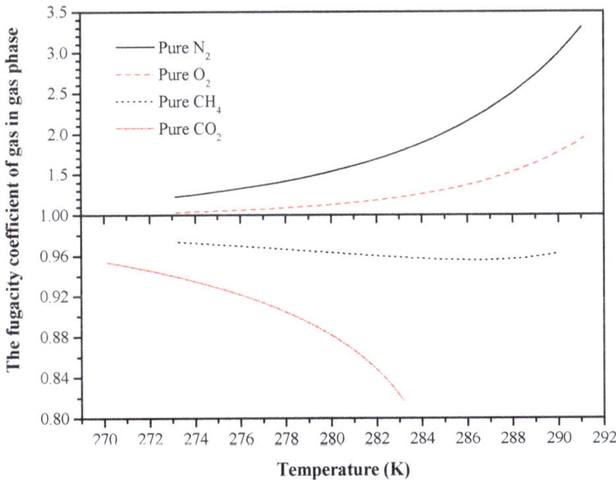

Figure 2. The fugacity coefficient of CH_4, CO_2, N_2 and O_2 in vapor phase.

The mole fraction of CH_4, CO_2, N_2 and O_2 in aqueous phase under phase equilibrium condition is shown in Figures 3–5. In this work, the gas mole fraction was considered as one of the factors affecting the phase equilibrium. It represents the ratio of the number of gas molecules maintaining the three-phase equilibrium to the number of all molecules in aqueous phase.

In Figures 3–5, under phase equilibrium condition, the gas mole fraction in aqueous phase decrease with temperature increment, and all the changing ranges are less than 1×10^{-3}. Therefore, there may exist a threshold value for the gas mole fraction in aqueous phase. In other words the hydrate will form when the gas mole fraction in aqueous phase reaches a certain threshold value. Furthermore, for methane hydrate, the results in this work are in agreement with the views of Walsh et al. [47] and Guo and Rodger [48]. Walsh et al. suggested that the threshold value of gas mole fraction triggering hydrate formation calculated by the molecular dynamics (MD) simulation was 1.5×10^{-3}. The threshold value is a reasonable explanation for reducing the temperature or increasing the pressure, which could effectively promote the formation of hydrate. This is because lowering the temperature or increasing the pressure will enhance the gas dissolution, which in turn causes the gas mole fraction in aqueous phase exceeding the threshold value, then hydrate forms.

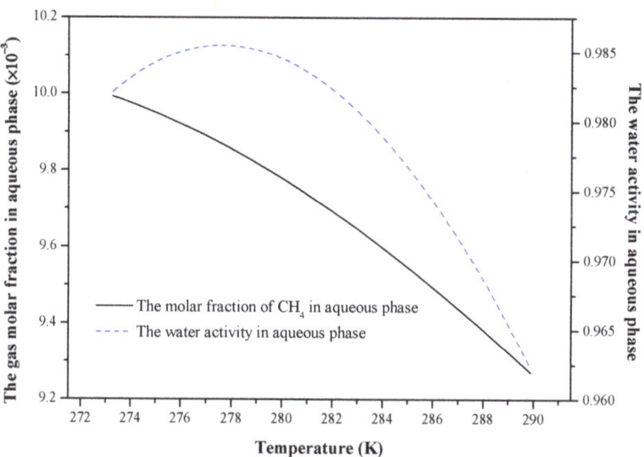

Figure 3. The mole fraction of CH_4 and water activity in aqueous phase.

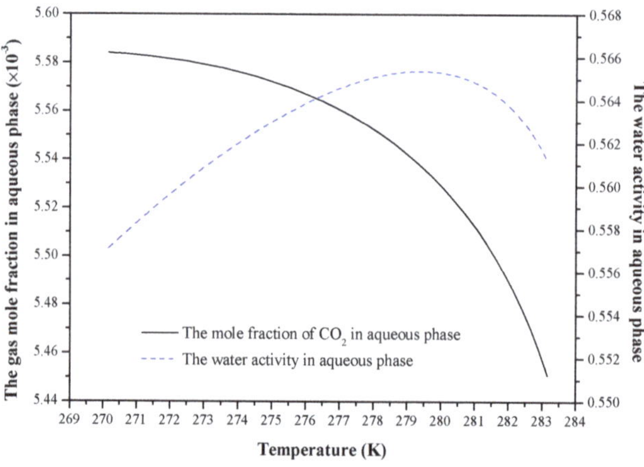

Figure 4. The mole fraction of CO_2 and water activity in aqueous phase.

In particular, although carbon dioxide has a large solubility in water, the mole fraction of carbon dioxide calculated in this work is not very large under phase equilibrium, as shown in Figure 4. It implied that the total number of carbon dioxide gas molecules in aqueous phase for maintaining the three-phase equilibrium is not large, and it may be much smaller than the sum of the gas molecules dissolved in the water. This is mainly because part of carbon dioxide gas molecules dissolved in water turn into carbonic acid, thus reducing the amount of carbon dioxide gas molecules existing in water. Meanwhile, the pH of aqueous phase will be changed, and affecting the activity of the water and phase equilibrium conditions.

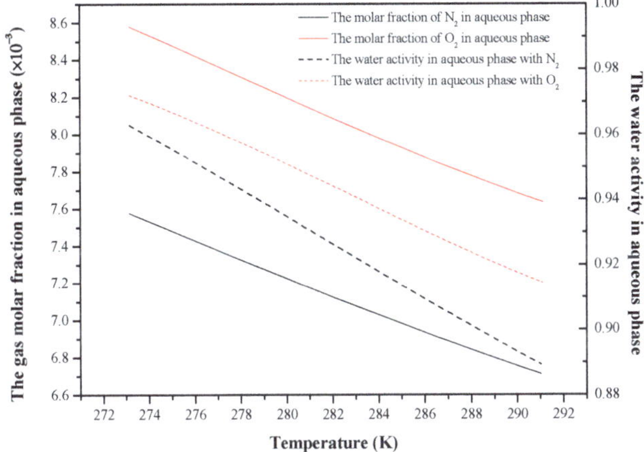

Figure 5. The mole fraction of N_2/O_2 and water activity in aqueous phase.

In Figure 3, water activity increases with the increase of temperature, in the methane system, reaching a maximum value of 0.985 when the temperature is about 278 K, and then decreases rapidly with the increase of temperature. The general variation trend of the water activity in carbon dioxide system is similar to that of methane system, as seen in Figure 4. In addition, the water activity of the CO_2 system reached its maximum value at about 279.5 K. However, the maximum water activity in the CO_2 system is only about 0.5658, which is probably because of the effect of the carbonic acid. Nevertheless, the activity of water in aqueous phase decreases almost linearly with temperature increase in nitrogen and oxygen systems, as shown in Figure 5.

Figures 6 and 7 show the experimental and predicted phase equilibrium conditions for the single gas hydrate systems. The temperature range, pressure range and AADP (%) are listed in Table 5.

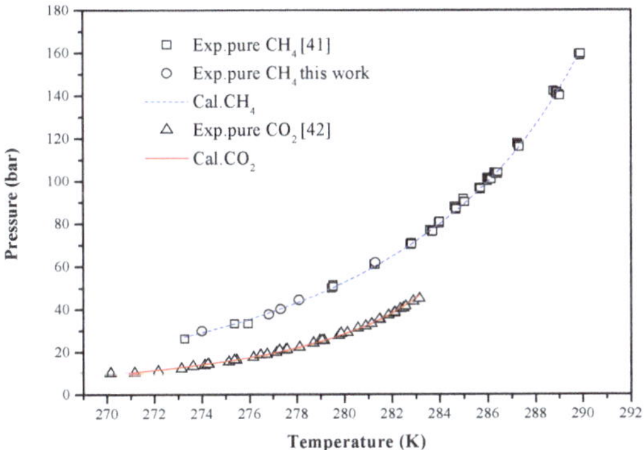

Figure 6. Experimental and predicted phase equilibrium conditions for CH_4/CO_2 + water systems. Sloan [41], squares □; Ma et al. [42], triangles Δ; this work, circles ○.

Figure 6 shows the experimental and predicted phase equilibrium pressures for CH_4 and CO_2. It can be seen the predicted results for all the gas systems are in excellent agreement with the experimental data. It should be noted that the type of carbon dioxide hydrate structure was set to sI, and, because the carbon dioxide gas molecule is too big to be encaged in the linked cavities, the filling rate of the gas molecules in the linked cavities, θ_j, was set to 0, as described by Chen and Guo [16].

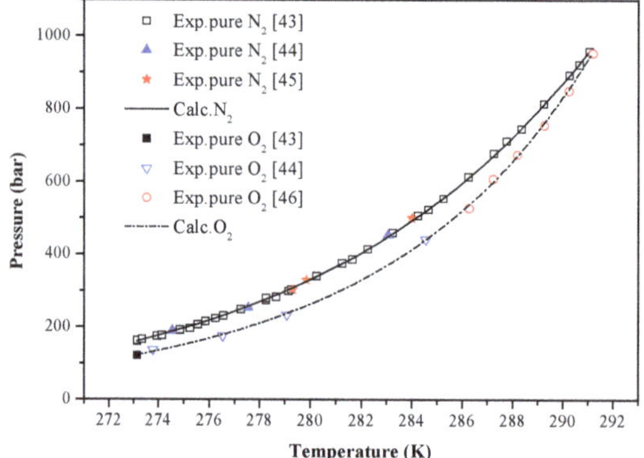

Figure 7. Experimental and predicted phase equilibrium conditions for N_2/O_2 + water systems. van Cleeff and Diepen [43], squares □, ■; Mohammadi et al. [44], triangles ▲,▽; Duc et al. [45], stars ★; van Cleeff and Diepen [46], circles ○.

The experimental and predicted phase equilibrium pressures for N_2 and O_2 are displayed in Figure 7. The predicted phase equilibrium pressures are in good agreement with the experiment. It is especially noteworthy that, when calculating oxygen and nitrogen hydrate, the hydrate structure was set to sII, which was based on the ideas proposed by Chen and Guo [16]. This is because the gas molecules of N_2 and O_2 are small and have a high filling rate in the connected cavities.

The gas mole fraction in Figure 8 was obtained by inverse phase equilibrium data using the framework proposed in this work. The gas mole fraction threshold value for maintaining the three-phase equilibrium state is different to the critical gas concentration. The gas mole fraction threshold value calculated in this work does not contradict the critical gas concentration proposed by Zhang et al. [49]. They pointed out that there is a critical gas concentration in aqueous phase that can spontaneously nucleate in the induction period, and the critical gas concentration is calculated by the total amount of carbon dioxide consumed in vapor phase until hydrate nucleation.

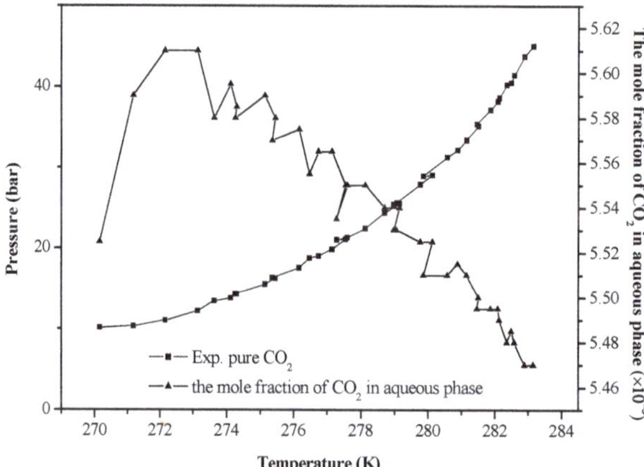

Figure 8. The experimental data and the mole fraction of carbon dioxide in aqueous phase for CO_2 + water systems under the phase equilibrium. The experimental data were reported by Ma et al. [42].

However, the phase equilibrium data cited in this work were recorded at the end of decomposition rather than in the preliminary stage of nucleation. Gas molecules entrapped in hydrate cage cannot be released totally, which was owing to memory effect [50]. Moreover, there theoretically exists a concentration difference as a force in mass transfer during hydrate nucleation and decomposition. Therefore, the gas mole fraction threshold value calculated in this work is less than the critical gas concentration.

In Figure 8, the threshold value of gas mole fraction achieves a maximum of 5.61×10^{-3} at 0 °C. A possible reason is that part of carbon dioxide molecules in aqueous phase react with water to form carbonic acid. When the temperature is above 0 °C and below 0 °C, the pressure increment and the temperature decrement become a dominant factor that results in more stability for the carbonic acid and less solubility of carbon dioxide, respectively. However, this analysis should be proved by further study.

Furthermore, since the correlation of gas mole fraction fitted in this work is a multivariate linear form, the trend of the gas mole fractions in Figures 4 and 8 are different. Therefore, a large number of accurate and reliable experiment data can effectively improve the prediction accuracy of the model in this work.

5. Conclusions

In this work, the Chen-Guo model coupled with the PSRK method were employed to predict phase equilibrium conditions of CH_4, CO_2, N_2 or O_2 in pure water systems. The gas mole fraction in aqueous phase is one of the factors that affect the phase equilibrium of gas hydrate proposed in this work. The gas mole fraction threshold value maintaining the three-phase equilibrium was obtained by reversed phase equilibrium data. Meanwhile, in order to obtain the water activity in aqueous phase, the correlation of the gas mole fraction threshold value in aqueous phase was fitted though UNIFAC model. The calculated water activity can effectively improve the accuracy of the prediction results, and the predicted results of this work are in good agreement with the experimental data reported in the references.

Author Contributions: conceptualization, W.Z. and Y.Z.; methodology, W.Z. and H.W.; software, H.W.; data curation, X.G.; project administration, Y.Z.; resources, Y.Z., writing—original draft preparation, W.Z..; writing—review and editing, J.W. and R.W.

Funding: This research was funded by the Young Teacher Capacity Improvement Fund of Henan Polytechnic University, grant number TM2017/02.

Conflicts of Interest: The authors declare no conflict of interest.

References

1. Sloan, E.D.; Koh, C.A. *Clathrate Hydrates of Natural Gas*, 3rd ed.; Chemical Industries, CRC Press, Taylor & Francis Group: Boca Raton, FL, USA, 2008.
2. Carroll, J.J. *Natural Gas Hydrates: A Guide for Engineers*; Gulf Professional Publishing: Houston, TX, USA, 2009.
3. Abdi-Khanghah, M.; Adelizadeh, M.; Naserzadeh, Z.; Barati, H.; Zhang, Z. Methane hydrate formation in the presence of ZnO nanoparticle and SDS: Application to transportation and storage. *J. Nat. Gas Sci. Eng.* **2018**, *54*, 120–130. [CrossRef]
4. Aaron, D.; Tsouris, C. Separation of CO_2 from Flue Gas: A Review. *Sep. Sci. Technol.* **2005**, *40*, 321–348. [CrossRef]
5. Ogawa, T.; Ito, T.; Watanabe, K.; Tahara, K.-I.; Hiraoka, R.; Ochiai, J.-I.; Ohmura, R.; Mori, Y.H. Development of a novel hydrate-based refrigeration system: A preliminary overview. *Appl. Therm. Eng.* **2006**, *26*, 2157–2167. [CrossRef]
6. Wang, X.; Dennis, M.; Hou, L. Clathrate hydrate technology for cold storage in air conditioning systems. *Renew. Sustain. Energy Rev.* **2014**, *36*, 34–51. [CrossRef]
7. Xu, C.-G.; Li, X.-S. Research progress of hydrate-based CO_2 separation and capture from gas mixtures. *RSC Adv.* **2014**, *4*, 18301–18316. [CrossRef]

8. Pan, Z.; Liu, Z.; Zhang, Z.; Shang, L.; Ma, S. Effect of silica sand size and saturation on methane hydrate formation in the presence of SDS. *J. Nat. Gas Sci. Eng.* **2018**, *56*, 266–280. [CrossRef]

9. Khan, M.N.; Warrier, P.; Peters, C.J.; Koh, C.A. Review of vapor-liquid equilibria of gas hydrate formers and phase equilibria of hydrates. *J. Nat. Gas Sci. Eng.* **2016**, *35*, 1388–1404. [CrossRef]

10. Taylor, F.W. The greenhouse effect and climate change. *Rep. Prog. Phys.* **1991**, *54*, 881. [CrossRef]

11. Zhao, W.-L.; Zhong, D.-L.; Yang, C. Prediction of phase equilibrium conditions for gas hydrates formed in the presence of cyclopentane or cyclohexane. *Fluid Phase Equilib.* **2016**, *427*, 82–89. [CrossRef]

12. van der Waals, J.H.; Platteeuw, J.C. Clathrate solutions. In *Advances in Chemical Physics*; Interscience Publishers Inc.: New York, NY, USA, 1959.

13. Nasrifar, K.; Moshfeghian, M.; Maddox, R.N. Prediction of equilibrium conditions for gas hydrate formation in the mixtures of both electrolytes and alcohol. *Fluid Phase Equilib.* **1998**, *146*, 1–13. [CrossRef]

14. Haghighi, H.; Chapoy, A.; Burgess, R.; Mazloum, S.; Tohidi, B. Phase equilibria for petroleum reservoir fluids containing water and aqueous methanol solutions: Experimental measurements and modelling using the CPA equation of state. *Fluid Phase Equilib.* **2009**, *278*, 109–116. [CrossRef]

15. Martin, A.; Peters, C.J. Thermodynamic Modeling of Promoted Structure II Clathrate Hydrates of Hydrogen. *J. Phys. Chem. B.* **2009**, *113*, 7548–7557. [CrossRef] [PubMed]

16. Chen, G.J.; Guo, T.M. Thermodynamic modeling of hydrate formation based on new concepts. *Fluid Phase Equilib.* **1996**, *122*, 43–65. [CrossRef]

17. Chen, G.J.; Guo, T.M. A new approach to gas hydrate modelling. *Chem. Eng. J.* **1998**, *71*, 141–151. [CrossRef]

18. Sun, C.Y.; Chen, G.J. Modelling the hydrate formation condition for sour gas and mixtures. *Chem. Eng. Sci.* **2005**, *60*, 4879–4885. [CrossRef]

19. Klauda, J.B.; Sandler, S.I. Phase behavior of clathrate hydrates: A model for single and multiple gas component hydrates. *Chem. Eng. Sci.* **2003**, *58*, 27–41. [CrossRef]

20. Ma, Q.L.; Chen, G.J.; Sun, C.Y.; Yang, L.Y.; Liu, B. Predictions of hydrate formation for systems containing hydrogen. *Fluid Phase Equilib.* **2013**, *358*, 290–295. [CrossRef]

21. Ma, Q.L.; Chen, G.J.; Guo, T.M. Modelling the gas hydrate formation of inhibitor containing systems. *Fluid Phase Equilib.* **2003**, *205*, 291–302. [CrossRef]

22. Liu, H.; Guo, P.; Du, J.; Wang, Z.; Chen, G.; Li, Y. Experiments and modeling of hydrate phase equilibrium of $CH_4/CO_2/H_2S/N_2$ quaternary sour gases in distilled water and methanol-water solutions. *Fluid Phase Equilib.* **2017**, *432*, 10–17. [CrossRef]

23. Abrams, D.S.; Prausnitz, J.M. Statistical thermodynamics of liquid mixtures: A new expression for the excess Gibbs energy of partly or completely miscible systems. *Aiche J.* **2010**, *21*, 116–128. [CrossRef]

24. Weidlich, U.; Gmehling, J. A modified UNIFAC model. 1. Prediction of VLE, hE, and .gamma..infin. *Ind. Eng. Chem. Res.* **1987**, *26*, 1372–1381. [CrossRef]

25. Delavar, H.; Haghtalab, A. Prediction of hydrate formation conditions using GE-EOS and UNIQUAC models for pure and mixed-gas systems. *Fluid Phase Equilib.* **2014**, *369*, 1–12. [CrossRef]

26. Delavar, H.; Haghtalab, A. Thermodynamic modeling of gas hydrate formation conditions in the presence of organic inhibitors, salts and their mixtures using UNIQUAC model. *Fluid Phase Equilib.* **2015**, *394*, 101–117. [CrossRef]

27. Dehaghani, A.H.S.; Karami, B. A new predictive thermodynamic framework for phase behavior of gas hydrate. *Fuel* **2018**, *216*, 796–809. [CrossRef]

28. Klauda, J.B.; Sander, S.T. A Fugacity Model for Gas Hydrate Phase Equilibria. *Ind. Eng. Chem. Res.* **2000**, *39*, 3377–3386. [CrossRef]

29. Heidemann, R.A.; Prausnitz, J.M. Equilibrium Data for Wet-Air Oxidation. Water Content and Thermodynamic Properties of Saturated Combustion Gases. *Ind. Eng. Chem. Process Des. Dev.* **1977**, *16*, 375–381. [CrossRef]

30. Sander, B.; Skjold-Jørgensen, S.; Rasmussen, P. Gas solubility calculations. I. Unifac. *Fluid Phase Equilib.* **1983**, *11*, 105–126. [CrossRef]

31. Fredenslund, A.; Gmehling, J.; Rasmussen, P. Vapor-liquid equilibria using UNIFAC: A group contribution method. *Fluid Phase Equilib.* **1977**, *1*, 317.

32. Fredenslund, A.; Jones, R.L.; Prausnitz, J.M. Group contribution estimation of activity coefficients in nonideal liquid mixtures. *AIChE J.* **1975**, *21*, 1086–1099. [CrossRef]

33. Gmehling, J.; Li, J.; Fischer, K. Further development of the PSRK model for the prediction of gas solubilities and vapor-liquid-equilibria at low and high pressures II. *Fluid Phase Equilib.* **1997**, *141*, 113–127. [CrossRef]

34. Fischer, K.; Gmehling, J. Further development, status and results of the PSRK method for the prediction of vapor-liquid equilibria and gas solubilities. *Fluid Phase Equilib.* **1996**, *121*, 185–206. [CrossRef]

35. Holderbaum, T.; Gmehling, J. PSRK: A group contribution equation based on UNIFAC. *Fluid Phase Equilib.* **1991**, *70*, 251–265. [CrossRef]

36. Sloan, E.D., Jr. *Clathrate Hydrates of Natural Gases*, 2nd ed.; CRC Press: New York, NY, USA, 1998.

37. Holder, G.D.; Corbin, G.; Papadopoulos, K.D. Thermodynamic and Molecular Properties of Gas Hydrates from Mixtures Containing Methane, Argon, and Krypton. *Ind. Eng. Chem. Fundam.* **1980**, *19*, 282–286. [CrossRef]

38. Soave, G. Equilibrium constants from a modified Redlich-Kwong equation of state. *Chem. Eng. Sci.* **1972**, *27*, 1197–1203. [CrossRef]

39. Sun, Q.; Guo, X.; Chapman, W.G.; Liu, A.; Yang, L.; Zhang, J. Vapor–hydrate two-phase and vapor–liquid–hydrate three-phase equilibrium calculation of THF/CH_4/N_2 hydrates. *Fluid Phase Equilib.* **2015**, *401*, 70–76. [CrossRef]

40. Krichevsky, I.R.; Kasarnovsky, J.S. Thermodynamical Calculations of Solubilities of Nitrogen and Hydrogen in Water at High Pressures. *J. Am. Chem. Soc.* **1935**, *57*, 2168–2171. [CrossRef]

41. Sloan, E.D. *Clathrate Hydrates of Natural Gas*; M. Dekker: New York, NY, USA, 2007.

42. Ma, Z.W.; Zhang, P.; Bao, H.S.; Deng, S. Review of fundamental properties of CO_2 hydrates and CO_2 capture and separation using hydration method. *Renew. Sustain. Energy Rev.* **2016**, *53*, 1273–1302. [CrossRef]

43. Van Cleeff, A.; Diepen, G.A.M. Gas hydrate of nitrogen and oxygen. *Recueil des Travaux Chimiques des Pays-Bas* **1960**, *79*, 582–586. [CrossRef]

44. Mohammadi, A.H.; Tohidi, B.; Burgass, R.W. Equilibrium Data and Thermodynamic Modeling of Nitrogen, Oxygen, and Air Clathrate Hydrates. *J. Chem. Eng. Data* **2003**, *48*, 612–616. [CrossRef]

45. Duc, N.H.; Chauvy, F.; Herri, J.-M. CO_2 capture by hydrate crystallization—A potential solution for gas emission of steelmaking industry. *Energy Convers. Manag.* **2007**, *48*, 1313–1322. [CrossRef]

46. Van Cleeff, A.; Diepen, G.A.M. Gas Hydrates of Nitrogen and Oxygen. II. *Recueil des Travaux Chimiques des Pays–Bas* **1965**, *84*, 1085–1093. [CrossRef]

47. Walsh, M.R.; Beckham, G.T.; Koh, C.A.; Sloan, E.D.; Wu, D.T.; Sum, A.K. Methane Hydrate Nucleation Rates from Molecular Dynamics Simulations: Effects of Aqueous Methane Concentration, Interfacial Curvature, and System Size. *J. Phys. Chem. C* **2011**, *115*, 21241–21248. [CrossRef]

48. Guo, G.J.; Rodger, P.M. Solubility of aqueous methane under metastable conditions: Implications for gas hydrate nucleation. *J. Phys. Chem. B* **2013**, *117*, 6498–6504. [CrossRef] [PubMed]

49. Zhang, P.; Wu, Q.; Mu, C.; Chen, X. Nucleation Mechanisms of CO_2 Hydrate Reflected by Gas Solubility. *Sci. Rep.* **2018**, *8*, 10441. [CrossRef] [PubMed]

50. Wu, Q.; Zhang, B. Memory effect on the pressure-temperature condition and induction time of gas hydrate nucleation. *J. Nat. Gas Chem.* **2010**, *19*, 446–451. [CrossRef]

Article

Calculation Model and Rapid Estimation Method for Coal Seam Gas Content

Fakai Wang [1], Xusheng Zhao [2], Yunpei Liang [1,*], Xuelong Li [1,*] and Yulong Chen [3,*]

[1] State Key Laboratory of Coal Mine Disaster Dynamics and Control, College of Resources and Environmental Science, Chongqing University, Chongqing 400044, China; wangfakai316@126.com
[2] Chongqing Research Institute of China Coal Technology and Engineering Group Crop., Chongqing 400037, China; cq_zxs@163.com
[3] State Key Laboratory of Hydroscience and Engineering, Tsinghua University, Beijing 100084, China
* Correspondence: liangyunpei@126.com (Y.L.); lixlcumt@126.com (X.L.); chenyl@tsinghua.edu.cn (Y.C.)

Received: 27 September 2018; Accepted: 12 November 2018; Published: 14 November 2018

Abstract: Coalbed gas content is the most important parameter for forecasting and preventing the occurrence of coal and gas outburst. However, existing methods have difficulty obtaining the coalbed gas content accurately. In this study, a numerical calculation model for the rapid estimation of coal seam gas content was established based on the characteristic values of gas desorption at specific exposure times. Combined with technical verification, a new method which avoids the calculation of gas loss for the rapid estimation of gas content in the coal seam was investigated. Study results show that the balanced adsorption gas pressure and coal gas desorption characteristic coefficient (K_t) satisfy the exponential equation, and the gas content and K_t are linear equations. The correlation coefficient of the fitting equation gradually decreases as the exposure time of the coal sample increases. Using the new method to measure and calculate the gas content of coal samples at two different working faces of the Lubanshan North mine (LBS), the deviation of the calculated coal sample gas content ranged from 0.32% to 8.84%, with an average of only 4.49%. Therefore, the new method meets the needs of field engineering technology.

Keywords: gas pressure; gas content; gas basic parameters; rapid estimation technology

---

1. Introduction

The basic parameters of coalbed gas are the foundation for preventing and controlling coal and gas outbursts [1]. As one of the most important basic parameters [2], coalbed gas content is employed to calculate the coal seam gas reserves, predicting gas emission from mines, and evaluating coal and gas outburst risk. Measuring coalbed gas content accurately is required to reduce the occurrence of mine disasters and the cost of mine gas hazard prevention [3]. At present, several methods have been proposed to determine the coalbed gas content, which can be roughly divided into direct and indirect methods. The indirect method is primarily used to calculate the gas content of coal through the Langmuir equation. However, this method has the disadvantages of in situ measurement process complexity and poor accuracy. Therefore, downhole direct measurement methods [4] have been widely used.

Many works have been conducted on theoretical and experimental studies of the estimation of gas content directly in the downhole. Jin and Firoozabadi [5] have studied phase behavior and flow in nanopores using density functional theory and various molecular simulations. Zhao et al. [1] present adsorption and desorption isotherms of methane, ethane, propane, n-butane, and isobutane, as well as carbon dioxide, for two shales and isolated kerogens determined by a gravimetric method. Bertard et al. [6] found that the early adsorption diffusion process of gas was proportional to the square of the time and formed a direct measurement method of coal seam gas content. This method predicts the gas loss through the gas desorption data and derivation equation, thus laying the foundation for determining the gas loss in the sampling process. McCulloch et al. [7] simplified the Bertard method and proposed a United States Bureau of Mines (USBM) direct gas content estimation method that is the square root calculations of desorption time ($\sqrt{t}$), and is proportional to the cumulative desorption. Ulery and Hyman [8] proposed a modified determination method (MDM) based on the measurement of various gas pressures, then the ideal gas law is used to calculate the desorption of gases under standard temperature and pressure conditions (STP). Mavor et al. [9,10] established a process for the estimation of gas content based on the USBM direct assay. Saghafi et al. [11] showed that the initial desorption of coal gas has an exponential relation with time. Smith and Williams [12] proposed a technique for the direct estimation of coal sample gas content from exposed rotary boreholes. Chase [13] determined the coal gas content by plotting the gas desorption rate curve and the cumulative desorption curve using the least squares method. Sawyer et al. [14] found that it is difficult to obtain more accurate gas desorption amounts and the residual gas volume by prematurely breaking the coal sample during desorption. Chen et al. [15] found that using an equation or method that does not have a large correlation with the gas desorption feature to calculate the gas content is usually more error-prone. Chen et al. [16] concluded that the negative exponential equation method is more consistent with the gas desorption law in the initial stage of tectonic coal. Lei et al. [17] proposed a new method of improvement based on the Barrer equation method to improve the accuracy of gas content measurement in coal seams. Zhang et al. [18] determined the gas desorption law of coal samples in different gas pressure conditions by experiments and proved that the $\sqrt{t}$ method is more consistent with the gas desorption law in the initial stage. Li and Yang [19] calculated and compared the gas loss in the sampling process using the graphic method and least squares method. The "Direct Gas Content Measurement Device (DGC)" [20–27], which was developed based on an empirical equation, can determine the gas content of coal. The calculation of the gas content is based on the law of gas desorption. The empirical equations of gas desorption in coal, which have been proposed by scholars throughout the world, are listed in Table 1.

Table 1. Coal gas desorption equations.

Equations	Q_t (mL/g)	Applicable Conditions
Barrer [28]	$\frac{Q_t}{Q_\infty} = \frac{2s}{V}\sqrt{\frac{D_t}{\pi}}$	$0 \leq \sqrt{t} \leq \frac{V}{2s}\sqrt{\frac{\pi}{D}}$
Winter [29]	$Q_t = \frac{v_1}{1-k_1}t^{1-k_1}$	$0 < k_t < 1$
Wang [30]	$Q_t = \frac{ABt}{1+Bt}$	
Sun [31]	$Q_t = at^i$	$0 < i < 1$
Exponential [32]	$Q_t = \frac{v_0}{b}\left(1 - e^{-bt}\right)$	

Barrer [28] concluded that adsorption-desorption is a reversible process, and the cumulative amount of gas absorbed or desorbed is proportional to the square root of time. Winter et al. [29] found that the change in the amount of desorbed gas over time can be expressed as a power equation. Wang et al. [30] believed that the gas desorption of coal with time is consistent with the Langmuir adsorption equation. Sun [31] confirmed that gas desorption in coal is mainly a diffusion process, and the change of desorption gas content with time can be expressed by a power equation. Others [32] believe that the decay process of coal gas desorption with time conforms to the exponential equation. Due to the extensive variety of research objects, the empirical equation has both reasonable and unreasonable components in revealing the law of gas desorption in coal [33]. Compared with the equations in Table 1, calculating the gas compensation amount is difficult due to three shortcomings. First, the definition of the gas desorption start time (zero time, as shown in Figure 1) is ambiguous. Second, the percentage of lost gas and original gas in the coal sample varies with different degrees of metamorphism and damage types. The model that is established by the derivation equation and the basic theory of the model have different assumptions, and the existence of these hypotheses may be very contradictory to the real environment. Therefore, the best method for measuring the gas content is to avoid the calculation of the gas loss.

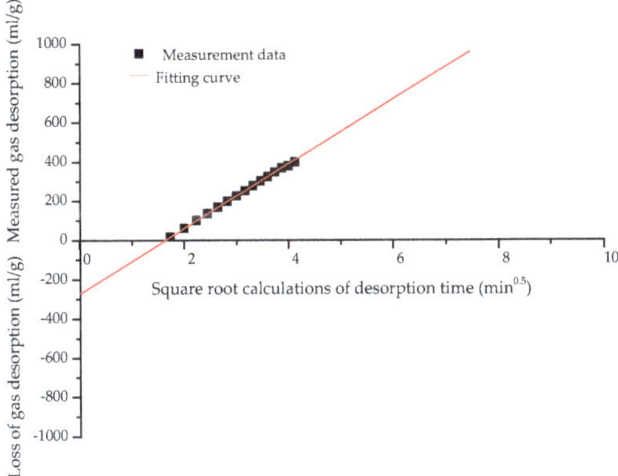

Figure 1. Diagram of gas desorption loss estimation.

In this paper, an analysis of the simulation results in gas desorption was performed, and a combination of numerical analysis and field tests were conducted. In order to improve the coalbed gas content measuring accuracy, a rapid method for determining the gas content in coal seams to avoid the calculation of gas loss was proposed for on-site measurement of the gas content in coal seams, which can accurately provide the basic parameters for mine safety production.

2. Experimental Study

2.1. Experimental Apparatus

To simulate the gas desorption process of a coal sample, a set of simulation equipment was designed and developed. A schematic of the experimental setup is shown in Figure 2.

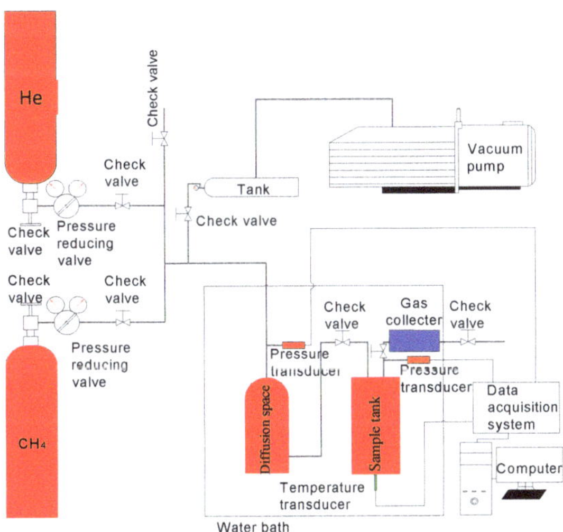

Figure 2. Schematic of the experimental setup.

The experimental device consisted of four systems, the details of each system are listed as follows:

(1)   Vacuum system: This system consisted of a composite vacuum gauge, a vacuum pump, a vacuum tube, a vacuum gauge, and a glass three-way valve.

(2)   Constant temperature system: This system consisted of a constant temperature water bath, a coal sample tank, a diffusion tank, a precision pressure gauge, and a high-purity methane gas source.

(3)   Adsorption balance system: This system consisted of precision pressure gauges, methane gas sources, inflatable tanks, coal sample tanks, and valves.

(4)   Desorption measurement control system: This system consisted of a pressure control valve and a homemade gas desorption analyzer.

To eliminate the influence of temperature on the simulation results, the device could achieve a constant ambient temperature in the coal sample gas adsorption-desorption process.

2.2. Coal Sample Preparation

The coal sample was taken from the No. 3 coal seam in the Lubanshan North Mine (LBS), which is located in Yibin City of Sichuan Province, and primarily consisted of lean coal. The location of the mine is shown in Figure 3. The No. 3 coal seam of LBS is located in the lower part of the Shanxi Formation, with an average distance of 58.04 m from No. 9 and an average thickness of 6.71 m. Figure 4 shows a general sketch of the coal seams and a diagram of the coal seam gas pressure measurement. The roof of the coal seam is mudstone and sandy mudstone, and the bottom slab is sandy mudstone and dark gray sandstone.

Figure 3. Geographical location of the LBS.

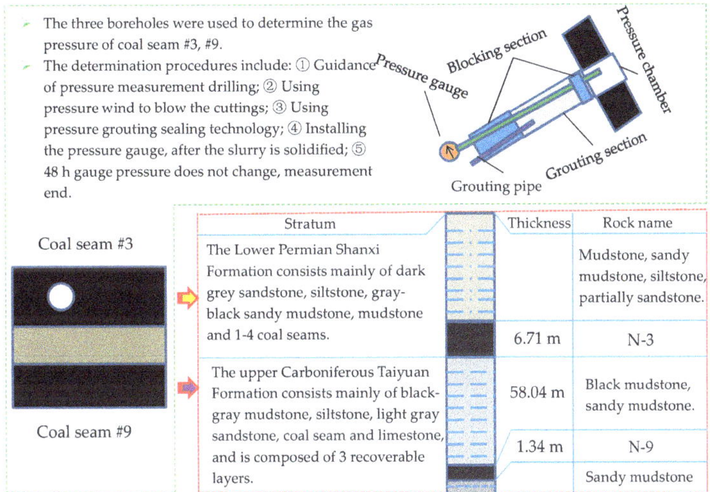

Figure 4. General sketch of the coal seam and a diagram of coal seam gas pressure measurement.

Coal seam No. 3 was sampled and marked as N-3. According to the "Sampling of coal seams" [34], coal samples were taken from the same coal seam at the same location. Five samples of coal with different damage types were collected, each with a mass of 5 kg, and were sealed and sent to a laboratory for the preparation of experimental coal samples. According to the experimental requirements, the parameters, such as the hardiness coefficient, true relative density, and proximate analysis of the coal, need to be separately determined. Therefore, the coal sample collected at the site must be processed into a sample that satisfies these requirements. In addition, the gas pressure (1.2 MPa) and coal seam temperature (35 °C) in the N-3 coal seam were measured on site, as shown in Figure 4.

According to "Methods for determining coal hardiness coefficient" [35], a sample for the estimation of the coal hardiness coefficient was prepared as follows: A coal sample of 1000 g was crushed and screened using standard sieves with apertures of 20 mm and 30 mm. Next, 50 g of the prepared sample was weighed into 1 part, with one set for every 5 parts, and a total of 3 groups were weighed. The coal hardiness coefficient to be measured was applied.

According to "Proximate analysis of coal" [36], samples for proximate analysis and estimation were prepared as follows: 500 g of raw coal was crushed and sieved to create samples of coal with a size less than 0.2 mm, and placed in a ground jar for sealing. Three samples were prepared; each sample's weight exceeded 50 g.

According to "Methods for determining the block density of coal and rock" [37], samples for density analysis and determination were prepared as follows: 50 g of raw coal was crushed and sieved to create samples of granular coal with a size less than 0.2 mm, and kept in a grinding jar sealed for use. Three samples were prepared; each sample's weight exceeded 2 g. Figure 5 shows the coal sample processing flow and related test equipment.

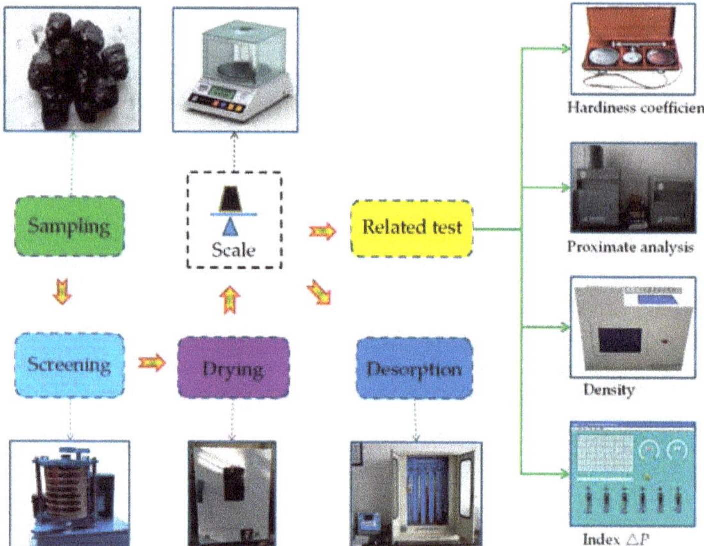

Figure 5. Coal sample preparation processes.

The preparation process of the desorption coal sample is described as follows: The raw coal was crushed and sieved to create 200 g of granules with sizes that ranged between 1 mm to 3 mm sieves. All samples were placed in a dryer at 105 °C for 3 h. After cooling, the coal samples were placed in a container isolated from air and sealed for subsequent use. Details of the coal samples are listed in Table 2.

Table 2. Preparation of coal samples with different specifications.

Term	Particle Size (mm)	Quality (g)	Quantity (parts)
Hardiness coefficient	20~30	50	15
Proximate analysis	<0.2	50	3
Density	<0.2	2	3
Adsorption constant	0.2~0.25	50	1
Desorption property	1~3	200	1

The preparation of coal samples with different specifications was used to determine relevant parameters and the coal sample gas desorption characteristic coefficient (K_t). Based on these indicators, the coal seam outburst risk assessment and coal seam classification could be carried out, and the index K_t could be calculated.

2.3. Experimental Procedure

The coal sample gas desorption process simulation was conducted by employing the experimental device shown in Figure 2. Since the gas desorption environment of the sample was always maintained at a temperature of 30 ± 1 °C and a gas outlet pressure of 0.1 MPa during the measurement process, the gas desorption of the sample could be considered to be an isothermal and isostatic desorption process. Dried coal samples with a weight and particle size ranging from 60 g to 80 g and 1 mm to 3 mm, respectively, were firstly loaded into the coal sample tank. After loading the coal sample, the sample tank was sealed and vacuumed with a water bath temperature of 35 °C until the pressure was less than 20 Pa. Then, methane with a purity of 99.9% was inlet into the diffusion tank with a defined pressure. After that, the sample tank and diffusion tank were connected to permit methane gas flow into the sample tank and begin the adsorption process. The adsorption process was considered finished only when the pressure in both the diffusion and sample tanks stayed constant, and this process usually continued for nearly 48 h. The gas desorption process began after the system pressure remained constant. The amount of desorption gas should be recorded every 30 s, and the test should be stopped after desorption for 30 min. The gas desorption capacity needs to be converted to the standard condition volume, and the conversion equation [38] is as follows:

$$W_t = \frac{273.2}{101325(273.2 + t_w)}(P_{atm} - 9.81h_w - P_S) \cdot W_t' \tag{1}$$

where W_t is the total amount of gas desorption in the standard state (mL), W_t' is the total gas desorption measured in the experimental environment (mL), t_w is the water temperature in the tube ($^\circ$C), P_{atm} is atmospheric pressure (Pa), h_w is the height of the water column in the measuring tube (mm), and P_S is the saturated water vapor pressure (Pa).

3. Experimental Results

3.1. Related Parameter

Proximate analysis is the main indicator for understanding the characteristics and the basic basis for evaluating the metamorphism of coal [36]. Elemental analysis is an important indicator for studying the degree of metamorphism of coal and estimating its carbonized product. It is also the basis for calorific calculation for coal as a fuel in industry. The proximate analysis indexes and elemental analysis indexes of the coal samples are listed in Table 3.

Table 3. Result of proximate analysis indexes and elemental analysis indexes.

Proximate Analysis Indexes							Elemental Analysis Indexes		
M_{ad} (%)	A_{ad} (%)	V_{daf} (%)	$S_{t,d}$ (%)	C_{daf} (%)	$Q_{b,d}$ (MJ/kg)	$G_{R,I}$	H_{daf} (%)	O_{daf} (%)	N_{daf} (%)
1.15	17.67	16.12	0.31	65.06	29.41	11.2	4.30	2.53	1.50

Notes: M_{ad} is the air dry basis moisture (%). A_{ad} is the air dry basis ash (%). V_{daf} is the dry ash-free basis of volatile content (%). $S_{t,d}$ is the true relative density (g/cm^3). $Q_{b,d}$ is the calorific value (MJ/kg). $G_{R,I}$ is the clean coal bond index (dimensionless). C_{daf} is the fixed carbon content (%). H_{daf} is the dry ash-free basis hydrogen content (%). O_{daf} is the dry ash-free basis oxygen content (%). N_{daf} is the dry ash-free basis nitrogen content (%).

In practice, the outstanding predictive index is an important indicator for the identification of outburst-prone coal seams [39]. The characteristic indicator and the adsorption constants are key indicators for quantifying the adsorption-desorption characteristic of coal. Table 4 lists the measured data of the relevant outstanding indicators of N-3 coal, the characteristic indexes, and the coal gas adsorption constants.

Table 4. Relevant indicator measured data.

Outstanding Predictive Indicators				Characteristics Indicators			Adsorption Constants	
f	ΔP (mmHg)	D_{cf}	P (MPa)	TRD (g/cm^3)	ARD (g/cm^3)	n (%)	a_{ac} (cm^3/g)	b_{ac} (MPa^{-1})
0.395	29.1	IV	1.20	1.32	1.23	6.8	29.6786	1.3236

Notes: f is the coal hardiness coefficient (dimensionless). ΔP is the initial velocity of diffusion of coal gas (mmHg). D_{cf} is the degree of coal fracturing (dimensionless), as shown in Table 5. P is the measured coal seam gas pressure (MPa). TRD is the true relative density of the coal sample (g/cm^3). ARD is the apparent relative density of the coal sample (g/cm^3). n is the ratio of the total volume of tiny voids to the total volume of coal (%). a_{ac} and b_{ac} are Langmuir adsorption constants; a_{ac} is the maximum gas adsorption capacity (cm^3/g) and b_{ac} is the adsorption constant (MPa^{-1}).

Table 5. Classification of the degree of coal fracturing.

Class	I	II	III	IV	V
Degree of coal fracturing	Massive coal	Slightly fractured coal	Severely fractured coal	Pulverized coal	Completely pulverized coal

The coal hardiness coefficient (f) reflects the ability of coal to resist damage, and can be employed to predict the ability to resist breakage and it's stability after drilling. When the index (f) exceeds 0.5, the coal has a strong ability to resist outburst. A comparison of Tables 4–6 reveals that f is 0.395 < 0.5, which indicates that coal N-3 is relatively easily destroyed under the gas pressure of 0.74 MPa.

Table 6. Thresholds of four indicators for the identification of outburst-prone coal seams.

Term	D_{cf}	ΔP (mmHg)	f	P (MPa)
Thresholds	III, IV, V	≥10	≤0.5	≥0.74

The index (ΔP) of the initial velocity of the diffusion of coal gas is also one of the indicators for predicting the risk of coal and gas outburst [22,39], which can reflect the degree of gas emission from gas-filled coal bodies. ΔP is 29.1 mmHg > 10 mmHg, which indicates that the coal seam has a rapid dispersion and a strong destruction capability.

Porosity (n) refers to the ratio of the mass of a certain substance contained in a certain volume to the mass and volume of the same substance at a specified temperature. The porosity of coal is not only an important index for measuring the development status of pores and cracks in coal, but also an important factor that affects the adsorption and infiltration capacity of coal.

The adsorption constants a_{ac} and b_{ac} were measured using a high-pressure volumetric method to determine the coalbed methane adsorption constants a_{ac} and b_{ac}. The adsorption constants a_{ac} and b_{ac} are calculated from the Langmuir adsorption equation [32,33] as follows:

$$Q = \frac{a_{ac}b_{ac}P}{1 + b_{ac}P} \tag{2}$$

where Q is the adsorption gas quantity (mL/g), P is the adsorption equilibrium gas pressure (MPa), and a_{ac} and b_{ac} are the Langmuir adsorption constants.

a_{ac} and b_{ac} are determined by the amount of coal gas sample adsorbed under different pressures. Therefore, the gas adsorption constant of the coal is an indicator of coal gas adsorption capacity. The physical meaning of a_{ac} is the maximum gas adsorption capacity of coal.

3.2. Gas Desorption Process

The gas adsorption capacity of the coal samples was calculated based on the gas adsorption equilibrium pressures and the results for the related parameters, by using the Langmuir adsorption equation. The simulation results are shown in Figure 6.

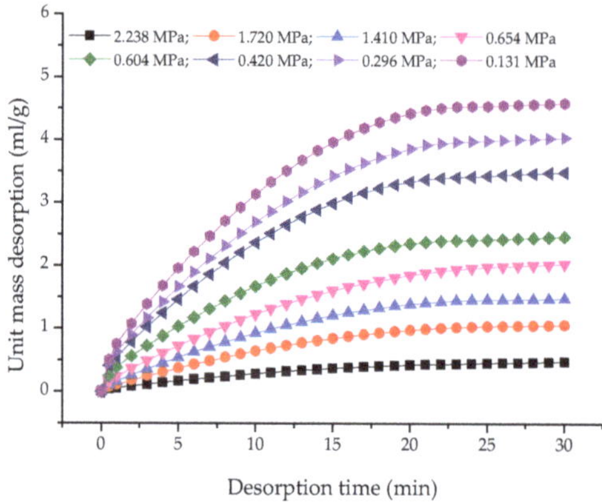

Figure 6. Desorption curves under different adsorption equilibrium gas pressures.

As shown in Figure 6, with an increase in the gas pressure, the coal samples with a unit mass are located at the same time point. The amount of desorption gas also gradually increases, but its increase decelerates, which mainly resulted from the primary limitation of the internal pores of the coal body. In addition, the slope of the gas desorption curve decreases and gradually flattens as the exposure times of the coal samples increase, which can be attributed to the decrease of adsorption gas and the complexity of the pore structure of coal [14,16]. During the initial stage of exposure, the gas concentration in the large pores of the coal body was released outward, and the resistance to gas migration was large; whereas in the later stage, the primary factor for determining the gas release rate was the large number of pores and fractures in the coal. The diffusion coefficient of gas will be reduced by one to two orders of magnitude.

Figure 7 shows the relationship between the gas desorption amount and the time for different gas pressures.

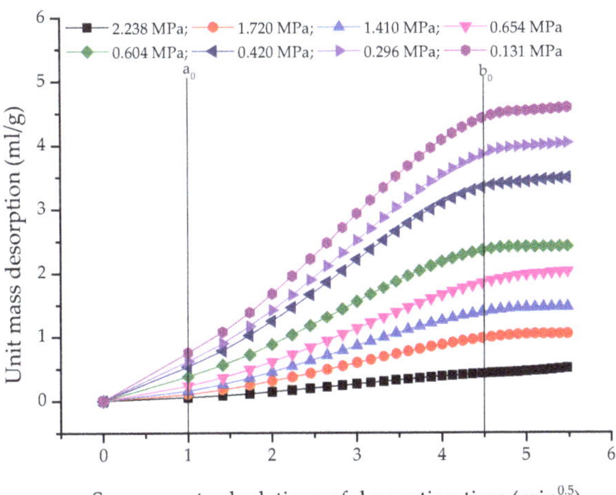

Figure 7. Relationship between the gas desorption amount and the square root of desorption time.

As shown in Figure 7, the gas desorption amount and the square root of the gas desorption time are linear. However, with the extension of the exposure time of the coal sample, the slope of the straight line will slightly decrease. The square root of the distinct turning point at the gas desorption time is $1 \text{ min}^{0.5}$ and $4.5 \text{ min}^{0.5}$; since the start of desorption, the slope of the line gradually increased and then decreased. The square root of the gas desorption time exceeds $4.5 \text{ min}^{0.5}$. The slope of the straight line is less than the slope of the straight line at the initial stage and gradually decreases; the square root of the gas desorption time falls between $1 \text{ min}^{0.5}$ and $4.5 \text{ min}^{0.5}$. As the gas pressure in the coal seam increases, the slope of the straight line increases. According to the calculation model of Winter [29], the change in the gas desorption rate with time can be expressed by an exponential equation [31] for certain other conditions as follows:

$$V_t - V_a \left(\frac{t}{t_a} \right)^{K_t} \tag{3}$$

Then, Equation (4) can be derived as follows:

$$K_t = \frac{Ln V_a - Ln V_t}{Ln t - Ln t_a} \tag{4}$$

Here, K_t is the gas desorption characteristic coefficient whose exposure time ranges from 1 min to 5 min $(mL/(g \cdot min^{0.5}))$. V_t and V_a are the gas desorption speed of coal samples with unit mass at the time t and t_a, respectively (cm^3/min). t and t_a are the gas desorption time and time in min, respectively.

The results were converted to the amount of gas desorbed from the coal sample, and the index (K_t) was determined by the least squares method. The results are listed in Table 7.

Table 7. Gas adsorption capacity and gas desorption characteristic coefficient (K_t).

P (MPa)	Q (mL/g)	K_1 (mL/(g·min$^{0.5}$))	K_2 (mL/(g·min$^{0.5}$))	K_3 (mL/(g·min$^{0.5}$))	K_5 (mL/(g·min$^{0.5}$))
0.131	4.72	0.1467	0.1380	0.1268	0.1202
0.296	6.92	0.2895	0.2644	0.2582	0.2292
0.420	8.20	0.3660	0.3568	0.3533	0.3274
0.604	9.70	0.4891	0.4434	0.4059	0.3852
0.653	10.04	0.5405	0.5067	0.4764	0.4326
1.410	13.40	0.7702	0.7176	0.6783	0.6283
1.720	14.20	0.8428	0.7897	0.7148	0.6470
2.238	15.19	0.9395	0.8846	0.8555	0.7991

Notes: P is the gas pressure (MPa). Q is the adsorption gas quantity (mL/g). K_t is the gas desorption characteristic coefficient whose exposure time ranges from 1 min to 5 min (mL/(g·min$^{0.5}$)).

As shown in Table 7, under the same adsorption-balanced gas pressure, different desorption characteristic coefficient values gradually decrease from K_1 to K_5 with the exposure time of the experimental coal sample; that is, $K_1 > K_2 > K_3 > K_4 > K_5$. First, this result is attributed to the gradual decrease of adsorbed gas and the decrease of the amount of available desorption gas. Second, with the accumulation of the amount of desorption gas in fixed space, the gas pressure in the fixed space and the pressure gradient in the coal gas gradually decrease. For the same gas desorption characteristic coefficient, K_t gradually increases with an increase in the adsorption equilibrium gas pressure. The larger the adsorption equilibrium gas pressure, the larger the amount of gas adsorbed by the coal sample under the larger adsorbed gas pressure gradient. When the gas is desorbed into the fixed space, the larger the gas pressure gradient between the fixed space and the coal sample, the larger the amount of gas desorption per unit time.

As the exposure time increases, the gas pressure gradients between adsorbed-balance gas and fixed-space cumulative gas gradually decrease, and the amount of desorption gas will gradually decrease during the unit exposure time. Therefore, K_t can be considered to be a reflection of the physical quantity of the gas desorption speed at different times.

4. Discussion

4.1. Relationship between Gas Pressure and K_t

The relationships between gas pressure and gas desorption characteristic coefficients are shown in Figure 8.

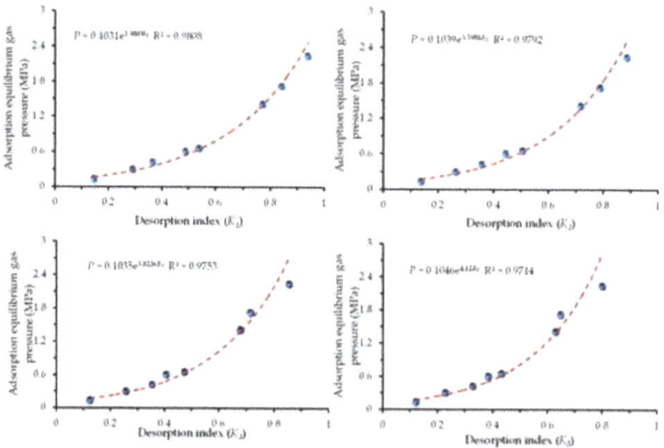

Figure 8. Relationships between gas pressure and gas desorption characteristic coefficients.

As shown in Figure 8, the gas desorption characteristic coefficient (K_t) also increases, and the increase in amplitude is gradually increased with gas pressure. Because coal is a natural adsorbent, the larger the adsorption pressure, the larger the amount of gas adsorption and the larger the amount of gas that is desorbed [31]. For the same gas desorption characteristic coefficient, the index (K_t) gradually increases with an increase in the adsorption equilibrium gas pressure. According to the adsorption theory of Langmuir [33], under the action of the larger adsorption equilibrium gas pressure, the coal sample absorbs a larger amount of gas. When the gas is desorbed into the fixed space, a larger pressure gradient of desorption gas is generated between the fixed space and the coal sample to promote coal adsorption equilibrium gas desorption.

For the coal seam adsorption equilibrium gas pressure and the different K_t respectively, trend fitting is available. The adsorption equilibrium gas pressure and the different K_t are exponential equation relations and have good correlation, the coefficient of determination (R^2) being higher than 0.97.

A comparison of the correlation curve of the gas pressure and K_t at different exposure times is shown in Figure 8. The results indicate that R^2 decreases to a minor extent with the exposure time of the coal sample due to the deviation of the gas desorption amount error caused by the increase in the exposure time [16,40]. However, the R^2 of the coal sample gas desorption regression fitting curve remains greater than 0.97, the correlation of regression fitting curve is higher, and the result is reliable. Therefore, the gas pressure can be expressed as follows:

$$P = A_c e^{B_c K_t} \tag{5}$$

where A_c and B_c are the constants that correspond to different desorption times, which are dimensionless; P is the adsorption equilibrium gas pressure (MPa); and K_t is the gas desorption characteristic coefficient that corresponds to different desorption times (mL/(g·min$^{0.5}$)).

4.2. Relationship between Gas Content and K_t

The relationships between gas content and gas desorption characteristic coefficients are shown in Figure 9.

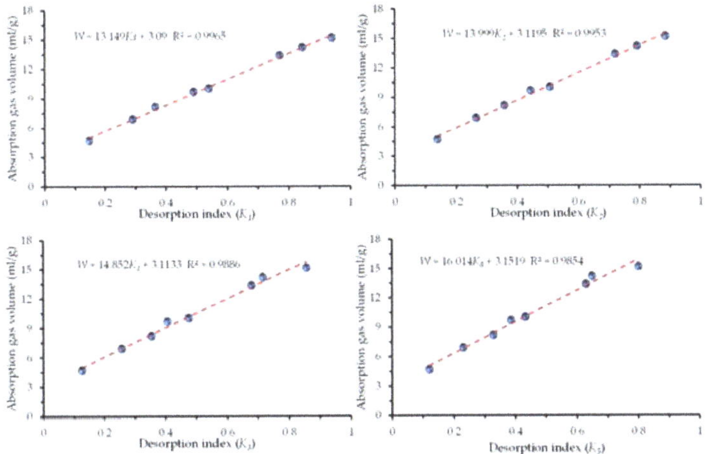

Figure 9. Relationships between gas content and gas desorption characteristic coefficients.

As shown in Figure 9, the gas content increases as K_t increases. Coal is a natural adsorbent with double pores and fissures. The larger the gas content, the larger K_t is. According to the adsorption theory equation of Langmuir [32], the larger the gas content, the larger the gas pressure, and the larger the amount of adsorbed coal gas. The larger the amount of adsorption gas in the coal sample, the larger the index K_t.

The coal seam gas content and K_t have a linear equation relationship and an excellent correlation. The correlation coefficient of the regression fitting curve showed a slight decrease with the exposure time. With an increase in desorption time, the deviation of the desorption amount of the coal sample gas gradually increases. However, R^2 remains greater than 0.98, which means the regression fitting curve has higher correlation and the result is reliable. Therefore, the relationship between the gas content and K_t can be expressed by Equation (6) as follows:

$$W = \partial K_t + \beta \tag{6}$$

where ∂ and β are the constants that correspond to different desorption times (dimensionless), W is the gas adsorption amount (mL/g), and K_t is the gas desorption characteristic coefficients that correspond to different desorption times (mL/(g·min$^{0.5}$)).

According to the adsorption theory equation of Langmuir [32], the gas pressure and gas content are not linear. To analyze the accuracy of the relationship between the coal seam gas pressure, and gas content and K_t for different exposure times, R^2 at different times is compared and listed in Table 8.

Table 8. Comparison of the coefficient of determination (R^2) under different exposure times.

Term	$P = A_c e^{BcKt}$	$W = \partial K_t + \beta$	Exposure Time (min)
	0.98079	0.99653	1
R^2	0.97922	0.99531	2
	0.97535	0.98858	3
	0.97140	0.98542	5
average of R^2	0.97669	0.99146	

As shown in Table 8, for R^2, the K_t used to describe the coal sample gas content and gas pressure at different exposure times can reach a high accuracy, especially when K_t is used to describe the gas content. The maximum R^2 is 0.99146.

Although, with the extension of exposure time, the K_t to describe coal gas pressure and gas content has a certain decrease; the decrease range is very small within 5 min, and the effect on the accuracy of the results is negligible.

4.3. Technical Verification

4.3.1. Experimental Verification

Based on the experimental data, the method proposed in this paper was used to calculate and revise the gas content of the coal samples under two different pressures after exposure for 1 min, 2 min, 3 min, and 5 min. The calculation results of the gas content and the calculation deviation are listed in Table 9.

Table 9. Calculated values and deviation of gas content.

Gas Pressure (MPa)	Gas Content (mL/g)	Exposure Time (min)	K_t (mL/(g·min$^{0.5}$))	Calculated Gas Content (mL/g)	Deviation (%)
		1	0.4634	9.18	−7.73
0.64	9.95	2	0.4232	9.04	−9.14
		3	0.4114	9.22	−7.33
		5	0.3860	9.33	−6.23
		1	0.6309	11.39	−1.04
0.912	11.51	2	0.5351	10.61	−7.81
		3	0.5137	10.74	−6.68
		5	0.5262	11.58	0.60

According to Table 9, the gas content calculated deviation range was from −9.14% to 0.6%, with an average of only 3.23%, which indicates that the new method can accurately calculate the gas content of coal samples. In addition, for the conditions of different adsorption equilibrium gas pressures, the calculated and measured gas content values are the smallest when the coal sample is exposed for 5 min, which indicates that the longer the exposure time, the closer the calculated gas content is to the measured value.

4.3.2. Field Verification

To test the reliability of the new method, the gas content of the N-3 coal seam in the LBS was measured.

The test sites were the E2305 inlet and the northern inlet of the LBS. The downhole gas desorption apparatus was used to directly measure the downhole desorption section of the coal sample. The underground gas desorption operation flow chart is shown in Figure 10. A total of six groups were tested, of which E2305 entered the air in 5 samples and the north inlet entered the air in 1 sample. The gas content measurement and calculation results are listed in Table 10.

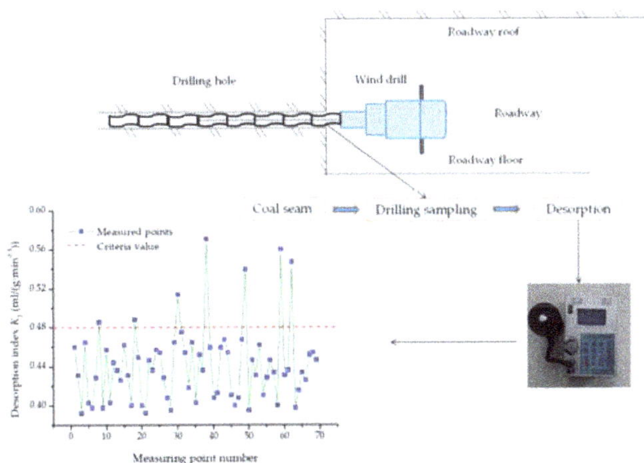

Figure 10. On-site gas desorption flow chart.

Table 10. Comparison of measured and calculated values of gas content.

Number	K_5 (mL/(g·min$^{0.5}$))	Calculated Values (mL/g)	Measured Values (mL/g)	Deviation (%)
1	0.4153	9.80	10.10	−2.97
2	0.3925	9.43	9.40	0.32
3	0.3611	8.93	9.41	−5.1
4	0.4385	10.17	10.71	−5.0
5	0.5186	11.45	10.93	4.76
6	0.40225	9.59	10.52	−8.84

Note: coal sample exposure time is 5 min.

As shown in Table 10, using the new method to measure and calculate the gas content of the coal seam at two different working faces of the LBS, the deviation of the calculated gas content ranged from 0.32% to 8.84%, with an average of only 4.49%. The main reasons for this finding are as follows: The coal sample is impure when it is collected by the method of powdered coal through holes due to coal expansion. Therefore, as the exposure time of the shallow coal bodies increases and the desorption rate of gas decreases, the K_5 calculated from the desorption law is also smaller than the actual K_5, which directly causes the calculated gas content of the new method to be smaller than the real gas content. When the indirect method is used to determine the gas content, the desorption rate that is measured at the site is less than the real desorption rate and will only affect the calculation of the loss and a part of the desorption amount, and the influence on the gas content of the raw coal is small.

Temperature has a significant effect on the gas adsorption and desorption of coal. The isothermal adsorption-desorption experiment is performed in the condition of the coal seam temperature, which is easily controlled. In field applications, there is a temperature difference between the coal seam and the roadway due to the cooling effect of the roadway; that is, the temperature of the coal sample changed when it was removed from the coal seam. This weak temperature change will have a certain influence on the coal gas content measurement results. From the point of view of field applications, the temperature difference between the coal seam and the roadway has a minimal effect on the final result of the new method.

The difference between the measured value and the calculated value using the new method of the gas content of coal sample exposure within 5 min is not distinct. The results conclude that the new method can accurately estimate the gas content of the coal seam in a field application, and the accuracy satisfies engineering needs. Therefore, the "calculation model and rapid estimation method of coal seam gas content" can be implemented in the field.

5. Conclusions

This paper is based on the analysis of simulation results in gas desorption and applies a field application for the investigation. A set of independently developed experimental apparatus was used to measure the gas desorption process of coal with a particle size of 1–3 mm in the N-3 coal seam of the LBS to research the relationship between the gas desorption law and the gas content. According to the specific exposure time of the gas desorption, the eigenvalues of the rules, and the establishment of a method for the rapid estimation of gas content in coal seams, the following main conclusions are obtained:

(1) The gas desorption amount and the square root of the gas desorption time are linear, and the slope of the straight line will slightly decrease with the extension of the exposure time. The slope of the straight line is less than the slope of the straight line at the initial stage and gradually decreases; the square root of the gas desorption time falls between $1~\text{min}^{0.5}$ and $4.5~\text{min}^{0.5}$. As the gas pressure increases, the slope of the straight line increases.

(2) Simulation and verification of the on-site gas desorption law verified that the gas pressure and gas content of coal seams and K_t have an exponential equation and a linear equation relationship, respectively. Using this equation relationship, a new method for accurately calculating the gas content of underground coal seams is constructed.

(3) Simulation experiments determined that the exposure time of the coal sample should be controlled within 5 min when using the new method to calculate gas content in a coal seam. The calculation equations at 1 min, 2 min, 3 min, and 5 min were given. The method can be used to calculate the coal seam gas content, and the deviation is within the allowable range of the project. Thus, this method can satisfy the needs of rapid gas content estimation at the site.

In this paper, the method for the calculation model and the rapid estimation of the gas content is simple and concise with respect to the operation and measuring accuracy. Changes in the ambient temperature of the test site will have an impact on the accuracy of the final result during field application, but this effect is negligible.

Author Contributions: Conceptualization, Y.L.; Methodology, X.Z.; Validation, F.W.; Formal Analysis, X.L.; Data Curation, Y.C.; Writing-Original Draft Preparation, F.W.; Writing-Review and Editing, X.L.; Funding Acquisition, Y.L.

Funding: This research received no external funding.

Acknowledgments: This study is financially supported by the National Science and Technology Major Project of China (Grant No. 2016ZX05043005), the State Key Research Development Program of China (Grant No. 2016YFC0801404 and 2016YFC0801402), and the National Natural Science Foundation of China (51674050), which are gratefully acknowledged. The authors also thank the editor and anonymous reviewers for their valuable advice.

Nomenclature

Q_t	cumulative amount of desorbed gas from time $t = 0$ to time t, mL/g
Q_∞	ultimate adsorption-desorption gas amount, mL/g
S	unit mass sample outer surface area, cm^2/g
V	unit mass volume of coal sample, mL/g
D	diffusion coefficient, cm^2/min
v_1	gas desorption speed at $t = 1$ min, mL/(g·min)
k_1	characteristic coefficient of gas desorption speed change
v_0	gas desorption speed at $t = 0$ min, mL/(g·min)
A	cumulative gas desorption amount from start to time t, mL/g
B	desorption constant, dimensionless
a, i	constants related to the gas content and structure of coal, dimensionless
b	gas desorption speed decay coefficient with time, dimensionless
W_t	total amount of gas desorption in the standard state, mL/g
W_t'	total gas desorption measured in the experimental environment, mL/g
t_w	water temperature in the tube, °C
P_{atm}	atmospheric pressure, MPa
h_w	height of the water column in the measuring tube, mm
P_S	saturated water vapor pressure, MPa
V_t, V_a	gas desorption speeds of the coal samples with unit mass at time t and t_a
t, t_a	gas desorption time and time in min
K_t	gas desorption characteristic coefficient whose exposure time ranges from 1 min to 5 min
M_{ad}	air dry basis moisture, %
A_{ad}	air dry basis ash, %
V_{daf}	dry ash-free basis of volatile content, %
$S_{t,d}$	true relative density, g/cm^3
$Q_{b,d}$	calorific value, MJ/kg
$G_{R,I}$	clean coal bond index, dimensionless
C_{daf}	fixed carbon content, %
H_{daf}	dry ash-free basis hydrogen content, %
O_{daf}	dry ash-free basis oxygen content, %
N_{daf}	dry ash-free basis nitrogen content, %
f	coal hardiness coefficient, dimensionless
ΔP	initial velocity of diffusion of coal gas, mmHg
D_{cf}	degree of coal fracturing, dimensionless
P	measured coal seam gas pressure, MPa
TRD	true relative density of the coal sample, g/cm^3
ARD	apparent relative density of the coal sample, g/cm^3
n	ratio of the total volume of tiny voids to the total volume of coal, %
a_{ac}	maximum gas adsorption capacity, cm^3/g
b_{ac}	adsorption constant, MPa^{-1}
Q	adsorption gas quantity, mL/g

References

1. Zhao, H.; Lai, Z.; Firoozabadi, A. Sorption Hysteresis of Light Hydrocarbons and Carbon Dioxide in Shale and Kerogen. *Sci. Rep.* **2017**, *7*, 16209. [CrossRef] [PubMed]
2. Dejam, M.; Hassanzadeh, H.; Chen, Z. Semi-analytical solutions for a partially penetrated well with wellbore storage and kkin effects in a double-porosity system with a gas cap. *Transp. Porous Media* **2013**, *100*, 159–192. [CrossRef]
3. Zou, Y.H.; Zhang, Q.H. Technical progress of direct determination of coal seam gas content in coal mines in China. *Min. Saf. Environ. Prot.* **2009**, *36*, 180–183. [CrossRef]

4. Jiang, W.Z.; Qin, Y.J.; Zou, Y.H.; Dong, J.; Xue, J.F.; Wu, Q.; Zhang, Q.H.; Zhang, S.T. *The Direct Method of Determining Coalbed Gas Content in the Mine*; The General Administration of Quality Supervision, Inspection and Quarantine of the People's Republic of China & the National Standardization Administration of China: Beijing, China, 2009; pp. 3–11.

5. Jin, Z.; Firoozabadi, A. Phase behavior and flow in shale nanopores from molecular simulations. *Fluid Phase Equilib.* **2016**, *430*, 156–168. [CrossRef]

6. Bertard, C.; Bruyet, B.; Gunther, J. Determination of desorbable gas concentration of coal direct method. *Rock Mech. Min. Sci.* **1970**, 43–65. [CrossRef]

7. Mcculloch, C.M.; Levine, J.R.; Kissell, F.N.; Deul, M. *Measuring the Methane Content of Bituminous Coalbed*; Department of the Interior Bureau of Mines Ri: Pittsburgh, PA, USA, 1975; p. 8043.

8. Ulery, J.P.; Hyman, D.M. The Modified Direct Method of Gas Content Determination: Application and Results. In Proceedings of the Coalbed Methane Symposia, The University of Alabama, Tuscaloosa, AL, USA, 17–21 May 1991; pp. 489–500.

9. Mavor, M.J.; Pratt, T.J. *Improved Methodology for Determining Total Gas Content, Vol. II. Comparative Evaluation of the Accuracy of Gas-in-Place Estimates and Review of Lost Gas Models*; Gas Research Inst., Topical Rep: Chicago, IL, USA, 1996.

10. Abate, A.F.; Nappi, A.; Narducci, F.; Ricciardi, S. A mixed reality system for industrial environment: An evaluation study. *CAAI Trans. Intell. Technol.* **2017**, *2*, 227–245.

11. Saghafi, A.; Williams, D.J.; Roberts, D.B. *Deternination of Coal Gas Content by Quick Crushing Method*; CSIRO Investigation Report: Canberra, Australia, 1995.

12. Smith, D.M.; Williams, F.L. A new technique for determining the methane content of coal. In Proceedings of the 16th Intersociety Energy Conversion Engineering Conference, New York, NY, USA, 9–14 August 1981; pp. 1272–1277.

13. Chase, R.W. *A Comparison of Methods Used for Determining the Natural Gas Content of Coalbeds from Exploratory Cores*; US Department of Energy: Washington, DC, USA, 1979.

14. Sawyer, W.K.; Zuber, M.D.; Kuuskraa, V.A.; Homer, D.M. Using reservoir simulation and field data to define mechanisms controlling coalbed methane production. In Proceedings of the Coalbed Methane Symposia, The University of Alabama, Tuscaloosa, AL, USA, 16–19 November 1987; pp. 295–307.

15. Chen, Y.; Yang, D.; Tang, J.; Li, X.; Jiang, C. Determination method of initial gas desorption law of coal based on flow characteristics of convergent nozzle. *J. Loss Prev. Process Ind.* **2018**, *54*, 222–228. [CrossRef]

16. Chen, J.W.; Zhang, R.L.; Chai, L. Analysis of sampling loss in the measurement of original gas content in coal seam with downhole desorption method. *Saf. Coal Mines* **2010**, *41*, 84–86. [CrossRef]

17. Lei, X.R.; Li, H.L.; Hou, H.H. Discussion on the calculation of desulfuration in detecting gas content in coal seam by desorption method. *Coal Mine Mod.* **2012**, *1*, 57–59. [CrossRef]

18. Zhang, X.Y.; Guo, M.Z.; Song, C.Y.; Gao, L.Q. Comparison and Analysis of Three Methods for Calculating Gas Loss in Coal Seam Gas Content Determination by Desorption Method. *Saf. Coal Mines* **2012**, *43*, 177–185. [CrossRef]

19. Li, Y.C.; Yang, S.Q. Comparative Analysis of Calculation Methods of Gas Loss. *Saf. Coal Mines* **2012**, *43*, 166–168. [CrossRef]

20. Wang, J.; Kong, B.; Mei, T.; Wei, H. A lane detection algorithm based on temporal–spatial information matching and fusion. *CAAI Trans. Intell. Technol.* **2017**, *2*, 275–291.

21. Scott, A.R. Hydrogeologic factors affecting gas content distribution in coal beds. *Int. J. Coal Geol.* **2002**, *50*, 363–387. [CrossRef]

22. Li, X.L.; Wang, E.Y.; Li, Z.H.; Liu, Z.T.; Song, D.Z.; Qiu, L.M. Rock burst monitoring by integrated microseismic and electromagnetic radiation methods. *Rock Mech. Rock Eng.* **2016**, *49*, 4393–4406. [CrossRef]

23. Liu, Y.; Wang, D.; Hao, F.; Liu, M.J.; Mitri, H.S. Constitutive model for methane desorption and diffusion based on pore structure differences between soft and hard coal. *Int. J. Min. Sci. Technol.* **2017**, *27*, 937–944. [CrossRef]

24. Cao, J.; Sun, H.T.; Dai, L.C.; Sun, D.L.; Wang, B.; Miao, F.T. Simulation research on dynamic effect of coal and gas outburst. *J. China Univ. Min. Technol.* **2018**, *47*, 113–120. [CrossRef]

25. Li, Z.Q.; Cheng, Q.; Liu, Y.W.; Duan, Z.P.; Song, D.Y. Research on gas diffusion model and experimental diffusion characteristic of cylindrical coal. *J. China Univ. Min. Technol.* **2017**, *46*, 1033–1040. [CrossRef]

26. Tang, J.; Jiang, C.; Chen, Y.; Li, X.; Wang, G.; Yang, D. Line prediction technology for forecasting coal and gas outbursts during coal roadway tunneling. *J. Nat. Gas Sci. Eng.* **2016**, *34*, 412–418. [CrossRef]

27. Yan, G.Q.; Wang, G.; Xin, L.; Du, W.Z.; Huang, Q.M. Direct fitting measurement of gas content in coalbed and selection of reasonable sampling time. *Int. J. Min. Sci. Technol.* **2017**, *27*, 299–305. [CrossRef]

28. Barrer, R.M. Gas flow in solids. *Philos. Mag.* **1939**, *28*, 148–162. [CrossRef]

29. Winter, K.; Janas, H. Gas emission characteristics of coal and methods of determining the desorbable gas content by means of desorbometers. In Proceedings of the 16th International Conference on Coal Mine Safety Research, Washington, DC, USA, 22–26 September 1996.

30. Wang, Y.A.; Yang, S.J. Some characteristic of coal seams with hazard of outburst. *J. Chin. Chem. Soc.* **1980**, *1*, 47–53. [CrossRef]

31. Kang, J.N. Application of direct determination technology of coal seam content in gas drainage effect evaluation of Zhongliang coal mine. *Min. Saf. Environ. Prot.* **2010**, *37*, 31–33. [CrossRef]

32. Yu, Q.X. *Coal Mine Gas Control*; China University of Mining and Technology Press: Xuzhou, China, 2012; pp. 34–37. ISBN 978-7-5646-1607-6.

33. Zou, Q.L.; Lin, B.Q.; Liu, T.; Zhou, Y.; Zhang, Z.; Yan, F.Z. Variation of methane adsorption property of coal after the treatment of hydraulic slotting and methane pre-drainage: A case study. *J. Nat. Gas Sci. Eng.* **2014**, *20*, 396–406. [CrossRef]

34. Sun, G.; Li, Y.H.; Wu, K.H.; Gong, X.L.; Tian, X.H.; Pan, M.G. *Sampling of Coal Seams*; The General Administration of Quality Supervision, Inspection and Quarantine of the People's Republic of China & the National Standardization Administration of China: Beijing, China, 2008; pp. 1–4.

35. Qi, Q.X.; Li, J.Q.; Mao, D.B.; Fu, J.X.; Zhang, X.L. *Methods for Determining Coal Hardiness Coefficient*; The General Administration of Quality Supervision, Inspection and Quarantine of the People's Republic of China & the National Standardization Administration of China: Beijing, China, 2010; pp. 2–3.

36. Han, L.T.; Lin, Y.J.; Chen, K.Q. *Proximate Analysis of Coal*; The General Administration of Quality Supervision, Inspection and Quarantine of the People's Republic of China & the National Standardization Administration of China: Beijing, China, 2008; pp. 1–9.

37. Qi, Q.X.; Li, J.Q.; Mao, D.B.; Fu, J.X.; Zhang, X.L. *Methods for Determining the Block Density of Coal and Rock*; The General Administration of Quality Supervision, Inspection and Quarantine of the People's Republic of China & the National Standardization Administration of China: Beijing, China, 2009; pp. 1–6.

38. Zhang, Q.L.; Zhang, S.A. *Experimental Method of High-Pressure Adsorption Isothermal to Coal—Capacity Method*; The General Administration of Quality Supervision, Inspection and Quarantine of the People's Republic of China & the National Standardization Administration of China: Beijing, China, 2008; pp. 1–5.

39. Hu, Q.T.; Zhao, X.S.; Zou, Y.H.; Li, Q.L.; Kang, J.N.; Zhang, Q.H.; Lei, H.Y. *Specification for Identification of Coal and Gas Outburst Mine*; State Administration of Work Safety: Beijing, China, 2006; pp. 1–6.

40. Li, X.L.; Li, Z.H.; Wang, E.Y.; Liang, Y.P.; Li, B.L.; Chen, P.; Liu, Y.J. Pattern recognition of mine microseismic (MS) and blasting events based on wave fractal features. *Fractals* **2018**, *26*, 1850029. [CrossRef]

Article

The Influence of Sorption Pressure on Gas Diffusion in Coal Particles: An Experimental Study

Xin Yang [1,2,3], Gongda Wang [2,3,4,*], Junying Zhang [2,3] and Ting Ren [4]

[1]   Faculty of Resources and Safety Engineering, China University of Mining and Technology Beijing, Beijing 100083, China; tbp1600101026@student.cumtb.edu.cn
[2]   Mine Safety Technology Branch of China Coal Research Institute, Beijing 100013, China; zhjy369@163.com
[3]   State Key Laboratory of Coal Mining and Clean Utilization, China Coal Research Institute, Beijing 100013, China
[4]   School of Civil, Mining and Environmental Engineering, University of Wollongong, Wollongong, NSW 2500, Australia; rtxuow@hotmail.com
*   Correspondence: wanggongda521@163.com

Received: 10 February 2019; Accepted: 9 April 2019; Published: 16 April 2019

Abstract: Gas pressure changes during the process of coal mine gas drainage and CBM recovery. It is of great importance to understand the influence of sorption pressure on gas diffusion; however, the topic remains controversial in past studies. In this study, four samples with different coal ranks were collected and diffusion experiments were conducted under different pressures through the adsorption and desorption processes. Three widely used models, i.e., the unipore diffusion (UD) model, the bidisperse diffusion (BD) model and the dispersive diffusion (DD) model, were adopted to compare the applicability and to calculate the diffusion coefficients. Results show that for all coal ranks, the BD model and DD model can match the experimental results better than the UD model. Concerning the fast diffusion coefficient D_{ae} of the BD model, three samples display a decreasing trend with increasing gas pressure while the other sample shows a V-type trend. The slow diffusion coefficient D_{ie} of BD model increases with gas pressure for all samples, while the ratio β is an intrinsic character of coal and remains constant. For the DD model, the characteristic rate parameter k_Φ does not change sharply and the stretching parameter α increases with gas pressure. Both D_{ae} and D_{ie} are in proportion to k_Φ, which reflect the diffusion rate of gas in the coal. The impacts of pore characteristic on gas diffusion were also analyzed. Although pore size distributions and specific surface areas are different in the four coal samples, correlations are not apparent between pore characteristic and diffusion coefficients.

Keywords: gas diffusion; gas pressure; unipore diffusion model; bidisperse diffusion model; dispersive diffusion model

1. Introduction

During the process of coal mine gas drainage and CBM recovery, the gas flow process can be divided into two stages. First, driven by the concentration gradient force, the gas adsorbed on the surface of coal matrix desorbs and then diffuses into the fracture/cleat system of coal. Second, the dissociative state gas permeates to the surface well or the underground borehole driven by the pressure gradient force. Therefore, two key factors that affect net gas movement result are the gas diffusion coefficient and the gas permeability in the fracture. The diffusion coefficient represents the essential parameter of diffusibility and related studies show that it could be affected by temperature [1,2], moisture [3], pressure [4,5], gas type [6–8], sample size [6,9,10], and coal sample features [11,12]. It should be noted that the coal seam gas pressure is in the dynamic condition during gas extraction. Hence, it is of great significance to understand the impact of pressure on the gas diffusion coefficient. Several research papers on this topic have been conducted but arguments can be found on how gas pressure impacts the diffusion coefficients.

For example, some scholars believe that the diffusion coefficient is in direct proportion to gas pressure. Charrière et al. [2] used CH_4 and CO_2 to conduct the adsorption kinetics experiments when the pressure is equal to 0.1 MPa and 5 MPa respectively. They found that the diffusion coefficient increases with gas pressure. Pan et al. [3] performed CH_4 adsorption/desorption diffusion test within 0~4 MPa pressures range, and results show a direct ratio between diffusion coefficient and gas pressure. Jian et al. [13] carried out the desorption experiments within 0~4.68 MPa pressure range and the conclusion remains the same. However, some scholars reckon that the diffusion coefficient decreases with the increase in pressure. Cui et al. [8] found that the diffusion coefficient of CO_2 reduces when gas pressure is smaller than 3.6 MPa. Staib et al. [4] conducted the adsorption kinetics experiments and analyzed the results using the BD model. It was found that the diffusion coefficient D_a lowers when the pressure increases. Shi et al. [14] tested the influence of CO_2 injection on microporous diffusion coefficient after the adsorption of CH_4 was balanced. Findings show that the increasing injection pressure of CO_2 would cause the reduction of micropore diffusion coefficient. There are also a few scholars who concluded that gas pressure has small effects on the diffusion coefficient. Nandi et al. [15] conducted CH_4 adsorption/desorption experiments on bituminous and anthracite coals and they did not find an apparent relationship between pressure and gas sorption rate. To summarize, the research outcomes are listed in Table 1.

Table 1. Summary of diffusion coefficient changing trend with the increase in pressure.

Author	Coal Sample	Gas Category	Diffusion Model	Pressure Range	Diffusion Coefficient Changing Trend
Delphine Charrière	Bituminous coal	CO_2	UD model	0.1 MPa, 5 MPa	Increase
Pan Zhejun	Bituminous coal	CH_4, CO_2	BD model	0~4 MPa	CH_4 increases and CO_2 remains unchanged
Jian Xin	Bituminous coal	CO_2	UD model	0~4.68 MPa	Increase
Cui X.J	Bituminous coal	CH_4, CO_2, N_2	UD model	<3.6 MPa	Decrease
Shi JQ		CH_4-CO_2	UD model	4.2 MPa	D_i decreases
Gregory Staib	Bituminous coal	CO_2	UD model, BD model, FDR model	0~4.5 MPa	D_a decreases and small impact on D_i
Satyendra P. Nandi	Bituminous coal and anthracite coal	CH_4	UD model	0~2.76 MPa	Small impact

By reviewing the previous studies, it can be concluded that the effect of pressure on gas diffusion coefficient remains controversial till now. It is difficult to compare the research outputs horizontally because of the diversified calculation models and experimental methods, such as experimental apparatus, gas pressure and gas type. Moreover, most of the coal samples used in the studies was bituminous coal, because gas diffusibility and diffusion coefficient in coal are closely correlated with the types of coals. It is unknown whether the results stand for coal with different metamorphic grades.

In the present study, we aim at investigating the influence of sorption pressure on gas diffusion and examining which previous finding is more convincible. To guarantee the comparability of the results, all the experiments are carried out under similar gas pressure range. Both adsorption and desorption kinetics test are conducted. Three widely used diffusion models are adopted to analyze the results and eliminate the possible differences induced by diffusion models. Four coal samples with different ranks are collected from typical mining areas in China, the test results are cross-compared to understand does the coal rank have impacts on the relationship between gas pressure and diffusion coefficients.

2. The Diffusion Models

Three widely used diffusion models, i.e., the unipore diffusion (UD) model, the bidisperse diffusion (BD) model and the dispersive diffusion (DD) model, will be used in this study. The expressions and the assumptions are introduced as follows.

2.1. The Unipore Diffusion Model

The UD [13,16,17] model assumes that the coal particle has only one type of pore and the gas diffuses under the concentration gradient between exterior and interior of the coal particle. The UD model is illustrated in Figure 1. Both UD model and BD model follow the following assumptions: (a) the diffusion system is isothermal; (b) the geometric shape of the particle coal is the standard sphere; (c) coal and pore system are isotropic and homogeneous; (d) the pores are incompressible; (e) gas follows the linear adsorption rule; (f) gas diffusion in pores is in line with Fick's Second Law. It can be expressed as [18],

$$\frac{m_t}{m_\infty} = 1 - \frac{6}{\pi^2} \sum_{n=1}^{\infty} \frac{1}{n^2} \exp\left(-\frac{Dn^2\pi^2 t}{r^2}\right) \tag{1}$$

m_t in the equation refers to the total gas adsorption/desorption quantity at time t, m_∞ is the total quantity after the gas adsorption/desorption is balanced, r represents the radius of spherical coal particle, D refers to the diffusion coefficient (m^2/s) and the value of $\frac{D}{r^2}$ is defined as the effective diffusion coefficient D_e (1/s).

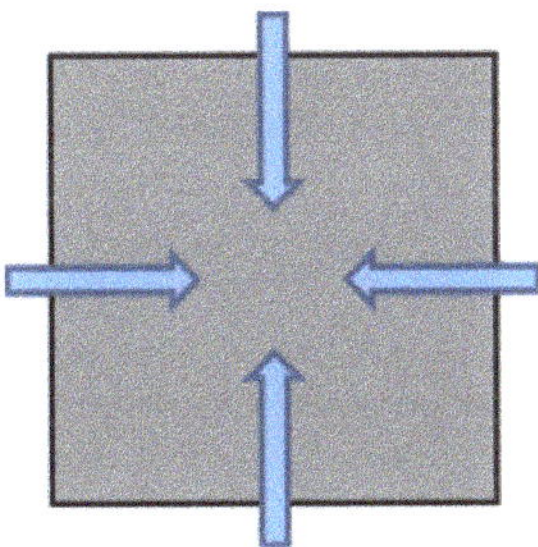

Figure 1. Concepts of gas diffusion under unipore diffusion (UD) Model [19].

2.2. The Bidisperse Diffusion Model

The BD model [5,8,9,14,20,21] assumes that the coal particle includes independent macropore and micropore systems, which are represented by D_a and D_i respectively. The gas diffusion under the two systems are driven by the concentration gradients between exterior and interior of the coal particle. The BD model is illustrated in Figure 2. The simplified BD model includes the fast macropore diffusion stage and the slow micropore diffusion stage [5,22].

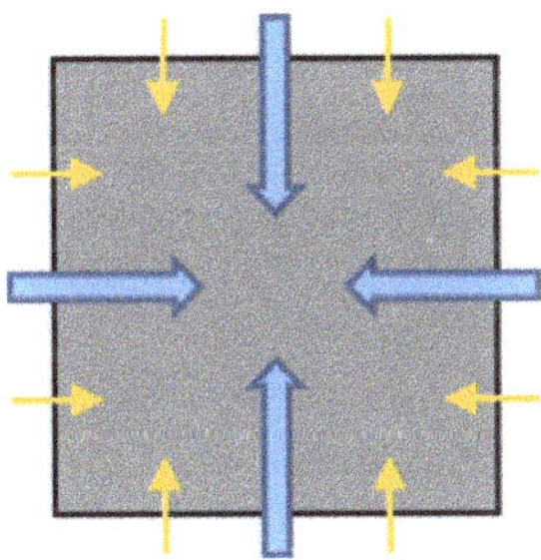

Figure 2. Concepts of gas diffusion under bidisperse diffusion (BD) Model [19].

Concerning the fast macropore diffusion stage, the diffusion model is denoted as,

$$\frac{m_a}{m_{a\infty}} = 1 - \frac{6}{\pi^2}\sum_{n=1}^{\infty}\frac{1}{n^2}\exp\left(-\frac{D_a n^2 \pi^2 t}{r_a^{\,2}}\right) \tag{2}$$

m_a in the equation refers to the total gas adsorption/desorption quantity at time t in the macropore, r_a and D_a represent the radius of macropore spherical coal particle and macropore diffusion coefficient (m²/s), respectively. The value of $\frac{D_a}{r_a^2}$ is defined as the effective diffusion coefficient D_{ae} (1/s).

Concerning the lower micropore diffusion stage, the diffusion model is denoted as,

$$\frac{m_i}{m_{i\infty}} = 1 - \frac{6}{\pi^2}\sum_{n=1}^{\infty}\frac{1}{n^2}\exp\left(-\frac{D_i n^2 \pi^2 t}{r_i^{\,2}}\right) \tag{3}$$

m_i in the equation refers to the total gas adsorption/desorption quantity in the micropore at time t, r_i and D_i represent the radius of micropore spherical coal particle and micropore diffusion coefficient (m²/s), respectively. The value of $\frac{D_i}{r_i^2}$ is defined as the effective diffusion coefficient D_{ie} (1/s).

The BD model is expressed as,

$$\frac{m_t}{m_\infty} = \frac{m_a + m_i}{m_{a\infty} + m_{i\infty}} = \beta\frac{m_a}{m_{a\infty}} + (1-\beta)\frac{m_i}{m_{i\infty}} \tag{4}$$

$\beta = \frac{m_{a\infty}}{m_{i\infty} + m_{a\infty}}$ is the ratio of macropore adsorption/desorption to the total adsorption/desorption.

2.3. The Dispersive Diffusion Model

In recent years, the dispersive diffusion model was developed and it assumes that a distribution of characteristic times for diffusion. The diffusion is dispersed and represents the wide distribution of diffusion feature time. Therefore, theoretically, the DD model can avoid the simplification of pore structure and reflect the real physical experimental process. The DD model is expressed as,

$$\frac{m_t}{m_\infty} = 1 - \exp\left[-\left(k_\varphi t\right)^\alpha\right] \tag{5}$$

m_t in the equation refers to the total gas adsorption/desorption quantity, m_∞ is the total quantity after the gas adsorption/desorption is balanced, k_φ is the characteristic rate parameter, α is the stretching parameter ($0 < \alpha < 1$). The research of Staib et al. [23] shows that α is an intrinsic property of coal and is greatly influenced by the coal pore characteristic.

3. Diffusion Experiments

To carry out the diffusion experiments and analyze the impact of pressure on the methane diffusion, the iSorb HP (Quantachrome) instrument was used. The set maximum adsorption pressure is 6 MPa, the coal sample quality is 40 g and the experimental temperature is 315 K. Coal samples were collected from the HuiChun long frame coal at Jilin Province, Hedong coking coal at Shanxi Province, Xinmi lean coal at Henan Province, and Qinshui meager coal at Shanxi Province (Figure 3), with the four coal samples are denoted as HC, HD, XM and QS, respectively. The coal samples were ground into 0.2~0.25 mm particle samples and the prepared coal particles were dried under the 100 °C vacuum state for 24 h to remove moisture. The proximate analysis results are shown in Table 2.

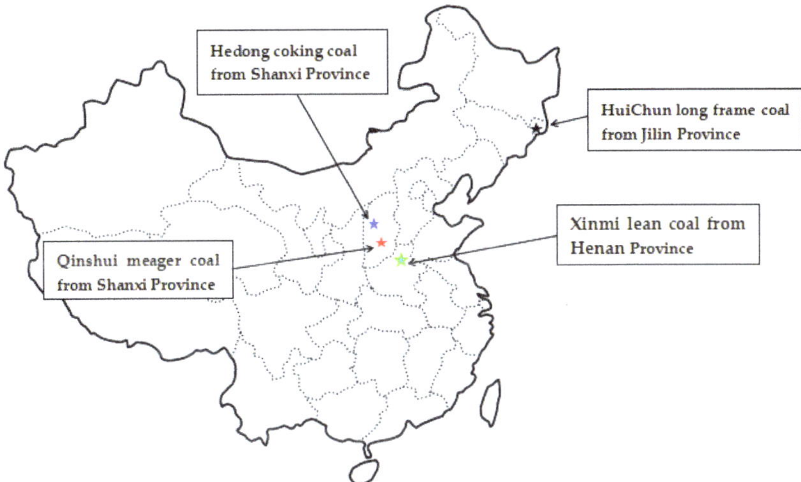

Figure 3. The diagram of coal samples collection places.

Table 2. Proximate analysis results.

Coal Sample	M_{ad} (%)	A_d (%)	V_{daf} (%)	F_c (%)
HC	8.40	32.47	47.40	11.73
HD	0.25	5.32	23.52	70.91
XM	2.65	7.83	17.25	72.27
QS	0.73	18.53	15.80	64.94

Manometric method is used in the experiments and the methane isothermal adsorption and diffusion kinetics are tested [24]. The gas state equation that implies the void volume of reference tank and coal samples tank is,

$$V = \frac{ZRTN_{He}}{P_{He}} \tag{6}$$

Four series data were recorded by pressure sensor in the experiments; (a) gas pressure in the reference tank before the gas is injected into the coal samples tank, P_{m1}; (b) gas pressure in the reference tank after the gas is injected into the coal samples tank, P_{m2}; (c) gas pressure in the coal samples tank before the reference tank gas is injected, P_{c1}; (d) gas pressure in the coal samples tank after the reference tank gas is injected, P_{c2}. P_{m1}, P_{m2} and P_{c1} are constant while P_{c2} is flexible.

The adsorption gas quantity at the i pressure step and time t in the adsorption diffusion process is,

$$N_{ti} = (N_{m1} - N_{m2}) - (N_{c2} - N_{c1}) = \left(\frac{P_{m1}V_m}{Z_{m1}RT} - \frac{P_{m2}V_m}{Z_{m2}RT}\right) - \left(\frac{P_{c2}V_c}{Z_{c2}RT} - \frac{P_{c1}V_c}{Z_{c1}RT}\right) \tag{7}$$

Therefore, the adsorption diffusion ratio of each pressure step is,

$$\frac{m_t}{m_\infty} = \frac{N_{ti}M_{CH_4}}{N_{\infty i}M_{CH_4}} = \frac{N_{ti}}{N_{\infty i}} \tag{8}$$

∞ in the equation represents the required time when the i pressure step is balanced.

While the total adsorption quantity at the pressure balance point is,

$$Q_{ad} = \frac{22.4 \times 1000 \times \sum_{i=1}^{n} N_{\infty i}}{m_{CH_4}} \tag{9}$$

Therefore, $\frac{m_t}{m_\infty}$ is equal to the gas adsorption quantity at the *i* pressure step and time *t* divided by the gas adsorption quantity when the *i* pressure step is balanced. Formula (9) is used to calculate the adsorption gas quantity at each balanced gas point and the adsorption/desorption isothermal lines [19] of coal samples are shown in Figure 4.

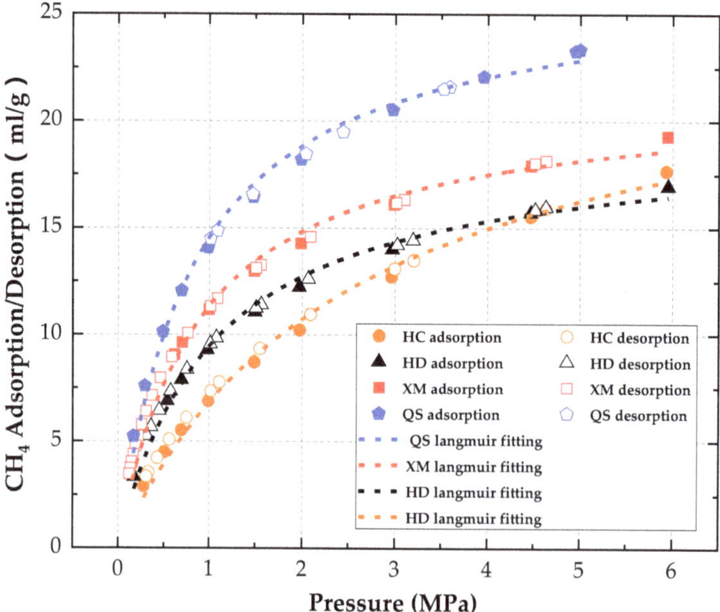

Figure 4. Adsorption-desorption isotherm of coal samples.

Langmuir model (Equation (10)) is used to fit the adsorption and desorption data of CH_4 and the correlation R^2 is listed in Table 3. It can be seen that the adsorption and desorption characteristics of CH_4 are represented well by the Langmuir model. The adsorption characteristic parameters are calculated in Table 3.

$$X = \frac{aP}{b + P} \tag{10}$$

a, b in the equation are the adsorption characteristic parameters. *a* represents the Langmuir adsorption quantity and *b* refers to the Langmuir adsorption pressure. *X* is the adsorbed gas quantity and *P* refers to the gas pressure.

Table 3. Adsorption characteristic parameters of coal samples.

Coal Sample	R^2	a/(mL/g)	b/MPa
HC	0.9935	24.81	2.62
HD	0.9891	19.33	1.04
XM	0.9930	21.37	0.88
GH	0.9933	25.46	0.67

Overall, similar to the previous findings [25,26], no apparent hysteresis phenomenon is found in the absorption/desorption process, in other words, the absorption/desorption process of CH_4 can be reversed.

4. Analysis and Discussion

4.1. Model Applicability Analysis

Based on the Equation (8), the diffusion ratio can be calculated at any moment. The approximate method was used to fit the experimental results when applying the UD and BD models. In the process of fitting the UD and BD models, findings show that the calculation results are adequately convergent when the infinite series n is expanded to the fifth term. This can ensure the accuracy of model fitting results and further reduces the calculation difficulties. Therefore, all data was processed by setting n expand to five as the standard for calculation.

Taking the gas balance pressure increases from 0.7 MPa to 1 MPa as an example, the experimental results and the fitting diffusion lines of coal samples are shown in Figure 5.

It is shown in Figure 5 that the fitting line by the UD model is below the experimental line before some critical moment regardless of the coal samples, indicating that the fitting value is smaller than the actual value. After a certain time, the fitting line keeps above the experimental line and implies that fitting value is larger than the actual value. Therefore, the experimental results cannot be restored by the fitting line regardless of moderating the UD coefficient. The fitting effect of the BD model is superior to the UD model and the fitting line of the DD model is closer to the actual line. It suggests that the whole gas adsorption/desorption process cannot be accurately described by the UD model due to the complicated pore structures. The fitting line through the BD model includes the double structure of macropore and micropore and thus shows a higher coincidence degree with the experimental results. The DD model shows the best coincidence degree with the real experimental results. Therefore, the BD model and DD model are selected to calculate the gas diffusion characteristic parameter.

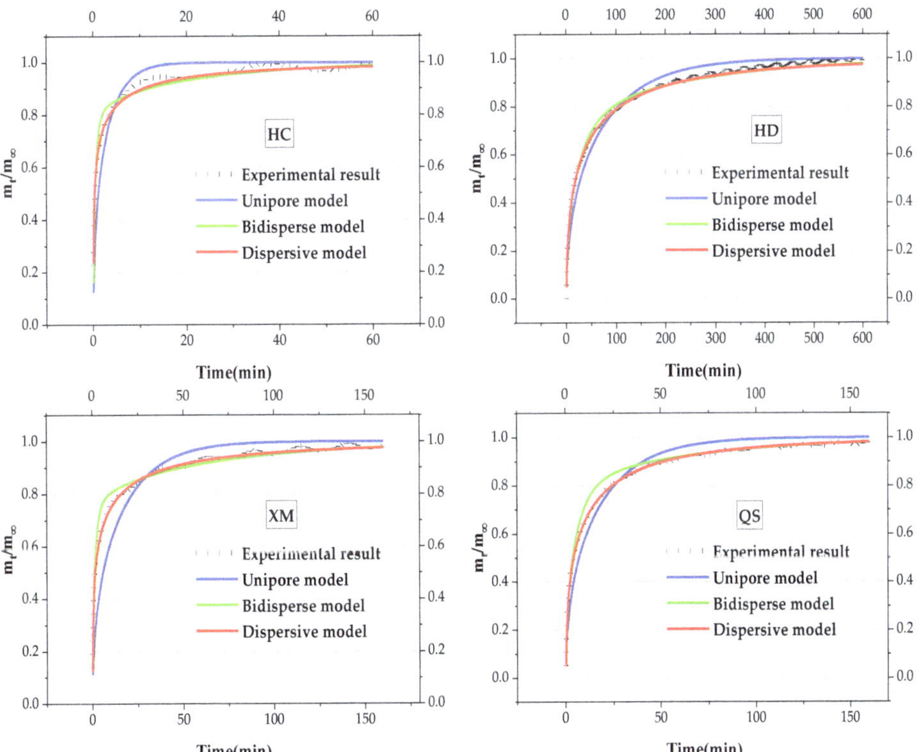

Figure 5. Experimental results and fitting diffusion lines of coal samples.

4.2. Analysis of Pressure's Effect on the Gas Diffusion

(1) The BD Model

Equation (4) implies that the BD model includes three unknown parameters, including fast effective diffusion coefficient D_{ae}, slow effective diffusion coefficient D_{ie} and the ratio of macropore adsorption/desorption to the total adsorption/desorption β. Using Equation (4) to calculate the BD characteristic parameters and analyze the impact of pressure on D_{ae} (Figure 6) and quadratic polynomial is to fit the results, the fitting goodness and calculated coefficient are shown in Table 4. As can be seen from Figure 6, the macropore diffusion coefficient D_{ae} decreases with the increase in pressure in three out of four sample coals (i.e, HC, XM and QS). Concerning the HD, D_{ae} shows a V-shape trend, which first decreases and then increases as the increases in pressure. Figure 6 also shows that the impact law of pressure on D_{ae} is better illustrated by the quadratic polynomial. When comparing the values of $\overline{D}_{ae}$, in both the adsorption and desorption processes, $\overline{D}_{ae}(\text{HC}) > \overline{D}_{ae}(\text{XM}) > \overline{D}_{ae}(\text{QS}) > \overline{D}_{ae}(\text{HD})$. The difference of $\overline{D}_{ae}$ in the absorption versus the desorption process becomes larger from HC to QS. No significant increasing trend of HC, XM and QS is found when the pressure increases, It is suspected that the set maximum pressure is not in the threshold level.

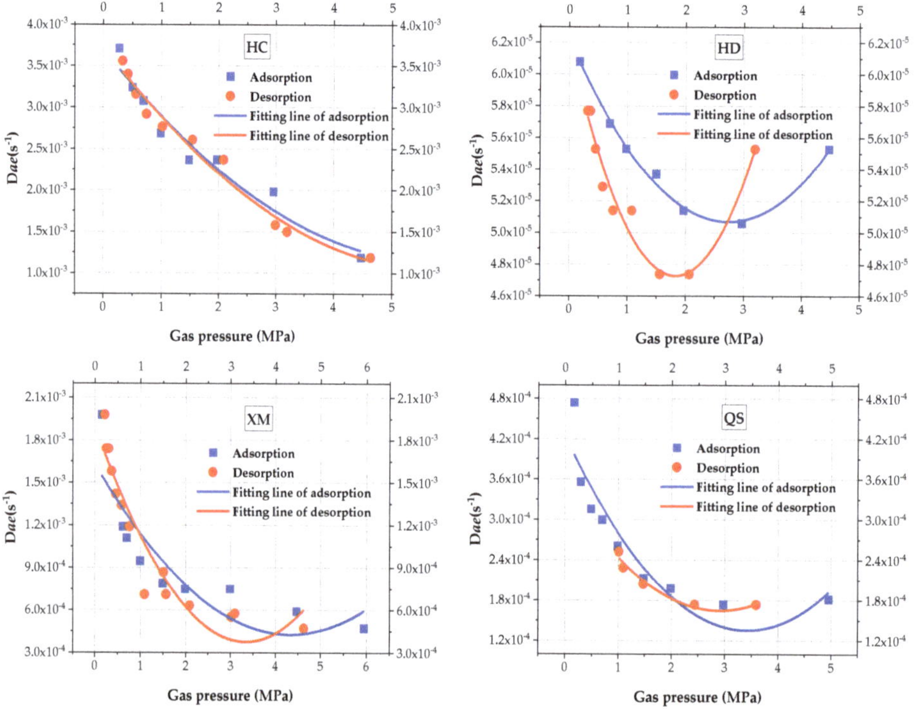

Figure 6. The diagram of the variation of macro-diffusion coefficients with pressure.

The impact of gas pressure on D_{ie} is analyzed and is shown in Figure 7. Linear regression is used to fit the results and, results of the fitting goodness and calculated coefficient are given in Table 5. It can be clearly seen that the slow efficient diffusion coefficient D_{ie} increases with the increase in pressure for all four samples. The impact law of pressure on D_{ie} is better explained by the linear regression. When comparing the values of $\overline{D}_{ie}$, the order is $\overline{D}_{ie}(\text{HC}) > \overline{D}_{ie}(\text{XM}) > \overline{D}_{ie}(\text{QS}) > \overline{D}_{ie}(\text{HD})$ in the adsorption process and $\overline{D}_{ie}(\text{HC}) > \overline{D}_{ie}(\text{QS}) > \overline{D}_{ie}(\text{XM}) > \overline{D}_{ie}(\text{HD})$ in the desorption process.

Table 4. Goodness of fit and diffusion coefficient.

Coal Sample	R^2		$\overline{D}_{ae}$	
	Adsorption	Desorption	Adsorption	Desorption
HC	0.950	0.977	2.58×10^{-3}	2.50×10^{-3}
HD	0.991	0.948	5.49×10^{-5}	5.29×10^{-5}
XM	0.764	0.906	9.53×10^{-4}	1.11×10^{-3}
QS	0.873	0.961	2.75×10^{-4}	2.07×10^{-4}

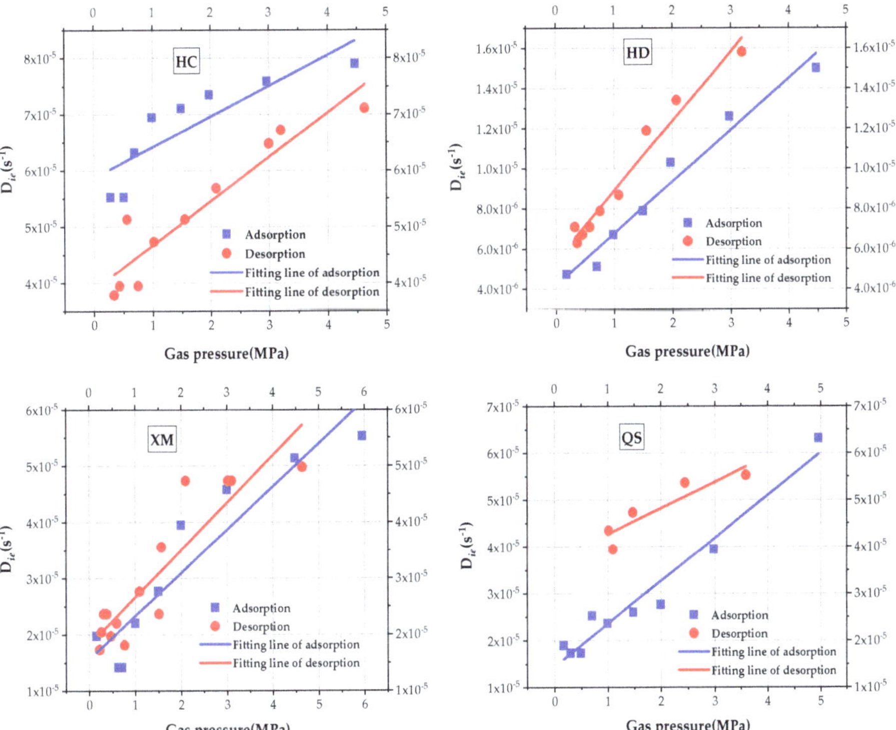

Figure 7. The diagram of variation of micro-diffusion coefficient with pressure.

Table 5. Goodness of fit and diffusion coefficient.

Coal Sample	R^2		$\overline{D}_{ie}$	
	Adsorption	Desorption	Adsorption	Desorption
HC	0.747	0.886	6.79×10^{-5}	5.27×10^{-5}
HD	0.968	0.967	8.92×10^{-6}	9.44×10^{-6}
XM	0.875	0.824	3.22×10^{-5}	3.03×10^{-5}
QS	0.955	0.828	2.88×10^{-5}	4.79×10^{-5}

The calculation results show that β is 0.74~0.76 for HC, 0.58~0.6 for HD, 0.67~0.69 for XM and 0.69~0.7 for QS, respectively, implying that the diffusion characteristic parameter β keeps constant in the adsorption/desorption process. This further indicates that β which represents the intrinsic property would not show a significant fluctuation with the change in pressure.

(2) The DD Model

The DD model includes two unknown characteristic parameters, the characteristic rate parameter k_Φ and the stretching parameter α. The influencing law of pressure on the k_Φ and α are re-analyzed, and calculated based on the gas diffusion experimental results and Equation (5). The results are shown in Figures 8 and 9, respectively.

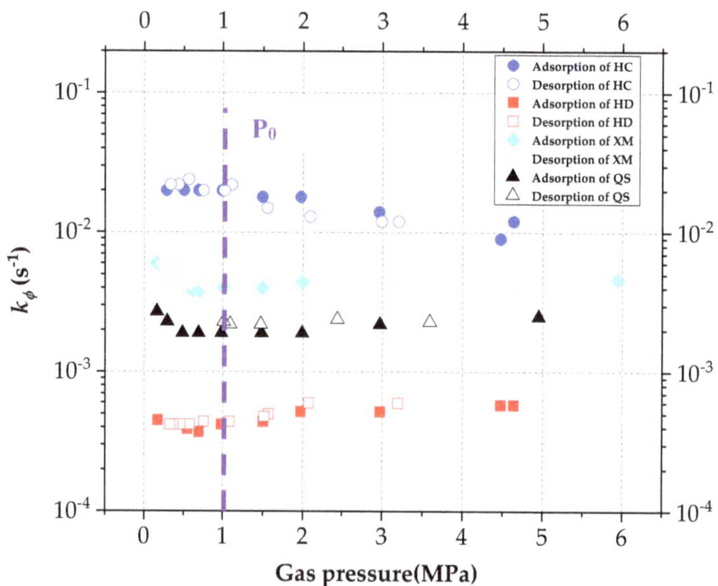

Figure 8. The diagram of variation of characteristic rate parameter k_Φ with pressure.

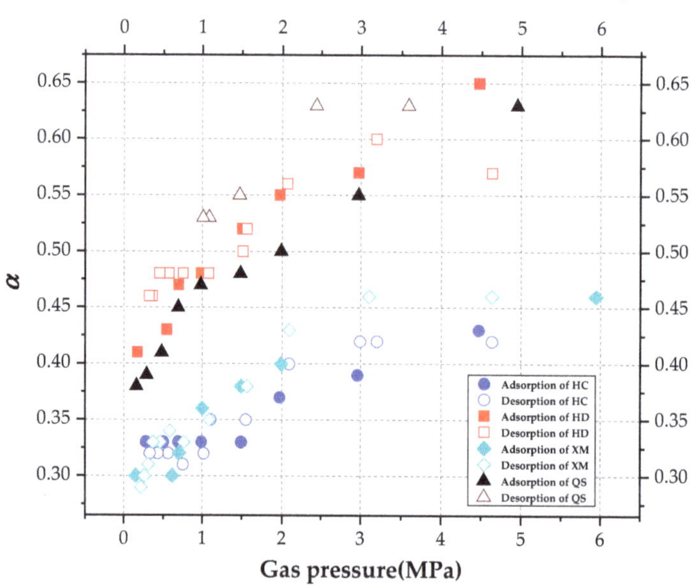

Figure 9. The diagram of variation of stretching parameter α with pressure.

Gregory Staib et al. [23,27] found that k_Φ decreases with the increase in pressure which ranges from 0~3 MPa in their studies. In terms of the vitrinite-rich coal samples, α increases with gas pressure while for the inertinite-rich coal samples, no significant changing trend is found for α. Figure 8 shows that in our study, k_Φ keeps unchanged in the pressure fluctuation process. Concerning XM and QS, k_Φ slightly decreases with the increase in pressure when the pressure is less than P_0, but it keeps constant while the pressure is larger than P_0.

As shown in Figure 9, α increases with pressure. The mean values of α were calculated in Table 6. The mean value of α is ordered as $\overline{\alpha}(HD) > \overline{\alpha}(QS) > \overline{\alpha}(XM) > \overline{\alpha}(HC)$ in the absorption process, while the order is $\overline{\alpha}(QS) > \overline{\alpha}(HD) > \overline{\alpha}(HC) > \overline{\alpha}(HM)$ in the desorption process.

Table 6. Stretching parameter $\overline{\alpha}$.

Coal Sample	Adsorption	Desorption	R^2
HC	0.355	0.356	0.876
HD	0.510	0.506	0.825
XM	0.376	0.353	0.908
QS	0.473	0.560	0.898

(3) Analysis of the correlation of diffusion characteristics parameters

The five diffusion characteristics parameters of the BD and DD models are treated by the homogenization procedure and the results are shown in Table 7. It can be seen that $\overline{D}_{ae}$, $\overline{D}_{ie}$ and $\overline{k}_\Phi$ are the largest for HC, in the middle for XM and QS, and the smallest for HD. The linear regression results of $\overline{k}_\Phi$ on $\overline{D}_{ae}$ and $\overline{k}_\Phi$ on $\overline{D}_{ie}$ are shown in Figures 10 and 11, respectively.

Table 7. Average gas diffusion parameters of experimental coal samples.

Coal Sample	$\overline{D}_{ae}$ (s^{-1})	$\overline{D}_{ie}$ (s^{-1})	$\overline{k}_\Phi$ (s^{-1})	$\overline{\alpha}$	$\overline{\beta}$
HC	2.54×10^{-3}	5.94×10^{-5}	1.78×10^{-2}	0.36	0.75
HD	5.38×10^{-5}	9.21×10^{-6}	4.73×10^{-4}	0.51	0.59
XM	1.05×10^{-3}	3.11×10^{-5}	4.69×10^{-3}	0.35	0.68
QS	2.51×10^{-4}	3.56×10^{-5}	2.16×10^{-3}	0.51	0.70

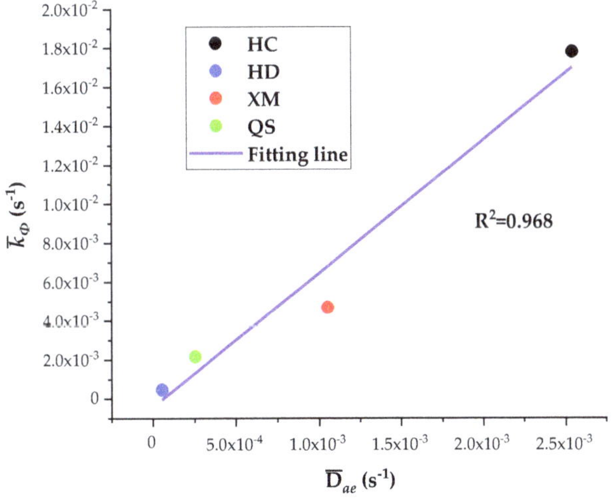

Figure 10. The linear regression of $\overline{k}_\Phi$ on $\overline{D}_{ae}$.

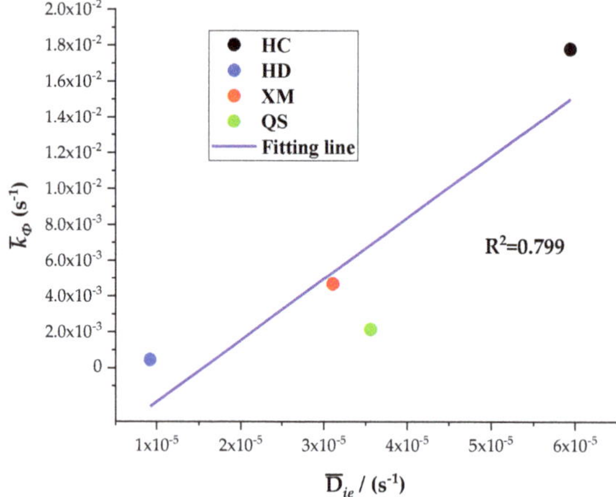

Figure 11. The linear regression of $\overline{k}_\Phi$ on $\overline{D}_{ie}$.

The results show a good linear correlation of $\overline{D}_{ae}$, $\overline{D}_{ie}$ and $\overline{k}_\Phi$ in our experimental results, and the goodness of fit is the best for $\overline{D}_{ae}$ and $\overline{k}_\Phi$. It suggests that both the diffusion coefficients D_{ae} and D_{ie} and characteristic rate parameter k_Φ are suitable for describing the coal gas diffusion rate. The analysis above suggests that the fast diffusion coefficient D_{ae} decreases with the increase in pressure while the slow diffusion coefficient D_{ie} increases with the increase in pressure. k_Φ keeps fixed and thus may be considered as a combined effect of D_{ae} and D_{ie}.

4.3. Analysis of the Relationship between Pore Structure Characteristics and Gas Diffusion

By analyzing and summarizing the impact, law of CH_4 diffusion under different pressures, we found the diversity of gas diffusion coefficients in both absorption and desorption process for different coal samples. Because the coal pore structures might directly affect the diffusion process of gas [28], experiments on the low temperature nitrogen absorption and mercury penetration were conducted to test the characteristics of coal pore structure.

The Quadrasorb instrument is used for the low temperature nitrogen absorption experiment and the PoreMaster60 mercury porosimeter instrument is applied for the mercury penetration. coal samples particles with 1~3 mm in size are prepared and dried. The low temperature nitrogen absorption method is suitable for testing the distribution of coal micropore ranging from 0~25 nm, which determines the coal specific surface area [28]. Because the mercury penetration method is not accurate in testing the micropore, it is only suitable for analyzing the pores which are bigger than 25 nm. Therefore, in this study, the computation of pore volume is calculated by the low temperature nitrogen absorption method when the pore size ranges from 0~25 nm and by the mercury penetration method if the pore size is bigger than 25 nm. The specific surface area and pore volume are given in Tables 8 and 9, respectively.

To further understand the impact of coal pore structure characteristics on the gas diffusion, we run the linear regressions of $\overline{D}_{ae}$, $\overline{D}_{ie}$ and $\overline{k}_\Phi$ on pore volume. As shown in Figure 12, the correlation between the pore volume and the diffusion coefficients is, largest for k_Φ ($R^2 = 0.912$), middle for D_{ae} ($R^2 = 0.793$) and smallest for D_{ie} ($R^2 = 0.722$).

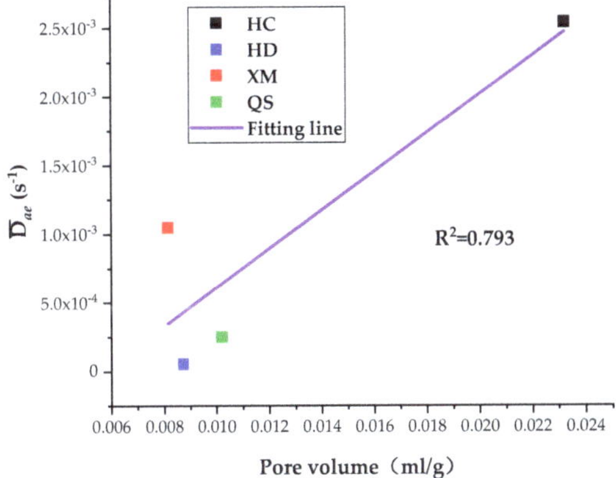

a) The impact of pore volume on $\overline{D}_{ae}$

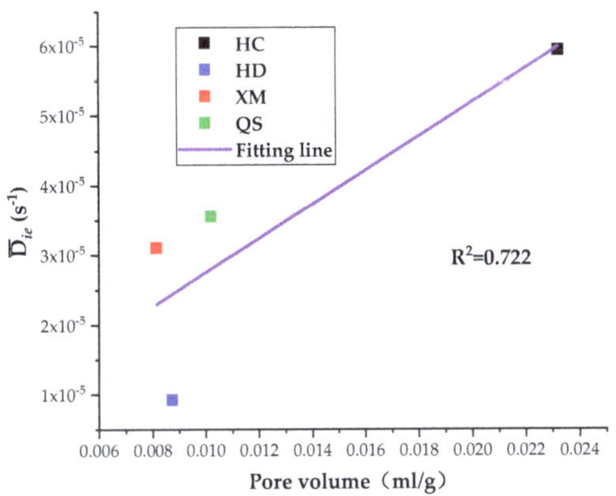

b) The impact of pore volume on $\overline{D}_{ie}$

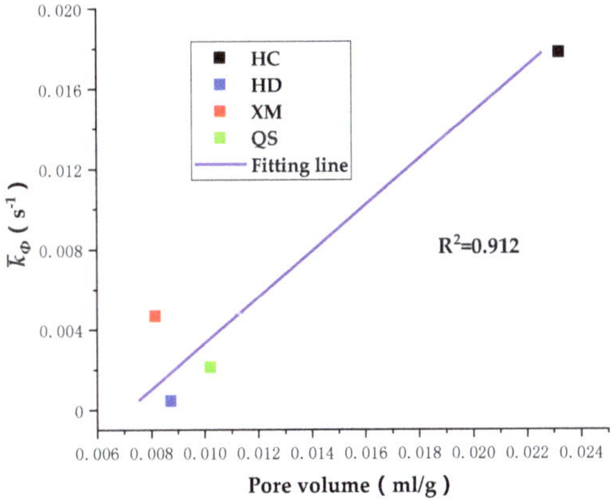

c) The impact of pore volume on $\overline{k}_{\Phi}$

Figure 12. The impact of pore volume on $\overline{D}_{ae}$, $\overline{D}_{ie}$ and $\overline{k}_{\Phi}$.

Table 8. Specific surface area of coal samples.

Coal Sample	Pore Size/nm	(0~25)
HC		11
HD	Specific surface area/(m^2/g)	0.234
XM		0.380
QS		0.177

Table 9. Pore volume of coal samples.

Coal Sample	Pore Size Ranges/nm	(0~10)	(10~100)	(100~1000)	(>1000)	Total
HC		1.21×10^{-2}	8.10×10^{-3}	2.50×10^{-3}	5.00×10^{-4}	2.32×10^{-2}
HD	Pore volume/(mL/g)	3.89×10^{-4}	4.52×10^{-3}	2.50×10^{-3}	1.30×10^{-3}	8.71×10^{-3}
XM		7.55×10^{-4}	3.27×10^{-3}	2.60×10^{-3}	1.50×10^{-3}	8.13×10^{-3}
QS		3.14×10^{-4}	6.25×10^{-3}	2.60×10^{-3}	1.00×10^{-3}	1.02×10^{-2}

Table 8 shows that the specific surface area is larger in HC relatively to other coal samples, indicating that the porosity in HC is well developed than other coal samples. It is shown in Table 7 that $\overline{D}_{ae}$, $\overline{D}_{ie}$ and $\overline{k}_{\Phi}$ of CH$_4$ is the largest in HC, suggesting the porosity development level is correlated with the diffusion rate. However, Figure 13a shows that $\overline{D}_{ae}$ of HD, QS and XM significantly increases when $\overline{D}_{ae}$ is smaller than 1.6×10^{-11} while the specific surface area keeps unchanged. Figure 13b,c show that the impact of specific surface area on $\overline{D}_{ie}$ and $\overline{k}_{\Phi}$ is small in all coal samples excluding HC. It is worth to mention that our experimental results can only be considered as reference due to the small number of coal samples. The impact of coal structure on the diffusion parameters requires further study. In conclusion, the fluctuation of diffusion coefficients with respect to the gas pressure is correlated to the variation of pore characteristics, but the reason is still mysterious due to lack of evidence.

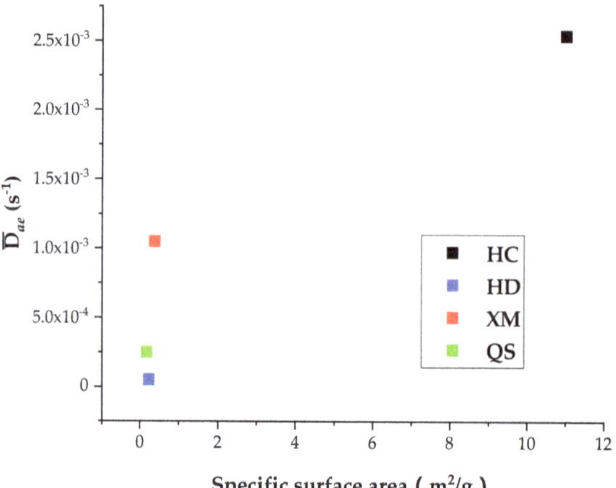

a) The impact of specific surface area on $\overline{D}_{ae}$

Figure 13. *Cont.*

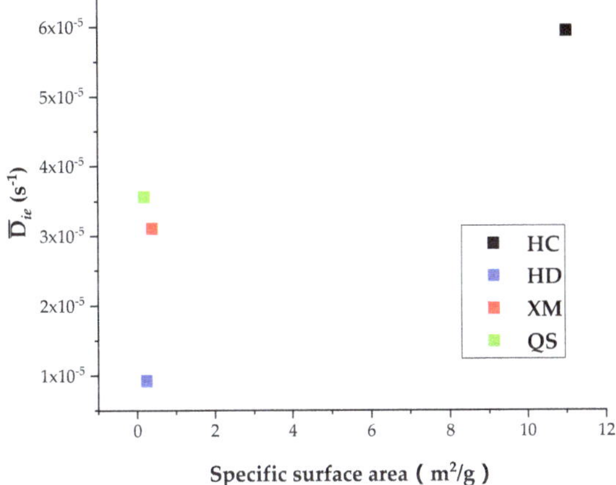

b) The impact of specific surface area on $\overline{D}_{ie}$

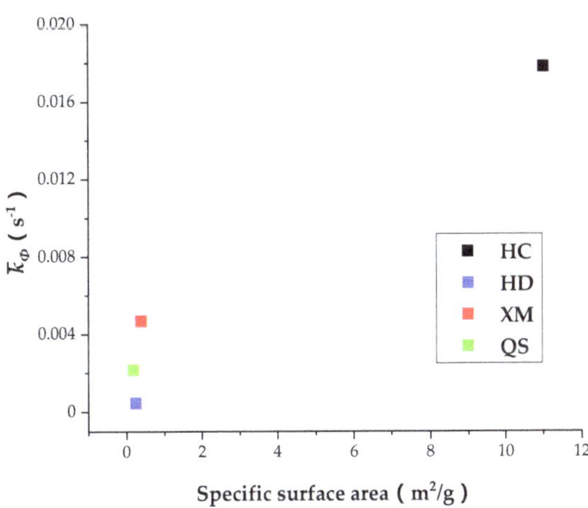

c) The impact of specific surface area on $\overline{k}_\Phi$

Figure 13. The impact of specific surface area on $\overline{D}_{ae}$, $\overline{D}_{ie}$, and $\overline{k}_\Phi$.

4.4. Discussion on the Influence of Inconstant Diffusion Coefficients on CBM Recovery

Previous studies have demonstrated that the BD diffusion cannot be overlooked and replaced by UD diffusion if diffusion is a constraint of gas production, especially for the coal seam with relatively large cleat spacing [24]. In this study, we found the BD and DD models are more accurate in describing the diffusion process, while pressure has apparent influence on the diffusion coefficients. From this point of view, the inconstant diffusion coefficients will have impacts on the CBM recovery rate. In terms of BD coefficients, most samples show an increase of fast diffusion coefficient D_{ae} but a decrease of slow diffusion coefficient D_{ie} during the drop of coal seam pressure. The increase or decrease of diffusion coefficient will certainly accelerate or hinder gas flow, but these two effects might be compromised for the BD model and the ultimate effect depends on the net value of these two effects. For the DD model, k_Φ keeps at a stable level, this phenomenon proves the above speculation as k_Φ can be seen as a combination of D_{ae} and D_{ie}. However, the stretching parameter α decreases during pressure dropping, which indicates the CBM recovery rate will be reduced due to the change of diffusion coefficient.

5. Conclusions

(1) Compared with the UD model, the BD and DD models are more accurate in describing the whole gas adsorption/desorption process.

(2) The fast efficient diffusion coefficient D_{ae} decreases with the increase in pressure for three out of four coal samples (i.e, HC, XM and QS) while it shows a V-shape with the increasing pressure for HD. The slow efficient diffusion coefficient D_{ie} is positively correlated with the pressure for all coal samples. The diffusion characteristic parameter β keeps constant in the adsorption and desorption process for all coal samples.

(3) k_Φ keeps fixed when the pressure changes and the stretching parameter α increases with the increase in pressure.

(4) Both the effective diffusion coefficient D_{ae} and D_{ie} and the characteristic rate parameter k_Φ can be used to describe the gas diffusion rate. The impact of pore volume on D_{ae}, D_{ie} and k_Φ differs in the four coal samples while D_{ae}, D_{ie} and k_Φ are slightly affected by the specific surface area. The influence of pore structure characteristics on gas diffusion ability still requires further study.

Author Contributions: Conceptualization, G.W. and X.Y.; methodology, G.W.; validation, J.Z.; formal analysis, X.Y. and G.W.; data curation, G.W.; writing—original draft preparation, X.Y. and G.W.; writing—review and editing, G.W., J.Z. and T.R.; visualization, X.Y. and G.W.; supervision, J.Z. and T.R.

Funding: This work was supported by National Natural Science Foundation of China (51604153), National Science and Technology Major Project (2016ZX05045-004-006), National key research and development Project (2018YFB0605601).

Acknowledgments: We sincerely thank assistant professor Chunling Xia from Queen Mary University of London for improving the language of this paper.

Conflicts of Interest: The authors declare no conflicts of interest.

References

1. Zhang, Y.; Xing, W.L.; Liu, S.Y.; Liu, Y.; Yang, M.J.; Zhao, J.F.; Song, Y.C. Pure methane, carbon dioxide, and nitrogen adsorption on anthracite from China over a wide range of pressures and temperatures: Experiments and modeling. *RSC Adv.* **2015**, *5*, 52612–52623. [CrossRef]

2. Charrière, D.; Pokryszka, D.; Behra, P. Effect of pressure and temperature on diffusion of CO_2 and CH_4 into coal from the Lorraine basin (France). *Int. J. Coal Geol.* **2010**, *81*, 373–380. [CrossRef]

3. Pan, Z.J.; Connell, L.D.; Camilleri, M.; Connelly, L. Effects of matrix moisture on gas diffusion and flow in coal. *Fuel* **2010**, *89*, 3207–3217. [CrossRef]

4. Staib, G.; Richard, S.; Gray, E.M.A. A pressure and concentration dependence of CO_2 diffusion in two Australian bituminous coals. *Int. J. Coal Geol.* **2013**, *116*, 106–116. [CrossRef]

5. Clarkson, C.R.; Bustin, R.M. The effect of pore structure and gas pressure upon the transport properties of coal: A laboratory and modelling study. 2. Adsorption rate modelling. *Fuel* **1999**, *78*, 1345–1362. [CrossRef]

6. Han, F.S.; Busch, A.; Krooss, B.M.; Liu, Z.Y.; Yang, J.L. CH_4 and CO_2 sorption isotherms and kinetics for different size fractions of two coals. *Fuel* **2013**, *108*, 137–142. [CrossRef]

7. Li, D.Y.; Liu, Q.F.; Weniger, P.; Gensterblum, Y.; Busch, A.; Krooss, B.M. High-pressure sorption isotherms and sorption kinetics of CH_4 and CO_2 on coals. *Fuel* **2010**, *89*, 569–580. [CrossRef]

8. Cui, X.J.; Bustin, R.M.; Dipple, G. Selective transport of CO_2, CH_4, and N_2 in coals: Insights from modeling of experimental gas adsorption data. *Fuel* **2004**, *83*, 293–303. [CrossRef]

9. Busch, A.; Gensterblum, Y.; Krooss, B.M.; Littke, R. Methane and carbon dioxide adsorption–diffusion experiments on coal: Upscaling and modelling. *Int. J. Coal Geol.* **2004**, *60*, 151–168. [CrossRef]

10. Zhang, J. Experimental study and modeling for CO_2 diffusion in coals with different particle sizes: Based on gas absorption (imbibition) and pore structure. *Energy Fuels* **2016**, *30*, 531–543. [CrossRef]

11. Karacan, C.Ö. Heterogeneous sorption and swelling in a confined and stressed coal during CO_2 injection. *Energy Fuels* **2003**, *17*, 1595–1608. [CrossRef]

12. Crosdale, P.J.; Beamish, B.B.; Valix, M. Coalbed methane sorption related to coal composition. *Int. J. Coal Geol.* **1998**, *35*, 147–158. [CrossRef]

13. Jian, X.; Guan, P.; Zhang, W. Carbon dioxide sorption and diffusion in coals: Experimental investigation and modeling. *Sci. China Earth Sci.* **2012**, *55*, 633–643. [CrossRef]

14. Shi, J.Q.; Durucan, S. A bidisperse pore diffusion model for methane displacement desorption in coal by CO_2 injection. *Fuel* **2003**, *82*, 1219–1229. [CrossRef]

15. Nandi, S.P.; Walker, P.L. Activated diffusion of methane from coals at elevated pressures. *Fuel* **1975**, *54*, 81–86. [CrossRef]

16. Švábová, M.; Weishauptová, Z.; Prˇibyl, O. The effect of moisture on the sorption process of CO_2 on coal. *Fuel* **2012**, *92*, 187–196. [CrossRef]

17. Pone, J.D.N.; Halleck, P.M.; Mathews, J.P. Sorption capacity and sorption kinetic measurements of CO_2 and CH_4 in confined and unconfined bituminous coal. *Energy Fuels* **2009**, *23*, 4688–4695. [CrossRef]

18. Crank, J. *The Mathematics of Diffusion*, 2nd ed.; Oxford University Press: New York, NY, USA, 1975; ISBN 0198533446.

19. Wang, G.D.; Ren, T.; Qi, Q.X.; Lin, J.; Liu, Q.Q.; Zhang, J. Determining the diffusion coefficient of gas diffusion in coal: Development of numerical solution. *Fuel* **2017**, *196*, 47–58. [CrossRef]

20. Siemons, N.; Wolf, K.H.A.A.; Bruining, J. Interpretation of carbon dioxide diffusion behavior in coals. *Int. J. Coal Geol.* **2007**, *72*, 315–324. [CrossRef]

21. Wang, G.D.; Ren, T.; Qi, Q.X.; Zhang, L.; Liu, Q.Q. Prediction of Coalbed Methane (CBM) Production Considering Bidisperse Diffusion: Model Development, Experimental Test, and Numerical Simulation. *Energy Fuels* **2017**, *31*, 5785–5797. [CrossRef]

22. Ruckenstein, E.; Vaidyanathan, A.S.; Youngquist, G.R. Sorption by solids with bidisperse pore structures. *Chem. Eng. Sci.* **1971**, *26*, 1305–1318. [CrossRef]

23. Staib, G.; Sakurovs, R.; Gray, E.M.A. Dispersive diffusion of gases in coals. Part I: Model development. *Fuel* **2015**, *143*, 612–619. [CrossRef]

24. Busch, A.; Gensterblum, Y. CBM and CO_2-ECBM related sorption processes in coal: A review. *Int. J. Coal Geol.* **2011**, *87*, 49–71. [CrossRef]

25. Goodman, A.L.; Busch, A.; Duffy, G.J.; Fitzgerald, J.E.; Gasem, K.A.M.; Gensterblum, Y.; Krooss, B.M.; Levy, J.; Ozdemir, E.; Pan, Z.; et al. An Inter-laboratory comparison of CO_2 isotherms measured on Argonne Premium Coal samples. *Energy Fuels* **2004**, *18*, 1175–1182. [CrossRef]

26. Wang, G.D. *Adsorption and Desorption Hysteresis of Coal Seam Gas and Its Influence on Gas Permeability [D]*; China University of Mining and Technology: Beijing, China, 2015. (In Chinese)

27. Staib, G.; Sakurovs, R.; Gray, E.M.A. Dispersive diffusion of gases in coals. Part II: An assessment of previously proposed physical mechanisms of diffusion in coal. *Fuel* **2015**, *143*, 620–629. [CrossRef]

28. Liu, H.H.; Mou, J.H.; Cheng, Y.P. Impact of pore structure on gas adsorption and diffusion dynamics for long-flame coal. *Nat. Gas Sci. Eng.* **2015**, *22*, 203–221. [CrossRef]

Article

Theoretical Methodology of a High-Flux Coal-Direct Chemical Looping Combustion System

Xiaojia Wang [1,*], Xianli Liu [1,2], Zhaoyang Jin [1], Jiewen Zhu [1] and Baosheng Jin [1]

[1]   Key Laboratory of Energy Thermal Conversion and Control of Ministry of Education,
     School of Energy and Environment, Southeast University, Nanjing 210096, China; liuxl_seu@163.com (X.L.);
     zhaoyangjinseu@163.com (Z.J.); jiewenzhuseu@163.com (J.Z.); bsjin@seu.edu.cn (B.J.)
[2]   China Energy Engineering Group Jiangsu Power Design Institute Co. Ltd., Nanjing 211102, China
*   Correspondence: xiaojiawang@seu.edu.cn; Tel.: +86-025-8379-2811

Received: 26 October 2018; Accepted: 30 November 2018; Published: 4 December 2018

Abstract: This study, as an extension of our previous experimental tests, presented a mechanism analysis of air reactor (AR) coupling in a high-flux coal-direct chemical looping combustion (CDCLC) system and provided a theoretical methodology to the system optimal design with favorable operation stability and low gas leakages. Firstly, it exhibited the dipleg flow diagrams of the CDCLC system and concluded the feasible gas–solid flow states for solid circulation and gas leakage control. On this basis, the semi-theoretical formulas of gas leakages were proposed to predict the optimal regions of the pressure gradients of the AR. Meanwhile, an empirical formula of critical sealing was also developed to identify the advent of circulation collapse so as to ensure the operation stability of the whole system. Furthermore, the theoretical methodology was applied in the condition design of the cold system. The favorable gas–solid flow behaviors together with the good control of gas leakages demonstrated the feasibility of the theoretical methodology. Finally, the theoretical methodology was adopted to carry out a capability assessment of the high-flux CDCLC system under a hot state in terms of the restraint of gas leakages and the stability of solid circulation.

Keywords: coal-direct chemical looping combustion; coupling mechanism; theoretical methodology; high-flux; gas leakage; pressure gradient

---

1. Introduction

Coal-direct chemical looping combustion (CDCLC) has been demonstrated as an attractive combustion technology of coal with the inherent feature for CO_2 capture [1,2]. The CDCLC concept is typically implemented in two interconnected reactors, the so-called fuel reactor (FR) and the air reactor (AR), with an oxygen carrier (OC) circulating in between to transfer oxygen and heat. Specifically, in the FR, the fuel is first devolatilized and gasified by the gasification agent steam, and then the gasification products (mainly CO, H_2, and CH_4) are further oxidized to CO_2 and H_2O by the OC. In the AR, the reduced OC from the FR is oxidized by the air for regeneration, and then will be recirculated back to the FR. By means of the OC particles that deliver oxygen from the AR to FR, the direct mixing of the fuel and air can be avoided, and further highly purified CO_2, without the dilution of N_2, can be acquired at the outlet of the FR via the condensation of steam [3–12].

Alternatively, a characterized CDCLC system consisting of a high-flux circulating fluidized bed (HFCFB) riser as the FR and a counter-flow moving bed (CFMB) as the AR was proposed in our previous studies [13–15], as shown in Figure 1. The main advantages of this design are that the HFCFB FR can provide high solids concentration over the whole reactor height for favorable gas–solid contact efficiency and reaction performance, and that the CFMB AR possesses steady solids flow and low-pressure drop. Besides, an inertial separator, connecting the two reactors, was specially designed as the carbon stripper to separate the coarse OC particles off the FR into the AR for regeneration, and also recirculate the fine particles of the unconverted coal char back to the FR for further conversion.

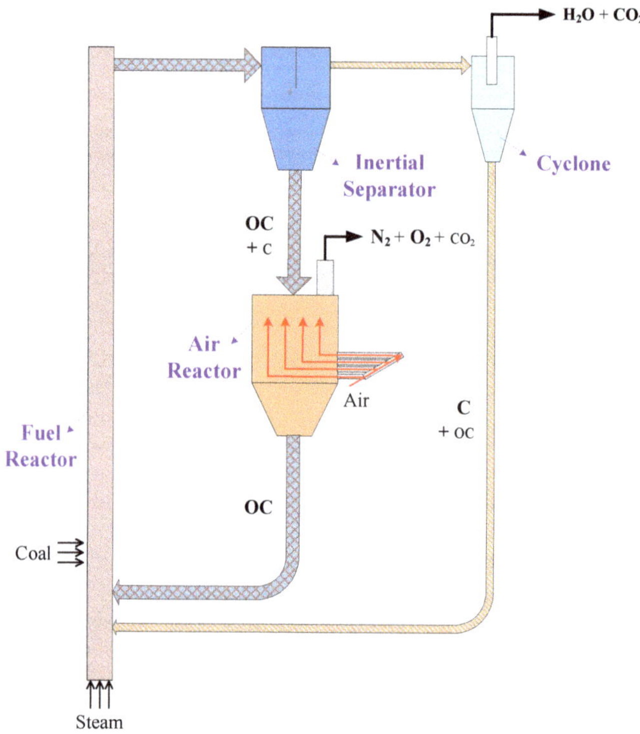

Figure 1. Schematic of the high-flux coal-direct chemical looping combustion (CDCLC) system (OC: oxygen carrier).

Up to now, we have successively developed cold [14] and pilot-scale hot [15] experimental systems of this high-flux CDCLC concept, preliminarily realizing the whole-system stable operation with acceptable gas–solid reaction performance under certain conditions. Similar feasibility studies have been experimentally conducted in different pilot-scale CDCLC units, e.g., the 10 kW_{th} [3] and 100 kW_{th} [9] units at Chalmers University of Technology (Sweden), the 10 kW_{th} [4] and 50 kW_{th} [5] units at Southeast University (China), the 1 MW_{th} unit from Technische Universität Darmstadt (Germany) [10], the 25 kW_{th} unit at Hamburg University of Technology (Germany) [11], the 25 kW_{th} unit from Ohio State University (America) [12], the 50 kW_{th} unit at Instituto de Carboquimica (ICB-CSIC) (Spain) [16], and the 5 kW_{th} CDCLC reactor at Huazhong University of Science and Technology (China) [17]. However, despite promising experimental results obtained in pilot-scale units, the CDCLC technology for CO_2 capture has to be further developed towards large-scale commercial applications. In this aspect, it is essential to develop theoretical methodologies, beside experimental studies, for a better understanding of hydrodynamic and reaction mechanisms in CDCLC processes, which can provide vital references to the design, operation, and process optimization of the future large-scale CDCLC power plants. By far, compared to the extensive experimental studies, few studies are available in the literature on the development of theoretical methodologies in terms of hydrodynamics and/or reaction mechanisms for CDCLC processes. Su et al. [18], based on the hydrodynamic equations for fluidized beds and the reaction kinetics, simulated the CDCLC process in a dual circulating fluidized bed (DCFB) system. Ohlemüller et al. [19] developed a process simulation model to predict the flow and reaction performances of a 1 MW_{th} unit at Technische Universität Darmstadt.

In our previous experimental tests, we have found that the coupling of the CFMB AR into the downcomer of the HFCFB FR makes the hydrodynamic mechanism of the whole system much more complicated, and hence leads to crucial effects on the operation independence of the two reactors (i.e., FR and AR) in terms of solid circulation stability and gas leakages [20]. In this context, it is necessary to carry out an in-depth mechanism investigation of this high-flux CDCLC system coupled by a CFMB AR, which is significant to the design and operation processes of the future CDCLC applications. Therefore, the objective of this study is to develop a theoretical methodology to illustrate the fundamental hydrodynamics of our high-flux CDCLC system, extending from the previous experimental studies. The main contributions of this work are listed as follows: (1) the screening of the feasible gas–solid flow states for solid circulation and gas leakage control on the strength of the dipleg flow diagrams of the CDCLC system; (2) the development of the semi-theoretical formulas of gas leakages to predict the optimal regions of the pressure gradients of the AR; (3) the development of the empirical formula of critical sealing to identify the advent of circulation collapse so as to ensure the whole-system operation stability; (4) the feasibility validation of the theoretical methodology through its application in the cold-state condition design; (5) the successful application of the theoretical methodology into the capability assessment of the high-flux CDCLC system under a hot state, in terms of the restraint of gas leakages and the stability of solid circulation.

2. Materials and Methods

2.1. Visualization Experimental Device

The visualization experimental device of the high-flux CDCLC system consists primarily of a FR, an inertial separator, a downcomer, an AR, and a J-valve. During the operation process, the FR, with an inner diameter of 60 mm and a height of 5.8 m, was operated in dense suspension upflow (DSU) regime with high solid circulation flux and solids holdup. An inertial separator was installed following the FR, and was used as the carbon stripper to separate the gas stream and elutriated particles off the FR. The particle outlet of the inertial separator was connected with the downcomer, and further the AR which was operated in moving bed regime with an inner diameter of 418 mm and a height of 0.7 m. After leaving the AR, the OC particles were transported back to the FR with the help of the J-valve. The drawing presented in Figure 2 schematically shows how the different sections of the visualization experimental setup are interconnected. The more detailed description can be found in our previous experimental studies with this system [14,20].

The tracer gas (99.99% CO) concentrations were continuously measured with a gas analyzer (MRU, Neckarsulm, Germany) at the outlets of the two reactors. The pressures were measured with pressure gages and a multi-channel differential pressure transducer. The gas flow rates were adjusted and measured by calibrated rotameters (Changzhou shuanghuan thermal instrument co., LTD, Changzhou, China) and then normalized to the standard state (labeled with a subscript *sta*). Specifically, $Q_{1,sta}$, $Q_{2,sta}$, $Q_{3,sta}$, and $Q_{4,sta}$ represent the inlet air flow rate of the FR, the fluidizing air flow rate of the J-valve, the aeration air flow rate of the J-valve, and the inlet air flow rate of the AR, respectively. $Q_{a,sta}$ and $Q_{b,sta}$ represent the outlet air flow rates of the FR and the AR, respectively.

2.2. Materials

The OC material used in this study was a natural iron ore from Harbin, China with an average particle diameter of 0.43 mm and bulk density of 1577 kg/m^3. The minimum fluidization gas velocity under the cold condition was 0.187 m/s [20].

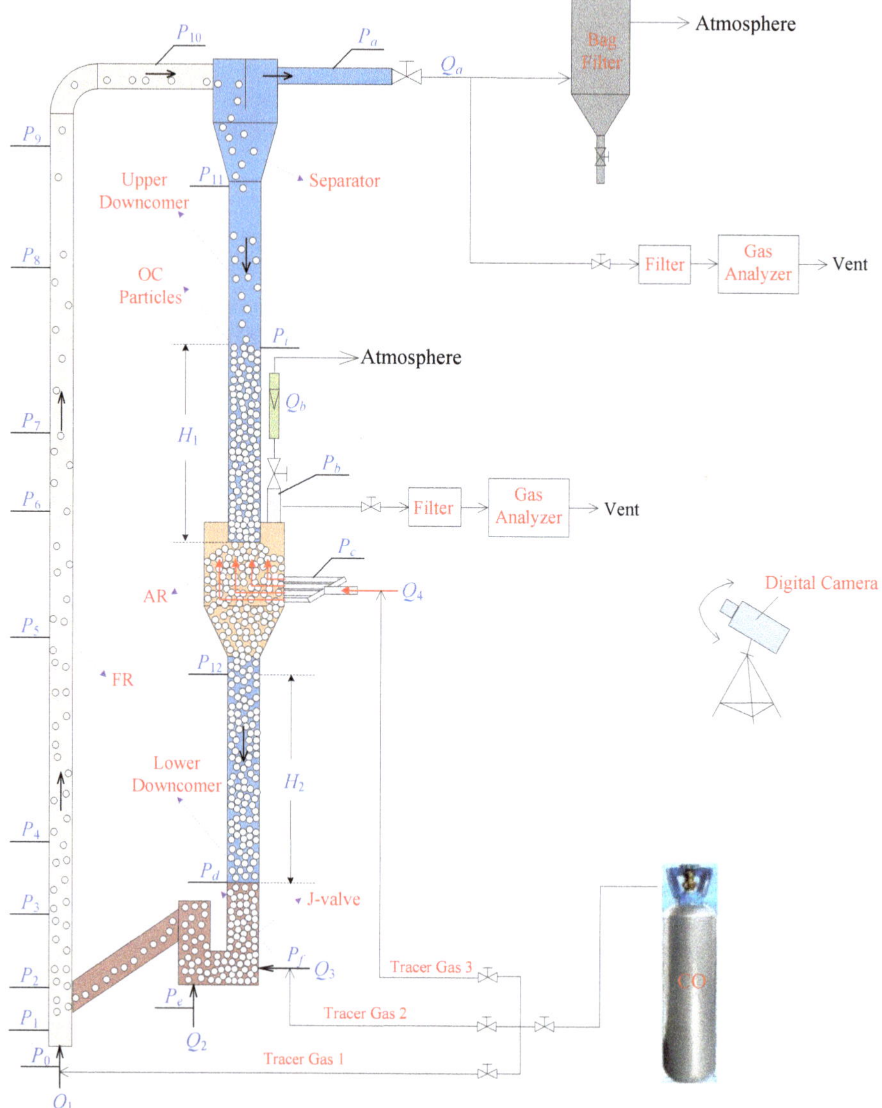

Figure 2. Schematic diagram of the cold-state experimental device of the high-flux CDCLC system. *P*: pressure; *Q*: gas flow; AR: air reactor; FR: fuel reactor.

2.3. Performance Indicators

The upper pressure gradient ($\Delta P_1/H_1$) represents the pressure gradient between the AR and the carbon stripper, which was expressed as Equation (1). The lower pressure gradient ($\Delta P_2/H_2$) represents the pressure gradient between the J-valve and the AR, which was expressed as Equation (2) [20].

$$\Delta P_1/H_1 = \left(\frac{p_b + p_c}{2} - p_i\right)/H_1 \tag{1}$$

$$\Delta P_2/H_2 = (p_d - p_{12})/H_2 \tag{2}$$

Solid circulation flux, G_s, represents the solid circulation ratio (kg/s) per unit area of the FR, which was estimated by [20,21]

$$G_s = \frac{\rho_b u_s A_{ud}}{A_f} = \frac{\rho_b A_{ud}}{A_f}(\Delta H/t) \tag{3}$$

The solids holdup in the FR, ε_s, could be estimated according to the local pressure drop [13,14,21–24].

$$\Delta P_Z/\Delta Z \approx [\rho_s \varepsilon_s + \rho_g(1 - \varepsilon_s)]g \tag{4}$$

The FR leakage ratio, f_1, represents the gas leakage ratio from the FR to the AR. During the experimental process, the FR leakage ratio was measured by using tracer gas 1 [14,20].

$$f_1 = -\frac{Q_{b,sta}x_{b,CO}}{Q_{a,sta}x_{a,CO} + Q_{b,sta}x_{b,CO}} \tag{5}$$

In this study, the upward direction is defined as the positive direction, and hence the FR leakage ratio should be negative.

The AR leakage ratio, f_2, represents the gas leakage ratio from the AR to the FR, which could be measured by using tracer gas 2 [14,20].

$$f_2 = \frac{Q_{a,sta}x'_{a,CO}}{Q_{a,sta}x'_{a,CO} + Q_{b,sta}x'_{b,CO}} \tag{6}$$

The J-valve leakage ratio, f_3, represents the gas leakage ratio of the J-valve aeration air into the AR, which was measured by using tracer gas 3 [20].

$$f_3 = \frac{Q_{b,sta}x''_{b,CO}}{Q_{a,sta}x''_{a,CO} + Q_{b,sta}x''_{b,CO}} \tag{7}$$

The detailed meanings of the symbols can be found in the nomenclature.

3. Results and Discussion

3.1. Gas–Solid Flow Characteristics

In typical circulating fluidized bed (CFB) reactors, the downcomer dipleg plays a critical role in solid circulation and gas seal. In order to drive particles, the whole dipleg is usually kept in a state of negative pressure gradient (i.e., a decrease of pressure with the increase of downcomer height) [25–31]. However, the coupling of the AR into the downcomer of our high-flux CDCLC system together with the special requirements of gas leakages makes the operation mechanism of the dipleg and even the whole system much more complicated. As shown in Figure 2, the existence of the AR divides the dipleg into the upper dipleg and the lower dipleg. During the CDCLC process, the lower dipleg stays at a negative pressure gradient state with the J-valve owning the maximum pressure so as to drive the solids for circulation. But the upper dipleg can situate at a pressure region across the positive and negative pressure gradient states, which has crucial effects on the circulation stability and gas leakage ratios. Therefore, a systematic study is necessary to improve the understanding of AR coupling effects on the hydrodynamic mechanism of this CDCLC system in terms of solid circulation stability and gas leakage controllability.

Figure 3 shows the possible gas–solid flow states in the upper dipleg during the high-flux CDCLC process, where the upward direction is defined as the positive direction [32]. As the abovementioned discussions, the upper dipleg flow can be categorized into seven flow states, according to the differential pressure between the two ends of the upper dipleg. In the first state, the top pressure of the upper dipleg is much larger than that at the bottom, and the gas flow moves downward with a much higher velocity than that of the solids. The big velocity difference between the gas–solid phases means the dramatic CO_2 leakage from the carbon stripper into the AR, and hence the great reduction of CO_2 capture efficiency. In the second state, the positive differential pressure between the top and bottom becomes much smaller, and the gas velocity is only slightly higher than the solids velocity, indicating a modest gas leakage from the FR to the AR. In the third state, when the differential pressure between the two ends of the upper dipleg becomes zero, the solids downward flow is controlled by gravity and the gas–solid relative velocity becomes zero. Then in the fourth state, the bottom pressure of the upper dipleg starts to outpace the top pressure, and the gas–solid flow becomes negative pressure gradient flow, leading to a further reduction of the downward velocity of gas phase. When the downward gas velocity further decreases to zero, it comes to the fifth state, so-called the ideal sealing state. At this point, the gas–solid relative velocity is equal to the absolute value of the solids descending velocity, indicating the ideal suppression of the gas leakages between the FR and AR. In the sixth state, with the further enhancement of the negative pressure gradient, the gas begins to flow upward with a low velocity, indicating a small amount of gas leakage from the AR to the FR. Finally, when a large amount of gas flow moves upward in terms of visible large bubbles, the upper dipleg will enter into the last state, so-called the critical sealing state, meaning that the whole-system particle circulation is about to be broken together with a dramatical leakage of N_2 from the AR into FR. In general, States 1 and 2 belong to the positive pressure gradient flow, State 3 belongs to the zero pressure gradient flow, and States 4 to 7 belong to the negative pressure gradient flow.

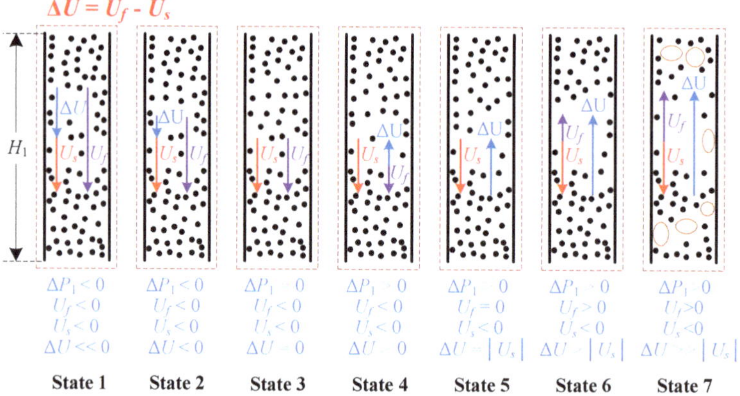

Figure 3. Gas–solid flow diagrams in the upper dipleg under different pressure gradient conditions.

Our previous experimental studies found that with the increase of the upper pressure gradient $\Delta P_1/H_1$, the FR leakage ratio f_1 had a linear drop until extinction while the AR leakage ratio f_2 firstly stayed at zero and then had a linear increase [20]. Thus, the variations of gas–solid flow state in the upper dipleg corresponding to the upper pressure gradient could be further deduced, as shown in Figure 4. It can be found that the gas flow direction in the upper dipleg changed from downward to upward. By referring to Figure 3, it can be concluded that the gas–solid flow in the upper dipleg had gone through States 2 to 6, demonstrating the feasibility of the selection of optimal operation region for the gas leakage control and solid circulation by means of the adjustment of the upper pressure gradient $\Delta P_1/H_1$. Consistent with the experimental studies [20], we set -3% and 3% as the limit values of the two gas leakages f_1 and f_2, respectively, and as the selection criteria of the upper pressure gradient. Thus, we can get the optimal region of $\Delta P_1/H_1$ corresponding to States 2 to 6 under the involved operation conditions: State 2 (-2.1 kPa/m $< \Delta P_1/H_1 < 0$ kPa/m), State 3 ($\Delta P_1/H_1 = 0$ kPa/m), State 4 (0 kPa/m $< \Delta P_1/H_1 < 1.6$ kPa/m), State 5 ($\Delta P_1/H_1 = 1.6$ kPa/m), and State 6 (1.6 kPa/m $< \Delta P_1/H_1 <$ 3.0 kPa/m).

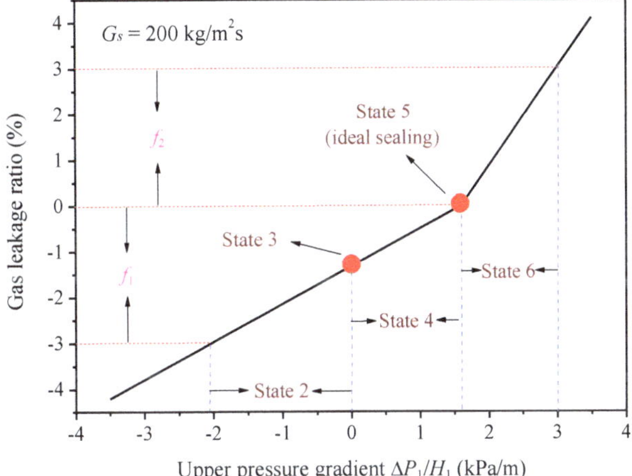

Figure 4. Variations of gas–solid flow state in the upper dipleg with the upper pressure gradient.

On the other hand, as mentioned above, because the J-valve is the driving source for solid circulation, it has the maximum pressure of the whole system. Hence, the lower dipleg necessarily stays at a negative pressure gradient region, i.e., States 4 to 7 as shown in Figure 3. However, considering that the gas leakage from the AR to the J-valve will result in the mixing of N_2 into the FR, and further the reduction of CO_2 capture concentration, State 4 should also better be avoided during the CDCLC process. Moreover, an excess gas leakage (i.e., State 7 shown in Figure 3) will cause serious damage on the stability of the solids downward flow, and further the solid circulation. Therefore, only States 5 and 6 were regarded as the preferred gas–solid flow states in the lower dipleg. Consistent with our previous experimental studies [20], we set 20% as the upper limit value of the gas leakage f_3, and as the selection criterion of optimal region under the involved operation conditions.

3.2. Theoretical Methodology for Gas Leakage Restraint

3.2.1. Semi-Theoretical Formulas of the Upper Pressure Gradient

From the analyses made above, we can find that the coupling of the CFMB AR makes the gas–solid flow in the upper dipleg very complicated, which covers a diversity of flow structures from positive pressure gradient to negative pressure gradient states. In a real CDCLC application, the optimal operation region should better locate among States 2 to 6 in order to acquire stable solid circulation and favorable restraint of gas leakages. Fortunately, we found that the optimal region exhibited a relatively symmetrical distribution rule with the ideal sealing state (i.e., State 5) as the core. Thus, it provides a possibility for us to propose an empirical equation applied to the high-flux CDCLC system based on the ideal sealing theory.

The modified Ergun equation was attempted to be applied in the stable moving bed flow of the upper dipleg under the ideal sealing state, which had the form of Equation (8) [33–35]. Meanwhile, according to the relationship between the solids velocity and solid circulation flux (see Equation (9)), we further got the correlation equation of the upper pressure gradient under the ideal sealing state with the solid circulation flux (i.e., Equation (10)).

$$(\Delta P_1 / H_1)_i = 150 \frac{(1-\varepsilon)^2 \mu_g |U_s|}{(\varepsilon \varphi_s d_s)^2} + 1.75 \frac{(1-\varepsilon)\rho_g U_s^2}{\varepsilon \varphi_s d_s} \tag{8}$$

$$|U_s| = \left(\frac{D_f}{D_{ud}}\right)^2 \frac{G_s}{\rho_s(1-\varepsilon)} \tag{9}$$

$$(\Delta P_1/H_1)_i = \left[150\frac{(1-\varepsilon)\mu_g}{\rho_s(\varepsilon\varphi_s d_s)^2}\left(\frac{D_f}{D_{ud}}\right)^2\right]G_s + \left[1.75\frac{\rho_g}{\varepsilon\varphi_s d_s(1-\varepsilon)\rho_s^2}\left(\frac{D_f}{D_{ud}}\right)^4\right]G_s^2 \qquad (10)$$

Figure 5 illustrates the comparison of predicted and experimental upper pressure gradients under the ideal sealing state (i.e., State 5). Table 1 lists the main parameters required for the calculation of the ideal upper pressure gradient. It can be seen that, with a solid circulation flux of 200 kg/m^2·s, the value of the theoretical ideal pressure gradient was about 1.6 kPa/m which was almost the same with the experimental value. Then, when the solid circulation flux increased to 300 kg/m^2·s, the theoretical and measured values of the ideal pressure gradient were increased to about 2.5 and 2.8 kPa/m, respectively. In general, the relative errors between the measured and predicted values of the ideal pressure gradient were kept to be lower than 15%, demonstrating the application feasibility of the modified Ergun equation in the prediction of the ideal pressure gradient of the high-flux CDCLC system.

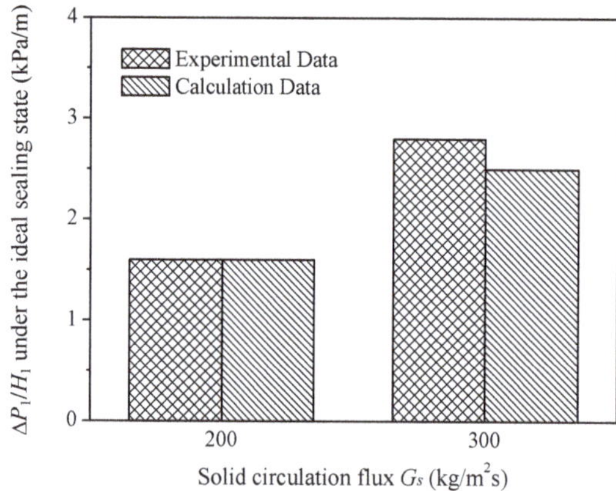

Figure 5. Comparison of predicted and experimental upper pressure gradients under ideal sealing states with different solid circulation fluxes.

Table 1. Parameters for the calculation of the ideal upper pressure gradient (OC: oxygen carrier; FR: fuel reactor).

Description	Value
Density of air ρ_g (kg/m^3)	1.29
Dynamic viscosity of air μ_g (Pa·s)	1.78×10^{-5}
Apparent density of the OC ρ_s (kg/m^3)	3015
Void fraction in the downcomer ε (-)	0.477
Mean diameter of the OC d_s (mm)	0.43
Sphere coefficient of the OC φ_s (-)	0.7
Diameter of the FR D_f (m)	0.06
Diameter of the upper downcomer D_{ud} (m)	0.10

In addition, from Figure 4, we can get the optimal region of the upper pressure gradient $\Delta P_1/H_1$ under a high-flux condition of 200 kg/m^2·s, which ranged between -2.1 kPa/m and 3.0 kPa/m. Thus, by associating the optimal region with the ideal pressure gradient (1.6 kPa/m), a semi-theoretical formula of gas leakages between the two reactors (i.e., Equation (11)) could be deduced, which includes two conterminal linear equations with the ideal pressure gradient chosen as the boundary point. This formula successfully established the important mapping relationships between the gas–solid flow states in the upper dipleg and the upper pressure gradient, which should be important coupling criteria of selecting design parameters and operating conditions.

$$f_i = \begin{cases} f_1 = \alpha_1[(\Delta P_1/H_1)_t - (\Delta P_1/H_1)_i]((\Delta P_1/H_1)_t < (\Delta P_1/H_1)_i, \alpha_1 \approx 0.8) \\ f_2 = \alpha_2[(\Delta P_1/H_1)_t - (\Delta P_1/H_1)_i]((\Delta P_1/H_1)_t \geq (\Delta P_1/H_1)_i, \alpha_2 \approx 2.2) \end{cases} \tag{11}$$

3.2.2. Semi-Theoretical Formulas of the Lower Pressure Gradient

From the analyses shown in Section 3.1, we can find that the optimal operation region of the lower dipleg should better locate between State 5 (i.e., the ideal sealing state) and State 6 in order to ensure stable solid circulation with acceptable gas leakage.

The correlation equation of the lower pressure gradient under the ideal sealing state was also derived from the modified Ergun equation [33–35], as shown in Equation (12). Here, it should be noted that the lower downcomer was designed to be cuboid shaped (0.1 m length $\times$ 0.1 m width) and the upper downcomer was cylinder shaped, which made the form of Equation (12) a bit different from that of Equation (10). Thus, the theoretical value of the ideal lower pressure gradient could be deduced to be about 1.3 kPa/m. Further, according to our previous experimental results [20], the optimal region of $\Delta P_2/H_2$ under a high-flux condition of 200 kg/m^2·s should be limited within 6.0 kPa/m in order to guarantee the J-valve leakage ratio lower than 20%. Thus, by associating the optimal region of the lower pressure gradient with the ideal pressure gradient, a semi-theoretical formula of J-valve gas leakage (i.e., Equation (13)) could be deduced, in which the coefficient β was used as the slope. Similarly, with Equation (11), this formula established the mapping relationships between the J-valve gas leakage and the lower pressure gradient, enabling a coupling criterion of selecting design parameters and operating conditions during the CDCLC process.

$$(\Delta P_2/H_2)_i = \left[150\frac{\pi}{4}\frac{(1-\varepsilon)\mu_g}{\rho_s(\varepsilon\varphi_s d_s)^2}\left(\frac{D_f}{L_{ld}}\right)^2\right]G_s + \left[1.75\left(\frac{\pi}{4}\right)^2\frac{\rho_g}{\varepsilon\varphi_s d_s(1-\varepsilon)\rho_s^2}\left(\frac{D_f}{L_{ld}}\right)^4\right]G_s^2 \tag{12}$$

$$f_3 = \beta[(\Delta P_2/H_2)_t - (\Delta P_2/H_2)_i]((\Delta P_2/H_2)_t \geq (\Delta P_2/H_2)_i, \beta \approx 4.3) \tag{13}$$

3.3. Theoretical Methodology for Circulation Stability

In a real CDCLC application, a critical sealing state (i.e., State 7 shown in Figure 3) can be used to identify the advent of circulation collapse. Therefore, it is also necessary to understand the critical sealing ability of the CDCLC system so as to prevent the emergency situation of operation instability. Here, an empirical formula of critical sealing proposed by Chang et al. [32] (see Equation (14)) was attempted to be applied in this high-flux CDCLC system, in which the coefficient γ was between 0.6–0.7.

$$\left|\frac{\Delta P}{H}\right|_c = \gamma g_c[\rho_s(1-\varepsilon) - 136] \tag{14}$$

Figure 6 exhibits the comparison of predicted and experimental upper pressure gradients under the critical sealing state. It can be found that, with a solid circulation flux of 250 kg/m^2·s, the experimental value of the critical sealing gradient was 10.7 kPa/m under an upper dipleg height of 1.07 m [20]. In order to ensure the accuracy of test measurement, another dipleg height (0.87 m) was adopted for the measure of the critical sealing gradient while the other operating conditions were kept constant. It can be seen that these two experimental results (10.9 kPa/m for 0.87 m height, and 10.7 kPa/m for 1.07 m height) were very close to each other, demonstrating the constancy of the critical sealing gradient. On the other hand, the calculation value of the critical sealing gradient based on Equation (14) was between 8.5 kPa/m and 9.9 kPa/m. Thus, the relative error between the measured and predicted values of the critical sealing gradient could be further calculated to be lower than 21% for γ of 0.6, and 8% for γ of 0.7, demonstrating the application feasibility of Chang et al. [32] equation in the prediction of the critical pressure gradient of the high-flux CDCLC system. Moreover, the value of 0.7 for the coefficient γ seems to be more suitable for this system, in view of the least relative error with the experimental values. Therefore, the semi-theoretical formula for the circulation stability of this high-flux system can be finally expressed as Equation (15).

$$(\Delta P/H)_c = 0.7g_c[\rho_s(1-\varepsilon) - 136]$$

(15)

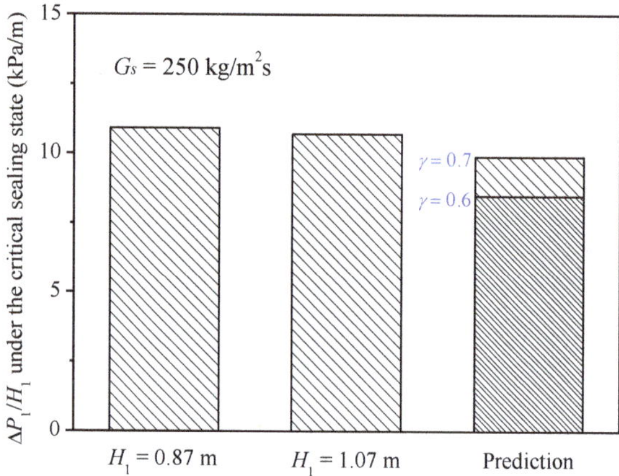

Figure 6. Comparison of predicted and experimental upper pressure gradients under the critical sealing state.

3.4. Theoretical Methodology Application to Condition Designs of the Cold System

Based on the theoretical methodology for the gas leakages and solid circulation, we proposed an optimal operation condition of the cold CDCLC system. Firstly, considering the feature and requirement of the high-flux operation, we selected a higher value of 300 kg/m^2·s as the solid circulation flux G_s while the corresponding FR superficial gas velocity $U_{f,sta}$ and the inlet air flow rate of the AR $Q_{4,sta}$ were set to be 10.7 m/s and 44 m^3/h, respectively. Thus, according to Equations (10) and (12), the theoretical ideal pressure gradients of the upper dipleg and the lower dipleg should be about 2.5 kPa/m and 1.9 kPa/m, respectively. Then, on the basis of the above semi-theoretical formulas (i.e., Equations (11) and (13)), we could further deduce that the optimal regions for gas leakage restraint were about −1.3 to 3.9 kPa/m for the upper pressure gradient $\Delta P_1/H_1$, and about 1.9 to 6.6 kPa/m for the lower pressure gradient $\Delta P_2/H_2$. Under this premise, we selected 3.8 kPa/m and 5.2 kPa/m as the proposed values of $\Delta P_1/H_1$ and $\Delta P_2/H_2$, respectively.

Figure 7 exhibits the whole-system pressure profile and the apparent solids holdup along the FR height under the proposed operation condition. As shown in Figure 7a, the pressures of each part under the high-flux condition were smoothly connected with each other, demonstrating the operation stability and the favorable coupling between each component. Besides, the high-pressure drop in the FR and low-pressure drop in the AR, they successfully exhibited the operation features of HFCFB and CFMB. From Figure 7b, we further observed the high solids holdup along the whole FR height, indicating the positive effect of high solid circulation flux on the efficiencies of gas–solid contact and reaction [14]. In addition, Table 2 summarizes the pressure gradients and gas leakage ratios under the proposed operation condition. It can be found that, although the upper pressure gradient $\Delta P_1/H_1$ close to the upper limit of the optimal region, the FR leakage ratio f_1 (−0.1%) and the AR leakage ratio f_2 (2.5%) can still be limited within their limits (i.e., −3% for f_1 and 3% for f_2), demonstrating the feasibility of the semi-theoretical Equation (11) for the prediction of the optimal region for gas leakage control. On the other hand, the lower pressure gradient $\Delta P_2/H_2$ was located at a value of 5.2 kPa/m, in which the J-valve leakage ratio f_3 (15.2%) could be kept within a proposed region (<20%) in order to ensure a favorable solid circulation.

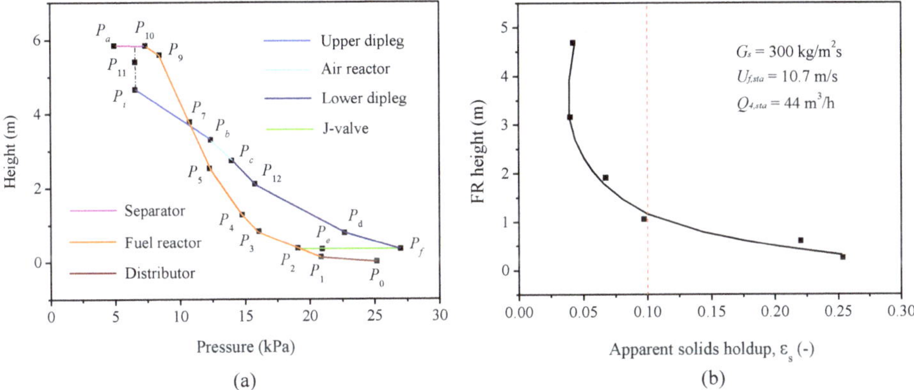

Figure 7. The pressure profile of the whole system and the apparent solids holdup along the FR under the proposed pressure gradient condition: (**a**) pressure profile, and (**b**) apparent solids holdup.

Table 2. Pressure gradients and gas leakage ratios under the proposed operation condition (AR: air reactor).

Description	Parameter	Measured Value (%)	Calculation Value (%)	Relative Error (%)
$\Delta P_1/H_1 = 3.8$ kPa/m	FR leakage ratio f_1	−0.1	0	-
	AR leakage ratio f_2	2.5	2.9	14
$\Delta P_2/H_2 = 5.2$ kPa/m	J-valve leakage ratio f_3	15.2	14.2	7

In general, the system operation stability, the high-flux feature, and particularly the gas leakage restraint were successfully achieved in this proposed operation condition, indicating the application feasibility of the semi-theoretical methodology to the system optimal design.

3.5. Hot State Application Assessment of the Theoretical Methodology

From the theoretical methodology of AR coupling principle with the high-flux CDCLC system, we could further carry out a capability assessment of the system in terms of the restraint of gas leakages and the stability of solid circulation under hot states. To facilitate the analysis and comparison, the structure parameters and OC material of the cold system were also adopted in the hot system. Besides, the solid circulation flux in the hot state was also selected as 300 kg/m^2·s so as to keep consistent with the proposed cold-state operation condition mentioned above. The only difference was that under the hot state, the operating temperature was as high as 1243 K with the corresponding dynamic viscosity $\mu_{g,hot} = 4.7 \times 10^{-5}$ Pa·s. Table 3 details the parameters for the calculation of ideal pressure gradients under the hot state.

According to Equation (10), the theoretical ideal pressure gradient of the upper dipleg under the hot state should be about 6.4 kPa/m. Similarly, according to Equation (12), the theoretical ideal pressure gradient of the lower dipleg under the hot state was about 5.0 kPa/m. Figure 8 shows the comparison of the ideal pressure gradients between the hot and cold states. We can observe the ideal pressure gradients of the two diplegs under the hot state were about 2.6 times of those under the cold state. This implies a lower requirement of the sealing height in the hot state, which should be beneficial for the spatial arrangement of the hot-state system. In addition, on the basis of the semi-theoretical formulas for gas leakage restraint (i.e., Equations (11) and (13)), the optimal regions were further calculated to be about 2.7 to 7.8 kPa/m for $\Delta P_1/H_1$, and about 5 to 9.7 kPa/m for $\Delta P_2/H_2$. On the other hand, according to Equation (15), the critical pressure gradient for the circulation stability could be deduced to be about 9.9 kPa/m. It can be found that the upper limit of the optimal region of $\Delta P_2/H_2$

(9.7 kPa/m) for gas leakage restraint was very close to the critical pressure gradient for the circulation stability (9.9 kPa/m), demonstrating the rationality of the choice of 20% as the upper limit standard of the J-valve leakage. Certainly, it should be noted that the approach of the optimal pressure gradients for gas leakages to the critical pressure gradient for circulation stability also means the increase in the risk of circulation collapse during the hot-state operation process.

Table 3. Parameters for the calculation of the ideal pressure gradients under the hot state.

Description	Value
Temperature (K)	1243
Solid circulation flux G_s (kg/m^2·s)	300
Gas dynamic viscosity under the hot state $\mu_{g,hot}$ (Pa·s)	4.7×10^{-5}
Apparent density of the OC ρ_s (kg/m^3)	3015
Void fraction in the downcomer ε (-)	0.477
Mean diameter of the OC d_s (mm)	0.43
Sphere coefficient of the OC φ_s (-)	0.7
FR diameter D_f (m)	0.06
Upper downcomer diameter D_{ud} (m)	0.10
Side length of the lower downcomer L_{ld} (m)	0.10

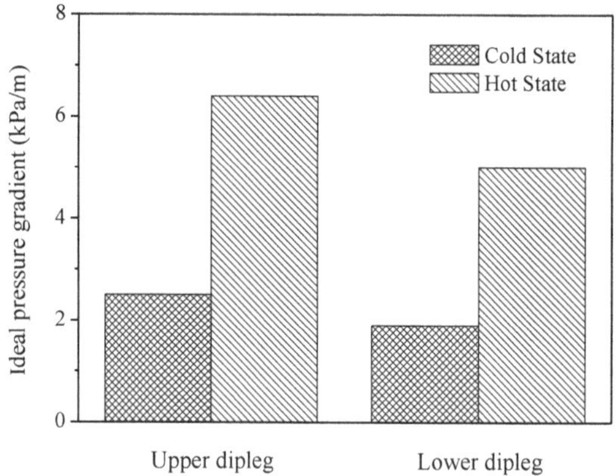

Figure 8. Comparison of the ideal pressure gradients between the hot and cold states.

For real high-flux CDCLC applications, we can first get the optimal sizes of the reactors and the downcomer based on the ideal pressure gradient equations and the solid circulation flux range in the design process. Then during the operating process, the relevant parameters (i.e., the solid-seal heights in the downcomer, the pressures of the FR and AR, the solid circulation flux, and the gas flow rates) can be adjusted flexibly and optimally to make sure the pressure gradients within the optimal regions for a favorable performance of operation and reaction.

4. Conclusions

Built upon the previous experimental studies of a high-flux CDCLC system, the objective of this study is to further investigate the fundamental coupling mechanism of the AR, and develop a comprehensive theoretical methodology to the system optimal design with favorable operation stability and low gas leakages. The following conclusions can be drawn from the present study:

(1)   During the CDCLC process, the dipleg flow can situate at a pressure region across the positive and negative pressure gradients, which can be categorized into seven flow states. Considering the gas leakages and the circulation stability, the upper dipleg of the AR was recommended

to be operated among State 2 to 6 while the lower dipleg of the AR should better run between States 5 and 6.

(2)   The gas leakages between the two reactors were expressed as two conterminal linear equations with the ideal pressure gradient chosen as the boundary point, which can be used to predict the optimal regions of the upper pressure gradient. Similarly, the J-valve leakage within the optimal region was expressed as a linear function of the lower pressure gradient of the AR. In addition, an empirical formula of critical sealing was developed for this high-flux CDCLC system, which can be used to identify the advent of circulation collapse so as to guarantee the operation stability.

(3)   The theoretical methodology for gas leakages and solid circulation was successfully applied to the condition design and operation of the cold system, achieving favorable gas–solid flow and circulation together with good control of gas leakages in the whole system.

(4)   The theoretical methodology was adopted to carry out a capability assessment of the high-flux CDCLC system under a hot state in terms of the restraint of gas leakages and the stability of solid circulation. The ideal pressure gradients under the hot state of 1243 K were about 2.6 times than those under the cold state, implying a lower requirement of sealing height in the hot state. However, on the other hand, the increase of the ideal pressure gradients also led to the approach of the optimal pressure gradients for gas leakages to the critical pressure gradients for circulation stability, which would increase the risk of circulation collapse during the operation process.

Author Contributions: conceptualization, X.W.; investigation, X.W. and X.L.; formal analysis, X.W. and X.L.; supervision, B.J.; writing—original draft preparation, X.W., X.L., and Z.J.; writing—review and editing, J.Z.

Funding: This research was funded by the National Natural Science Foundation of China (grant numbers 51806035, 51741603), the Natural Science Foundation of Jiangsu Province (grant number BK20170669), the Fundamental Research Funds for the Central Universities (grant number 2242018K40117), and the Guangdong Provincial Key Laboratory of New and Renewable Energy Research and Development (grant number Y707s41001).

Conflicts of Interest: The authors declare no conflict of interest.

Nomenclature

A_{ud}	sectional area of the upper downcomer (m^2)
A_f	sectional area of the upper FR (m^2)
d_s	mean diameter of the OC particles (mm)
D_f	FR diameter (m)
D_{ud}	upper downcomer diameter (m)
f_1	FR leakage ratio
f_2	AR leakage ratio
f_3	J-valve leakage ratio
f_i	gas leakage ratio between the two reactors ($i = 1$ or 2)
g	acceleration due to gravity (9.8 m/s^2)
g_c	conversion coefficient (9.8 N/kg)
G_s	solid circulation flux (kg/m^2·s)
H_1	solid-seal height of the upper dipleg of the AR (m)
H_2	solid-seal height of the lower dipleg of the AR (m)
ΔH	scale height in the upper dipleg of the AR (m)
L_{ld}	side length of the lower downcomer (m)
P_b	pressure of the AR outlet (kPa)
P_c	pressure of the AR inlet (kPa)
P_d	top pressure of the lower dipleg of the AR (kPa)
P_i	pressure at the interface of the dense phase and dilute phase of the upper downcomer (kPa)

P_{11}	pressure at the underside of the separator (kPa)
P_{12}	pressure at the top position of the lower dipleg (kPa)
ΔP_Z	local pressure drop at two adjacent elevations of the FR (kPa)
$(\Delta P/H)_c$	critical pressure gradient for circulation stability (kPa/m)
$\Delta P_1/H_1$	upper pressure gradient of the AR (kPa/m)
$(\Delta P_1/H_1)_i$	upper pressure gradient of the AR under the ideal sealing state (kPa/m)
$(\Delta P_1/H_1)_t$	transient upper pressure gradient of the AR (kPa/m)
$\Delta P_2/H_2$	lower pressure gradient of the AR (kPa/m)
$(\Delta P_2/H_2)_i$	lower pressure gradient of the AR under the ideal sealing state (kPa/m)
$(\Delta P_2/H_2)_t$	transient lower pressure gradient of the AR (kPa/m)
$Q_{1,sta}$	inlet air flow rate of the FR distributor (m^3/h)
$Q_{2,sta}$	fluidizing air flow rate of the J-valve (m^3/h)
$Q_{3,sta}$	aeration air flow rate of the J-valve (m^3/h)
$Q_{4,sta}$	inlet air flow rate of the AR (m^3/h)
$Q_{a,sta}$	outlet air flow rate of the FR (m^3/h)
$Q_{b,sta}$	outlet air flow rate of the AR (m^3/h)
t	measured duration of the OC particles passing through the scale height (s)
T	operation temperature (K)
u_s	velocity of the OC particles in the upper dipleg (m/s)
$U_{f,sta}$	FR superficial gas velocity (m/s)
U_s	solids velocity (m/s)
$x_{a,CO}$	concentration of tracer gas 1 measured at the outlet of the separator (ppm)
$x_{b,CO}$	concentration of tracer gas 1 measured at the AR outlet (ppm)
$x'_{a,CO}$	concentration of tracer gas 2 measured at the outlet of the separator (ppm)
$x'_{b,CO}$	concentration of tracer gas 2 measured at the AR outlet (ppm)
$x''_{a,CO}$	concentration of tracer gas 3 measured at the outlet of the separator (ppm)
$x''_{b,CO}$	concentration of tracer gas 3 measured at the AR outlet (ppm)
ΔZ	height difference between two adjacent elevations of the FR (m)
α_1	slope of the linear fitting equation of FR leakage ratio
α_2	slope of the linear fitting equation of AR leakage ratio
β	slope of the linear fitting equation of J-valve leakage ratio
γ	dimensionless coefficient of the fitting equation of critical sealing gradient
ε	void fraction in the downcomer
ε_s	cross-sectional average solids holdup in the FR
φ_s	sphere coefficient of the OC particles
ρ_b	bulk density of the OC particles (kg/m^3)
ρ_g	density of air (kg/m^3)
ρ_s	apparent density of the OC particles (kg/m^3)
μ_g	dynamic viscosity of air (Pa·s)

References

1. Kim, H.R.; Wang, D.; Zeng, L.; Bayham, S.; Tong, A.; Chung, E.; Kathe, M.V.; Luo, S.; McGiveron, O.; Wang, A.; et al. Coal direct chemical looping combustion process: Design and operation of a 25-kWth sub-pilot unit. *Fuel* **2013**, *108*, 370–384. [CrossRef]
2. Bayham, S.C.; Kim, H.R.; Wang, D.; Tong, A.; Zeng, L.; McGiveron, O.; Kathe, M.V.; Chung, E.; Wang, W.; Wang, A.; et al. Iron-based coal direct chemical looping combustion process: 200-h continuous operation of a 25-kWth subpilot unit. *Energy Fuels* **2013**, *27*, 1347–1356. [CrossRef]
3. Berguerand, N.; Lyngfelt, A. Design and operation of a 10 kW$_{th}$ chemical-looping combustor for solid fuels-testing with South African coal. *Fuel* **2008**, *87*, 2713–2726. [CrossRef]
4. Shen, L.H.; Wu, J.H.; Xiao, J. Experiments on chemical looping combustion of coal with a NiO based oxygen carrier. *Combust. Flame* **2009**, *156*, 721–728. [CrossRef]
5. Xiao, R.; Chen, L.; Saha, C.; Zhang, S.; Bhattacharya, S. Pressurized chemical-looping combustion of coal using an iron ore as oxygen carrier in a pilot-scale unit. *Int. J. Greenh. Gas Control* **2012**, *10*, 363–373. [CrossRef]

6. Abad, A.; Gayán, P.; de Diego, L.F.; García-Labiano, F.; Adánez, J. Fuel reactor modelling in chemical-looping combustion of coal: 1. Model formulation. *Chem. Eng. Sci.* **2013**, *87*, 277–293. [CrossRef]

7. García-Labiano, F.; de Diego, L.F.; Gayán, P.; Abad, A.; Adánez, J. Fuel reactor modelling in chemical-looping combustion of coal: 2-simulation and optimization. *Chem. Eng. Sci.* **2013**, *87*, 173–182. [CrossRef]

8. Wang, X.; Jin, B.; Zhang, Y.; Zhang, Y.; Liu, X. Three dimensional modeling of a coal-fired chemical looping combustion process in the circulating fluidized bed fuel reactor. *Energy Fuels* **2013**, *27*, 2173–2184. [CrossRef]

9. Markström, P.; Linderholm, C.; Lyngfelt, A. Operation of a 100 kW chemical-looping combustor with Mexican petroleum coke and Cerrejón coal. *Appl. Energy* **2014**, *113*, 1830–1835. [CrossRef]

10. Ströhle, J.; Orth, M.; Epple, B. Design and operation of a 1 MW_{th} chemical looping plant. *Appl. Energy* **2014**, *113*, 1490–1495. [CrossRef]

11. Thon, A.; Kramp, M.; Hartge, E.-U.; Heinrich, S.; Werther, J. Operational experience with a system of coupled fluidized beds for chemical looping combustion of solid fuels using ilmenite as oxygen carrier. *Appl. Energy* **2014**, *118*, 309–317. [CrossRef]

12. Bayham, S.; McGiveron, O.; Tong, A.; Chung, E.; Kathe, M.; Wang, D.; Zeng, L.; Fan, L.S. Parametric and dynamic studies of an iron-based 25-kW_{th} coal direct chemical looping unit using sub-bituminous coal. *Appl. Energy* **2015**, *145*, 354–363. [CrossRef]

13. Wang, X.; Jin, B.; Liu, X.; Zhang, Y.; Liu, H. Experimental investigation on flow behaviors in a novel in situ gasification chemical looping combustion apparatus. *Ind. Eng. Chem. Res.* **2013**, *52*, 14208–14218. [CrossRef]

14. Wang, X.; Jin, B.; Liu, H.; Wang, W.; Liu, X.; Zhang, Y. Optimization of in situ gasification chemical looping combustion through experimental investigations with a cold experimental system. *Ind. Eng. Chem. Res.* **2015**, *54*, 5749–5758. [CrossRef]

15. Wang, X.; Jin, B.; Zhu, X.; Liu, H. Experimental evaluation of a novel 20 kW_{th} in situ gasification chemical looping combustion unit with an iron ore as the oxygen carrier. *Ind. Eng. Chem. Res.* **2016**, *55*, 11775–11784. [CrossRef]

16. Adánez, J.; Abad, A.; Perez-Vega, R.; Luis, F.; García-Labiano, F.; Gayán, P. Design and operation of a coal-fired 50 kW_{th} chemical looping combustor. *Energy Procedia* **2014**, *63*, 63–72. [CrossRef]

17. Ma, J.; Zhao, H.; Tian, X.; Wei, Y.; Rajendran, S.; Zhang, Y.; Bhattacharya, S.; Zheng, C. Chemical looping combustion of coal in a 5 kW_{th} interconnected fluidized bed reactor using hematite as oxygen carrier. *Appl. Energy* **2015**, *157*, 304–313. [CrossRef]

18. Su, M.; Zhao, H.; Ma, J. Computational fluid dynamics simulation for chemical looping combustion of coal in a dual circulation fluidized bed. *Energy Convers. Manag.* **2015**, *105*, 1–12. [CrossRef]

19. Ohlemüller, P.; Alobaid, F.; Abad, A.; Adanez, J.; Ströhle, J.; Epple, B. Development and validation of a 1D process model with autothermal operation of a 1 MW_{th} chemical looping pilot plant. *Int. J. Greenh. Gas Control* **2018**, *73*, 29–41. [CrossRef]

20. Wang, X.; Liu, X.; Jin, B.; Wang, D. Hydrodynamic study of AR coupling effects on solid circulation and gas leakages in a high-flux in situ gasification chemical looping combustion system. *Processes* **2018**, *6*, 196. [CrossRef]

21. Wang, X.; Jin, B.; Zhong, W.; Zhang, M.; Huang, Y.; Duan, F. Flow behaviors in a high-flux circulating fluidized bed. *Int. J. Chem. React. Eng.* **2008**, *6*, A79. [CrossRef]

22. Issangya, A.S.; Bai, D.; Bi, H.T.; Lim, K.S.; Zhu, J.; Grace, J.R. Suspension densities in a high-density circulating fluidized bed riser. *Chem. Eng. Sci.* **1999**, *54*, 5451–5460. [CrossRef]

23. Namkung, W.; Kim, S.W.; Kim, S.D. Flow regimes and axial pressure profiles in a circulating fluidized bed. *Chem. Eng. J.* **1999**, *54*, 245–252. [CrossRef]

24. Li, Z.Q.; Wu, C.N.; Wei, F.; Jin, Y. Experimental study of high-density gas-solids flow in a new coupled circulating fluidized bed. *Powder Technol.* **2004**, *139*, 214–220. [CrossRef]

25. Jing, S.; Li, H. Study on the flow of fine powders from hoppers connected to a moving-bed standpipe with negative pressure gradient. *Powder Technol.* **1999**, *101*, 266–278. [CrossRef]

26. Nagashima, H.; Ishikura, T.; Ide, M. Flow characteristics of a small moving bed downcomer with an orifice under negative pressure gradient. *Powder Technol.* **2009**, *192*, 110–115. [CrossRef]

27. Wang, J.; Bouma, J.H.; Dries, H. An experimental study of cyclone dipleg flow in fluidized catalytic cracking. *Powder Technol.* **2000**, *112*, 221–228. [CrossRef]

28. Wang, Q.; Yang, H.; Wang, P.; Lu, J.; Liu, Q.; Zhang, H.; Wei, L.; Zhang, M. Application of CPFD method in the simulation of a circulating fluidized bed with a loop seal, part I—Determination of modeling parameters. *Powder Technol.* **2014**, *253*, 814–821. [CrossRef]

29. Wang, Q.; Yang, H.; Wang, P.; Lu, J.; Liu, Q.; Zhang, H.; Wei, L.; Zhang, M. Application of CPFD method in the simulation of a circulating fluidized bed with a loop seal Part II—Investigation of solids circulation. *Powder Technol.* **2014**, *253*, 822–828. [CrossRef]

30. Yin, S.; Jin, B.; Zhong, W.; Lu, Y.; Shao, Y.; Liu, H. Gas-solid flow behavior in a pressurized high-flux circulating fluidized bed riser. *Chem. Eng. Commun.* **2014**, *201*, 352–366. [CrossRef]

31. Wang, Z.; Zhao, T.; Yao, J.; Liu, K.; Takei, M. Influence of particle size on the exit effect of a full-scale rolling circulating fluidized bed. *Part. Sci. Technol.* **2018**, *36*, 541–551. [CrossRef]

32. Chang, G.; Yang, G.; Yang, S.; Li, S.; Li, H. Gas solid flow in a moving bed under negative pressure difference. *J. Chem. Ind. Eng.* **1980**, *31*, 229–240.

33. Ergun, S. Fluid flow through packed columns. *Chem. Eng. Prog.* **1952**, *48*, 89–94.

34. Leung, L.S.; Jones, P.J. Flow of gas—Solid mixtures in standpipes. A review. *Powder Technol.* **1978**, *20*, 145–160. [CrossRef]

35. Ozahi, E.; Gundogdu, M.Y.; Carpinlioglu, M.Ö. A modification on Ergun's correlation for use in cylindrical packed beds with non-spherical particles. *Adv. Powder Technol.* **2008**, *19*, 369–381. [CrossRef]

Article

Hydrodynamic Study of AR Coupling Effects on Solid Circulation and Gas Leakages in a High-Flux In Situ Gasification Chemical Looping Combustion System

Xiaojia Wang *, Xianli Liu, Baosheng Jin and Decheng Wang

Key Laboratory of Energy Thermal Conversion and Control of Ministry of Education, School of Energy and Environment, Southeast University, Nanjing 210096, China; liuxl_seu@163.com (X.L.); bsjin@seu.edu.cn (B.J.); dechengw76@126.com (D.W.)
* Correspondence: xiaojiawang@seu.edu.cn

Received: 27 September 2018; Accepted: 16 October 2018; Published: 18 October 2018

Abstract: In situ gasification chemical looping combustion (iG-CLC) is a novel and promising coal combustion technology with inherent separation of CO_2. Our previous studies demonstrated the feasibility of performing iG-CLC with a high-flux circulating fluidized bed (HFCFB) riser as the fuel reactor (FR) and a counter-flow moving bed (CFMB) as the air reactor (AR). As an extension of that work, this study aims to further investigate the fundamental effects of the AR coupling on the oxygen carrier (OC) circulation and gas leakages with a cold-state experimental device of the proposed iG-CLC system. The system exhibited favorable pressure distribution characteristics and good adaptability of solid circulation flux, demonstrating the positive role of the direct coupling method of the AR in the stabilization and controllability of the whole system. The OC circulation and the gas leakages were mainly determined by the upper and lower pressure gradients of the AR. With the increase in the upper pressure gradient, the OC circulation flux increased initially and later decreased until the circulation collapsed. Besides, the upper pressure gradient exhibited a positive effect on the restraint of gas leakage from the FR to the AR, but a negative effect on the suppression of gas leakage from the AR to the FR. Moreover, the gas leakage of the J-valve to the AR, which is directly related to the solid circulation stability, was exacerbated with the increase of the lower pressure gradient of the AR. In real iG-CLC applications, the pressure gradients should be adjusted flexibly and optimally to guarantee a balanced OC circulation together with an ideal balance of all the gas leakages.

Keywords: in situ gasification chemical looping combustion; high-flux circulating fluidized bed; counter-flow moving bed; gas leakage; coupling mechanism

1. Introduction

Chemical looping combustion (CLC), which possesses an inherent feature of isolating CO_2 during the combustion process, has been regarded as a promising novel combustion technology with a low energy penalty for carbon capture [1,2]. A conventional CLC system is mainly comprised of a fuel reactor (FR) and an air reactor (AR). The fuel introduced into the FR is oxidized to CO_2 and H_2O by a solid oxygen carrier (OC). The reduced OC particles are then transferred to the AR where they are re-oxidized upon contact with air. Compared to conventional combustion methods, the fuel will no longer mix with N_2 in a CLC process, by means of the circulation of OC between the two reactors. Therefore, the flue gas leaving the FR will only contain CO_2 and H_2O with a complete conversion of the fuel, which enables efficient and energy-saving CO_2 capture via the condensation of H_2O [3,4].

On the basis of fuel types, CLC technology can be broadly categorized into gas-fueled CLC and solid-fueled CLC. So far, studies on gas-fueled CLC have been extensively conducted in different prototype reactors [5–14]. In recent years, considering the price advantage of solid fuels (coal as the main representative) in comparison with those of gaseous fuels, solid-fueled CLC began to attract increasing attention [15–21]. One of the coal-fueled CLC approaches is the so-called in situ gasification chemical looping combustion (iG-CLC) which involves three reaction steps between two reactors [20,21]. As shown in Figure 1, two of the reaction steps occur in the FR, i.e., the coal gasification reaction (Step 1) and the subsequent oxidation reactions of the gasification gases by the OC (Step 2). The other reaction step proceeds in the AR, i.e., the reoxidation reaction of the OC by air (Step 3).

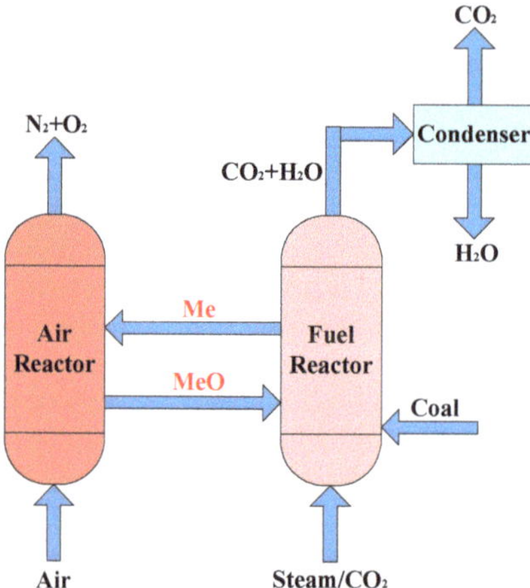

Figure 1. Schematic diagram of an in situ gasification chemical looping combustion (iG-CLC) system.

Up to now, some experimental studies on iG-CLC systems have been carried out with various reactor designs [16,18,22–28]. An iG-CLC configuration with a bubbling fluidized bed (BFB) as the FR and a circulating fluidized bed (CFB) as the AR was first proposed by Berguerand and Lyngfelt [16] from Chalmers University of Technology (10 kW_{th}), in view of the advantages of sufficient solid residence time in the BFB and favorable gas-solid contact in the CFB. Still, with CFBs as the ARs, some other FR designs have also been put forward. Shen et al. [18] from Southeast University adopted a spout-fluid bed (SFB) instead of the BFB as the FR (10 kW_{th}), in which the strong solid mixing enhanced the gas-solid contact and reaction efficiencies. Thon et al. [22] from Hamburg University of Technology designed a two-stage BFB as the FR (25 kW_{th}) for the purpose of enhancing the conversion of combustible gases. Bayham et al. [23] from Ohio State University utilized a counter-current moving bed as the FR (25 kW_{th}) with OC particles flowing downwards and gases flowing upwards, showing high competitiveness in the combustion performance and CO_2 purity. Adánez et al. [24] from CSIC proposed a CFB as the FR (20 kW_{th}) in order to achieve intense gas-solid contact and reaction. This kind of dual circulating fluidized bed (DCFB) design (i.e., both the FR and AR are CFBs) can also be found in the units by Markström et al. [25] from Chalmers University of Technology (100 kW_{th}), Ma et al. [26] from Huazhong University of Science and Technology (5 kW_{th}), and Ströhle et al. [27] from Technische Universität Darmstadt (1 MW_{th}). In addition, some other units with different design concepts have also been constructed and operated. Xiao et al. [28] from Southeast University developed a 50 kW_{th} iG-CLC unit where the AR and FR were designed as a BFB and a CFB, respectively. Pressurized conditions were achieved in their work, demonstrating the positive role of high-pressure operation in carbon conversion, CO_2 capture concentration, and combustion efficiency.

In our previous studies, we also proposed an iG-CLC system based on the high-flux circulating fluidized bed (HFCFB) technology [29–32]. Specifically, an HFCFB riser was designed as the FR because it can provide not only a favorable gas-solid contact but also sufficient solid holdups over the whole reactor height, which should be very beneficial to the interphase reaction efficiency. Specifically, compared to the common CFB FRs [24–27], the higher OC inventory in this HFCFB FR can potentially compensate for the low reactivity of OC to a certain degree, and hence this high-flux iG-CLC system will have a large advantage on the operation cost by virtue of the use of low-grade natural iron ores with a lower reactivity as the OCs. Certainly, the high-flux operation will inevitably lead to a higher pressure drop. Therefore, greater energy consumption is required to compensate for the pressure loss. However, compared to the cost reduction from the use of cheap OCs, the slight increase in energy consumption should be acceptable. A counter-flow moving bed (CFMB), in view of its advantages of steady solid flow, low pressure drop, and compact structure, was employed as the AR. Besides, it can be simply placed in the middle of the HFCFB downcomer to enhance the simplicity and stabilization of the whole system. Moreover, an inertial separator-based carbon stripper was specially designed to separate the reduced OC from the FR to the AR for reoxidation, and recirculate the unconverted char for further conversion.

Based on the design concept of this high-flux iG-CLC system, we successively built and tested a cold visualization experimental device operating at the ambient temperature [30,31] and a hot pilot-scale 20 kW$_{th}$ unit operating at the high-temperature conditions between 800–1000 °C [32]. After a series of tests, the steady operation with favorable gas-solid flow and reaction performance of the whole system could be realized under certain conditions, preliminarily verifying the feasibility of this design. With the deepening of the research, we have found that, compared to most iG-CLC systems with indirect serial structures of the two reactors [16,24–28], the direct coupling of the CFMB AR into the downcomer of the HFCFB FR in our system indeed contributes to the stabilization and controllability of the whole system. However, this coupling method of the AR also inevitably brings about greater challenges on the control of gas leakages between the two reactors, which will, in turn, affect the matching principle of reactors, such as the OC circulation characteristics.

The aim of this study is to investigate the fundamental effects of the AR coupling on the system operation stability and gas leakages. The pressure distribution characteristics of the whole system and the adaptability of solid circulation flux were first investigated with the cold-state visualization device for the proposed iG-CLC system. The effects of the upper and lower pressure gradients of the AR on the OC circulation and the gas leakages were further established. Moreover, by giving consideration to the AR coupling effects under various operational conditions, one optimal operating condition was recommended to demonstrate the adjustment feasibility for balanced solids circulation with low gas leakages during the iG-CLC process.

2. Materials and Methods

2.1. Experimental Device

Figure 2 gives a schematic representation of the experimental device for the proposed iG-CLC system under cold conditions. Here, only a brief description of the experimental setup is provided. A more detailed description can be found in our initial studies of this system [31].

2.1.1. Main Assembly

The main assembly predominantly consists of a FR (5), a carbon stripper (6), a downcomer (7), an AR (8), a J-valve (11), and a bag filter (12).

The FR (5) is an HFCFB riser with a height of 5.8 m and an inner diameter of 60 mm. The AR (8) is a CFMB, mainly consisting of a gas inlet (10), a gas outlet (9), a cylindrical channel (0.418 m inner diameter × 0.5 m height), and a cone channel (0.2 m height). The carbon stripper (6) is an inertial separator. For the purpose of visualization, some sections of the FR, the AR, and the downcomer are made of plexiglas.

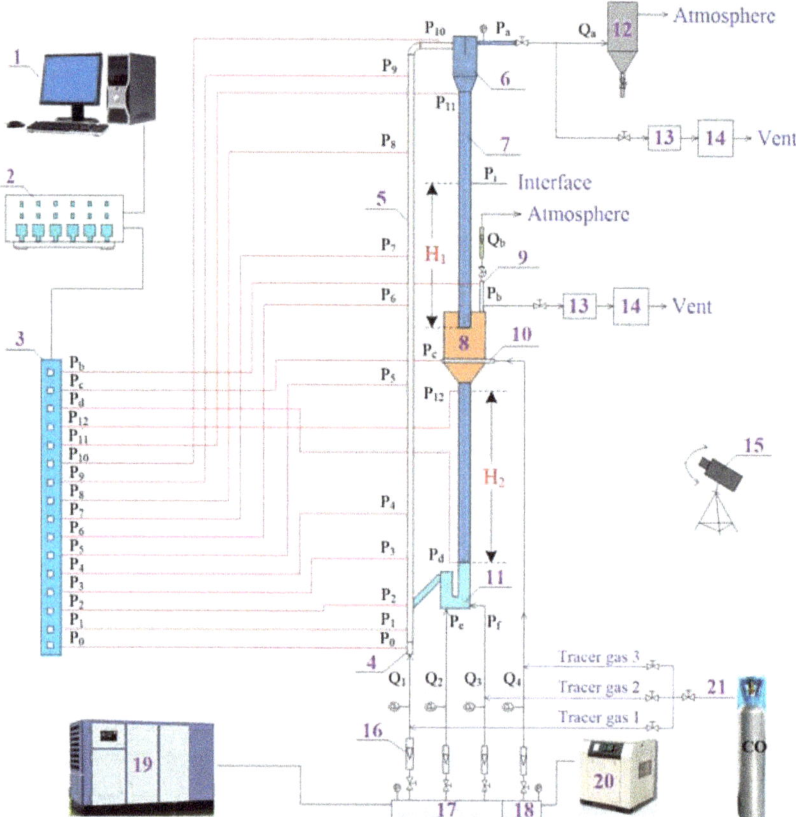

Figure 2. Schematic of the visualization experimental device of the iG-CLC system. 1—computer, 2—A/D converter, 3—differential pressure transducer, 4—riser gas chamber, 5—fuel reactor (FR), 6—carbon stripper, 7—downcomer, 8—air reactor (AR), 9—AR gas outlet, 10—AR gas inlet, 11—J-valve, 12—bag filter, 13—filter, 14—gas analyzer, 15—digital camera, 16—rotameter, 17—FR gas chamber, 18—AR gas chamber, 19—90 kW air compressor, 20—18 kW air compressor, 21—tracer gas. P—pressure gauge, Q—gas flow.

2.1.2. Gas Supply System

The air stream from a 90 kW air compressor (Nanjing Compressor Co. Ltd., Nanjing, China) (19) was introduced into the FR to circulate the OC particles. Another air stream from an 18 kW air compressor (Guangzhou Panyu JOYO Air Compressor Factory, Guangzhou, China) (20) was fed into the AR through a tube distributor.

A high-purity carbon monoxide (99.99%) stream was used as the tracer gas (21) which would be fed into the FR, the AR, and the J-valve in turn during the testing stages.

2.1.3. Data Acquisition System

Gas flow rates in the FR, AR, and J-valve were controlled and measured by calibrated rotameters (16).

The pressures of monitoring nodes were measured by pressure manometers and a multi-channel differential pressure transducer (3). During the experiments, the pressures of the two reactors could be adjusted by back-pressure regulators.

The tracer gas concentrations at the outlets of the two reactors were measured by a gas analyzer (MRU, Neckarsulm, Germany) (14).

A digital camera (15) was used to photograph the flow regimes and capture some special flow phenomena through the visualization sections during the experiments.

2.2. Material

The OC used in this study was a natural iron ore from Harbin, China. Prior to the experiments, the OC particles were crushed and sieved to a mean diameter of 0.43 mm. More details on this OC material can be found in Table 1.

Table 1. Main physical properties of the oxygen carrier.

Description	Value	
	Diameter (mm)	**Mass Fraction (%)**
	>1.25	0.05
	1–1.25	0.71
	0.6–1	2.79
Particle size distribution	0.45–0.6	22.94
	0.4–0.45	45.14
	0.3–0.4	22.94
	<0.3	5.43
Mean diameter d_p (mm)	0.43	-
Apparent density ρ_s (kg/m^3)	3015	-
Bulk density ρ_b (kg/m^3)	1577	-
Minimum fluidization gas velocity U_{mf} (m/s)	0.187	-

2.3. Experimental Procedures

The particles of the OC were first packed in the downcomer. Then, the gas stream from the 90 kW air compressor was introduced into the FR from the J-valve and FR distributor to drive the OC particles for circulation. After the particle circulation became balanced, another gas stream from the 18 kW air compressor was introduced into the AR and coupled into the original circulation system. As the system rebounded to a steady state, the effects of operating parameters began to be tested with the real-time monitoring of the system flow state, by means of pressure tracking, time sampling, gas tracer, and so on. A detailed explanation of the data processing can be found in Section 2.4.

2.4. Data Evaluation

2.4.1. Gas Flow Rates

The gas flow rates in this study were all normalized to the standard state with the subscript *sta*. According to the conservation of mass, the total gas flow rate of the system can be calculated as [31]

$$Q_{in,sta} = Q_{1,sta} + Q_{2,sta} + Q_{3,sta} + Q_{4,sta} = Q_{a,sta} + Q_{b,sta} = Q_{out,sta} \tag{1}$$

where $Q_{in,sta}$ and $Q_{out,sta}$ represent the total inlet air flow rate and the total outlet air flow rate of the system, respectively. $Q_{1,sta}$, $Q_{2,sta}$, $Q_{3,sta}$, and $Q_{4,sta}$ are the inlet air flow rate of the FR distributor, the fluidizing air flow rate of the J-valve, the aeration air flow rate of the J-valve, and the inlet air flow rate of the AR, respectively. $Q_{a,sta}$ and $Q_{b,sta}$ are the outlet air flow rates of the FR and the AR, respectively.

All the inlet/outlet gas flow rates mentioned above were measured by calibrated rotameters except $Q_{a,sta}$ which can be deduced from Equation (1).

$$Q_{a,sta} = Q_{1,sta} + Q_{2,sta} + Q_{3,sta} + Q_{4,sta} - Q_{b,sta} \tag{2}$$

The FR superficial gas velocity can be calculated as:

$$U_{f,sta} = \frac{Q_{f,sta}}{A_f}$$ (3)

where A_f is the sectional area of the FR. $Q_{f,sta}$ represents the total inlet air flow rate of the FR, which can be estimated as:

$$Q_{f,sta} = Q_{1,sta} + Q_{2,sta} + (1 - f_3)Q_{3,sta}$$ (4)

where f_3 represents the J-valve leakage ratio, and a detailed explanation of this parameter can be found in Section 2.4.3.

2.4.2. Solid Circulation Flux

The solid circulation flux can be estimated as:

$$G_s = \frac{\rho_b u_s A_d}{A_f} = \frac{\rho_b A_d}{A_f}(\Delta H/t)$$ (5)

where ρ_b, u_s, and A_d represent the bulk density of the OC, the downward flow velocity of the OC particles in the upper dipleg, and the sectional area of the upper downcomer, respectively. ΔH is a scale height in the upper dipleg for the measurement of solid circulation flux, and t is the measured duration of the traced OC particles passing through the scale height.

2.4.3. Gas Leakage Ratios

The distribution of the FR exhaust gas can be measured through the use of tracer gas 1. f_1, named as FR leakage ratio, represents the gas leakage ratio of the FR into the AR [31].

$$f_1 = \frac{Q_{b,sta} x_{b,CO}}{Q_{a,sta} x_{a,CO} + Q_{b,sta} x_{b,CO}}$$ (6)

where $x_{a,CO}$, and $x_{b,CO}$ are the concentrations of tracer gas 1 measured at the outlets of the separator and the AR, respectively.

Similarly, the distribution of the exhaust gas from the AR inlet can be investigated by using tracer gas 2. f_2, named as the AR leakage ratio, represents the gas leakage ratio of the AR into the FR [31].

$$f_2 = \frac{Q_{a,sta} x'_{a,CO}}{Q_{a,sta} x'_{a,CO} + Q_{b,sta} x'_{b,CO}}$$ (7)

where $x'_{a,CO}$ and $x'_{b,CO}$ are the concentrations of tracer gas 2 measured at the outlets of the separator and the AR, respectively.

Moreover, the distribution of the J-valve aeration air can be measured by the use of tracer gas 3. f_3, the so-called J-valve leakage ratio, represents the gas leakage ratio of the aeration air into the AR.

$$f_3 = \frac{Q_{b,sta} x''_{b,CO}}{Q_{a,sta} x''_{a,CO} + Q_{b,sta} x''_{b,CO}}$$ (8)

where $x''_{a,CO}$ and $x''_{b,CO}$ are the concentrations of tracer gas 3 measured at the outlets of the separator and the AR, respectively.

2.4.4. Pressure Gradients

The upper pressure gradient ($\Delta P_1/H_1$) represents the pressure gradient between the AR and the separator.

$$\Delta P_1/H_1 = \left(\frac{P_b + P_c}{2} - P_i\right)/H_1 \tag{9}$$

where P_b and P_c are the pressures of the AR outlet and inlet, respectively. P_i represents the pressure at the top position of the upper dipleg of the AR, i.e., the pressure at the interface of the dense phase and dilute phase of the upper downcomer. As the pressure loss is very small in the dilute phase region of the upper downcomer, the value of P_i can be approximated by the pressure at the underside of the separator (i.e., P_{11}). H_1 is the solid-seal height of the upper dipleg of the AR.

The lower pressure gradient ($\Delta P_2/H_2$) represents the pressure gradient between the J-valve and AR.

$$\Delta P_2/H_2 = (P_d - P_{12})/H_2 \tag{10}$$

where P_d is the pressure at the bottom position of the lower dipleg of the AR, and P_{12} the pressure at the top position of the lower dipleg. H_2, named as the solid-seal height of the lower dipleg, is represented by the height difference between the pressure monitoring nodes P_{12} and P_d.

3. Results and Discussion

As shown in Table 2, experiments were carried out with wide ranges of operating conditions (e.g., solid mass flux, superficial gas velocity, and pressure gradients) in order to present the primary flow behaviors of the proposed iG-CLC system. Thereinto, the operating condition of $G_s = 310 \text{ kg/m}^2\cdot\text{s}$, $U_{f,sta} = 10.7$ m/s, $Q_{4,sta} = 48 \text{ m}^3/\text{h}$ is defined as the reference condition which realized a balanced operation but involved relatively obvious gas leakage features to facilitate the observation and analysis.

Table 2. Main operation conditions of the iG-CLC tests.

Description	Range of Values
Solid circulation flux G_s (kg/m$^2\cdot$s)	170–480
FR superficial gas velocity $U_{f,sta}$ (m/s)	7–12.5
AR inlet air flow rate $Q_{4,sta}$ (m^3/h)	20–60
Upper pressure gradient $\Delta P_1/H_1$ (kPa/m)	−4–13
Lower pressure gradient $\Delta P_2/H_2$ (kPa/m)	2.0–7.0

3.1. Pressure Balance and Solid Circulation of the iG-CLC System

An appropriate pressure balance is vital to the balanced gas flow and solid circulation in an iG-CLC system; therefore, 19 pressure measuring nodes were mounted on the main assembly, the labels can be seen in Figure 2. Figure 3 shows the pressure profile of the system under the reference condition. It can be found that the sum of the pressure drop within a solid circulation loop was zero. This indicates the whole system was self-stabilizing and hence a small perturbation would not break the balance of solid circulation. The pressure drop along the height of the FR was conspicuous with the total pressure difference between P_1 at the bottom and P_{10} at the top reaching up to 12.8 kPa. In contrast, the pressure drop of the AR (i.e., P_c-P_b) was only about 1.5 kPa with a CFMB structure.

Due to the coupling of the AR in the middle of the downcomer, the dipleg of the downcomer was divided into two parts, the upper dipleg and the lower dipleg of the AR. Here, we adopted H_1, see Figure 2, as the solid-seal height of the upper dipleg and H_2, see Figure 2, as the solid-seal height of the lower dipleg. As shown in Figure 3, the pressures of the two diplegs both linearly increased from top to bottom, indicating that the solid flow structures in the two diplegs under the reference condition both belonged to the negative pressure differential flow. Moreover, the existence of the AR made the pressure distributions of the two diplegs relatively independent of each other. To be specific, the pressure distribution along the upper dipleg was mainly determined by the outlet pressures of the AR and FR, and the solid-seal height of the upper dipleg (i.e., H_1), while the pressure distribution along the lower dipleg was determined by the pressures of the AR and the J-valve together with the solid-seal height of the lower dipleg (i.e., H_2).

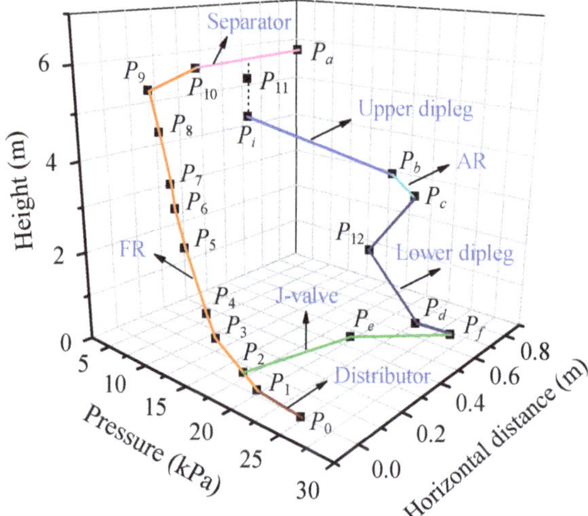

Figure 3. Pressure profile of the iG-CLC system under the reference condition.

The solid circulation rate is significant to the performance of an iG-CLC system, which determines the carrying capacity of oxygen and heat transferred by the OC from the AR to the FR. Figure 4 shows the distributions of the solid circulation flux G_s corresponding to the FR superficial gas velocity $U_{f,sta}$. It can be seen that a wide range of G_s from 170 to 480 kg/m^2·s had been achieved, indicating a good adaptability of this iG-CLC system on the OC circulation flux. This also demonstrates the positive role of the direct coupling method of the AR in the stabilization and controllability of the whole system. Thus, in the hot operation process, this iG-CLC system can achieve the feasible adjustment of oxygen and heat transfer according to actual situations. In particular, the capacity of high solid circulation flux ($G_s \geq 200$ kg/m^2·s) can greatly increase the OC inventory in the FR, and thus compensate for the possible low reactivity of the OC. Therefore, this iG-CLC system also provides the feasibility of the use of low-grade natural iron ores with lower reactivity as the OCs.

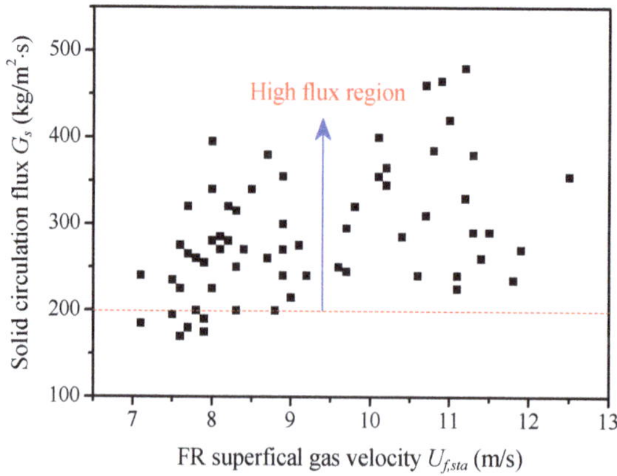

Figure 4. Distributions of the solid circulation flux corresponding to the FR superficial gas velocity.

3.2. Effect of the AR Coupling on the Solid Circulation

The direct coupling of the CFMB AR into the HFCFB system will inevitably affect the gas-solid flow behaviors of the system. In our previous work, we indeed notice that the OC circulation seemed to be affected during the adjustment process of the AR back pressure [31]. Therefore, it is significant to understand the fundamental effects of the AR coupling on the flux and further the stability of OC circulation.

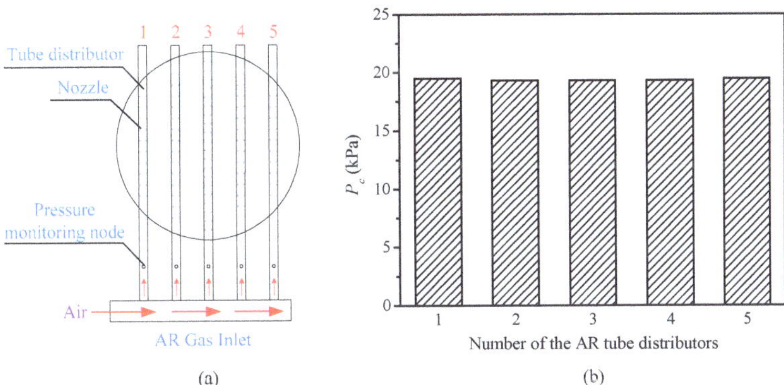

Figure 5. Pressure profiles of the parallel AR tube distributors under the reference condition: (**a**) schematic of the AR tube distributors, and (**b**) pressure profiles.

Figure 5 shows the pressure profiles of the parallel AR tube distributors under the reference condition. It can be seen that the inlet pressures of the five distributors were basically the same, demonstrating the realizability of even flow and distribution of gas-solid phases in the AR under certain conditions. In this context, a series of tests were carried out to investigate the fundamental effects of the AR coupling on the OC circulation flux and the system stability. We found that the OC circulation was actually determined by more factors that had interactions with each other (e.g., the pressures of the two reactors and the J-valve, the solid-seal heights above and under the AR), not just the AR back pressure. Here, we proposed a combined influence factor, the so-called upper pressure gradient $\Delta P_1/H_1$, which integrated the pressures of the two reactors and the solid-seal height of the upper dipleg, and hence should be able to more reasonably reflect the characteristics of OC circulation. Figure 6a shows the variations of the solid circulation flux with the upper pressure gradient $\Delta P_1/H_1$ while keeping all other parameters constant. It can be seen that the whole process could be divided into three sequential stages. In the first stage (-3.3 kPa/m $\leq \Delta P_1/H_1 \leq 8.9$ kPa/m), the solid circulation flux had a linear increase from 215 to 260 kg/m²·s with the increase of $\Delta P_1/H_1$, which was defined as the stage of circulation strengthening. In the second stage (8.9 kPa/m $< \Delta P_1/H_1 < 10.7$ kPa/m), the so-called transition stage, the solid circulation flux began to decrease. In the third stage ($\Delta P_1/H_1 \geq 10.7$ kPa/m), the solid circulation flux would drastically decline until the circulation collapsed completely, which was defined as the stage of circulation collapse.

The above results indicated that, with the increase in the upper pressure gradient $\Delta P_1/H_1$, the solid circulation flux would increase initially and later decrease until the circulation collapsed. First, in the stage of circulation strengthening, the material seal of the upper dipleg could overcome most of the gas leakage from the AR to the separator in spite of the increase of the upper pressure gradient. Better still, the increasing upper pressure gradient due to the increase of the AR back pressure also gave rise to the decrease of the lower pressure gradient $\Delta P_2/H_2$, and further, the decline of the aeration air leakage ratio from the J-valve to the AR (i.e., f_3), which indirectly enhanced the driving force of the J-valve aeration air for the OC circulation, and hence led to the increase of solid circulation flux. Then, starting from the transition stage, the material seal of the upper dipleg would be gradually destroyed by the increasing upper pressure gradient with the creation of large bubbles in the upper dipleg, as shown in Figure 6b (the bubbles were circled in red). The retrograde motion of the bubbles in the upper dipleg greatly blocked the downward flow of the OC, which was the reason the solid circulation of the whole system was broken. Worse still, the appearance of bubbles also meant the massive leakage of N_2 from the AR into the separator, and hence the great reduction in the CO_2 concentration. Given the above, under the operational conditions shown in Figure 6a, the upper pressure gradient $\Delta P_1/H_1$ should be limited to 8.9 kPa/m (i.e., within the stage of circulation strengthening), where the OC circulation was balanced and adjustable.

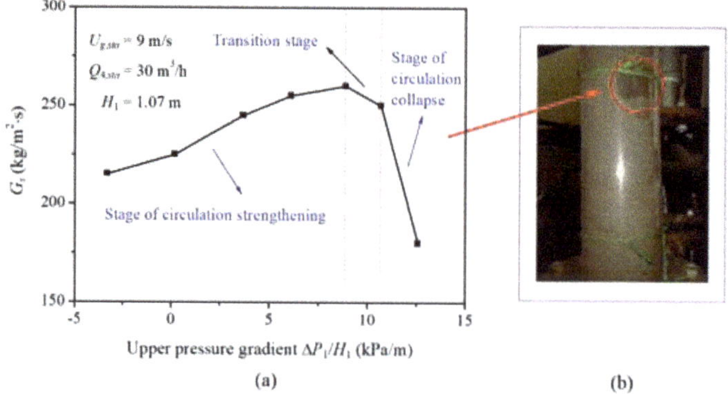

Figure 6. Variations of the solid circulation patterns with the upper pressure gradient: (**a**) variations of the solid circulation flux; and (**b**) snapshot of the state of circulation collapse.

3.3. Effect of the AR Coupling on the Gas Leakages

The gas leakages between reactors are one of the key factors affecting the performance of an iG-CLC system, such as the CO_2 capture efficiency, the CO_2 capture concentration, and even the solid circulation flux. As shown in Figure 7, there are three possible routes of gas leakages in our iG-CLC system: From FR to AR (i.e., FR leakage ratio f_1), from AR to FR (i.e., AR leakage ratio f_2), and from J-valve to AR (i.e., J-valve leakage ratio f_3). It can be found that the AR is the critical component associated with each leakage route. Hence, it is very meaningful to understand the influence mechanism of the AR coupling on the gas leakages in order to discover feasible solutions.

After a series of tests, the gas leakages were also found to be determined by the pressure gradients of the AR, which integrated the effects of the pressures of the two reactors and the J-valve, and the solid-seal heights of the upper and lower diplegs. Figure 8 shows the variations of the FR leakage ratio f_1 and the AR leakage ratio f_2 with the upper pressure gradient $\Delta P_1/H_1$ under a high solid flux condition (G_s = 200 kg/m²·s). It can be seen that the FR leakage ratio f_1 could be effectively reduced with an increase in the upper pressure gradient $\Delta P_1/H_1$. When $\Delta P_1/H_1$ exceeded 0.7 kPa/m, the value of f_1 would decline to about zero, indicating the gas leakage from the FR to the AR had almost disappeared. However, we also observed a contrary trend of the AR leakage ratio f_2 with the upper pressure gradient $\Delta P_1/H_1$. First, f_2 was kept near zero when $\Delta P_1/H_1$ increased from -2.2 to 1.6 kPa/m, indicating the upper dipleg could seal the AR gas stream perfectly within this range of $\Delta P_1/H_1$. When $\Delta P_1/H_1$ exceeded 1.6 kPa/m, f_2 would rise rapidly, indicating the material seal in the upper dipleg began to gradually lose its effectiveness.

In the commercial application, if the FR leakage ratio f_1 is excessive, the CO_2 capture efficiency will be greatly reduced. Moreover, more unreacted char will be carried by the leaked gas of the FR into the AR for combustion, which will increase the risk of OC sintering, and further influence the operation stability of the whole system. On the other hand, if the AR leakage ratio f_2 becomes too large, a large amount of N_2 from the AR will bypass into the separator and mix with the FR exhaust gas, resulting in a substantial decrease in the CO_2 concentration. Worse still, the gas leakage may also impede the downward flow of the solid in the upper dipleg and break the system circulation stability, as mentioned in Section 3.2. Therefore, both the FR leakage ratio f_1 and the AR leakage ratio f_2 must be limited to lower values. However, in view of the opposite effects of the upper pressure gradient $\Delta P_1/H_1$ on f_1 and f_2, a coordination control and optimized matching is inevitable. Under the involved operation conditions, the range of $\Delta P_1/H_1$ between -2.1 to 3.0 kPa/m should be an optimal region for adjustment, in which the values of f_1 and f_2 could be limited to 3%, together with a favorable solid circulation.

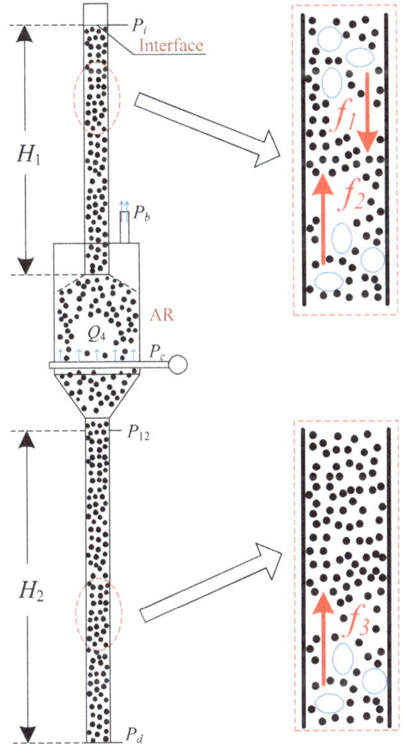

Figure 7. The possible routes of gas leakages in the proposed iG-CLC system.

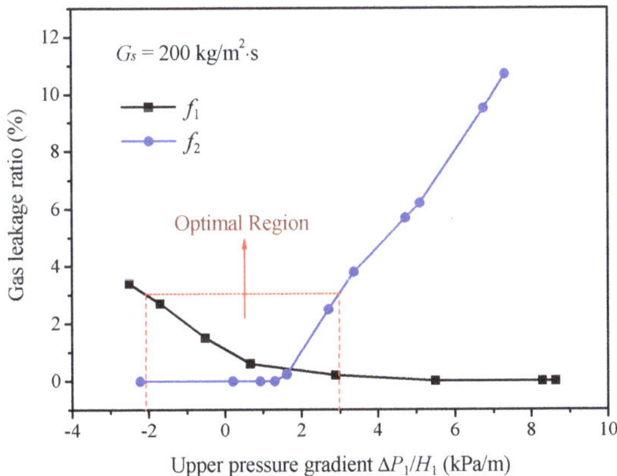

Figure 8. Variations of FR leakage ratio and AR leakage ratio with the upper pressure gradient.

On the other hand, the lower pressure gradient $\Delta P_2/H_2$ could be used to reflect the gas leakage ratio of the J-valve to the AR. Figure 9 presents the variations of the J-valve leakage ratio f_3 with the lower pressure gradient $\Delta P_2/H_2$. When the lower pressure gradient $\Delta P_2/H_2$ increased from 2.3 to 6.9 kPa/m, the J-valve leakage ratio f_3 had an obvious increase from 6.0% to 21.3% with a near-linear trend. This indicated a negative effect of the lower pressure gradient on the suppression of the gas leakage from the J-valve to the AR. In the iG-CLC application, a small amount of gas leakage from the J-valve to the AR will rarely affect the circulation stability of the system, and hence can be accepted. However, an excess gas leakage will cause the aeration gas stream of the J-valve to no longer contribute to the solid circulation, but impede the downward flow of the particles. Worse yet, the excess gas leakage will lower the air inflow to the AR, and hence reduce the thermal power of the proposed iG-CLC system. Therefore, under the involved operation conditions, the lower pressure gradient $\Delta P_2/H_2$ should be limited within 6.0 kPa/m, and thus the J-valve leakage ratio f_3 could also be controlled within a low value (20%) with a favorable solid circulation.

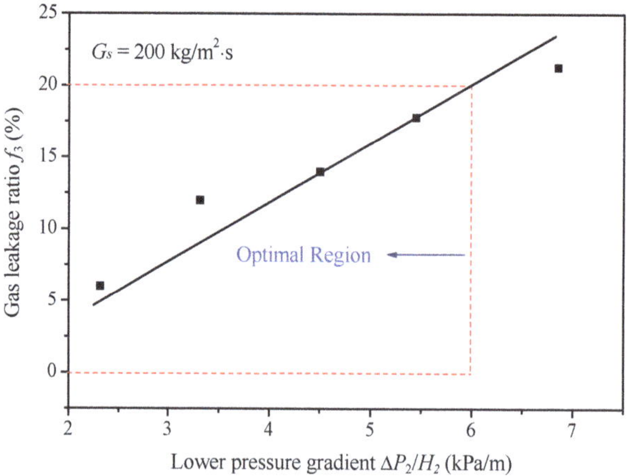

Figure 9. Variations of the J-valve leakage ratio with the lower pressure gradient.

3.4. Performance Optimization of the AR Coupling

Based on the above analysis, the coupling of the AR has important effects, by virtue of the upper pressure gradient $\Delta P_1/H_1$ and the lower pressure gradient $\Delta P_2/H_2$, on the characteristics of gas leakages and even the solid circulation stability. In the practical operation process, we can adjust the relevant parameters (i.e., the pressures of the two reactors and the J-valve, and the solid-seal heights in the downcomer) flexibly and optimally to guarantee the pressure gradients within the optimal ranges for an ideal performance of operation and reaction.

In order to better exhibit the effect of the AR coupling, we carried out a comparison of gas leakages between the reference condition and an optimal condition, as shown in Table 3. On the basis of the coupling criteria proposed in Section 3.3, the upper pressure gradient $\Delta P_1/H_1$ and lower pressure gradient $\Delta P_2/H_2$ in the optimal test were selected to be 3.5 kPa/m and 5.0 kPa/m, respectively. It could be found that for the reference condition, although achieving a balanced solid circulation in the whole system, an unsatisfactory gas leakage of the AR (i.e., f_2) was observed, indicating considerable mixing of N_2 from the AR into the FR exhaust gas stream, and hence an obvious reduction in the CO_2 concentration. However, the good news is that, with an optimization of the pressure gradients, the iG-CLC unit, under the optimal condition, inhibited the gas leakages, together with a favorable solid circulation. This demonstrates the significance of the study of the AR coupling mechanism in the high-flux iG-CLC system for the achievement of high CO_2 capture efficiency and

CO_2 capture concentration under a balanced system operation, which could provide vital information and experience for the design of future large-scale coal-fired CLC power plants.

Table 3. Comparison of gas leakages between the reference condition and optimal condition.

Description	Reference Condition	Optimal Condition
Solid circulation flux G_s (kg/m^2·s)	310	310
Upper pressure gradient $\Delta P_1/H_1$ (kPa/m)	8.0	3.5
Lower pressure gradient $\Delta P_2/H_2$ (kPa/m)	3.8	5.0
FR leakage ratio f_1 (%)	0	0
AR leakage ratio f_2 (%)	10.7	1.9
J-valve leakage ratio f_3 (%)	11.0	15.0

4. Conclusions

On the basis of the previous feasibility studies of a high-flux iG-CLC system, this work further investigated the AR coupling effects on the system operation stability and gas leakages with a cold-state visualization device, enabling the design parameters and operating conditions. The following conclusions can be drawn from the present study:

(1) The iG-CLC system exhibited favorable pressure distribution characteristics and good adaptability of solid circulation flux, demonstrating the positive role of the direct coupling method of the AR in the stabilization and controllability of the whole system.

(2) With the increase of the upper pressure gradient of the AR, the OC circulation flux would increase initially and later decrease until the circulation collapsed, demonstrating the crucial effect of the AR coupling on the OC circulation flux and further the circulation stability. In the real iG-CLC applications, the upper pressure gradient of the AR should be limited within the stage of circulation strengthening in order to guarantee a balanced and adjustable OC circulation.

(3) The gas leakage ratios of the FR and the AR were determined by the upper pressure gradient of the AR, while the gas leakage ratio of the J-valve was determined by the lower pressure gradient. In the iG-CLC applications, we can adjust the pressures of the two reactors and the solid-seal heights in the downcomer flexibly and optimally to ensure the two pressure gradients within the optimal ranges for an ideal balance of all the gas leakages.

(4) By giving consideration to the AR coupling effects under various operation conditions comprehensively, one operating condition with 3.5 kPa/m for the upper pressure gradient and 5.0 kPa/m for the lower pressure gradient was recommended. Under this condition, the gas leakages between the two reactors could be limited to 3%, and the gas leakage of the J-valve could also be below 20% to guarantee the solid circulation. This demonstrates the significance of the study of the AR coupling mechanism in the high-flux iG-CLC system for the achievement of high CO_2 capture efficiency and CO_2 capture concentration under a balanced system operation.

Author Contributions: X.W. conceived and designed the experiments; X.L. performed the experiments; X.W. and X.L. analyzed the data; B.J. supervised the research; X.W., X.L. and D.W. wrote the manuscript.

Funding: This research was funded by the National Natural Science Foundation of China (grant numbers 51741603, 51806035, 51676038), the Natural Science Fund project in Jiangsu Province (grant number BK20170669), the Fundamental Research Funds for the Central Universities (grant number 2242018K40117), and the Guangdong Provincial Key Laboratory of New and Renewable Energy Research and Development (grant number Y707s41001).

Conflicts of Interest: The authors declare no conflict of interest.

References

1. Fan, L.S.; Zeng, L.; Wang, W.; Luo, S. Chemical looping processes for CO_2 capture and carbonaceous fuel conversion—Prospect and opportunity. *Energy Environ. Sci.* **2012**, *5*, 7254–7280. [CrossRef]
2. Lyngfelt, A.; Leckner, B.; Mattisson, T. A fluidized-bed combustion process with inherent CO_2 separation; application of chemical-looping combustion. *Chem. Eng. Sci.* **2001**, *56*, 3101–3113. [CrossRef]

3. Abad, A.; Mattisson, T.; Lyngfelt, A.; Johansson, M. The use of iron oxide as oxygen carrier in a chemical-looping reactor. *Fuel* **2007**, *86*, 1021–1035. [CrossRef]
4. Mattisson, T.; García-Labiano, F.; Kronberger, B.; Lyngfelt, A.; Adánez, J.; Hofbauer, H. Chemical-Looping Combustion using syngas as fuel. *Int. J. Greenh. Gas Control* **2007**, *1*, 158–169. [CrossRef]
5. Ishida, M.; Jin, H.; Okamoto, T. A fundamental study of a new kind of medium material for chemical-looping combustion. *Energy Fuels* **1996**, *10*, 958–963. [CrossRef]
6. Jin, H.; Okamoto, T.; Ishida, M. Development of a novel chemical-looping combustion: Synthesis of a solid looping material of NiO/NiAl$_2$O$_4$. *Ind. Eng. Chem. Res.* **1999**, *38*, 126–132. [CrossRef]
7. Mattisson, T.; Järdnäs, A.; Lyngfelt, A. Reactivity of some metal oxides supported on alumina with alternating methane and oxygen application for chemical-looping combustion. *Energy Fuels* **2003**, *17*, 643–651. [CrossRef]
8. Cho, P.; Mattisson, T.; Lyngfelt, A. Comparison of iron-, nickel-, copper-and manganese-based oxygen carriers for chemical-looping combustion. *Fuel* **2004**, *83*, 1215–1225. [CrossRef]
9. Abad, A.; Mattisson, T.; Lyngfelt, A.; Rydén, M. Chemical-looping combustion in a 300 W continuously operating reactor system using a manganese-based oxygen carrier. *Fuel* **2006**, *85*, 1174–1185. [CrossRef]
10. De Diego, L.F.; García-Labiano, F.; Gayán, P.; Celaya, J.; Palacios, J.M.; Adánez, J. Operation of a 10 kWth chemical-looping combustor during 200 h with a CuO-Al$_2$O$_3$ oxygen carrier. *Fuel* **2007**, *86*, 1036–1045. [CrossRef]
11. Adánez, J.; Dueso, C.; de Diego, L.F.; García-Labiano, F.; Gayán, P.; Abad, A. Methane combustion in a 500 Wth chemical-looping combustion system using an impregnated Ni-based oxygen carrier. *Energy Fuels* **2008**, *23*, 130–142. [CrossRef]
12. Kolbitsch, P.; Bolhàr-Nordenkampf, J.; Pröll, T.; Hofbauer, H. Operating experience with chemical looping combustion in a 120 kW dual circulating fluidized bed (DCFB) unit. *Int. J. Greenh. Gas Control* **2010**, *4*, 180–185. [CrossRef]
13. Ma, J.; Zhao, H.; Tian, X.; Wei, Y.; Zhang, Y.; Zheng, C. Continuous Operation of Interconnected Fluidized Bed Reactor for Chemical Looping Combustion of CH$_4$ Using Hematite as Oxygen Carrier. *Energy Fuels* **2015**, *29*, 3257–3267. [CrossRef]
14. Diglio, G.; Bareschino, P.; Mancusi, E.; Pepe, F. Techno-Economic Evaluation of a small-scale power generation unit based on a Chemical Looping Combustion Process in Fixed Bed Reactor network. *Ind. Eng. Chem. Res.* **2018**, *57*, 11299–11311. [CrossRef]
15. Cao, Y.; Pan, W.P. Investigation of Chemical looping combustion by solid fuels: 1 Process analysis. *Energy Fuels* **2006**, *20*, 1836–1844. [CrossRef]
16. Berguerand, N.; Lyngfelt, A. Design and operation of a 10 kW$_{th}$ chemical-looping combustor for solid fuels-testing with South African coal. *Fuel* **2008**, *87*, 2713–2726. [CrossRef]
17. Leion, H.; Mattisson, T.; Lyngfelt, A. Solid fuels in chemical-looping combustion. *Int. J. Greenh. Gas Control* **2008**, *2*, 180–193. [CrossRef]
18. Shen, L.H.; Wu, J.H.; Xiao, J. Experiments on chemical looping combustion of coal with a NiO based oxygen carrier. *Combust. Flame* **2009**, *156*, 721–728. [CrossRef]
19. Fan, L.S.; Li, F. Chemical looping technology and its fossil energy conversion applications. *Ind. Eng. Chem. Res.* **2010**, *49*, 10200–10211. [CrossRef]
20. Abad, A.; Gayán, P.; de Diego, L.F.; García-Labiano, F.; Adánez, J. Fuel reactor modelling in chemical-looping combustion of coal: 1. Model formulation. *Chem. Eng. Sci.* **2013**, *87*, 277–293. [CrossRef]
21. García-Labiano, F.; de Diego, L.F.; Gayán, P.; Abad, A.; Adánez, J. Fuel reactor modelling in chemical-looping combustion of coal: 2-simulation and optimization. *Chem. Eng. Sci.* **2013**, *87*, 173–182. [CrossRef]
22. Thon, A.; Kramp, M.; Hartge, E.-U.; Heinrich, S.; Werther, J. Operational experience with a system of coupled fluidized beds for chemical looping combustion of solid fuels using ilmenite as oxygen carrier. *Appl. Energy* **2014**, *118*, 309–317. [CrossRef]
23. Bayham, S.; McGiveron, O.; Tong, A.; Chung, E.; Kathe, M.; Wang, D.; Zeng, L.; Fan, L.S. Parametric and dynamic studies of an iron-based 25-kW$_{th}$ coal direct chemical looping unit using sub-bituminous coal. *Appl. Energy* **2015**, *145*, 354–363. [CrossRef]
24. Adánez, J.; Abad, A.; Perez-Vega, R.; Luis, F.; García-Labiano, F.; Gayán, P. Design and Operation of a Coal-fired 50 kW$_{th}$ Chemical Looping Combustor. *Energy Procedia* **2014**, *63*, 63–72. [CrossRef]
25. Markström, P.; Linderholm, C.; Lyngfelt, A. Operation of a 100 kW chemical-looping combustor with Mexican petroleum coke and Cerrejón coal. *Appl. Energy* **2014**, *113*, 1830–1835. [CrossRef]

26. Ma, J.; Zhao, H.; Tian, X.; Wei, Y.; Rajendran, S.; Zhang, Y.; Bhattacharya, S.; Zheng, C. Chemical looping combustion of coal in a 5 kW$_{th}$ interconnected fluidized bed reactor using hematite as oxygen carrier. *Appl. Energy* **2015**, *157*, 304–313. [CrossRef]

27. Ströhle, J.; Orth, M.; Epple, B. Design and operation of a 1 MW$_{th}$ chemical looping plant. *Appl. Energy* **2014**, *113*, 1490–1495. [CrossRef]

28. Xiao, R.; Chen, L.; Saha, C.; Zhang, S.; Bhattacharya, S. Pressurized chemical-looping combustion of coal using an iron ore as oxygen carrier in a pilot-scale unit. *Int. J. Greenh. Gas Control* **2012**, *10*, 363–373. [CrossRef]

29. Wang, X.; Jin, B.; Zhang, Y.; Zhang, Y.; Liu, X. Three Dimensional Modeling of a Coal-Fired Chemical Looping Combustion Process in the Circulating Fluidized Bed Fuel Reactor. *Energy Fuels* **2013**, *27*, 2173–2184. [CrossRef]

30. Wang, X.; Jin, B.; Liu, X.; Zhang, Y.; Liu, H. Experimental investigation on flow behaviors in a novel in situ gasification chemical looping combustion apparatus. *Ind. Eng. Chem. Res.* **2013**, *52*, 14208–14218. [CrossRef]

31. Wang, X.; Jin, B.; Liu, H.; Wang, W.; Liu, X.; Zhang, Y. Optimization of in Situ Gasification Chemical Looping Combustion through Experimental Investigations with a Cold Experimental System. *Ind. Eng. Chem. Res.* **2015**, *54*, 5749–5758. [CrossRef]

32. Wang, X.; Jin, B.; Zhu, X.; Liu, H. Experimental Evaluation of a Novel 20 kW$_{th}$ in Situ Gasification Chemical Looping Combustion Unit with an Iron Ore as the Oxygen Carrier. *Ind. Eng. Chem. Res.* **2016**, *55*, 11775–11784. [CrossRef]

Article

Experimental Study on Spray Breakup in Turbulent Atomization Using a Spiral Nozzle

Ondřej Krištof [1], Pavel Bulejko [1,*] and Tomáš Svěrák [1,2]

[1] Heat Transfer and Fluid Flow Laboratory, Faculty of Mechanical Engineering, Brno University of Technology, Technická 2, 616 69 Brno, Czech Republic; ondrej.kristof@vut.cz (O.K.); sverak@fch.vutbr.cz (T.S.)

[2] Institute of Materials Chemistry, Faculty of Chemistry, Brno University of Technology, Purkyňova 464/118, 612 00 Brno, Czech Republic

* Correspondence: pavel.bulejko@vut.cz; Tel.: +420-541-144-912

Received: 15 September 2019; Accepted: 26 November 2019; Published: 3 December 2019

Abstract: Spiral nozzles are widely used in wet scrubbers to form an appropriate spray pattern to capture the polluting gas/particulate matterwith the highest possible efficiency. Despite this fact, and a fact that it is a nozzle with a very atypical spray pattern (a full cone consisting of three concentric hollow cones), very limited amount of studies have been done so far on characterization of this type of nozzle. This work reports preliminary results on the spray characteristics of a spiral nozzle used for gas absorption processes. First, we experimentally measured the pressure impact footprint of the spray generated. Then effective spray angles were evaluated from the photographs of the spray and using the pressure impact footprint records via Archimedean spiral equation. Using the classical photography, areas of primary and secondary atomization were determined together with the droplet size distribution, which were further approximated using selected distribution functions. Radial and tangential spray velocity of droplets were assessed using the laser Doppler anemometry. The results show atypical behavior compared to different types of nozzles. In the investigated measurement range, the droplet-size distribution showed higher droplet diameters (about 1 mm) compared to, for example, air assisted atomizers. It was similar for the radial velocity, which was conversely lower (max velocity of about 8 m/s) compared to, for example, effervescent atomizers, which can produce droplets with a velocity of tens to hundreds m/s. On the contrary, spray angle ranged from 58° and 111° for the inner small and large cone, respectively, to 152° for the upper cone, and in the measured range was independent of the inlet pressure of liquid at the nozzle orifice.

Keywords: spiral nozzle; gas absorption; spray atomization; droplet size; droplet velocity

---

1. Introduction

Gas–liquid absorption processes for gaseous contaminants removal are crucial in diverse industrial fields and are basic means for air pollution mitigation associated with large-scale industrial operations. To ensure the reduction of the gaseous pollutants released from such processes, many industrial facilities adopt a gas scrubbing system as a post-treatment of produced polluted air. Gas scrubbers are rather complicated devices, in which the polluted air is cleaned using a sprayed liquid, mostly various aqueous solutions depending on the gas to be removed. This includes, for example, CO_2 and VOC removal [1–3], flue gas desulphurization [4–6], and ammonia separation [7]. Last but not least, scrubbing systems have widely been recognized in separation of particulate matter [8–12], the release of which is due mostly to various combustion processes [13–15] and, for example, comminution technologies [16,17]. This is very important due to associated health and environmental concerns [18,19]. Comparing the ability to remove particulate and gaseous pollutants, gas scrubbers have lower efficiency in one cleaning cycle but are able to treat a significantly higher amount of polluted air compared to, for example, air filtration or membrane contactors [20–23]. In these applications, filter/membrane causes additional resistance to airflow, thus disabling processing of huge amounts of polluted air generated in large-scale industrial processes. On the other hand, gas scrubbers work with sprayed absorption liquid media, which must then be subjected to another processing step, such as regeneration (if possible) or disposal, which can be very expensive and energy demanding [24]. Due to these facts, there is a large field for development and optimization of these systems in terms of operational parameters, chemicals used or spraying arrangement including usage of extremely broad spectrum of nozzles. Nozzles are crucial in a plethora of industrial applications including dust control [25], spray cooling [26–30], hydraulic descaling [31], and play a vital role in gas scrubbers as they can provide an adequate spray pattern of the absorption liquid. This is necessary to create the largest possible contact surface with gas phase to ensure process intensification, appropriate mass transfer and thus separation efficiency. Such a type of nozzle, which has scarcely been studied in terms of the spray characteristics, is the spiral nozzle.

Spiral nozzles have been found in different applications including flue gas desulfurization and spray towers [4,5,32], spray drying [33], distillation [34], petrochemical industry [35], and fire suppression [36,37]. Several studies tried to focus on the basic spray characteristics including droplet-size distribution, the inlet liquid pressure–flowrate relationship, and mass spray density [38–41] at different hydrodynamic conditions. Some authors did a research on the spray surface geometry [42] and even air-assisted atomization using the spiral nozzle [43]. Important works related to the present study are further described in detail. Li et al. [4] investigated spray characteristics of spiral nozzles used in flue gas desulfurization. They observed the flowrate was linear with a square root of pressure and droplet diameter was a power function of pressure. The droplet diameter variations with pressure were similar for spiral nozzles with different orifice, while the spray angle varied slightly at pressures higher than 40 kPa. Zhang et al. [32] studied spray characteristics of spiral nozzles with different diameters using particle image velocimetry at different spray pressures. They observed both, the spray pressure and nozzle diameter, to have an influence on the spray angle, droplet diameter, droplet size uniformity, and sprayed area diameter. In a biomass pyrolysis experiments, they found a nozzle with a diameter of 5.6 mm and the liquid pressure of two bars to be ideal for the quenching of pyrolysis vapors. In another study, Zhou et al. [41] conducted experiments on the flow distribution characteristics of a low-pressure high-flux spiral nozzle using a flow distribution testing system. They analyzed the effect of nozzle size on the flow distribution by varying several parameters including nozzle dimensions (nozzle length to diameter ratio), radial flowrate, position of sprayed surface, and spraying angle. The results indicated that with increasing nozzle length to diameter ratio, the flowrate decreases and spraying angle increases. With increasing pressure, the relationships were the same as in the above mentioned works. Dong et al. [5] compared desulfurization performance of scrubbers with spiral and Dynawave nozzle. Li et al. [42] developed a spray surface geometry model of a spiral nozzle with involute atomization. They further simulated the model using MATLAB, which validated its effectiveness and provided a theoretical basis for designing and manufacturing the spiral nozzles. Wasik et al. [38] studied the influence of nozzle type (including a TF6 spiral nozzle) on a mass spray density. For other works on spiral nozzles, especially those with different application, refer to [33–35].

In this study, we focus on a spiral nozzle (Figure 1a), which generates an atypical cone-like spray consisting of three concentric hollow cones (Figure 1b), thus creating a full cone pattern (for further details refer to [44]). The pattern is formed via a complicated nozzle geometry without axis of symmetry and gradually narrowing orifice, in which the liquid is sprayed from several rebound surfaces at different rake angles. This nozzle is quite different compared to spiral channel nozzles [45] or spiral flow nozzles [46,47] with which it can be confused. Generally, a lot of works have been carried out on nozzles creating a full-cone spray in various application fields [48–58]. Conversely, the amount of works on nozzles with a spiral geometry is quite limited despite the fact that they possess several very interesting features [59]:

- They are made of one piece of material (no internal parts, hence resistant to clogging);
- They have high discharge coefficients (higher flowrates possible at lower pressure drops);
- They can provide fine atomization, wide range of flowrates and spray angles;

- A reduction in waste energy appearing as noise;
- Very wide effective patterns produced allow replacing several nozzles with one.

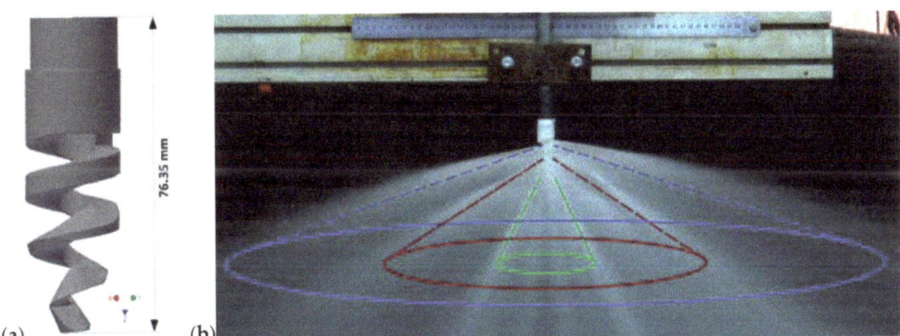

Figure 1. A 3D model of the spiral nozzle studied (**a**) and the concentric hollow cone pattern produced by the nozzle (**b**).

Therefore, we investigate the pressure impact footprint of the falling liquid (cross-sectional pattern of the spray), spray radial, and tangential velocity distribution using the laser Doppler anemometry (LDA). The pressure impact footprint of the spray is generally important in spray cooling applications; in gas absorption processes this is not typically used. However, this can be easily used for validation of CFD data, as it is practically simpler to measure pressure impact compared to using the LDA method which is not always available. Further, we aimed at the spray morphology. This involved droplet-size distribution, spray angle, and determining the area of primary and secondary breakup of the spray using a classical photography modified with an Nd:YAG (neodymium-doped yttrium aluminum garnet) pulse laser, and assess the breakup areas using selected dimensionless numbers, which has not been carried out before for similar type of nozzle. We also compared the droplet diameter with selected theoretical relationships used for prediction of droplet and ligament diameter. Finally, we tried to outline flow conditions for ideal nozzle operation.

2. Underlying Phenomena

Liquid atomization (i.e., a breakup of liquid jet into dispersed fine droplets) is a complex process involving several physical/chemical phenomena that take place simultaneously. The spray characteristics (e.g., morphology, droplet size distribution, etc.) are strongly dependent on the atomizer used, especially on its size and geometry. Further, it is dependent on the physical properties of the fluids involved (i.e., the atomized liquid and the environment into which the liquid is sprayed (usually ambient air)). The spray atomization is strongly influenced by the liquid density, viscosity, and surface tension. The effect of density is rather lower, as indicated by the experimental data [60]. Conversely, the influence of the surface tension is quite essential. Surface tension is a force acting against the formation of a new surface area. Atomization involves two main phases (i.e., primary and secondary atomization). In the primary atomization, the disruptive forces act against the consolidating forces and cause oscillations of the liquid. Once the disruptive forces are stronger than the consolidating ones (surface tension), the bulk liquid disintegrates into smaller formations (ligaments, larger drops). Then the secondary atomization occurs (i.e., larger droplets or ligaments split into smaller droplets in a gas caused by either greater relative velocity or turbulence) [61]. A governing parameter relating disruptive inertial and restorative surface tension forces is the non-dimensional Weber number:

$$We = \frac{\rho v^2 d}{\sigma},\qquad(1)$$

where ρ, v, d, and σ are the fluid density, absolute radial velocity of liquid, characteristic dimension, and surface tension, respectively. The larger the Weber number, the higher the tendency toward the liquid breakup [62]. Viscosity is another very important parameter affecting the droplet size distribution and mainly the flow mode inside the atomizer, thus influencing the spray pattern/morphology. The influence of viscosity on the flow in the nozzle is quite complex and depends on the type of atomizer. Generally, drop size increases with increasing viscosity and delays the liquid jet breakup [60]. Both surface tension and viscosity decrease the tendency of the jet/sheet to disintegrate, which is accounted for by the Ohnesorge number (i.e., the ratio of viscous to surface tension forces):

$$Oh = \frac{\mu_1}{\sqrt{\sigma \rho_1 d}}, \tag{2}$$

where ρ_1 and μ_1 are the liquid density and viscosity, respectively. Many spray nozzles form a liquid sheet from bulk liquid prior to the atomization itself. The liquid sheet exits the nozzle orifice and may oscillate, which results in the formation of liquid ligaments, which are then broken into droplets. The droplet size is mostly in the same order as the liquid sheet thickness [63]. The primary atomization is strongly dependent on the liquid jet Weber number:

$$We_{jet} = \frac{\rho_1 v_{jet}^2 d_o}{\sigma}, \tag{3}$$

where d_o is the nozzle orifice diameter and v_{jet} is the jet velocity (i.e., liquid velocity inside the nozzle prior to exiting the orifice). If the jet Weber number is lower, the surface tension forces impede the formation of a new surface area, thus preventing the liquid sheet breakup. Conversely at a larger jet Weber number, the breakup occurs due to the inertial forces to completely dominate over the surface tension forces causing the liquid sheet to tear into ligaments and droplets [64]. A critical value of the jet Weber number describes the onset of a decrease of the radial breakup distance. This value is typically around 1000 depending on the nozzle type [65]. After the first breakup phase, the secondary atomization may occur which is indicated by the gas Weber number:

$$We_g = \frac{\rho_g v_{AR}^2 d_D}{\sigma}, \tag{4}$$

where ρ_g is the ambient gas density. Droplet viscous forces are significant for $Oh > 0.1$. For Oh below this value, the breakup was observed to be independent of Oh. For $Oh < 0.1$, the transition We for individual breakup modes were practically constant as reported previously [66–68]. The individual modes are droplet deformation/vibrational breakup for $0 < We < 11$, bag breakup for $11 < We < 35$, multimode breakup for $35 < We < 80$, sheet thinning for $80 < We < 350$, and catastrophic breakup for $We > 350$. These are; however, not valid for conditions with higher Oh. The individual modes of secondary atomization are often depicted in We-Oh space, refer, for example, to [68–71]. To assess objectively which regime factually took place in the atomization process, several correlations for critical Weber number (We_c) were proposed for $Oh \rightarrow 0$. One of such correlation was suggested for $Oh < 4$ by Gelfand [72] as follows:

$$We_c = We_{cOh \rightarrow 0}\left(1 + 1.5Oh^{1.74}\right), \tag{5}$$

where $We_{cOh \rightarrow 0}$ is the critical We at low Oh, as listed above for individual modes.

The liquid sheet is formed via impinging the edge layer of the liquid jet on the surface of the helix as indicated in the CFD model (Figure 2). Therefore, we can expect an analogy with the radial spread of a liquid jet over a horizontal plane. This can be used to estimate the liquid sheet thickness (t_{sh}) based on the free-surface similarity boundary layer concept as developed by Watson [73] and used by Ren et al. [74] and Zhou and Yu [75]. Assuming the sheet flow on the helix surface be turbulent, the sheet thickness at the edge of the helix can be estimated as follows:

$$t_{sh} = \frac{d_o^2}{8r_h} + \frac{0.0245d_o^{1/5}r_h^{4/5}}{Re_o^{1/5}},$$ (6)

where d_o and r_h are the nozzle orifice diameter (11.25 mm) and width of the helix (6.45 mm), respectively, and Re is the Reynolds number at the nozzle orifice calculated as follows:

$$Re_o = \frac{d_o v_{jet} \rho_l}{\mu_l}$$ (7)

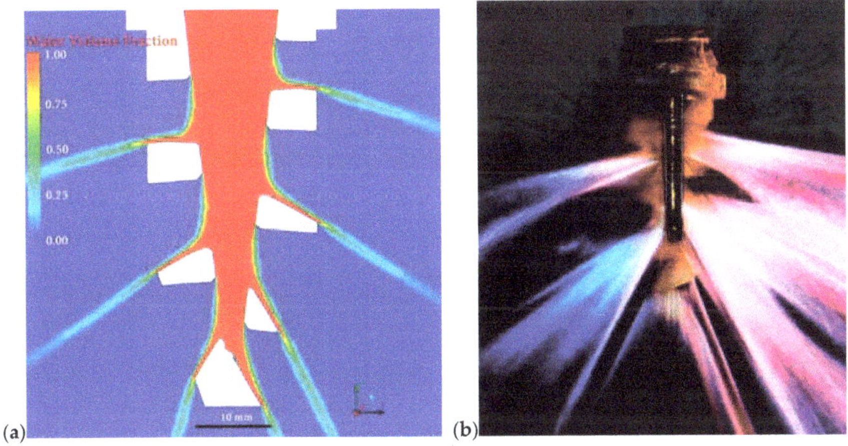

(**a**)   (**b**)

Figure 2. A CFD model illustration of the formation of the liquid sheet by impinging the peripheral jet layer onto the surface of the helix (**a**) and a photograph of the same with the applied screen filter (**b**).

To estimate the spray velocity in a given radial location (r), it is necessary to consider the effect of viscous interaction with the helix surface. Therefore, a non-dimensional sheet thickness is defined as the ratio of the actual thickness (t_{sh}) to an inviscid sheet thickness (t_{sh0}):

$$\delta = \frac{t_{sh}}{t_{sh0}} = 1 + \frac{0.196}{Re_o^{1/5}}\left(\frac{r}{d_o}\right)^{9/5}.$$ (8)

Note that t_{sh0} is the first term in Equation (6). The average sheet velocity at the edge of the helix can be calculated as [74]:

$$v_{sh} = \frac{K\sqrt{\Delta p}}{2\pi r_h t_{sh0}\delta},$$ (9)

where K is the flow factor, which is a characteristic constant of the nozzle and Δp is the pressure drop at the nozzle (inlet pressure of the liquid). There is a relationship between the volumetric flowrate (Q_V) and the inlet water pressure (Δp) as follows:

$$Q_V = K\sqrt{\Delta p}.$$ (10)

For the spiral nozzle used in this study, the value is about 75. Further breakup of the sheet into ligaments and droplets is due to inherent instabilities caused by the wave growth. The wavelength at the sheet breakup governs the size of the ligaments and ultimately the droplet diameter. Dombrowski and Johns [76] developed a theory to predict the wave instability of liquid sheets in an inviscid gas. The model assumes sinusoidal waves to be on the liquid sheet, and the force balance is performed considering the inertial, pressure, and surface tension forces be associated with the wave displacement. After simplification, the force balance can be expressed as follows:

$$\left(\frac{\partial f}{\partial \tau}\right)^2 + \frac{\mu_L \tilde{\omega}^2}{\rho_1}\left(\frac{\partial f}{\partial \tau}\right) - \frac{2\left(\rho_g \tilde{\omega} v_{sh}^2 - \sigma \tilde{\omega}^2\right)}{\rho_1 t_{sh}} = 0, \tag{11}$$

where τ is time and f and $\tilde{\omega}$ are the breakup parameter and wavenumber, respectively. The breakup parameter, also called dimensionless wave amplitude, was first investigated by Weber [77] who obtained a value of 12 and further confirmed by Dombrowski and Hooper [78] who stated that this value is constant regardless of the experimental conditions. However, in this study we better used the following correlation as it can be a function of the nozzle geometry [63]:

$$f = Re^{0.07} We^{0.37}, \tag{12}$$

where the Weber and Reynolds numbers are calculated for jet (i.e., liquid properties), liquid jet velocity prior to entering the nozzle orifice, and nozzle orifice diameter as the characteristic length. Another attempt was to estimate theoretically the ligament and droplet sizes based on the wave instabilities, which are the main cause of sheet disintegration. With an assumption of attenuating sheet and the formed ligaments be of cylindrical shape, the diameter estimate can be expressed as follows [76]:

$$d_L = 2\left(\frac{4}{3f}\right)^{1/3}\left(\frac{k^2\sigma^2}{\rho_g \rho_1 v_{sh}^2}\right)^{1/6}\left(1 + 2.6\mu_1 \sqrt[3]{\frac{k\rho_g^4 v_{sh}^8}{6f\rho_l^2\sigma^5}}\right)^{1/5}, \tag{13}$$

where k for a uniform velocity radiating sheet can be expressed as follows:

$$k = \frac{r t_{sh}}{v_{sh}}. \tag{14}$$

Accordingly, the estimate of the ligament breakup time $\tau_{L,bu}$ can be written as follows [74,77,79]:

$$\tau_{L,bu} = 24 \sqrt{\frac{2\rho_1}{\sigma}}\left(\frac{d_L}{2}\right)^{3/2}. \tag{15}$$

Based on the ligament size d_L, the droplet diameter d_D can be calculated [76,77]:

$$d_D = 1.882 d_L\left(1 + 3Oh_{sh}\right)^{1/6}, \tag{16}$$

where Oh_{sh} is the Ohnesorge number based on the sheet thickness.

3. Experimental

3.1. Spiral Nozzle

A TF-28 150 asymmetric spiral nozzle (BETE, USA) was used in this study. Figure 3 shows a 3D scan of the nozzle obtained using an ATOS Tripple Scan 8M camera and its profile. The nozzle is made of Teflon and has a narrowing spiral-like orifice. The spiral is divided into three sections based on the rake angle of the rebound surfaces related to the nozzle Z-axis. Liquid falling on the individual rebound surfaces of the helix forms a full cone. The full cone can be considered for a combination of three water curtains in the form of hollow concentric cones (Figures 1 and 4). Based on their position to each other, they can be called as the outer (upper) cone, inner large, cone and inner small cone, as indicated in Figure 4a.

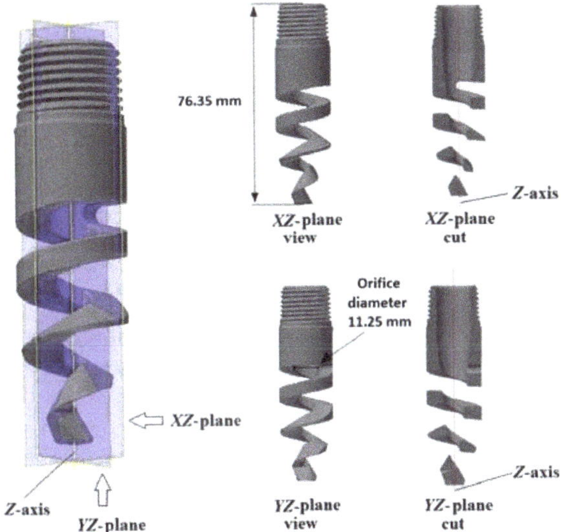

Figure 3. The BETE TF-28 150 spiral nozzle with designation of the longitudinal axis (*Z*-axis), views from different planes, and longitudinal cross sections (plane cuts).

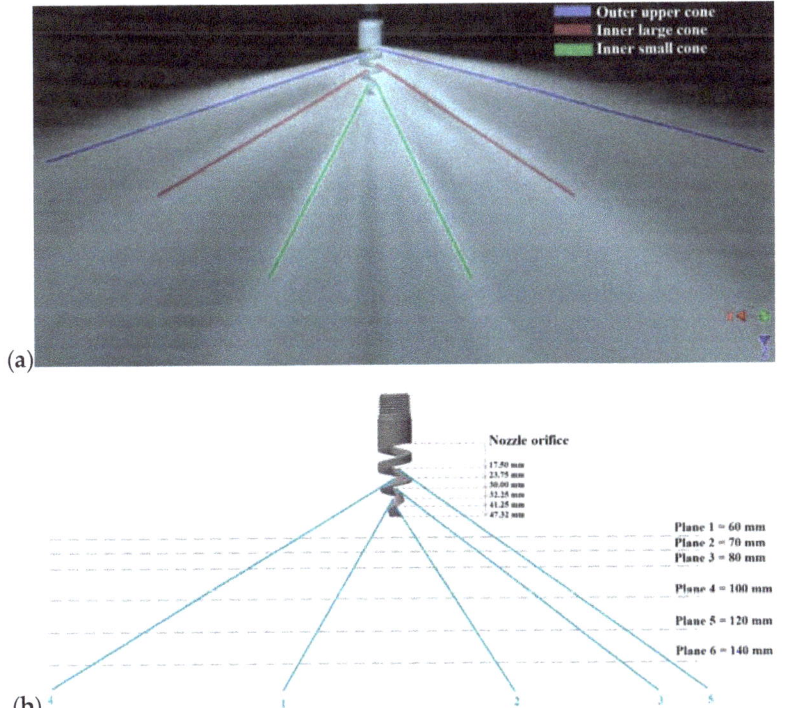

Figure 4. An illustration of the main streams of the spray generated (**a**) and a depiction of the sprayed water beams numbered (blue); the distances of the beginning of the water beams from the nozzle orifice and the distances of individual impact planes from the nozzle orifice (**b**).

3.2. Pressure Impact Footprint

Measurement of the pressure impact footprint was done to visualize the pressure patterns generated by the sprayed water. During spraying the water curtain fell down on surface connected to an electronic tensometric pressure sensor. This surface was placed on a position system, which was operated using a computer software. The real impact pressure values were recorded depending the sensor position related to the nozzle. The impact measurement was performed twice, each with a different arrangement of the impact surface. In each experiment, different hydrodynamic conditions and different distances of the nozzle from the impact surface (planes, Figure 4b) were adopted.

3.2.1. Measuring the Impact Pressure in Four Planes

In the first set, pressure impact footprint of the water curtain was measured. This was done in five planes (Figure 4b) (i.e., in the distances of 60, 80, 100, and 140 mm from the nozzle orifice). In each experiment, an inlet water pressure was 2 bars corresponding to a flowrate of 1.76 L/s. These are optimal hydrodynamic conditions for the operation of the nozzle tested. The experimental setup is shown in Figure 5. It was a metal plate (1) placed on an upper moving arm (5), in the middle of which was a dismountable metal casing (2) with a bevel-like top. At the top of the casing, an impact surface (3) was attached and connected to a tensometric pressure sensor placed inside the casing. The impact surface was circular with a diameter of 12 mm. The water falling down on the impact surface generates an impact pressure, which is taken by the sensor and the signal is recorded by the data acquisition system. The upper arm (5) was moved using a step motor in the direction of X-axis (Figure 5). This system was placed on the top of a bottom moving arm (6), which was moving in the direction of Y-axis (Figure 5). The area under the nozzle (4) was scanned in a length of 300 mm in the X-direction and 600 mm in the Y-direction, which was the maximum possible range of the experimental device used. The shift of both arms was set to 5 mm.

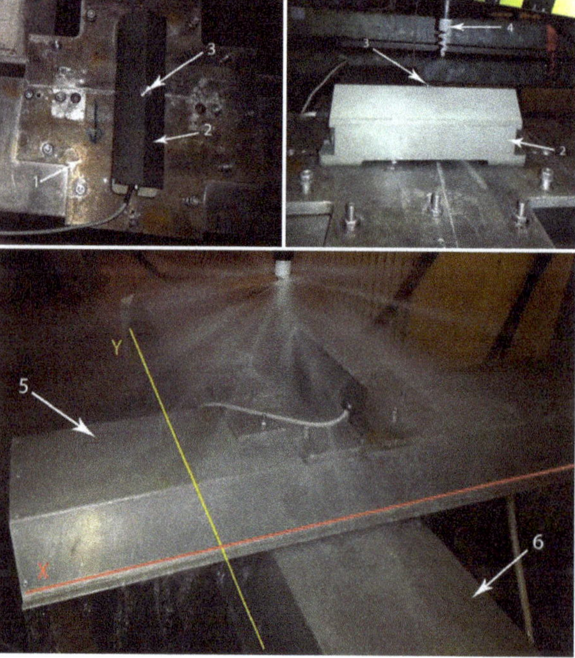

Figure 5. Experimental setup for the impact measurement: Bed plate (1), metal casing (2), impact surface (3), spiral nozzle (4), upper movable arm (5), lower movable arm (6).

3.2.2. Measuring the Impact Pressure in One Plane at Different Water Inlet Pressures

These experiments (Figure 6) were performed to clarify the effect of different inlet water pressures (the pressure of water measured prior to entering the nozzle orifice) on the impact footprint. The impact surface (2) was in the form of a circular plate with a diameter of 200 mm, in the center of which was an opening (3) with a diameter of 1 mm. The opening was connected to a closed space under the plate where the tensometric pressure sensor was placed (4). The space was completely filled up with water prior to the measurement to ensure the impact pressure be transferred to the sensor through the water fluctuations. The remaining parts of the experimental setup including the moving arms in the X and Y axes were identical with the previous measurements (Figure 5). The only difference was the extent of the area scanned, which was 240 mm in both directions with a step of 3 mm. All the measurements were performed for plane 2 (i.e., in a distance of 70 mm from the nozzle orifice; Figure 4b). The hydrodynamic conditions are shown in Table 1.

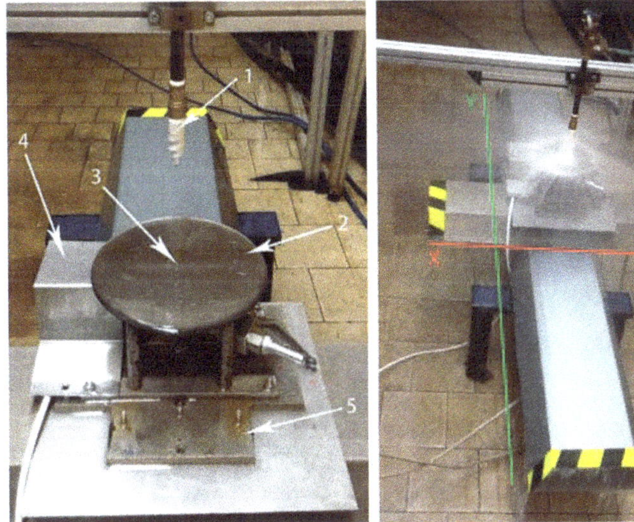

Figure 6. Experimental setup for the impact measurement with different water inlet pressures: Spiral nozzle (1), impact surface (2), pressure sensor opening (3), pressure sensor covering (4), bed plate of the impact surface (5).

Table 1. Experimental conditions at pressure impact measurement.

Inlet Pressure (bar)	1.00	1.25	1.5	1.75	2.00
Flowrate (L/s)	1.25	1.41	1.54	1.65	1.76

3.3. Spray Morphology

Spray morphology was observed inside a transparent laboratory-scale sprinkle chamber made of PVC glass (Figure 7a). The chamber with a water reservoir were fixed on a frame (Figure 7b) together with a centrifugal pump providing water circulation through the nozzle. The experiments were performed at 20 °C and an inlet pressure of 0.9 bar corresponding to a flowrate of 1.19 L/s. This flowrate was selected due to ideal properties of the spray for measuring in the experimental device due mainly to the size of the transparent chamber. Lower flowrate caused improper spray properties including very narrow span and practically no breakup of the liquid stream exiting the nozzle into ligaments/droplets. Higher flowrate caused significant rebound of the liquid stream from the chamber wall, which had a negative effect on the snapshots taking.

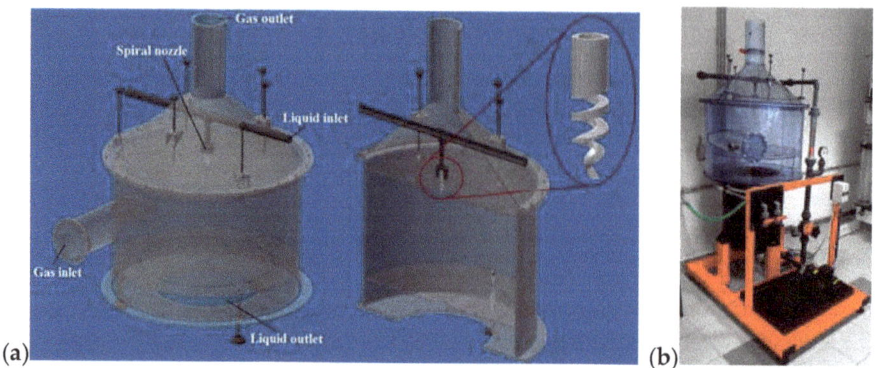

Figure 7. A 3D model of the laboratory sprinkle chamber (**a**) and a real view of the same (**b**).

For a detailed evaluation of the spray, an apparatus as shown in Figure 8 was used (for the real view refer to supplementary Figure S1). A classical photography with a modified type of lightning (i.e., an Nd:YAG pulse laser lightning (2) with a pulse length of 5 ns) was used (a spray snapshot using classical photography and modified with laser is shown in Figure S2). The images were taken using a Canon D70 camera (1) with a Canon EF 10 mm f/2.8 USM Macro objective. From the images, the areas of primary and secondary atomization of the spray were observed. The images were further used to measure droplet size using the Stream Motion software (Olympus Corporation, Shinjuku, Japan).

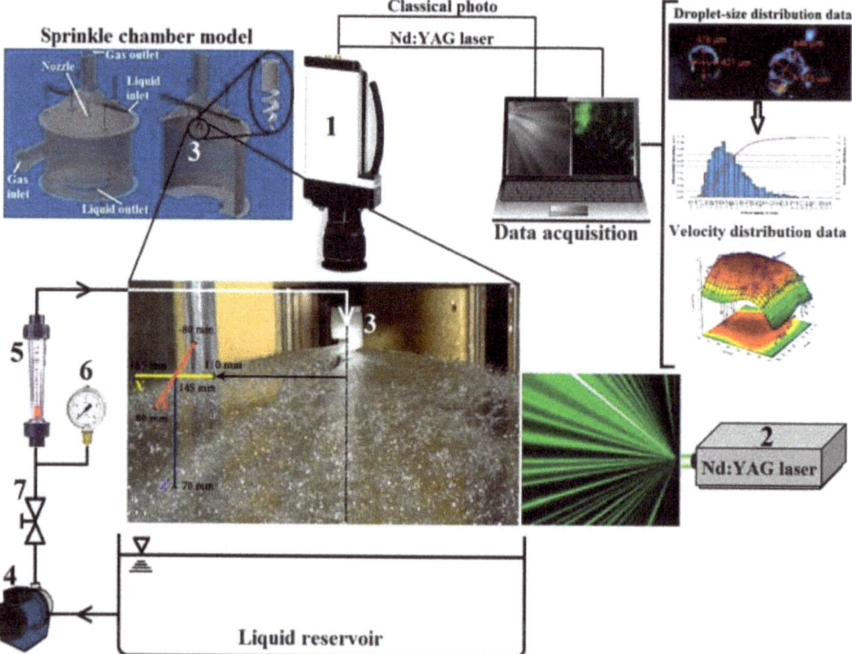

Figure 8. A scheme of the experimental setup for the assessment of the spray morphology/kinetics with the Canon D70 camera (1), Nd:YAG pulse laser (2), spiral nozzle (3), centrifugal pump (4), flowmeter (5), manometer (6), and valve (7).

From the measured droplet sizes, distribution curves were plotted and fitted with suitable distribution functions. The appropriateness of the fitting was then tested using the Kolmogorov–Smirnov test at a significance level of 0.05. Two density distribution functions, which are frequently used to describe droplet size in sprays were chosen (i.e., log-normal and Rosin-Rammler distribution) [80]. Log-normal density distribution can be expressed as follows:

$$f(D) = \frac{1}{\bar{\sigma}_D \sqrt{2\pi}} \exp\left[-\frac{\left(\ln D_i - \bar{\mu}_D\right)^2}{2\bar{\sigma}_D^2} \right],\tag{17}$$

where D_i, μ_D, and σ_D are the droplet diameter, mean, and standard deviation of the distribution, respectively. Rosin-Rammler density distribution is described by the following equation:

$$f(D) = \frac{\alpha}{\beta}\left(\frac{D_i}{\beta}\right)^{\alpha-1} \exp\left[-\left(\frac{D_i}{\beta}\right)^{\alpha} \right],\tag{18}$$

where α and β are empirical constants of the distribution. Further representative means of the droplet size were calculated (i.e., surface-weighted mean), which is used in the area of absorption processes where interphase area between two phases is important:

$$D_{20} = \left[\frac{\sum N_i D_i^2}{\sum N_i} \right]^{\frac{1}{2}}.\tag{19}$$

In addition, it is widely used Sauter mean diameter, which characterizes the spray fineness and is often used to calculate the efficiency and rate of mass transport in chemical reactions. It is a ratio of droplet volume to its surface area:

$$D_{32} = \frac{\sum N_i D_i^3}{\sum N_i D_i^2}.\tag{20}$$

Finally, volume weighted mean was calculated as follows:

$$D_{43} = \frac{\sum N_i D_i^4}{\sum N_i D_i^3}.\tag{21}$$

Another parameter studied was the spray angle, which was assessed using two methods. The first method was based on the pressure impact footprint curves, which were approximated using an Archimedean spiral. The other method was based on a visualization of the liquid spray in an open space to determine the spraying angle of individual cones using the Stream Motion software. Both methods are further explained in detail in the Results and Discussion section.

3.4. Spray Kinetics

Another method used was the laser Doppler anemometry (LDA), which provided us with information about the spray velocity distribution. The LDA setup was arranged on the same apparatus as for the experiments for the spray morphology evaluation (Figure 8). Thus, we obtained radial and tangential velocities from the upper and inner large cone. The LDA was performed for eight horizontal planes (Figure 9). In each plane, the spray velocity was measured in two perpendicular axes (the X-axis in yellow, and the Y-axis in red).

The beginning of the spray velocity measurement in the X-axis was at a distance of 110 mm from the nozzle center, as shown in Figure 9a (first position). Then, the spray velocity was measured after each 5 mm up to a distance of 165 mm from the nozzle center (12 positions in total). The spray velocity in the Y-axis was measured in a distance of 145 mm from the nozzle center and ranged from −80 to 80 mm (Figure 9a, detail shown in Figure 9b) perpendicular to the X-axis (33 positions in total). In each position, the measurement took 30 s or 30 thousand samples was taken. In the Z-axis, the point 0 mm corresponds to the position above the spray. The radial velocity was then measured up to 70 mm from this point in a step of 10 mm (8 positions in total; Figure 9b).

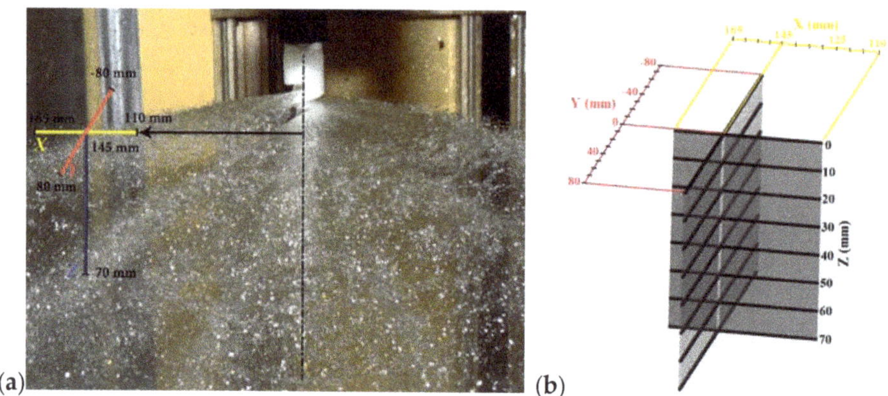

Figure 9. A representation of the measured position in relation to the spray (**a**) with a detailed demonstration/description of the measured positions (**b**).

4. Results and Discussion

4.1. Pressure Impact Footprint

Data from measurements in individual planes are presented in Figure 10. In the 3D graphs of Figure 10, the axes are not equidistant and the 2D graphs show the Cartesian coordinate system from the view above, turned according to the nozzle position in the experiments. Therefore, the 2D graphs can be considered for the spray cones' baselines and serve mainly as a visualization of the spray impact footprint of the two inner cones (the upper cone was out of the scanned area). The values of the impact pressure are very low (0.61 kPa at the highest). The reason for this is the size of the impact area, which is quite large for such a fine spray. The pressure impact intensity decreases with increasing distance from the nozzle orifice from about 0.61 kPa in the plane 1 to less than half (0.28 kPa) in the plane 6, which corresponds to the distances of individual planes from the orifice (60 and 140 mm of the plane 1 and 6, respectively). The effective spraying angle of the inner cones was further evaluated. This was done only for the pressure of 2 bars because with increasing inlet pressure, the effective spraying angle varied very slightly. Figure 10d shows the maximum impact pressure in relation to set inlet pressure of the liquid at the nozzle. We can see a linear increase up to an inlet pressure of 1.5 bar, then the impact pressure increase slowed down. The same is shown in supplementary Figure S3, which compares the varying intensity of the impact pressure at a distance of 70 mm (plane 2) from the nozzle in relation to changing increasing inlet pressure.

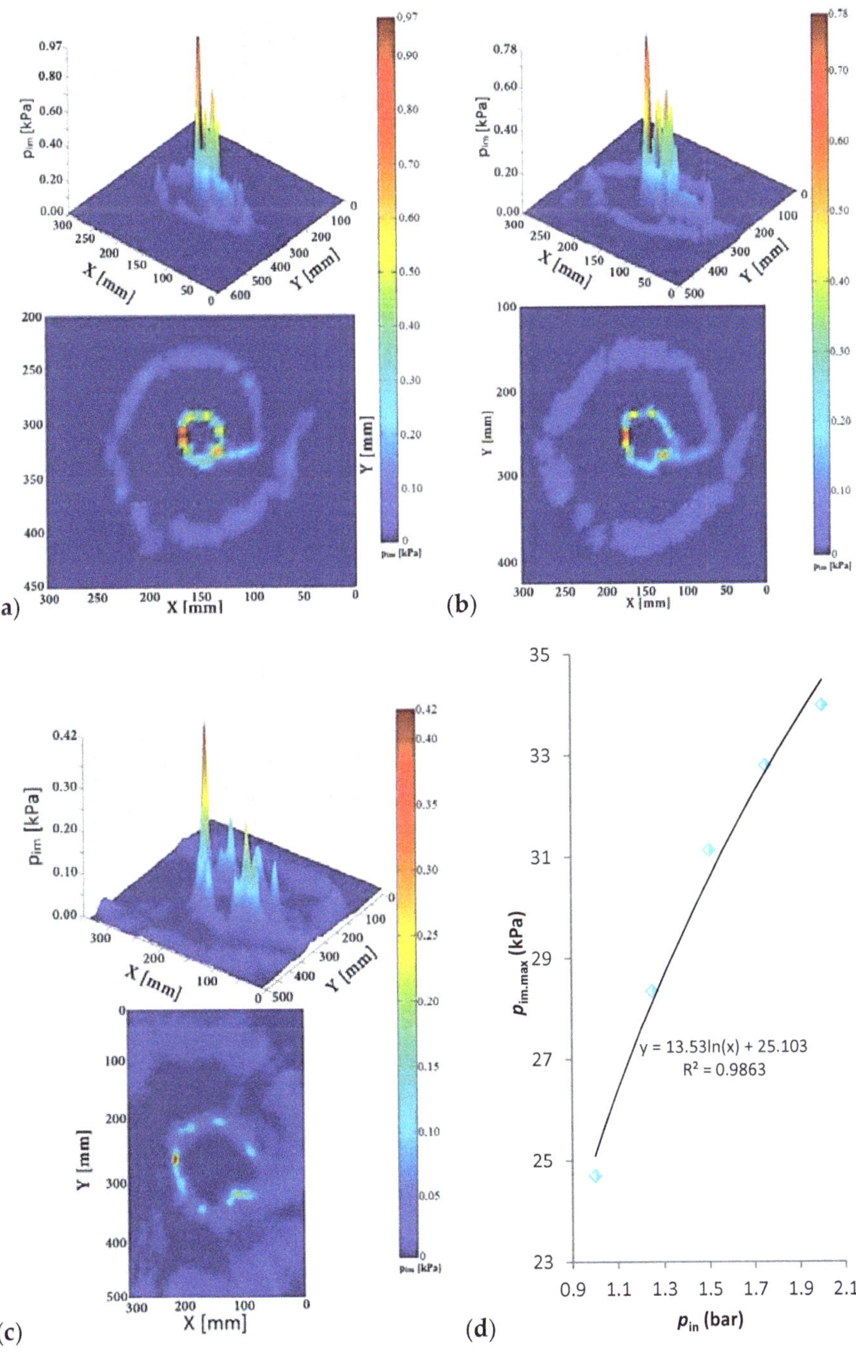

Figure 10. Pressure distribution pattern of the sprayed liquid measured in the plane 1 (**a**), 3 (**b**), 6 (**c**), and a relationship between maximum impact pressure and inlet pressure of liquid at nozzle (**d**).

4.2. Spray Angle

The base of the inner cones can be expressed using the parametric equation of the Archimedean spiral as follows:

$$x = a\theta \cos\theta + S_x, \tag{22}$$

$$y = a\theta \sin\theta + S_y, \tag{23}$$

where x and y are the points of the curve in the coordinate system, a is the parameter, S_x and S_y are the shifted centers of the spirals, and θ is the angle between half line describing the spiral trajectory and polar axis of the system. The values of the parameter a for the inner cones are shown in (Table S1). An example of the expression for plane 3 (Figure 11a) is for the small cone as follows:

$$x = 2.6\theta \cos\theta + 145, \tag{24}$$

$$y = 2.6\theta \sin\theta + 250. \tag{25}$$

In addition, for the large cone:

$$x = 9.0\theta \cos\theta + 145, \tag{26}$$

$$y = 9.0\theta \sin\theta + 250, \tag{27}$$

where for each plane $\theta \in (2.95\pi; 4.95\pi)$. The upper cone could not be evaluated using this method due to its large extent and was obtained using another method as explained further. The spiral curves delineate the trajectory of the impact footprint and well corresponds to experimental data (Figure 11a). An exception can be found in the fourth quadrant (beam number 5, Figure 11a). This is due to the nozzle geometry and transition between individual cones resulting in a local change of liquid flow.

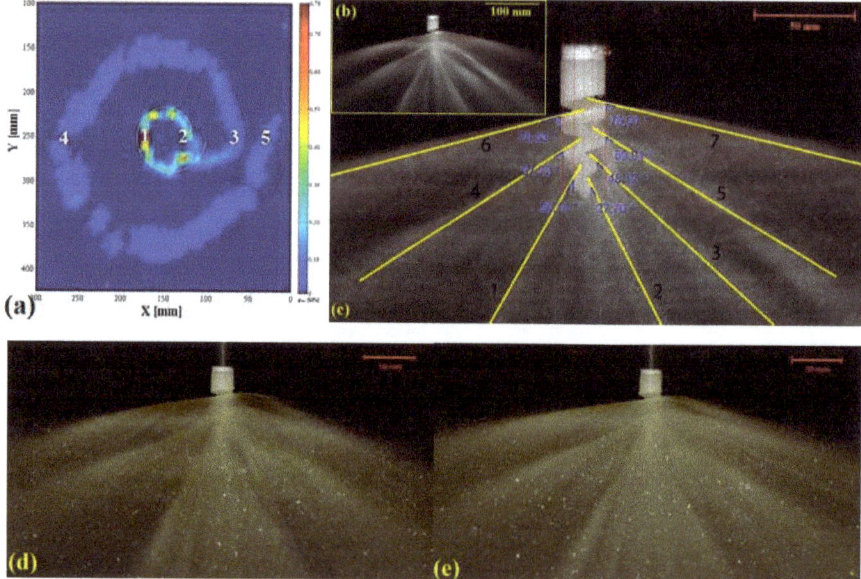

Figure 11. Pressure impact footprint with numbered water jets (individual numbers relates to individual water beams, refer to Figure 4b) and a graphical depiction of the spirals tracing the cones' baselines used for spraying angle assessment (in plane 3) (**a**), a photograph of the spray with an image filter applied to determine the main liquid streams (**b**), a visualization of the spray with assessment of inner

angles formed by the liquid beam and nozzle longitudinal axis (**c**), and comparison of the spray formed at an inlet liquid pressure of 1 bar (**d**) and 2 bars (**e**).

The extent of the small cone angle was calculated as a sum of the inner angles of the beams no. 1 and 2 (refer to Figure 11a). The extent of the large cone was a sum of the beam no. 4 and an arithmetic average of beams no. 3 and 5. The inner angles of individual water beams were determined in relation to the nozzle longitudinal axis (the Z-axis) in the same position as in experiments (i.e., in the XZ plane (refer to Figure 3)). Based on the extent of water beams of both cones, the effective spraying angle γ_{ef} was calculated. The average values of effective spraying angle were 61.7° and 115.5° for small and large cone, respectively, (for detailed results, refer to Table S2).

Another method to determine the spraying angle was using the photographs of the spray with applied screen filters to assess the main liquid beams (Figure 11b) and comparing with the photographs of the nozzle without spray (Figure 11c). Evaluation of the angles was done in the Stream Motion software (refer to Table S3). The small, large, and upper cone angles were 58.0°, 111.4°, and 152.3°, respectively. Comparing the results of the angles for the small and large cone with those obtained with the previous method (Table S2), we get values, which are smaller by 6% and 3.5%, respectively. This difference is mainly caused by a subjective assessment of the 2D photograph of a 3D spray, which may bring an error into the determination of the spraying angle. Both discrepancies are; however, very small and can be neglected.

Comparing the results of the spray angle with other studies, one significant difference can be observed (i.e., an independence of the spray angle on the inlet pressure (flowrate) of the liquid; Figure 11d,e), refer also to (Figure S4). This is in agreement with a previous study on characterization of a spiral nozzle, in which authors observed minimal variation of the spray angle at pressures above 0.4 bar [4]. On the contrary, such behavior is quite different compared, for example, to an air–water impinging jet atomizer [81], in which the spray angle increased with increasing pressure, or pressure swirl atomizer, in which the spray angle increased up to 7 bar and then decreased [82]. Some researchers also observed a decrease in spray angle with increasing inlet pressure in the whole measured range [48]. Another typical feature of the nozzle studied is its large spray angle, thus large spray coverage. This can be observed in nozzles mostly used for fire suppression applications [36]. However, this property is also required in the applications related to gas cleaning via absorption processes, especially for gas scrubbers with larger spray tower diameter.

4.3. Spray Breakup

Here we discuss the breakup of the outer upper cone only (Figure 4a). The other two inner cones are not evaluated as it was not allowed by the experimental setup. Moreover, the results are compared with mathematical models assuming the analogy with radial spread of a liquid jet over a horizontal plane. The condition of the horizontal plane is fulfilled for the upper cone only (Figure 2), which is formed by impinging the liquid jet on the first twist of the helix, which is horizontal (refer to CFD model at cross-section, Figure 2). The other helix surfaces (second and third twist) have different rake angles and are not horizontal, so the model could not be applied to these.

Individual atomization phases can be observed in the pictures (Figure 12) (i.e., primary/secondary atomization in various distances from the nozzle). The water sheet remains continuous up to a distance of about 80 mm (Figure 12a), even though some local perforations can be observed. Some perforations can be observed throughout the whole length of the liquid sheet (Figure 12b 2). The jet Weber number was higher than 22,000, which is far higher than the critical value of 1000 proposed by other researchers [65,83]. At this Weber number, we can expect very short radial breakup distance and the sheet to thin rapidly. The sheet thickness was estimated using the Equation (6) to be 2.47 mm at the edge of the helix and gradually decreased with increasing radial distance from the nozzle. At a distance of 78 mm, corresponding to the first sheet breakup into ligaments, the sheet thickness was estimated to 0.33 mm. The estimated sheet thickness also decreased with increasing liquid flowrate. At a radial distance of about 80 mm the sheet integrity is markedly disturbed (i.e., a primary atomization occurs up to 125 mm forming larger ligament structures; Figure 12b 1). The first ligaments observed (at the distance of 78 mm) were as large as (1.86 ± 0.57) mm with a breakup time estimate of 10.9 ms (according to Equation (15)) and attenuating down to (0.75 ± 0.19) mm with a breakup time of 2.8 ms (at a distance of approximately 120 mm). The estimate of the ligament diameter was 0.70 mm (based on Equation (13) using the correlation in Equation (12) for the dimensionless wave amplitude, at a radial distance of 120 mm). However, conducting experiments at different flowrates is necessary to assess further this comparison.

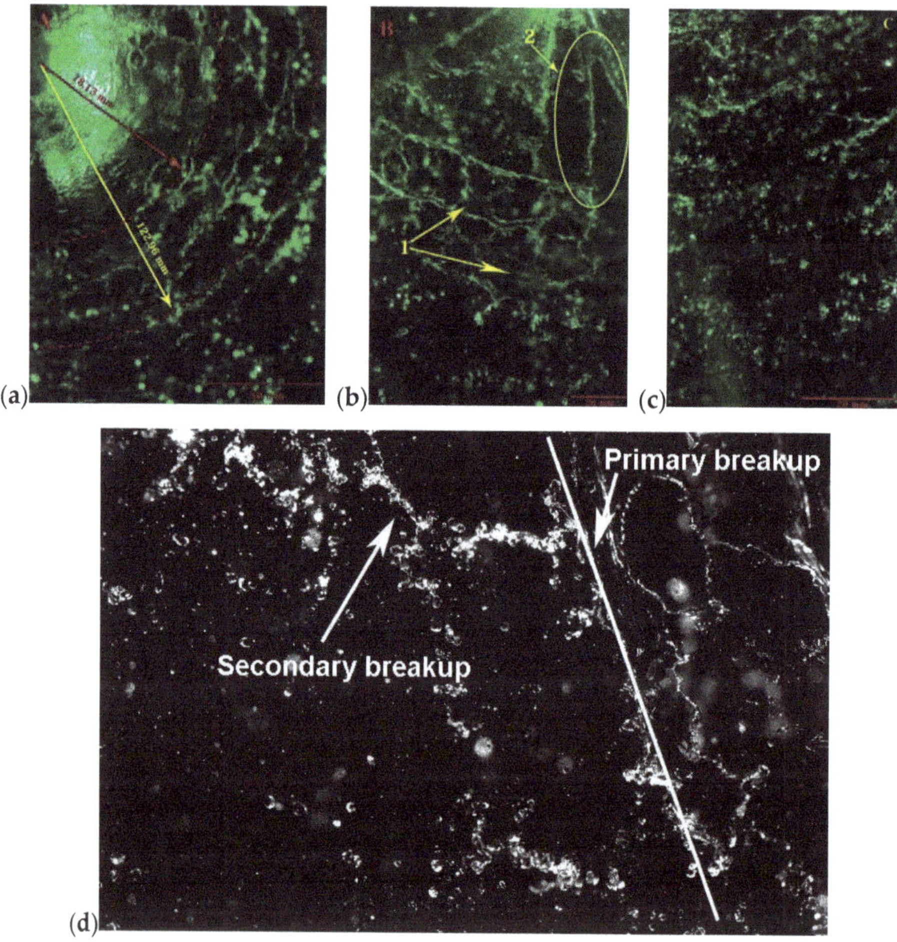

Figure 12. Individual phases of spray atomization (**a**), primary spray atomization to ligaments, 1–forming of ligaments, 2–liquid sheet rupture (**b**) and secondary atomization to individual droplets (**c**), and a detail of the boundary of both phases (**d**).

Subsequently a breakup to individual droplets (Figure 12c) occurred in a very narrow area. From this area the droplet-size distribution and structure has changed very slightly indicating area of secondary (or better quasi-secondary) atomization of the spray. A detail of the boundary between primary (sheet breakup into ligaments) and the quasi-secondary atomization of the spray (ligaments breaking into droplets) is shown in a modified photograph (Figure 12d). Generally, similar qualitative spray characteristics were very scarcely found in the literature. Qualitatively the most similar sprays were observed in studies focusing on characterization of fire sprinklers, refer, for example, to [64,74,75,84,85]. These sprinklers consist of a convergent nozzle with a cone-disc deflector placed under in a defined distance to create a circular liquid sheet. This is very similar to the spray generated by the spiral nozzle as the principle of forming the liquid sheet is practically the same (i.e., a rebound from a surface into the ambient space).

The gas Weber number calculated for the measured droplet sizes and radial velocities was always lower than 11. This may indicate, according to some studies, no secondary breakup [86–88], droplet deformation [67], or vibrational breakup [66]. The latter two may be true as indicated in obtained photographs (Figure 12d, also refer to an example shown in Figure S5). This is; however, a mere qualitative observation of the spray. As the liquid is sprayed into a quiescent air, it is more probable that the secondary atomization did not occur at the adopted experimental conditions. It is also important to say that the Weber number ranges defining individual secondary breakup modes were mostly derived for liquid jets. In liquid sheets or even conical sheets, the situation can be different and is rather a suggestion for future research. In general, vibrational breakup is not always observed. Oscillations at a natural frequency of the drop are typical and a formation of only a few fragments with comparable sizes as the original droplet are observable at this mode [68].

4.4. Droplet Size Distribution

Droplet size distribution was obtained from the modified classical photography of the upper cone in various distances normal to the nozzle axis. From detailed photographs of the secondary atomization area, droplet size was measured using the Stream Motion software. Prior to the measurement, a filter was applied to the photographs to better recognize the droplet edges. Spherical droplets' diameter was measured once, whereas the diameter of droplets of non-spherical shape was measured twice (two diameters normal to each other, refer to Figure S6), and then an arithmetic average was calculated. Using this method, 1773 droplets' diameters were evaluated. Droplet-size distribution curves were then plotted using the obtained data (Figure 13). The width of one size class was 0.07 mm. This was obtained by dividing the whole distribution width by square root of the number of measured droplet sizes. The agreement between experimental and theoretical distribution was assessed using the Kolmogorov–Smirnov test, which revealed the best fit only for the log-normal distribution for which the *P*-values were higher than the selected significance level of 0.05 (refer to Table S4).

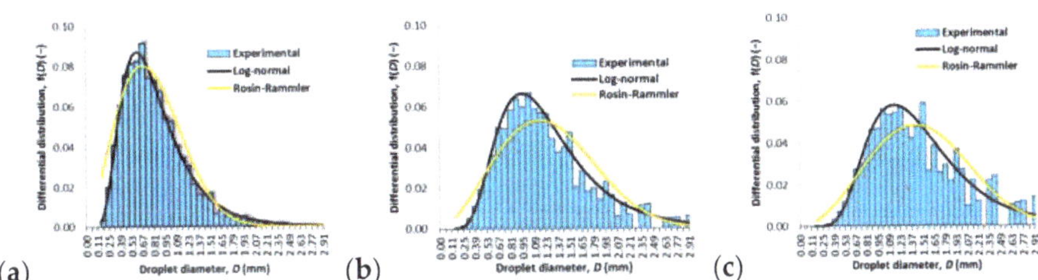

Figure 13. Number (**a**), surface (**b**), and volume weighted (**c**) droplet-size distribution fitted with selected theoretical distribution functions.

Droplet-size distribution of a spray can be considered monodisperse if the standard deviation is approximately less than 10% of the mean particle diameter [80]:

$$\frac{\overline{\sigma}_D}{\overline{\mu}_D} < 0.1. \tag{28}$$

In our case, according to the number droplet-size distribution of the spray, the ratio is as high as 0.48, confirming a polydisperse spray. The average droplet size was 0.815 mm, while the mode was around 0.587 mm with a portion of 8.7%. The surface-weighted average (D_{20}) was 0.997 mm and the Sauter mean diameter (D_{32}) was 1.201 mm. The largest amount of droplet surface (6.6%) was represented by a droplet diameter of 0.911 mm. For the volume-weighted distribution, the mean value is shifted to larger droplet diameters as large as 1.423 mm and a mode of 1.108 mm representing 5.8% of the liquid volume. The theoretical predictions of the droplet diameter were based on the theory of the liquid sheet instability, assuming an analogy with a rebound of liquid jet from a horizontal plane. A droplet diameter of 1.413 mm was calculated based on Equation (16), which is quite different from the average value obtained experimentally (0.815 mm), despite the expected analogy with the theory adopted for the diameter estimate. It is; therefore, necessary to conduct further research in this area as the description of conical sheets breakup mechanisms is rather scarce in the literature [89]. The measured droplet diameters are mostly much larger compared to different types of atomizers, which produce droplets in the range up to 100–150 µm in diameter, refer to, for example, [58,90–93], but in a similar order as compared to fire sprinklers [64,79,94]. Comparing the Sauter mean diameter empirical formulas developed for different atomizers we can also see rather significant differences. For example, the Sauter mean diameter for pressure-atomized sprays is a function of several parameters, as proposed in the correlation by Elkotb [95]:

$$D_{32} = 3.08\mu_l^{0.385}(\sigma\rho_l)^{0.737}\rho_g^{0.06}\Delta p^{-0.54}, \tag{29}$$

where Δp is the pressure drop at the nozzle (i.e., the pressure difference between pressure inside the nozzle and the ambient pressure of the space, into which the liquid is sprayed). This formula gives the Sauter mean diameter as large as 1.844 mm, which is more than 0.6 mm larger than that obtained experimentally.

4.5. Liquid Velocity Distribution

Using the LDA method, radial and tangential velocities in the outer upper cone (Figure 4a) were obtained. 3D surface graphs of absolute radial velocity in relation to position were created. Data was processed into velocity distribution histograms of radial and tangential velocity component at individual points (Figure 14). The absolute radial velocity (v_{AR}) was obtained through a vector sum of radial and tangential velocity at a given point. An example of the data processing into histograms for the position (140,0,40) mm (x,y,z) is shown in Figure 14. The velocities minus signs are due to the LDA experimental setup, which measured the radial velocities as negative. An average radial and tangential velocity at the point (140,0,40) mm was (7.78 ± 2.03) m s^{-1} and (0.17 ± 1.28) m s^{-1}, respectively. The vector sum of the both velocities (i.e., the representative absolute radial velocity of the liquid) is 7.78 m s^{-1}. This means that the contribution of the tangential velocity was very small with a very narrow distribution. This is in contrast with a spiral flow nozzle with an annular slit [47] using which a tangential velocity up to 8 m/s was observed (in relation to slit width). Looking at the geometry of the spiral nozzle studied, more noticeable tangential velocity contribution was expected. This is due to the centripetally-oriented impact surfaces of the spiral causing torque of the liquid at the outlet. This would probably be more noticeable at lower liquid velocities, at which the liquid sprayed copies the trajectory demarcated by the spiral. However, this effect is rather limited at high velocities due to high impact force causing a strong rebound of the liquid from the surface in the radial direction, thus eliminating the formation of the torqued flow.

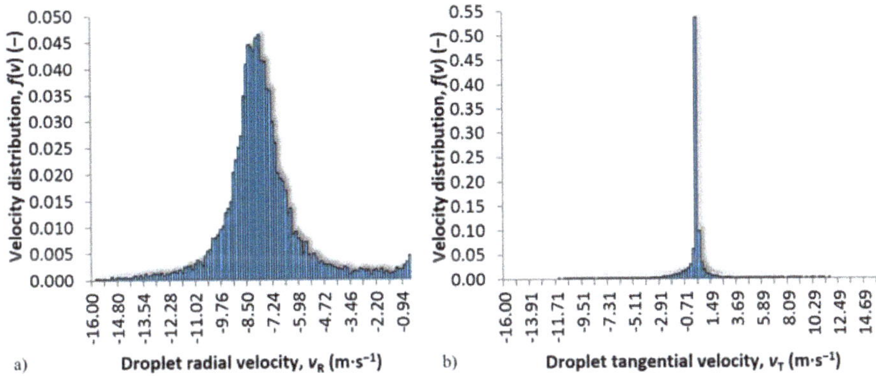

Figure 14. Droplet radial (**a**) and tangential (**b**) velocity distribution at a position (140,0,40) mm (*x,y,z*).

Using the treated data, 3D surface graphs of absolute radial velocity in relation to radial X- and Y-position at different axial distances (Z) were plotted and approximated using the distance weighted least squares algorithm in the TIBCO Statistica (USA) software. The 3D graphs are shown from two different orientations, and blue marks represent individual points approximated by the plane. Figure 15a shows a decrease in velocity with increasing distance. At the Z-axis 0 and 10 mm, the velocity was measured right above the liquid sheet; therefore, the measured velocities are low. The highest velocity is at the Z-axis between 40 and 50 mm, where the main liquid stream occurs and then slightly decrease (the same in 2D is shown in supplementary Figure S7). The highest radial velocity of 8.26 m s^{-1} was measured at a radial and axial distance of 110 and 40 mm, respectively. At the same axial distance was also the lowest measured velocity of 7.22 m s^{-1}, but at the position corresponding to a radial distance of 165 mm (the most distant position from the nozzle axis). This is due to the kinetic energy loss in the space with increasing distance from the nozzle orifice. However, at the 0 and 10 mm Z-distance, there is an obvious decrease of the velocity profile, which is higher than could be expected from the loss of kinetic energy. This is probably caused by the large spray angle (the liquid in the upper cone is sprayed from the nozzle almost horizontally) and a descent of the liquid sheet due to gravity. In a distance between 110 and 165 mm, the descent by 15 mm can be expected. The kinetic energy loss (liquid flow deceleration) is also due to the friction caused by the complicated nozzle orifice geometry. This can be expressed using the Euler number (*Eu*) (i.e., the ratio of pressure loss due to flow restriction and kinetic energy per volume of the flow):

$$Eu = \frac{2\Delta p}{\rho_l v^2},$$ (30)

where v is the liquid velocity inside the nozzle. Figure 16 shows the relationship between Euler number and Reynolds number at nozzle orifice (Re_o) of the flow inside the nozzle (Equation (7)).

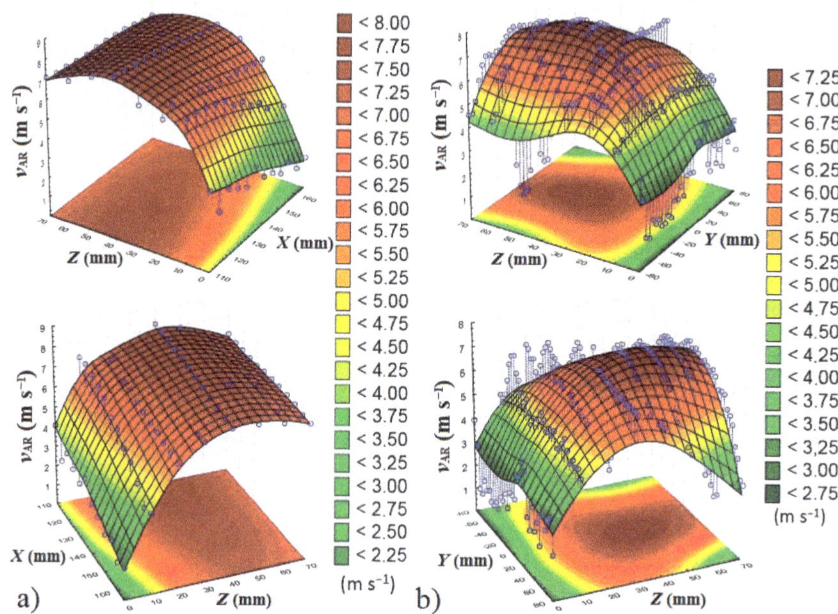

Figure 15. Absolute radial velocity in relation to X-distance (**a**) and Y-distance (**b**).

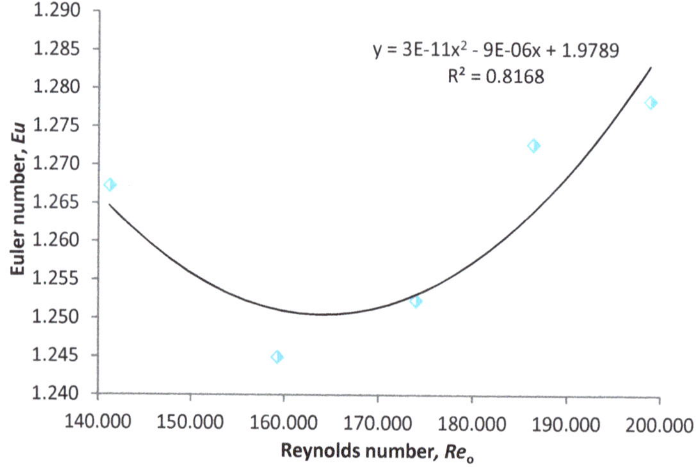

Figure 16. Euler number in relation to Reynolds number at nozzle.

We can see an atypical behavior of the nozzle studied. The friction losses expressed in terms of Eu are quite low (ideal frictionless flow corresponds to $Eu = 1$). However, Eu first decreases with increasing Re from approx. 1.4×10^5 to 1.6×10^5. Such a high Re corresponds to fully developed turbulent flow and any further increase is expected to rather increase pressure losses due to friction. Nonetheless, the extraordinary nozzle orifice geometry probably causes this slight anomaly and up to the point corresponding to Re of about 1.6×10^5 the friction losses slightly decreased. This can outline ideal operation conditions for the use of this nozzle in gas cleaning applications. At lower Re

the friction is higher but the turbulence is lower at the same time causing non-ideal conditions for the removal of the contaminative gases from air. With further increase of *Re*, *Eu* starts to also increase up to 1.28 corresponding to an inlet pressure of 2 bars (i.e., a flowrate of 1.76 L/s; refer to Table 1). It is further suggested for future work to measure a wider range of inlet pressures to obtain more complex idea about the nozzle behavior. However, we were limited to a pressure of 3 bars at the highest due to the nozzle made of Teflon (for higher pressures, metal nozzles are necessary). The nozzle behavior described in Figure 16 is quite different compared to a hollow-cone pressure swirl nozzle as studied by Nonnenmacher and Piesche [96]. The authors observed higher values of *Eu* increasing approx. from 50 to 80 with *Re* increasing approx. from 300 to 50,000.

5. Conclusions

This work tried to provide basic characteristics of a spray produced using a spiral nozzle and compare the spray properties with other atomizers. The spray produced using the spiral nozzle consists of three concentric liquid cones creating a full-cone pattern. Typical spray angle is quite large compared to other nozzles, the upper large cone having a spray angle as large as 150°. At the adopted conditions, the droplet-size distribution was larger. Sauter mean diameter was 1.201 mm, the same is true for velocity which was, conversely, lower compared to, for example, effervescent atomizers. Ideal operational conditions were found to be at a liquid flowrate of 1.41 L/s corresponding to an inlet pressure of 1.25 bar. Based on the initial comparison with the theory, we can see a gap in the mathematical description of the conical sheets. Therefore, future research is necessary to further understand the breakup behavior of the liquid jets on plates horizontal as well as at different angles. Future work will also focus on the description of the conical sheet breakup using detailed observations with optical methods, and basic RANS simulations (Reynolds-averaged Navier-Stokes equations) and large eddy simulations (LES) with adaptable computational mesh to catch as many of the smallest structures and droplets formed during the sheet atomization as possible. The presented results will also serve as the experimental verification of the developed CFD models.

Supplementary Materials: The following are available online at http://www.mdpi.com/2227-9717/7/12/911/s1, Figure S1: Representation of the measured position in relation to the spray (a) and a detailed demonstration/description of the same (b); Figure S2: A snapshot of classical (a) and modified (b) photography, Figure S3: Impact pressures of individual measurements in plane 2; Table S1: The parameters of the Archimedean spiral equation; Table S2: Effective spraying angles γ_{ef} of small and large cones for planes 1, 3, and 5; Table S3: Effective spraying angles γ_{ef} of small and large cones for planes 1, 3, and 5; Figure S4: Comparison of the spray at two different flowrates of 1.25 l/s corresponding to a pressure at the nozzle of 1 bar (A) and 1.76 l/s corresponding the a pressure at the nozzle of 2 bars (B); Figure S5: Deformation (yellow-circled) and vibrational breakup (red-circled) of droplets; Table S4: Results of the Kolmogorov–Smirnov test; Figure S6: Detail of individual droplets (a) and evaluation of their size (b); Figure S7: Velocity profile of the liquid sheet in relation to radial distance X at individual axial positions Z

Author Contributions: O.K. conceived, designed, and performed the experiments and evaluated data; P.B. evaluated the data and wrote the paper; T.S. conceived the work, provided the laboratory with the equipment and supervision.

Funding: This paper has been supported by the project "Computer Simulations for Effective Low-Emission Energy Engineering" funded as project no. CZ.02.1.01/0.0/0.0/16_026/0008392 by Operational Program Research, Development and Education, Priority axis 1: Strengthening capacity for high-quality research.

Conflicts of Interest: The authors declare no conflicts of interest.

References

1. Seyboth, O.; Zimmermann, S.; Heidel, B.; Scheffknecht, G. Development of a Spray Scrubbing Process for Post Combustion CO_2 Capture with Amine Based Solvents. *Energy Procedia* **2014**, *63*, 1667–1677. [CrossRef]
2. Vlasák, J.; Svěrák, T.; Dreveny, L.; Kalivoda, J. Air Purification from CO_2 Gas Using a Scrubber and Suggestions to Reach a Better Efficiency. *Mater. Sci. Forum* **2019**, *955*, 1–6. [CrossRef]
3. Dumont, É. Mass Transfer in Multiphasic Gas/Liquid/Liquid Systems. KLa Determination Using the Effectiveness-Number of Transfer Unit Method. *Processes* **2018**, *6*, 156. [CrossRef]

4. Li, Z.; Wang, S.; Wang, X. Spray characteristics of spiral nozzles used in wet flue gas desulfurization system. *J. Southeast Univ. Nat. Sci. Ed.* **2008**, *38*, 493–495.

5. Dong, H.; Ma, H.; Li, D.; Sun, G. Comparative study on scrubbing desulfurization performances of spiral nozzle and Dynawave nozzle. *Pet. Refin. Eng.* **2006**, *36*, 32–35.

6. Sadegh-Vaziri, R.; Amovic, M.; Ljunggren, R.; Engvall, K. A Medium-Scale 50 MWfuel Biomass Gasification Based Bio-SNG Plant: A Developed Gas Cleaning Process. *Energies* **2015**, *8*, 5287–5302. [CrossRef]

7. Byeon, S.-H.; Lee, B.-K.; Raj Mohan, B. Removal of ammonia and particulate matter using a modified turbulent wet scrubbing system. *Sep. Purif. Technol.* **2012**, *98*, 221–229. [CrossRef]

8. Zhou, J.; Zhou, S.; Zhu, Y. Characterization of Particle and Gaseous Emissions from Marine Diesel Engines with Different Fuels and Impact of After-Treatment Technology. *Energies* **2017**, *10*, 1110. [CrossRef]

9. Balas, M.; Lisy, M.; Kubicek, J.; Pospisil, J. Syngas Cleaning by Wet Scrubber. *WSEAS Trans. Heat Mass Transf.* **2014**, *9*, 195–204.

10. Nwokolo, N.; Mamphweli, S.; Makaka, G. Analytical and Thermal Evaluation of Carbon Particles Recovered at the Cyclone of a Downdraft Biomass Gasification System. *Sustainability* **2017**, *9*, 645. [CrossRef]

11. Avveduto, A.; Ferella, F.; De Giovanni, M.; Innocenzi, V.; Pace, L.; Tripodi, P. L'Aquila Smart Clean Air City: The Italian Pilot Project for Healthy Urban Air. *Environments* **2017**, *4*, 78. [CrossRef]

12. Adamec, V.; Schüllerová, B.; Hrabová, K.; Skeřil, R.; Kadlec, R.; Bulejko, P.; Adam, V. The nanoparticles concentration in the traffic loaded urban area. In Proceedings of the 9th International Conference on Nanomaterials—Research and Application, NANOCON 2017, Brno, Czech Republic, 18–20 October 2018; pp. 698–703.

13. Poláčik, J.; Šnajdárek, L.; Špiláček, M.; Pospíšil, J.; Sitek, T. Particulate Matter Produced by Micro-Scale Biomass Combustion in an Oxygen-Lean Atmosphere. *Energies* **2018**, *11*, 3359. [CrossRef]

14. Abdulwahid, A.A.; Situ, R.; Brown, R.J. Underground Diesel Exhaust Wet Scrubbers: Current Status and Future Prospects. *Energies* **2018**, *11*, 3006. [CrossRef]

15. Sitek, T.; Pospíšil, J.; Poláčik, J.; Špiláček, M.; Varbanov, P. Fine combustion particles released during combustion of unit mass of beechwood. *Renew. Energy* **2019**, *140*, 390–396. [CrossRef]

16. Svěrák, T.; Bulejko, P.; Krištof, O.; Kejík, P.; Kalivoda, J.; Horský, J. Covering ability of aluminum pigments prepared by milling processes. *Powder Technol.* **2017**, *305*, 396–404. [CrossRef]

17. Sverak, T. Sorption purpose lime hydrate grinding. *Int. J. Miner. Process.* **2004**, *74*, S379–S383. [CrossRef]

18. Bulejko, P.; Adamec, V.; Skeřil, R.; Schüllerová, B.; Bencko, V. Levels and Health Risk Assessment of PM10 Aerosol in Brno, Czech Republic. *Cent. Eur. J. Public Health* **2017**, *25*, 129–134. [CrossRef]

19. Bulejko, P.; Svěrák, T.; Dohnal, M.; Pospíšil, J. Aerosol filtration using hollow-fiber membranes: Effect of permeate velocity and dust amount on separation of submicron TiO_2 particles. *Powder Technol.* **2018**, *340*, 344–353. [CrossRef]

20. Sverak, T.; Bulejko, P.; Ostrezi, J.; Kristof, O.; Kalivoda, J.; Kejik, P.; Mayerova, K.; Adamcik, M. Separation of gaseous air pollutants using membrane contactors. *IOP Conf. Ser. Earth Environ. Sci.* **2017**, *92*, 012061. [CrossRef]

21. Bulejko, P. Numerical Comparison of Prediction Models for Aerosol Filtration Efficiency Applied on a Hollow-Fiber Membrane Pore Structure. *Nanomaterials* **2018**, *8*, 447. [CrossRef]

22. Ghasem, N. Modeling and Simulation of the Absorption of CO_2 and NO_2 from a Gas Mixture in a Membrane Contactor. *Processes* **2019**, *7*, 441. [CrossRef]

23. Bulejko, P.; Krištof, O.; Dohnal, M.; Svěrák, T. Fine/ultrafine particle air filtration and aerosol loading of hollow-fiber membranes: A comparison of mathematical models for the most penetrating particle size and dimensionless permeability with experimental data. *J. Membr. Sci.* **2019**, *592*, 117393. [CrossRef]

24. Bogacki, J.; Marcinowski, P.; Majewski, M.; Zawadzki, J.; Sivakumar, S. Alternative Approach to Current EU BAT Recommendation for Coal-Fired Power Plant Flue Gas Desulfurization Wastewater Treatment. *Processes* **2018**, *6*, 229. [CrossRef]

25. Gao, G.; Wang, C.; Kou, Z. Experimental Studies on the Spraying Pattern of a Swirl Nozzle for Coal Dust Control. *Appl. Sci.* **2018**, *8*, 1770. [CrossRef]

26. Chabičovský, M.; Hnízdil, M.; Tseng, A.A.; Raudenský, M. Effects of oxide layer on Leidenfrost temperature during spray cooling of steel at high temperatures. *Int. J. Heat Mass Transf.* **2015**, *88*, 236–246. [CrossRef]

27. Brozova, T.; Chabicovsky, M.; Horsky, J. Influence of the Surface Roughness on the Cooling Intensity During Spray Cooling. In Proceedings of the 25th Anniversary International Conference on Metallurgy and Materials (METAL), Ostrava, Czech Republic, 25–27 May 2016; Volume 2016, pp. 41–46.

28. Tseng, A.A.; Raudensky, M.; Lee, T.-W. Liquid Sprays for Heat Transfer Enhancements: A Review. *Heat Transf. Eng.* **2016**, *37*, 1401–1417. [CrossRef]

29. Stransky, M.; Brozova, T.; Raudensky, M.; Hnizdil, M.; Turon, R. Effect of Various Spray Cooling Configurations on Hardness Profile of Tubes. In Proceedings of the 24th International Conference on Metallurgy and Materials, Ostrava, Czech Republic, 3–5 June 2015; Volume 2015, pp. 880–885.

30. Lee, T.-W.; Hnizdil, M.; Chabicovsky, M.; Raudensky, M. Approximate Solution to the Spray Heat Transfer Problem at High Surface Temperatures and Liquid Mass Fluxes. *Heat Transf. Eng.* **2018**, *40*, 1649–1655. [CrossRef]

31. Votavová, H.; Pohanka, M.; Bulejko, P. Cooling homogeneity measurement during hydraulic descaling in spray overlapping area. In Proceedings of the 24th International Conference on Metallurgy and Materials, Brno, Czech Republic, 3–5 June 2015; pp. 265–270.

32. Zhang, D.; Li, Z.; Yi, W.; Wang, F. Test study of spray characteristics of spiral nozzle in the spray tower. *Acta Energiae Solaris Sin.* **2013**, *34*, 1969–1972.

33. Ren, Z.; Hao, Y.; Wang, B. Experimental Analysis on Spiral Pressure Nozzle and Spray Angle Control in the Spray Dryer. In Proceedings of the 2011 International Conference on Internet Computing and Information Services, Hong Kong, China, 17–18 September 2011; pp. 163–165.

34. Cong, H.; Li, X.; Li, Z.; Li, H.; Gao, X. Combination of spiral nozzle and column tray leading to a new direction on the distillation equipment innovation. *Sep. Purif. Technol.* **2016**, *158*, 293–301. [CrossRef]

35. Cheng, X.M.; Hashimoto, B.; Kage, S.; Matsumae, Y.; Horii, K.; Ohsumi, K. Flow in Spiral Nozzle for Rope-installation in Petrochemical Pipeline Repairs. *J. Jpn. Pet. Inst.* **1992**, *35*, 382–389. [CrossRef]

36. Tanner, G.; Knasiak, K.F. Spray Characterization of Typical Fire Suppression Nozzles. In Proceedings of the Third International Water Mist Conference, Madrid, Spain, 22–24 September 2003; pp. 1–16.

37. Jiang, L.; Wu, R.; Zhao, H.; Mei, P.; Zhang, Q.; Zhu, J.; Xiao, J.; Lei, B. Middle or low water pressure direct spiral double helix converging nozzle structure optimization and flow field analysis. *High Technol. Lett.* **2015**, *21*, 261–268.

38. Wąsik, W.; Walczak, A.; Węsierski, T. The impact of fog nozzle type on the distribution of mass spray density. *MATEC Web Conf.* **2018**, *247*, 00058. [CrossRef]

39. Liu, N.-L.; Zhang, X. Distribution of droplet diameters and the spline of their empirical equation for spiral nozzle. *J. Exp. Fluid Mech.* **2006**, *20*, 8–12.

40. Liu, N.; Zhang, X. Experimental research on spray and flow rate characteristics of spiral nozzle. *Trans. Chin. Soc. Agric. Mach.* **2006**, *37*, 79–82.

41. Zhou, H.; Guo, W.; Zhu, Y.; Ma, W. Flow Distribution Characteristics of Low-pressure High-flux Spiral Nozzles. *J. Chin. Soc. Power Eng.* **2017**, *37*, 577–583.

42. Li, C.; Wang, S.; Zhong, Z.; Zhang, X.; Kang, L. Research on spray surfaces geometry model of spiral nozzles with involute atomization. *Trans. Chin. Soc. Agric. Mach.* **2007**, *38*, 19.

43. Kalata, W.; Brown, K.J.; Schick, R.J. Air Assisted Atomization in Spiral Type Nozzles. In Proceedings of the 25th Annual Conference on Liquid Atomization and Spray Systems, ILASS Americas, Pittsburgh, PA, USA, 21–23 May 2013; p. 8.

44. Bete TF Spray Nozzle. Available online: https://www.johnbrooks.ca/product/bete-tf-spray-nozzle/ (accessed on 14 May 2019).

45. Han, H.; Wang, P.; Liu, R.; Li, Y.; Wang, J.; Jiang, Y. Experimental study on atomization characteristics of two common spiral channel pressure nozzles. *E3S Web Conf.* **2019**, *81*, 01022. [CrossRef]

46. Kim, T.H.; Matsuo, S.; Setoguchi, T.; Yu, S. A study on flow characteristics in a spiral flow nozzle. *Int. J. Turbo Jet Engines* **2006**, *23*, 129–136. [CrossRef]

47. Matsuo, S.; Kim, T.-H.; Setoguchi, T.; Kim, H.D.; Lee, Y.-W. Effect of nozzle geometry on the flow characteristics of spiral flow generated through an annular slit. *J. Therm. Sci.* **2007**, *16*, 149–154. [CrossRef]

48. Jain, M.; John, B.; Iyer, K.N.; Prabhu, S.V. Characterization of the full cone pressure swirl spray nozzles for the nuclear reactor containment spray system. *Nucl. Eng. Des.* **2014**, *273*, 131–142. [CrossRef]

49. Zacarías, A.; Venegas, M.; Lecuona, A.; Ventas, R.; Carvajal, I. Experimental assessment of vapour adiabatic absorption into solution droplets using a full cone nozzle. *Exp. Therm. Fluid Sci.* **2015**, *68*, 228–238. [CrossRef]

50. Valencia-Bejarano, M.; Langrish, T.A.G. Experimental investigation of droplet coalescence in a full-cone spray from a two-fluid nozzle using laser diffraction measurements. *At. Sprays* **2004**, *14*, 355–374. [CrossRef]
51. Tseng, A.A.; Bellerová, H.; Pohanka, M.; Raudensky, M. Effects of titania nanoparticles on heat transfer performance of spray cooling with full cone nozzle. *Appl. Therm. Eng.* **2014**, *62*, 20–27. [CrossRef]
52. Sada, E.; Takahashi, K.; Morikawa, K.; Ito, S. Drop size distribution for spray by full cone nozzle. *Can. J. Chem. Eng.* **1978**, *56*, 455–459. [CrossRef]
53. Naz, M.Y.; Sulaiman, S.A.; Ariwahjoedi, B.; Zilati, K. Visual characterization of heated water spray jet breakup induced by full cone spray nozzles. *J. Appl. Mech. Tech. Phys.* **2015**, *56*, 211–219. [CrossRef]
54. Naz, M.Y.; Sulaiman, S.A.; Ariwahjoedi, B.; Ku Shaari, K.Z. Investigation of vortex clouds and droplet sizes in heated water spray patterns generated by axisymmetric full cone nozzles. *Sci. World J.* **2013**, *2013*. [CrossRef]
55. Kohnen, B.T.; Pieloth, D.; Musemic, E.; Walzel, P. Characterization of full cone nozzles. *At. Sprays* **2011**, *21*, 317–325. [CrossRef]
56. Jiang, P.; Wang, Q.; Sabariman, I.; Specht, E. Experimental study on heat transfer of pressurized spray cooling on the heated plate by using 45° full cone nozzles. *Appl. Mech. Mater.* **2014**, *535*, 32–36. [CrossRef]
57. Bellerová, H.; Tseng, A.A.; Pohanka, M.; Raudensky, M. Heat transfer of spray cooling using alumina/water nanofluids with full cone nozzles. *Heat Mass Transf.* **2012**, *48*, 1971–1983. [CrossRef]
58. Jašíková, D.; Kotek, M.; Lenc, T.; Kopecký, V. The study of full cone spray using interferometric particle imaging method. *EPJ Web Conf.* **2012**, *25*, 01033. [CrossRef]
59. Standard Full Cone Spiral Nozzle—BETE Online Catalogue. Available online: http://www.spray-nozzle. co.uk/docs/default-source/spec-sheet-pdf$'$s/tf-full-cone-data-sheet.pdf?sfvrsn=5cb16e8d_5 (accessed on 1 May 2019).
60. Lefebvre, A.H.; McDonell, V.G. *Atomization and Sprays*; CRC Press: Boca Raton, FL, USA, 2017; ISBN 978-1-315-12091-1.
61. Rahman, M.A.; Heidrick, T.; Fleck, B.A. Correlations between the two-phase gas/liquid spray atomization and the Stokes/aerodynamic Weber numbers. *J. Phys. Conf. Ser.* **2009**, *147*, 012057. [CrossRef]
62. Guildenbecher, D.R.; López-Rivera, C.; Sojka, P.E. Droplet Deformation and Breakup. In *Handbook of Atomization and Sprays: Theory and Applications*; Ashgriz, N., Ed.; Springer: Boston, MA, USA, 2011; pp. 145–156, ISBN 978-1-4419-7264-4.
63. Ashgriz, N.; Li, X.; Sarchami, A. Instability of Liquid Sheets. In *Handbook of Atomization and Sprays: Theory and Applications*; Ashgriz, N., Ed.; Springer: Boston, MA, USA, 2011; pp. 75–95, ISBN 978-1-4419-7264-4.
64. Vegad, C.S.; Chakravarthy, S.R.; Kumar, A. Dynamics of a radially expanding circular liquid sheet and its atomization characteristics. *Fire Saf. J.* **2018**, *100*, 51–63. [CrossRef]
65. Taylor, G.I. The dynamics of thin sheets of fluid. III. Disintegration of fluid sheets. *Proc. R. Soc. Lond. Ser. Math. Phys. Sci.* **1959**, *253*, 313–321.
66. Pilch, M.; Erdman, C.A. Use of breakup time data and velocity history data to predict the maximum size of stable fragments for acceleration-induced breakup of a liquid drop. *Int. J. Multiph. Flow* **1987**, *13*, 741–757. [CrossRef]
67. Faeth, G.M.; Hsiang, L.-P.; Wu, P.-K. Structure and breakup properties of sprays. *Int. J. Multiph. Flow* **1995**, *21*, 99–127. [CrossRef]
68. Guildenbecher, D.R.; López-Rivera, C.; Sojka, P.E. Secondary atomization. *Exp. Fluids* **2009**, *46*, 371. [CrossRef]
69. Luo, K.; Shao, C.; Chai, M.; Fan, J. Level set method for atomization and evaporation simulations. *Prog. Energy Combust. Sci.* **2019**, *73*, 65–94. [CrossRef]
70. Zhao, H.; Liu, H. Breakup Morphology and Mechanisms of Liquid Atomization. *Adv. Jet Engines* **2019**.
71. Chryssakis, C.A.; Assanis, D.N.; Tanner, F.X. Atomization Models. In *Handbook of Atomization and Sprays: Theory and Applications*; Ashgriz, N., Ed.; Springer US: Boston, MA, USA, 2011; pp. 215–231, ISBN 978-1-4419-7264-4.
72. Gelfand, B.E. Droplet breakup phenomena in flows with velocity lag. *Prog. Energy Combust. Sci.* **1996**, *22*, 201–265. [CrossRef]
73. Watson, E.J. The radial spread of a liquid jet over a horizontal plane. *J. Fluid Mech.* **1964**, *20*, 481–499. [CrossRef]
74. Ren, N.; Blum, A.; Zheng, Y.; Do, C.; Marshall, A. Quantifying the Initial Spray from Fire Sprinklers. *Fire Saf. Sci.* **2008**, *9*, 503–514. [CrossRef]

75. Zhou, X.; Yu, H.-Z. Experimental investigation of spray formation as affected by sprinkler geometry. *Fire Saf. J.* **2011**, *46*, 140–150. [CrossRef]

76. Dombrowski, N.; Johns, W.R. The aerodynamic instability and disintegration of viscous liquid sheets. *Chem. Eng. Sci.* **1963**, *18*, 203–214. [CrossRef]

77. Weber, C. The break-up of liquid jets (Zum Zerfall eines Flüssigkeitsstrahles). *ZAMM J. Appl. Math. Mech. Z. Angew. Math. Mech.* **1931**, *11*, 136–154. [CrossRef]

78. Dombrowski, N.; Hooper, P.C. The effect of ambient density on drop formation in sprays. *Chem. Eng. Sci.* **1962**, *17*, 291–305. [CrossRef]

79. Wu, D.; Guillemin, D.; Marshall, A.W. A modeling basis for predicting the initial sprinkler spray. *Fire Saf. J.* **2007**, *42*, 283–294. [CrossRef]

80. Crowe, C.T.; Schwarzkopf, J.D.; Sommerfeld, M.; Tsuji, Y. *Multiphase Flows with Droplets and Particles*, 2nd ed.; CRC Press: Boca Raton, FL, USA, 2011; ISBN 978-1-4398-4050-4.

81. Xia, Y.; Khezzar, L.; Alshehhi, M.; Hardalupas, Y. Droplet size and velocity characteristics of water-air impinging jet atomizer. *Int. J. Multiph. Flow* **2017**, *94*, 31–43. [CrossRef]

82. Liu, C.; Liu, F.; Yang, J.; Mu, Y.; Hu, C.; Xu, G. Experimental investigations of spray generated by a pressure swirl atomizer. *J. Energy Inst.* **2019**, *92*, 210–221. [CrossRef]

83. Huang, J.C.P. The break-up of axisymmetric liquid sheets. *J. Fluid Mech.* **1970**, *43*, 305–319. [CrossRef]

84. Jordan, S.J.; Ryder, N.L.; Repcik, J.; Marshall, A.W. Spatially-resolved spray measurements and their implications. *Fire Saf. J.* **2017**, *91*, 723–729. [CrossRef]

85. Myers, T.; Trouvé, A.; Marshall, A. Predicting sprinkler spray dispersion in FireFOAM. *Fire Saf. J.* **2018**, *100*, 93–102. [CrossRef]

86. Dai, Z.; Faeth, G.M. Temporal properties of secondary drop breakup in the multimode breakup regime. *Int. J. Multiph. Flow* **2001**, *27*, 217–236. [CrossRef]

87. Krzeczkowski, S.A. Measurement of liquid droplet disintegration mechanisms. *Int. J. Multiph. Flow* **1980**, *6*, 227–239. [CrossRef]

88. Cao, X.-K.; Sun, Z.-G.; Li, W.-F.; Liu, H.-F.; Yu, Z.-H. A new breakup regime of liquid drops identified in a continuous and uniform air jet flow. *Phys. Fluids* **2007**, *19*, 057103. [CrossRef]

89. Sirignano, W.A.; Mehring, C. Review of theory of distortion and disintegration of liquid streams. *Prog. Energy Combust. Sci.* **2000**, *26*, 609–655. [CrossRef]

90. Sun, Y.; Alkhedhair, A.M.; Guan, Z.; Hooman, K. Numerical and experimental study on the spray characteristics of full-cone pressure swirl atomizers. *Energy* **2018**, *160*, 678–692. [CrossRef]

91. Wittner, M.O.; Karbstein, H.P.; Gaukel, V. Air-Core-Liquid-Ring (ACLR) Atomization: Influences of Gas Pressure and Atomizer Scale Up on Atomization Efficiency. *Processes* **2019**, *7*, 139. [CrossRef]

92. Li, W.; Qian, L.; Song, S.; Zhong, X. Numerical Study on the Influence of Shaping Air Holes on Atomization Performance in Pneumatic Atomizers. *Coatings* **2019**, *9*, 410. [CrossRef]

93. Darwish Ahmad, A.; Abubaker, A.M.; Salaimeh, A.A.; Akafuah, N.K. Schlieren Visualization of Shaping Air during Operation of an Electrostatic Rotary Bell Sprayer: Impact of Shaping Air on Droplet Atomization and Transport. *Coatings* **2018**, *8*, 279. [CrossRef]

94. Ren, N.; Baum, H.R.; Marshall, A.W. A comprehensive methodology for characterizing sprinkler sprays. *Proc. Combust. Inst.* **2011**, *33*, 2547–2554. [CrossRef]

95. Elkotb, M.M. Fuel atomization for spray modelling. *Prog. Energy Combust. Sci.* **1982**, *8*, 61–91. [CrossRef]

96. Nonnenmacher, S.; Piesche, M. Design of hollow cone pressure swirl nozzles to atomize Newtonian fluids. *Chem. Eng. Sci.* **2000**, *55*, 4339–4348. [CrossRef]

Review

Carbon Mineralization by Reaction with Steel-Making Waste: A Review

Mohamed H. Ibrahim [1], Muftah H. El-Naas [1,*], Abdelbaki Benamor [1], Saad S. Al-Sobhi [2] and Zhien Zhang [3,*]

[1] Gas Processing Centre, College of Engineering, Qatar University, 2713 Doha, Qatar; m.ibrahim@qu.edu.qa (M.H.I.); benamor.abdelbaki@qu.edu.qa (A.B.)

[2] Department of Chemical Engineering, College of Engineering, Qatar University, 2713 Doha, Qatar; saad.al-sobhi@qu.edu.qa

[3] William G. Lowrie Department of Chemical and Biomolecular Engineering, The Ohio State University, Columbus, OH 43210, USA

* Correspondence: Muftah@qu.edu.qa (M.H.E.-N.); zhang.4528@osu.edu (Z.Z.); Tel.: +974-4403-7695 (M.H.E.-N.); Tel.: +1-419-210-6449 (Z.Z.)

Received: 9 February 2019; Accepted: 20 February 2019; Published: 24 February 2019

Abstract: Carbon capture and sequestration (CCS) is taking the lead as a means for mitigating climate change. It is considered a crucial bridging technology, enabling carbon dioxide (CO_2) emissions from fossil fuels to be reduced while the energy transition to renewable sources is taking place. CCS includes a portfolio of technologies that can possibly capture vast amounts of CO_2 per year. Mineral carbonation is evolving as a possible candidate to sequester CO_2 from medium-sized emissions point sources. It is the only recognized form of permanent CO_2 storage with no concerns regarding CO_2 leakage. It is based on the principles of natural rock weathering, where the CO_2 dissolved in rainwater reacts with alkaline rocks to form carbonate minerals. The active alkaline elements (Ca/Mg) are the fundamental reactants for mineral carbonation reaction. Although the reaction is thermodynamically favored, it takes place over a large time scale. The challenge of mineral carbonation is to offset this limitation by accelerating the carbonation reaction with minimal energy and feedstock consumption. Calcium and magnesium silicates are generally selected for carbonation due to their abundance in nature. Industrial waste residues emerge as an alternative source of carbonation minerals that have higher reactivity than natural minerals; they are also inexpensive and readily available in proximity to CO_2 emitters. In addition, the environmental stability of the industrial waste is often enhanced as they undergo carbonation. Recently, direct mineral carbonation has been investigated significantly due to its applicability to CO_2 capture and storage. This review outlines the main research work carried out over the last few years on direct mineral carbonation process utilizing steel-making waste, with emphasis on recent research achievements and potentials for future research.

Keywords: carbon capture; CO_2 sequestration; steel-making waste; steel slag

---

1. Introduction

Fossil fuels are used as the main source of energy globally, and now they supply over 80% of the world energy demand [1]. Fossil fuels are expected to remain the most used energy source for years to come. This is due to the ever-increasing demand for energy created by the thriving economies around the globe. International Energy Agency reported a total energy demand of 574 exajoules globally in 2014 [2]. Although there are multiple sources for atmospheric CO_2, human activities, such as transportation and electricity generation, which directly burn several kinds of fossil fuels (including coal, oil, and natural gas), release more CO_2 into the atmosphere. This leads to increases in the earth temperature and in turn causes global warming. Hence, mitigating CO_2 emissions is a key to decrease global warming and sustain a better future for humanity [3]. Carbon capture and storage serves as the main technology for mitigating carbon emissions. Numerous conventional CO_2 capture technologies based on a post-combustion approach are being used in the industry. Separation based methods, such as absorption, adsorption, and membrane separation, are the most utilized separation technologies available [4–7]. Mineral carbonation is one of few technologies that work as both capture and storage technologies [8]. It is based on the principles of natural rock weathering, where the CO_2 dissolved in rainwater reacts with alkaline rocks to form carbonate minerals. The active alkaline elements (Ca/Mg) are the fundamental reactants for mineral carbonation reaction. Although the reaction is thermodynamically favored, it takes place over a large time scale. The challenge of mineral carbonation is to offset this limitation by accelerating the carbonation reaction with minimal energy and feedstock consumption. Calcium and magnesium silicates are generally selected for carbonation due to their abundance in nature. Industrial waste residues emerge as an alternative source of carbonation minerals that have higher reactivity than natural minerals; they are also inexpensive and readily available in proximity to CO_2 emitters. In addition, the environmental stability of the industrial waste is often enhanced as they undergo carbonation. Recently, mineral carbonation has been investigated significantly, due to its applicability to CO_2 capture and storage. Despite the growing interest in mineral carbonation research, there have not been any focused reviews that assess the status of CO_2 sequestration using steel-making waste. In this review, mineral carbonation using steel-making waste is reviewed in the light of different process parameters and their effect on CO_2 uptake. Potentials for future research in the area are highlighted.

1.1. CO_2 Storage

Several storage techniques are used to store CO_2, and the most feasible option to do so is geological sequestration [1]. Literature work investigating geological CO_2 storage has seen a substantial increase in the last decade [2]. Practically, over 1 million ton CO_2 is being sequestered in 14 individual different locations around the globe [3]. Estimates of CO_2 storage capacity varies depending on the region at which the study has been conducted. Nonetheless, the capacity is in the range of 100–20,000 giga ton CO_2 worldwide [4]. One of the mature CO_2 storage techniques is to inject it into depleted gas or oil reservoirs. Carbon dioxide is used to increase reservoirs pressure to produce enough driving force to push the gas/oil out of it. In other words, it enhances oil recovery in active wells by extracting the residual oil left. Additionally, CO_2 can be used to recover natural gas (methane, CH_4) trapped in coal beds. The main premise behind the idea is that CH_4 can be quickly displaced from coal by carbon dioxide injection, allowing CO_2 to be stored in the porous structure of the coal bed [5]. Injecting CO_2 into saline aquifers is also a viable option that commercially exists with an acceptable capacity [6,7]. Carbon dioxide is usually injected in its supercritical conditions [8]. At these conditions, CO_2 is buoyant relative to porous rocks and saline aquifers. Thus, there is always a possibility that buoyant CO_2 could leak to the surface and cause catastrophic environmental impacts. Most critically, monitoring programs for post-injection are limited and do not provide long-term detectability of the gas that can potentially escape from the storage medium [9]. Hence, these approaches cannot be taken for granted and considered as permanent and safe CO_2 storage solutions. Mineral carbonation is one approach that can provide long-term storage solution in addition to being CO_2 leak-free. This is due to the fact that carbonates are in a lower energy state than CO_2 [10]. More importantly, it possesses extremely large sequestration capacity compared to other geological storage options, as indicated in Table 1.

Table 1. Storage capacities for several geological storage options [11].

Reservoir Type	Estimated Range of Storage Capacity (GtCO$_2$)
Mineral carbonation	Very large (>10,000) [a]
Saline aquifers	1000–10,000
Oil and gas fields	675–900 [b]
Coal beds	3–200

[a] No specific number can be given, however, massive potential to sequester CO$_2$ exists. [b] Including fields that are not economically viable to inject carbon dioxide into.

1.2. Mineral Carbon Sequestration

Mineral carbon sequestration is based on the principles of the natural carbonation process of natural rocks, where the CO$_2$ dissolved in rainwater forms a weak carbonic acid. Consequently, alkali and alkaline earth metals (i.e., Ca and Mg) neutralize the acid to from insoluble carbonate minerals [12,13]. Sequestration happens in several alkaline minerals, such as calcite (CaCO$_3$), dolomite (Ca/Mg(CO$_3$)$_2$), magnesite (MgCO$_3$), siderite (FeCO$_3$), and serpentine (Mg$_3$Si$_2$O$_5$(OH)$_4$) [14]. For mineral carbonation, having a sufficient amount of a certain natural mineral is an essential factor. Hence, magnesium-based silicates are utilized since they are available in considerable amounts globally [11]. However, increasing CO$_2$ levels led to more CO$_2$ absorption by the oceans, hence increasing its acidity by 30% since the industrial era started [15]. Hence, this limits the natural carbonation process. The formed carbonates are in solid form resulting from exothermic reaction (Equation (1)) and a certain amount of heat is released, depending on the type of metal oxide reacting.

$$Mg_3Si_2O_5(OH)_4 + 3CO_2 \rightarrow 3MgCO_3 + 2SiO_2 + 2H_2O + 64\ kJ/mol \tag{1}$$

Carbonates require a high amount of energy to decompose back into CO$_2$. Hence, carbonates can be considered as thermodynamically stable CO$_2$ sink [16]. CO$_2$ will be fixed permanently without further monitoring to check its stability [17]. Table 2 shows the composition of different minerals rocks and the mass of CO$_2$ that can be sequestered by a unit mineral mass (mass CO$_2$/mass mineral). This ratio is based on the theoretical basis and considered as the maximum potential carbonation capacity for the specific mineral. Different minerals have different ratios according to their alkali metal content. Whether or not the maximum capacity can be reached, it is subject to the carbonation process and different operating parameters.

Table 2. Carbonation potential for different naturally occurring minerals [18].

Mineral/Formula	U (mass CO$_2$/mass mineral)
Serpentine (Mg$_3$Si$_2$O$_5$(OH)$_4$)	0.40
Serpentine ((Mg,Fe)$_3$Si$_2$O$_5$(OH)$_4$)	0.48
Wollastonite (CaSiO$_3$)	0.36
Olivine (Fe$_2$SiO$_4$)	0.36
Olivine (Mg$_2$SiO$_4$)	0.56

Mineral sequestration technique was first proposed by Seiftriz [19]. The proposed idea suggested introducing high purity CO$_2$ to accelerate the carbonation process. This ensured that carbonation time can be shortened from geological time scale to hours or minutes. Since then, the literature work has expanded greatly. However, it is clear that the research progress is facing challenges in enhancing the carbonation process to be viable to deploy on a large scale, as shall be demonstrated in the following sections. Nonetheless, the technique possesses several advantages over other sequestration techniques such as ocean and geologic sequestration, due to concerns over long term carbon leakage, as described previously [20]. Mineral carbonation produces more stable products that have the potential to be profitable and usually produced in fewer steps than other techniques. Additionally, the heat of the reaction can be further utilized as a source of energy.

Mineral sequestration techniques are often divided into in situ or ex situ manner. In situ sequestration requires injecting CO_2 into underground reservoirs to start a reaction with the existing underground minerals to form carbonates. Ex situ sequestration is related to carbonation process above the ground, where the raw natural mineral needs to be mined and treated before it undergoes the carbonation process. The scope of this work focuses on ex situ mineral carbonation and its related mechanisms and applications.

Although naturally occurring minerals have the potential to sequester huge amounts of CO_2 due to their abundance, it is not practically feasible due to the cost of extracting and pretreatment of the minerals and the impacts associated with it [21]. In addition to numerous process challenges in terms of carbonation efficiency and energy intensity (temperature and pressure). Alkaline industrial waste rich with Mg^{2+} and Ca^{2+} is an attractive alternative for CO_2 sequestration. It can be used to imitate mineral carbonation without the additional mining cost associated with natural rocks. Even so, alkaline waste is available in less amounts than natural minerals. It is available at lower cost, higher reactivity, and uptake capacity, and less pretreatment is required. Table 3 summarizes the most studied industrial alkaline waste and their alkali earth metal composition (Mg and Ca), in addition to their production rate per year and CO_2 emissions associated with their production. Examples of the industrial wastes include fly ash, such as coal and shale oil ashes, cement industry waste dust, and steel slag.

Table 3. Alkaline solids studied in the literature, their composition and global production per year and the CO_2 emission.

Alkaline Solid Waste	Production Per Year (t)	CO_2 Emissions Per Year (t) [a]	Examples	Composition (wt.%)	Ref.
Steel slags	315–420 [a]	171 [a]	Basic Oxygen Furnace (BOF) Electric Arc Furnace (EAF) Blast Furnace Slag (BFS)	Ca: 45–55, Mg: 2–5 Ca: 40–46, Mg: 1–6.5 Ca: 35–43, Mg: 4–7	[17,22–24]
Waste cement	1100 [a]	62 [a]	Cement kiln dust	Ca: 35–50, Mg: 0–2	[25]
Fly ash	600 [a]	12,000 [a]	Coal fly ash Oil shale ash	Ca: 35–53, Mg: 0–3 Ca: 10–25	[26–34]
Air pollution control (APC) residues	1.2 [b]	N/A	Waste incineration plant residues	Ca: 35–38, Mg: 0–1	[35,36]
Red mud	1.25 [a]	3.6 [a]	Red gypsum	Ca: 1–6, Mg: 1–5	[37]

[a] Retrieved from [10]; [b] Retrieved from [38].

Globally, cement industries account for 5% of the total CO_2 emissions [39,40]. Furthermore, the steel industry accounts for 7% of CO_2 emissions globally [41]. There are four main types of steelmaking slags, including blast furnace (BF), basic oxygen furnace (BOF), electric arc furnace (EAF), and ladle furnace (LF) slags. The slags consist of several oxides, primarily calcium, iron, and magnesium oxides that are present in different phases. On average, manufacturing 1 ton of steel produces approximately 420 kg of BOF and 180 kg of EAF [42]. There are two main approaches for ex situ alkaline waste carbonation: direct and indirect; each one has several sub-classifications based on the carbonation technique.

Direct carbonation technique implies that the carbonation process happens in one single step. On the other hand, indirect carbonation has more than one step (two or more) that usually involves pre-treatment of used minerals. Typically, mineral ores undergo a pre-treatment process where the reactive chemical components (i.e., alkali earth metals) in the rocks are separated for the mineral core. Pre-treatment usually involves mining, grinding, and activation of the rock minerals (Table 4). The end product of the pre-treatment process is almost pure carbonate form of the mineral. Then, the mineral carbonate is reacted with carbon dioxide in a separate step [12]. Table 5 shows the CO_2 uptake capacity using natural rocks. It lists numerous studies that investigated the carbonation efficiency of minerals using natural minerals (i.e., olivine and serpentine) by dry and aqueous carbonation routes. The table includes details about the process parameters and carbonation conversion.

Table 4. Description of pretreatment process.

Pre-Treatment Method	Description	Advantages/Disadvantages	Ref.
Grinding (reduction in size)	Rock minerals undergo grinding process to reduce particle size of the minerals to less than 63 µm (smaller particles size, more surface area available)	Advantages: • Surface area increase • Carbonation efficiency increase Disadvantages: • Energy intensive (releases additional CO_2 emissions)	[43,44]
Heat (thermal) activation	Naturally occurring minerals contains water molecules bounded to its chemical structure i.e., up to 13% of serpentine is water. By heating the mineral up to 600 °C water is removed and more mineral is available for carbonation	Advantages: • Specific surface area of mineral is increased by removal of water molecules Disadvantages: • Economically not feasible	[45,46]
Surface activation	Increasing the mineral surface area by treating it with steam or extraction acids	Advantages: • Specific surface area of mineral is increased Disadvantages: • Increasing CO_2 capture cost • Some of alkali earth metal elements can be reduced due to leaching	[47]
Magnetic separation	The presence of iron element in mineral rocks can decrease the carbonation efficiency due to the formation iron oxides layers on the surface of the mineral. Separating iron compounds magnetically before carbonation can solve this issue	Advantages: • Increases carbonation capacity • Separating iron bearing compounds as marketable products Disadvantages: • Increasing CO_2 capture cost	[48]
Sonication (ultrasound)	Ultrasound waves are used with extraction acid in conjunction to enhance the rate of mineral dissolution in the acid. The waves forms bubbles in the liquid that can enhance the mass transfer and the mineral dissolution rate.	Advantages: • Increases carbonation capacity by enhancing the dissolution rate Disadvantages: • Only tested on lab scale studies	[49]

1.3. Indirect Carbonation

The term indirect mineral carbonation refers to the carbonation process that happens in two or more stages. Extraction of the reactive elements (Mg and Ca) form the mineral solid matrix is an essential step in indirect carbonation. Typically, strong acids are used as extracting agents. The extracted mineral then undergoes the carbonation process by reacting with CO_2. One of the main advantages of using extraction is that it allows the production of almost pure mineral, as other impurities available in the natural mineral core can be removed after the reactive metal is extracted [50]. Numerous methods exist that can achieve mineral extraction, such as using acids, molten salts, caustic soda, and bioleaching. Table 6 lists all extraction methods studied in the literature with their associated extraction reactions. Every extraction technique possesses its intrinsic advantage and disadvantages. For example, using HCl produces pure alkali earth metal; however, it is significantly energy intensive if the recovery of HCl is required. Inversely, molten salts are less energy intensive in terms of regeneration. Nonetheless, molten salts are more corrosive compared to HCl [12]. Another approach of indirect carbonation is pH swing. pH swing refers to the extraction of mineral carbonate from the solid matrix at low pH condition in the first stage. In a second stage, the pH of the extraction solution is raised to improve carbonate formation [51]. Although acids are able to extract significant amounts of calcium and magnesium ions from the feedstock, as explained previously, pH plays a great role in the precipitation of calcium and magnesium carbonates. Therefore, increasing the solution pH to approximately 10 in the second stage of mineral extracting helps to increase the rate of carbonate precipitation [10]. Hence, due to the added cost of implementing a second process step and the extensive use of extraction agents, direct carbonation is a more practical mineral carbonation option. In this review, direct aqueous carbonation using steel-making waste is being reviewed.

Table 5. Mineral carbonation studies for natural minerals.

Material	Carbonation Method	Composition (wt.%)	Maximum CO_2 Uptake %	Reactor Type	Conditions/Remarks	Ref.
Serpentine	Direct aqueous carbonation	MgO: 40	30%	Batch	Temperature: 300 °C CO_2 Partial pressure: 335.4 atm	[52]
Olivine	Direct aqueous carbonation	CaO: 0.07 SiO_2: 41.4 MgO: 49.7 Al_2O_3: 0.21 Fe_2O_3: 2.7	91%	CSTR	Temperature: 155 °C Process pressure: 185 atm	[53]
Serpentine	Indirect using pH swing	-	42%	Batch	Temperature: 70 °C Process pressure: 1 atm	[51]
Serpentine	Direct aqueous carbonation	CaO: 0.15 SiO_2: 39.5 MgO: 38.7 Al_2O_3: 0.35 Fe_2O_3: 4.86	7%	Batch	Temperature: 25 °C Process pressure: 125 atm	[46]
Wollastonite $CaSiO_3$	Direct aqueous carbonation	-	69%	Batch	Temperature: 200 °C Process pressure: 20 bar Particle size: <38 μm L/S: 2	[54]
Serpentinite	Dry carbonation	-	50%	Fluidized bed	Temperature: 500 °C Process pressure: 20 bar	[55]
Wollastonite $CaSiO_3$	Direct aqueous carbonation	-	83.5%	Batch	Temperature: 150 °C Process pressure: 40 bar Particle size: <30 μm	[56]
Serpentinite	Dry carbonation	MgO: 35.3	0.0075 gCO_2/g serpentinite	Fluidized bed	Temperature: 90 °C Moist CO_2 Process pressure: 1 bar	[57]

Table 6. Description of different indirect carbonation methods.

Indirect Carbonation	Description/Reactions	Ref.
Acid extraction	Numerous acids are investigated in the literature as an extraction agent. Examples of these acids are: acetic acid, nitric acid, formic acid and hydrochloric acid. Acid extraction is achieved through multiple routes. The most straightforward extraction method includes mixing the mineral and the extracting agent in a stirred reactor or vessel at a certain temperature and pressure to extract the minerals, followed by carbonation process of the extracted mineral according to the following reactions: $Mg_3Si_2O_5(OH)_{4(s)} + 6HCl_{aq} + H_2O \xrightarrow{100^\circ C} 3MgCl_2 6H_2O_{aq)} + 2SiO_{2(s)}$ $MgCl_2 6H_2O_{aq)} \xrightarrow{250^\circ C} MgCl(OH)_{(aq)} + HCl_{aq} + 5H_2O$ $2MgCl(OH)_{(aq)} \xrightarrow{Cooling} Mg(OH)_{2(s)} + MgCl_{2(aq)}$	[46]
Molten salt	The molten salt process is aimed to reduce energy requirements resulting from HCl extraction. It shares many similarities with HCl extraction except the molten salt, $3MgCl_2 \cdot 3.5H_2O_{(aq)}$, is being used as an extracting agent. $Mg_3Si_2O_5(OH)_{4(s)} + 3MgCl_2 \cdot 3.5H_2O_{(aq)} \xrightarrow{200^\circ C} 6Mg(OH)Cl + 2SiO_{2(s)} + 9.5H_2O$	[12]
Ammonia	Using NH_4Cl as extraction agent according to the following reaction: $4NH_4Cl_{aq} + 2CaO \cdot SiO_2 \rightarrow 2CaCl_{2(aq)} + 2SiO_2 + 4NH_3 + 2H_2O$ $2CaCl_{2(aq)} + 2CO_{2(g)} + 4NH_{3(aq)} + 2H_2O \rightarrow 2CaCO_{3(s)} + 4NH_4Cl_{aq}$	[51]
Caustic soda	Using sodium hydroxide as an extracting agent: $CaSiO_{3(s)} + NaOH_{aq} \xrightarrow{200^\circ C} NaCaSiO_3(OH)_s$ $2NaOH_{aq} + CO_{2(aq)} \xrightarrow{200^\circ C} Na_2CO_{3(aq)} + H2O$ $Na_2CO_{3(aq)} + 3NaCaSiO_3(OH)_s + H_2O \xrightarrow{200^\circ C} 4NaOH_{aq} + CaCO_3 + NaCa_2Si_3O_8(OH)_S$	[12]
Bioleaching	Bioleaching is defined as the process of using bacteria to extract minerals from natural rocks, can be applied for the extraction of Ca & Mg oxides from silicates.	[58]

2. Direct Carbonation

Direct carbonation includes the reaction of CO_2 with a suitable feedstock or Calcium/Magnesium rich solid residue in a single step. It is relatively easier to implement compared to indirect carbonation. Hence, it has the potential to be used in industrial scale. The following sections explain the working principles of direct carbonation, discussing the operational parameters that have the most impact on the carbonation capacity.

2.1. Gas-Solid Carbonation

The reaction of gaseous CO_2 with solid minerals is the most basic and straightforward approach of direct carbonation, first studied by Lackner et al [52]. In the case of olivine carbonation:

$$Mg_2SiO_{4(s)} + 2\,CO_{2(g)} \rightarrow MgCO_{3(s)} + SiO_{2(s)} \tag{2}$$

The reaction suffers from very slow reaction rates. Hence, it is usually carried over an elevated temperature and pressure. However, due to thermodynamics limitations, the temperature is restricted between 170–400 °C for most natural minerals, as equilibrium shifts to the reactant side with increasing temperature [52]. Hence, the maximum carbonation temperature is a function of CO_2 partial pressure and the specific mineral used (Table 7). The process high temperature requirement can be further utilized to generate steam that will be used to produce electricity [12]. Nevertheless, the process slow kinetics is still the main obstacle hindering further progress even at high temperature and pressure.

Table 7. Maximum carbonation temperature for several minerals at CO_2 partial pressure of 1 bar [59].

Mineral	Maximum Carbonation Temperature (°C)
Olivine (Mg_2SiO_4)	241
Wollastonite ($CaSiO_3$)	280
Calcium oxide (CaO)/Calcium hydroxide ($Ca(OH)_2$)	887
Magnesium oxide (MgO)/Magnesium hydroxide ($Mg(OH)_2$)	406

Thus, the experimentally obtained carbonation rate of direct dry rock minerals carbonation is insignificant, even at elevated temperature and pressure [12]. Kwon et al. [60] reported that introducing moisture to the flue gas in dry carbonation process can increase the carbonation rate significantly. However, due to low CO_2 sequestering efficiency, 8 tons of olivine would be required to capture 1 ton of CO_2. This renders the process practically not viable and reduces its wide scale applicability. Hence, focus shifted on dry carbonation of pure magnesium and calcium oxides [52]. Nevertheless, the limited availability of the used minerals hindered the research progress. However, alkali earth metals can be extracted from the mineral rocks and industrial wastes. This adds another process step (extraction) to the overall process scheme. Hence, the process becomes an indirect carbonation process, which is discussed previously.

2.2. Direct Aqueous Carbonation

As explained in the introduction section, natural carbonation occurs when CO_2 is dissolved in rain water according the following equation:

$$CO_{2\,(g)} + H_2O_{(l)} \leftrightarrow HCO_{3(aq)}^- + H_{(aq)}^+ \leftrightarrow CO_{3(aq)}^{2-} + 2H_{(aq)}^+ \tag{3}$$

The aqueous solution becomes more acidic due to the presence of protons ($H_{(aq)}^+$) resulting from CO_2 solubility in water. Hence, by imitating natural carbonation, carbonation of natural minerals could take place in aqueous media in a single stage process. When the rock mineral is placed in aqueous solution, calcium or magnesium element in the solid matrix leaches from mineral ore according to Equation (4):

$$(Ca/Mg)SiO_4 + 2H^+ \rightarrow (Ca^{2+}/Mg^{2+}) + SiO_2 + H_2O \tag{4}$$

Eventually, mineral carbonate is formed when the mineral reacts with dissolved CO_2 (bicarbonate proton)

$$\left(Ca^{2+}/Mg^{2+}\right) + HCO_{3(aq)}^- \rightarrow (Ca/Mg)CO_3 + H_2O \tag{5}$$

When studying direct aqueous carbonation, a good practice is to study the process parameters, such as temperature, pressure, and solution medium, that can maximize CO_2 uptake capacity. For aqueous carbonation specifically, increasing process pressure enhances CO_2 solubility in the aqueous medium. Therefore, according to Equation (5), the reaction will shift towards the products side, which is highly desirable. On the other hand, increasing the reaction rates can be enhanced by elevating temperature. However, this applies to a certain extent due to the decline in CO_2 solubility in the solution with increasing temperature. In other words, CO_2 solubility in a certain solution dictates the upper limit at which the process temperature can be elevated. Carbonation conversion is a way to measure the carbonation efficiency and is defined according to the following equation:

$$\eta_{(Carbonation)}\% = \frac{\text{Quantatity of Mg or Ca converted to carbonate}}{\text{Quantatity of Mg or Ca avialable in mineral}} \times 100 \tag{6}$$

Hence, the efficiency is reported on the basis of magnesium and calcium content of the mineral, not the total quantity of the used mineral.

Studies investigating direct aqueous carbonation reaction mechanism revealed that aqueous carbonation proceeds in two distinct steps, as opposed to 1 step in dry carbonation [61]:

(1)   Dissolution of alkali earth element into the solution (leaching step).
(2)   Formation of mineral carbonate (carbonation step).

Typically, leaching of alkali metal into aqueous solution is the rate limiting step. Nonetheless, altering the process parameters, such as temperature and pressure, can make the carbonation step the rate limiting step [62]. This is explained by the formation of the carbonation products on the surface of the minerals that will increase the mass transfer resistance between the dissolved CO_2 and mineral core. Hence, controlling the dissolution rate and finding the best process parameters is a must to ensure sufficient carbonation efficiency. Different minerals have different dissolution rates [63]. The rate depends on the morphology of the mineral (surface area and structure) [64]. Thus, pre-treatment techniques stated in Table 4 can be used to enhance the dissolution rate, hence, the carbonation efficiency.

3. Steelmaking Waste Mineral Carbonation

Solid industrial wastes are generally alkaline and rich in Ca/Mg and can therefore be applied as an additional feedstock for mineral CO_2 sequestration. The main advantages of industrial waste are that they are available at low to no costs in proximity to industrial emitters, almost no pre-treatment is needed, and they are more reactive in less energy intensive conditions. In addition, the end product of the sequestration can be used in several applications, i.e., as a construction material and in fertilizers. The fundamental working principles for mineral CO_2 sequestration apply for industrial waste in the same way. In fact, the major elements of e.g., steel slag (Mg, Ca, Si, and Fe) are present in a comparable concentration as in natural rocks. However, trace metals and soluble salt concentrations are available in more quantities compared to the average composition of natural rocks. Thus, steel industry waste can undergo the same direct and indirect carbonation techniques previously explained. Presently, the research is going towards optimizing the uptake capacity of CO_2 by modifying the operating parameters including pressure, temperature, liquid-to-solid ratio, CO_2 gas flowrate, solid particle size, and pretreatment. Table 8 presents a summary of steel making waste which have been tested as mineral carbonation in terms of feedstocks; feed composition, the experimental CO_2 capture capacity, and the different process conditions were investigated. The mineral carbonation uptake is a function of process temperature, CO_2 partial pressure, and steel waste surface area, which affect the carbon dioxide dissolution rate, the diffusion rate of ions through the reaction with steel slag. The pH value is an additional essential parameter in mineral carbonation process. Optimum pH for aqueous carbonation is achieved at pH of 10 [65]. pH of the process influences the carbonation reaction, as the reaction is more favorable in alkaline mediums. In addition, the pH decreases continuously as carbonation, due to CO_2 being dissociated into the solution. Eventually, the pH value remains unchanged at around 7 after the carbonation process ends. This signifies that the mineral carbonation process will not proceed in acidic mediums. Figure 1 summarizes the different aspects that affect the ex-situ mineral carbonation process, such as different reactor types and process parameters.

3.1. Temperature and Particle Size

Huijgen et al. [21] were among the first to utilized steel slag as feed stock for mineral carbonation. The authors studied parameters that could affect the carbonation rate, which include reaction temperature and steel slag particle size. An autoclave reactor was used to carry out the reactor, and a 450 mL of the slurry was used with a liquid to solid ratio of 20 kg/kg. A maximum conversion of 70% of the calcium in the feed stock was carbonated at a pressure of 19 bar and temperature of 100 °C was achieved. The authors reported that at higher temperatures leaching of calcium from steel slag components will proceed faster, hence increasing the reaction rate, but the solubility of CO_2 in the solution decreases. This was also observed by Han et al. [65]. To achieve this carbonation percentage, the particle size was reduced from <2 mm to <38 μm. Reducing the particle size will produce more surface area for the carbonation reaction to occur, hence increasing the conversion. Particle size and specific surface area are among the most important factors affecting the dissolution kinetics of any kind of material. Mineral particle size determines its reactive surface area in addition to its leaching mechanisms. Typically, grinding is used to achieve a specific particle size. However, it is an energy intensive process. Hence, determining the optimal particles size will help in reducing the process cost in addition to increasing its efficiency. Baciocchi et al. [43] reported that the parameter that most affected the CO_2 uptake of the slag was particle size, especially the specific surface of the particles. An increase in temperature also had a positive effect, achieving a maximum uptake of 130 g CO_2/kg slag. The authors reported that an average particle size of less than 150 micrometers is considered as optimum.

3.2. Liquid to Solid Ratio

Liquid to solid ratio is defined by the amount of steel slag that is being utilized in a certain amount of aqueous medium (mass/mass). Revathy et al. reported that the carbonation efficiency increased when the S/L ratio decreased. The results indicate that when L/S is increased from 5 to 10 g/g, the carbonation degree of steel slag also increases. A further L/S increase causes a decrease in the carbonation degree of steel slag. This is caused due to the presence of extra liquid that leads to dilution of calcium ion concentration in the aqueous medium [66]. In a similar manner, the sequestration capacity of slag water slurry increased with the L/S ratio from 2 to 10, after which it decreased. This is due to the fact that a high amount of water inside the reactor causes blocking in the diffusion of gas molecules in the slurry [67].

Table 8. Summary of the tested steel slag for mineral carbonation in terms of CO_2 capture capacity, feedstock composition and process parameters.

Material	Carbonation Method	Composition (wt.%)	Maximum CO_2 Uptake	Reactor Type	Conditions/Remarks	Year	Ref
Steel Slag	Direct aqueous carbonation	Fe_2O_3: 35.5 CaO: 31.7 SiO_2: 9.1 MgO: 6.0	74% of Ca content	Batch	CO_2 pressure: 19 bar Temperature:100 °C Particle size: <38 μm Reaction time: 30 min	2005	[21]
BSF	Indirect aqueous carbonation: (extraction) using acetic acid	CaO: 40.6 SiO_2: 34.1 MgO: 10.7 Al_2O_3: 9.4	0.23 g CO_2/g CaO	Stirred batch	3.6 liters of Acetic acid was used to produce carbonates by leaching	2008	[50]
Steel slag	Indirect carbonation	CaO: 32.1 SiO_2: 19.4 MgO: 9.4 Al_2O_3: 8.6 Fe_2O_3: 26.4	30%	Batch	Slag was leached in deionized water Ambient temperature and pressure. Increasing the leachate temperature from 60 °C enhanced the Ca-leaching	2008	[70]
LFS	Indirect aqueous carbonation	CaO: 58.1 SiO_2: 26.4 MgO: 6.2 Al_2O_3: 4.6 FeO: 4.30	0.247 g CO_2/g CaO	Stirred batch	L/S: 10 Temperature: 20 °C Process pressure: 1 bar CO_2: 15 vol.% CO_2 flowrate: 5 mL/min Rotational speed: 200 rpm	2008	[71]
Steel slag	Indirect aqueous using pH swing using NH_4Cl	CaO: 44.5 SiO_2: 9.28 MgO:7.6 Fe_2O_3: 19.1 Al_2O_3: 2.3	70% of Ca content	Stirred batch	CO_2: 13 vol.% Temperature: 80 °C Pressure: 1 atm Rotational speed: 300 rpm	2008	[72]
APC	Direct aqueous carbonation	CaO: 35 SiO_2: 1.01 Al_2O_3: 0.21 MgO: 0.84	0.25 g CO_2/g CaO	Batch	CO_2: 100 vol.% L/S: 0.2 Temperature: 30 °C Process pressure: 3 bar	2009	[36]
EAF Slag	Indirect aqueous carbonation (extraction) using nitric acid	CaO: 41.6 SiO_2: 18.8 MgO: 8.0 Al_2O_3: 3.4	0.359 g CO_2/g CaO & MgO	Batch	L/S: 0.2 Temperature: 22 °C	2010	[73]
Industrial wastes from acetylene production	Carbonation by atmospheric CO_2	CaO: 41.6 SiO_2: 18.8	0.476 g CO_2/g waste	N/A	L/S: 0.33	2010	[74]

Table 8. *Cont.*

Material	Carbonation Method	Composition (wt.%)	Maximum CO_2 Uptake	Reactor Type	Conditions/Remarks	Year	Ref
BFS	Indirect carbonation (extraction) using nitric acid	CaO: 51.1 SiO_2: 11.5 MgO: 4.2 Al_2O_3: 1.5 Fe_2O_3: 24.1	0.27 g CO/g CaO	Slurry	L/S: 10 Temperature: 70 °C CO_2 Partial pressure: 101.3 kPa CO_2 flowrate: 0.1 L/min Particles size: <44 μm	2011	[75]
Steel slag	Direct aqueous carbonation	CaO: 38.84 MgO: 10.36 Al_2O_3: 3.91 Fe_2O_3: 32.8	93% based on CaO content	high-gravity rotating packed bed	Rotational speed: 750 rpm Temperature: 65 °C Process pressure: 1 bar L/S: 20	2012	[76]
BOFS	Direct aqueous carbonation	CaO: 36.37 MgO: 7 Al_2O_3: 1.89 Fe_2O_3: 10.36	99% based on CaO content	Rotating packed bed	Rotational speed: 1000 rpm Temperature: 25 °C L/S: 20 mg/L Process pressure: 1 bar CO_2: 30 vol.% CO_2 flowrate: 1.8 L/min	2013	[77]
BOFS	Direct aqueous carbonation	CaO: 41.15 SiO_2: 10.59 MgO: 9.21 Al_2O_3: 2.24 Fe_2O_3: 24.41 MnO: 2.75	89.4%	Slurry reactor	Temperature: 25 °C L/S: 20 CO_2 pressure: 1 bar CO_2 flowrate: 1 L/min Slurry volume: 350 mL	2013	[78]
BSF	Indirect aqueous carbonation (extraction) using EDTA	CaO: 47.15 SiO_2: 31.08 MgO: 3.34 Al_2O_3: 13.81 Fe_2O_3: 0.378 MnO: 0.71	0.09 g CO_2/g slag	Batch	Temperature: 25 °C Process pressure: 1 bar CO_2 flowrate: 1.5 L/min	2013	[17]
Steel slag	Dry carbonation	-	0.0449 g CO_2/g slag	Batch	Temperature: 600 °C CO_2%: 10 vol.% CO_2 flowrate: 1.5 L/min	2014	[79]
BOFS	Direct aqueous carbonation	CaO: 43 SiO_2: 12.9 Fe_2O_3: 28.7	0.16 g CO_2/g CaO	rotating packed bed	Rotational speed: 541 rpm Temperature: 25 °C L/S: 10 Process pressure: 1 bar	2014	[80]

Table 8. *Cont.*

Material	Carbonation Method	Composition (wt.%)	Maximum CO_2 Uptake	Reactor Type	Conditions/Remarks	Year	Ref
Steel slag	Indirect aqueous carbonation (extraction) using NH_4SO_4	CaO: 38.98 SiO_2: 12.13 MgO: 8.96 Al_2O_3: 2.74 Fe_2O_3: 22.53 MnO: 3.58	74%	Batch	Temperature: 65 °C L/S: 15 g/L Process pressure: 1 bar	2014	[24]
Steel slag	Direct aqueous carbonation	CaO: 41.3 SiO_2: 20.9 MgO: 6.2 Al_2O_3: 2.3 Fe_2O_3: 20.7	0.264 g CO_2/g CaO	Batch	Temperature: 60 °C L/S: 10 CO_2 flowrate: 0.6 L/min Process pressure: 10 bar CO_2%: 100 vol.%	2015	[66]
BOFS	Direct aqueous carbonation	CaO: 23 SiO_2: 6 MgO: 3.8 Al_2O_3: 1.1 Fe_2O_3: 25	0.403 g CO_2/g CaO	Batch	Temperature: 100 °C L/S: 5 L/kg CO_2: 100 vol.% Process pressure: 10 bar Particle size: <150 μm	2015	[81]
EAFS	Dry carbonation	CaO: 42.8 SiO_2: 4.49 MgO: 4.96 Al_2O_3: 0.28 Fe_2O_3: 42.8	0.657 g/g CaO	Slurry		2015	[82]
BOFS	Direct aqueous carbonation	CaO: 51.1 SiO_2: 11.2 MgO: 4.2 Al_2O_3: 1.2 Fe_2O_3: 24	57%	Batch	Temperature: 50 °C L/S: 20 mL/g CO2 flowrate: 0.1 L/min Process pressure: 1 bar	2016	[83]
Steel slag	Dry carbonation	CaO: 28.27 SiO_2: 15.4 MgO: 7.88 Al_2O_3: 1.01 Fe_2O_3: 24.25	0.011 g CO_2/g slag	Batch	Temperature: 50 °C CO_2: 100 vol.%	2016	[67]
EAF	-	CaO: 33.19 SiO_2: 16.71 MgO: 9.43 Al_2O_3: 6.73 Fe_2O_3: 38.19	0.052 g CO_2/g slag	Batch	Temperature: 25 °C Process pressure: 10.68 bar L/S: 10	2016	[68]

Table 8. *Cont.*

Material	Carbonation Method	Composition (wt.%)	Maximum CO$_2$ Uptake	Reactor Type	Conditions/Remarks	Year	Ref
BOF	Direct aqueous carbonation	CaO: 31 SiO$_2$: 5.1 MgO: 7.5 Fe$_2$O$_3$: 27	0.536 g CO$_2$/g slag	Batch	Temperature: 83.7 °C Process pressure: 5.9 bar L/S: 5 L/kg CO$_2$: 60.6 vol.%	2016	[84]
BFS	Direct aqueous carbonation	CaO: 42.5 SiO$_2$: 31.9 MgO: 4.81 Al$_2$O$_3$: 13 Fe$_2$O$_3$: 0.34	0.0295 g CO$_2$/g slag	Batch	Temperature: 50 °C Process pressure: 5 bar L/S: 3 CO$_2$: 100 vol.%	2017	[69]

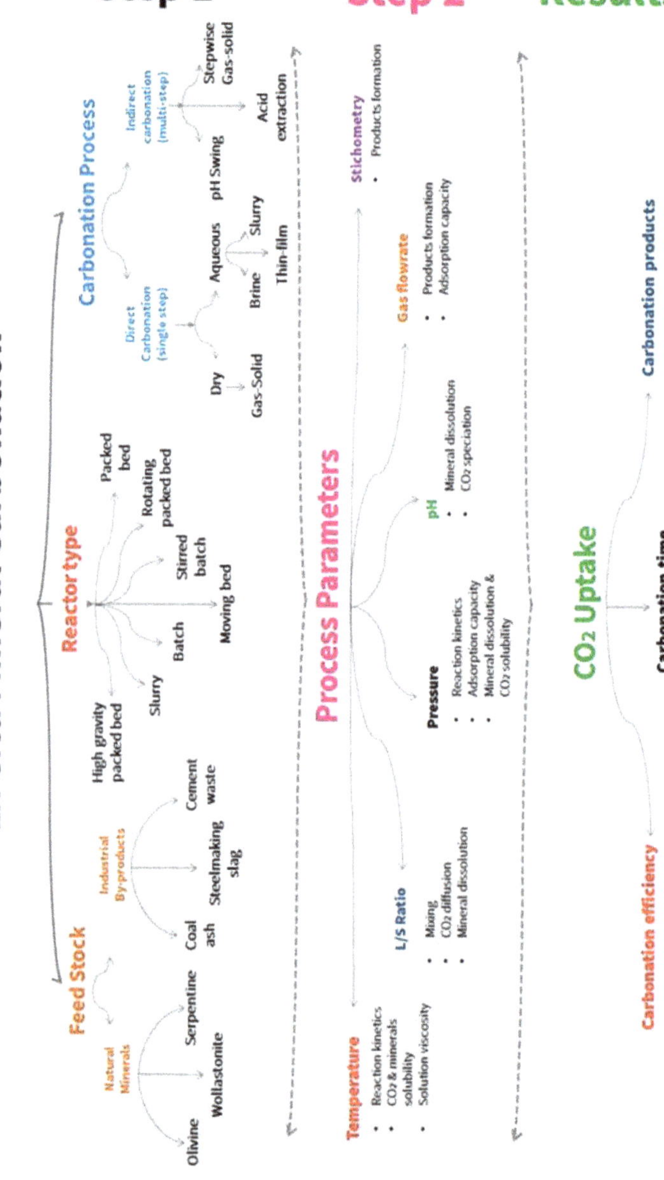

Figure 1. Ex-situ mineral carbonation literature summary.

3.3. Pressure

At constant temperature, CO_2 gas solubility increases along with pressure according to Henry's law. Hence, CO_2 molecules that are involved in the carbonation process will be more as the pressure is elevated. The effect of pressure on CO_2 uptake was tested at 10, 50, 100, and 150 bar under the same condition (50 °C, L/S = 1) by Han et al. [65]. The carbonation conversion was found to be 21% and 50.2% for 10 and 150 bar, respectively. Similarly, Ghacham et al. [68] reported that a higher CO_2 partial pressure caused more CO_2 to be soluble in aqueous medium, forming carbonic acid and consequently increasing the bicarbonate ions formation. Therefore, more bicarbonates will react with calcium ions. Hence, higher CO_2 uptake. Fagerlund et al. [55] reported that that the carbonation rate and degree might increase exponentially with time, as long as a high enough CO_2 pressure could be maintained. Additionally, high pressure will cause the reaction time to be shorter, hence, having lower carbonation time. Similarly, Eloneva et al. [69] reported shorter reaction times as the partial pressure of CO_2 is increased.

4. Summary and Future Prospective

Carbon capture and sequestration can be achieved through different techniques that have the potential to capture substantial amounts of CO_2 and help reduce its emissions. Mineral carbonation is evolving as a possible candidate to sequester CO_2 from medium-sized emissions point sources. The process of natural carbonation forms the basis of mineral carbonation process. Active alkaline elements (Ca and Mg) are the fundamental reactants for mineral carbonation reaction. Industrial alkaline wastes, such as steel-making waste, are rich with these alkaline compounds, especially calcium and magnesium oxides. Hence, they are studied in the literature as a possible mineral carbonation process feedstock. Several parameters govern the carbonation process, including process temperature, pressure, and liquid to solid ratio. There is still a room for improvement by targeting higher CO_2 uptake value. This can be achieved by using a different aqueous medium to carry out the carbonation process, i.e., reject brine and the development of reactor systems that minimize mass transfer limitations. Optimizing the interactions between process parameters, such as the interplay between temperature, pressure, and the degree of mixing, will contribute to the carbonation process. In addition, studying the adsorption behavior of CO_2 on other elements, such as iron oxide, will give more insights into increasing CO_2 uptake.

Author Contributions: Conceptualization, M.H.E.-N. and M.H.I.; Formal Analysis, M.H.I. and A.B.; Investigation, A.B., M.H.E.-N. and M.H.I.; Resources, S.S.A. and Z.Z.; Writing-Original Draft Preparation, M.H.E.-N. and M.H.I.

Funding: This research received no external funding.

Conflicts of Interest: The authors declare no conflict of interest.

References

1. Leung, D.Y.C.; Caramanna, G.; Maroto-Valer, M.M. An overview of current status of carbon dioxide capture and storage technologies. *Renew. Sustain. Energy Rev.* **2014**, *39*, 426–443. [CrossRef]
2. Gale, J.; Abanades, J.C.; Bachu, S.; Jenkins, C. Special Issue commemorating the 10th year anniversary of the publication of the Intergovernmental Panel on Climate Change Special Report on CO_2 Capture and Storage. *Int. J. Greenh. Gas Control* **2015**, *40*, 1–5. [CrossRef]
3. Bui, M.; Adjiman, C.S.; Bardow, A.; Anthony, E.J.; Boston, A.; Brown, S.; Fennell, P.S.; Fuss, S.; Galindo, A.; Hackett, L.A.; et al. Carbon capture and storage (CCS): The way forward. *Energy Environ. Sci.* **2018**, *11*, 1062–1176. [CrossRef]
4. Chunbao Charles, X.U.; Cang, D.Q. A brief overview of low CO_2 emission technologies for iron and steel making. *J. Iron Steel Res. Int.* **2010**, *17*, 1–7.
5. White, C.M.; Strazisar, B.R.; Granite, E.J.; Hoffman, J.S.; Pennline, H.W. Separation and capture of CO_2 from large stationary sources and sequestration in geological formations—Coalbeds and deep saline aquifers. *J. Air Waste Manag. Assoc.* **2003**, *53*, 645–715. [CrossRef] [PubMed]

6. Torp, T.A.; Gale, J. Demonstrating storage of CO_2 in geological reservoirs: The Sleipner and SACS projects. *Energy* **2004**, *29*, 1361–1369. [CrossRef]

7. Maldal, T.; Tappel, I.M. CO_2 underground storage for Snøhvit gas field development. *Energy* **2004**, *29*, 1403–1411. [CrossRef]

8. Matter, J.M.; Kelemen, P.B. Permanent storage of carbon dioxide in geological reservoirs by mineral carbonation. *Nat. Geosci.* **2009**, *2*, 837–841. [CrossRef]

9. Michael, K.; Golab, A.; Shulakova, V.; Ennis-King, J.; Allinson, G.; Sharma, S.; Aiken, T. Geological storage of CO_2 in saline aquifers-A review of the experience from existing storage operations. *Int. J. Greenh. Gas Control* **2010**, *4*, 659–667. [CrossRef]

10. Azdarpour, A.; Asadullah, M.; Mohammadian, E.; Hamidi, H.; Junin, R.; Karaei, M.A. A review on carbon dioxide mineral carbonation through pH-swing process. *Chem. Eng. J.* **2015**, *279*, 615–630. [CrossRef]

11. Metz, B.; Davidson, O.; de Coninck, H.; Loos, M.; Meyer, L. *IPCC Special Report on Carbon Dioxide Capture and Storage*; Cambridge University Press: Cambridge, UK, 2005.

12. Bobicki, E.R.; Liu, Q.; Xu, Z.; Zeng, H. Carbon capture and storage using alkaline industrial wastes. *Prog. Energy Combust. Sci.* **2012**, *38*, 302–320. [CrossRef]

13. Kunzler, C.; Alves, N.; Pereira, E.; Nienczewski, J.; Ligabue, R.; Einloft, S.; Dullius, J. CO_2 storage with indirect carbonation using industrial waste. *Energy Procedia* **2011**, *4*, 1010–1017. [CrossRef]

14. Olajire, A.A. A review of mineral carbonation technology in sequestration of CO_2. *J. Pet. Sci. Eng.* **2013**, *109*, 364–392. [CrossRef]

15. Harrould-kolieb, E.R. A governing framework for international ocean acidification policy. *Mar. Policy* **2019**, *102*, 10–20. [CrossRef]

16. Power, I.M.; Harrison, A.L.; Dipple, G.M.; Wilson, S.A.; Kelemen, P.B.; Hitch, M.; Southam, G. Carbon Mineralization: From Natural Analogues to Engineered Systems. *Rev. Mineral. Geochem.* **2013**, *77*, 305–360. [CrossRef]

17. Mun, M.; Cho, H. Mineral carbonation for carbon sequestration with industrial waste. *Energy Procedia* **2013**, *37*, 6999–7005. [CrossRef]

18. Sanna, A.; Uibu, M.; Caramanna, G.; Kuusik, R.; Maroto-Valer, M.M. A review of mineral carbonation technologies to sequester CO_2. *Chem. Soc. Rev.* **2014**, *43*, 8049–8080. [CrossRef] [PubMed]

19. Seifritz, W. *CO_2 Disposal by Means of Silicates*; Nature Publishing Group: London, UK, 1990.

20. Gunning, P.J.; Hills, C.D.; Carey, P.J. Accelerated carbonation treatment of industrial wastes. *Waste Manag.* **2010**, *30*, 1081–1090. [CrossRef] [PubMed]

21. Huijgen, W.J.J.; Witkamp, G.J.; Comans, R.N.J. Mineral CO_2 sequestration by steel slag carbonation. *Environ. Sci. Technol.* **2005**, *39*, 9676–9682. [CrossRef] [PubMed]

22. Liu, Q.; Liu, W.; Hu, J.; Wang, L.; Gao, J.; Liang, B.; Yue, H.; Zhang, G.; Luo, D.; Li, C. Energy-efficient mineral carbonation of blast furnace slag with high value-added products. *J. Clean. Prod.* **2018**, *197*, 242–252. [CrossRef]

23. Lee, S.; Kim, J.W.; Chae, S.; Bang, J.H.; Lee, S.W. CO_2 sequestration technology through mineral carbonation: An extraction and carbonation of blast slag. *J. CO_2 Util.* **2016**, *16*, 336–345. [CrossRef]

24. Dri, M.; Sanna, A.; Maroto-Valer, M.M. Mineral carbonation from metal wastes: Effect of solid to liquid ratio on the efficiency and characterization of carbonated products. *Appl. Energy* **2014**, *113*, 515–523. [CrossRef]

25. Sanna, A.; Dri, M.; Hall, M.R.; Maroto-Valer, M. Waste materials for carbon capture and storage by mineralisation (CCSM)—A UK perspective. *Appl. Energy* **2012**, *99*, 545–554. [CrossRef]

26. Dananjayan, R.R.T.; Kandasamy, P.; Andimuthu, R. Direct mineral carbonation of coal fly ash for CO_2 sequestration. *J. Clean. Prod.* **2016**, *112*, 4173–4182. [CrossRef]

27. Nyambura, M.G.; Mugera, G.W.; Felicia, P.L.; Gathura, N.P. Carbonation of brine impacted fractionated coal fly ash: Implications for CO_2 sequestration. *J. Environ. Manag.* **2011**, *92*, 655–664. [CrossRef] [PubMed]

28. Mayoral, M.C.; Andrés, J.M.; Gimeno, M.P. Optimization of mineral carbonation process for CO_2 sequestration by lime-rich coal ashes. *Fuel* **2013**, *106*, 448–454. [CrossRef]

29. Ji, L.; Yu, H.; Wang, X.; Grigore, M.; French, D.; Gözükara, Y.M.; Yu, J.; Zeng, M. CO_2 sequestration by direct mineralisation using fly ash from Chinese Shenfu coal. *Fuel Process. Technol.* **2017**, *156*, 429–437. [CrossRef]

30. Mazzella, A.; Errico, M.; Spiga, D. CO_2 uptake capacity of coal fly ash: Influence of pressure and temperature on direct gas-solid carbonation. *J. Environ. Chem. Eng.* **2016**, *4*, 4120–4128. [CrossRef]

31. Wee, J.H. A review on carbon dioxide capture and storage technology using coal fly ash. *Appl. Energy* **2013**, *106*, 143–151. [CrossRef]

32. Reddy, K.J.; John, S.; Weber, H.; Argyle, M.D.; Bhattacharyya, P.; Taylor, D.T.; Christensen, M.; Foulke, T.; Fahlsing, P. Simultaneous capture and mineralization of coal combustion flue gas carbon dioxide (CO_2). *Energy Procedia* **2011**, *4*, 1574–1583. [CrossRef]

33. Jo, H.Y.; Kim, J.H.; Lee, Y.J.; Lee, M.; Choh, S.J. Evaluation of factors affecting mineral carbonation of CO_2 using coal fly ash in aqueous solutions under ambient conditions. *Chem. Eng. J.* **2012**, *183*, 77–87. [CrossRef]

34. Uibu, M.; Velts, O.; Kuusik, R. Developments in CO_2 mineral carbonation of oil shale ash. *J. Hazard. Mater.* **2010**, *174*, 209–214. [CrossRef] [PubMed]

35. Araizi, P.K.; Hills, C.D.; Maries, A.; Gunning, P.J.; Wray, D.S. Enhancement of accelerated carbonation of alkaline waste residues by ultrasound. *Waste Manag.* **2016**, *50*, 121–129. [CrossRef] [PubMed]

36. Baciocchi, R.; Costa, G.; Polettini, A.; Pomi, R.; Prigiobbe, V. Comparison of different reaction routes for carbonation of APC residues. *Energy Procedia* **2009**, *1*, 4851–4858. [CrossRef]

37. Gadikota, G.; Fricker, K.J.; Jang, S.-H.; Park, A.-H.A. Carbonation of silicate minerals and industrial wastes and their potential use as sustainable construction materials. In *Advances in CO_2 Capture, Sequestration, and Conversion*; American Chemical Society: Washington, DC, USA, 2015; pp. 115–137.

38. Gomes, H.I.; Mayes, W.M.; Rogerson, M.; Stewart, D.I.; Burked, I.T. Alkaline residues and the environment: A review of impacts, management practices and opportunities. *J. Clean. Prod.* **2016**, *112*, 3571–3582. [CrossRef]

39. Xi, F.; Davis, S.J.; Ciais, P.; Crawford-Brown, D.; Guan, D.; Pade, C.; Shi, T.; Syddall, M.; Lv, J.; Ji, L.; et al. Substantial global carbon uptake by cement carbonation. *Nat. Geosci.* **2016**, *9*, 880–883. [CrossRef]

40. Damiani, D.; Litynski, J.T.; McIlvried, H.G.; Vikara, D.M.; Srivastava, R.D. The US Department of Energy's R&D program to reduce greenhouse gas emissions through beneficial uses of carbon dioxide. *Greenh. Gases Sci. Technol.* **2012**, *2*, 9–19.

41. Pan, S.-Y. CO_2 Capture by accelerated carbonation of alkaline wastes: A review on its principles and applications. *Aerosol Air Qual. Res.* **2012**, *12*, 770–791. [CrossRef]

42. Said, A.; Laukkanen, T.; Järvinen, M. Pilot-scale experimental work on carbon dioxide sequestration using steelmaking slag. *Appl. Energy* **2016**, *177*, 602–611. [CrossRef]

43. Baciocchi, R.; Costa, G.; Polettini, A.; Pomi, R. Influence of particle size on the carbonation of stainless steel slag for CO_2 storage. *Energy Procedia* **2009**, *1*, 4859–4866. [CrossRef]

44. Jung, W.M.; Kang, S.H.; Kim, W.S.; Choi, C.K. Particle morphology of calcium carbonate precipitated by gas-liquid reaction in a Couette-Taylor reactor. *Chem. Eng. Sci.* **2000**, *55*, 733–747. [CrossRef]

45. McKelvy, M.J.; Chizmeshya, A.V.G.; Diefenbacher, J.; Béarat, H.; Wolf, G. Exploration of the role of heat activation in enhancing serpentine carbon sequestration reactions. *Environ. Sci. Technol.* **2004**, *38*, 6897–6903. [CrossRef] [PubMed]

46. Maroto-Valer, M.M.; Fauth, D.J.; Kuchta, M.E.; Zhang, Y.; Andrésen, J.M. Activation of magnesium rich minerals as carbonation feedstock materials for CO_2 sequestration. *Fuel Process. Technol.* **2005**, *86*, 1627–1645. [CrossRef]

47. Kleiv, R.A.; Thornhill, M. Mechanical activation of olivine. *Miner. Eng.* **2006**, *19*, 340–347. [CrossRef]

48. Veetil, S.P.; Mercier, G.; Blais, J.F.; Cecchi, E.; Kentish, S. Magnetic separation of serpentinite mining residue as a precursor to mineral carbonation. *Int. J. Miner. Process.* **2015**, *140*, 19–25. [CrossRef]

49. Said, A.; Mattila, O.; Eloneva, S.; Järvinen, M. Enhancement of calcium dissolution from steel slag by ultrasound. *Chem. Eng. Process. Process Intensif.* **2015**, *89*, 1–8. [CrossRef]

50. Eloneva, S.; Teir, S.; Salminen, J.; Fogelholm, C.J.; Zevenhoven, R. Fixation of CO_2 by carbonating calcium derived from blast furnace slag. *Energy* **2008**, *33*, 1461–1467. [CrossRef]

51. Park, A.H.A.; Fan, L.S. CO_2 mineral sequestration: Physically activated dissolution of serpentine and pH swing process. *Chem. Eng. Sci.* **2004**, *59*, 5241–5247. [CrossRef]

52. Lackner, K.S.; Butt, D.P.; Wendt, C.H. Progress on binding CO_2 in mineral substrates. *Energy Convers. Manag.* **1997**, *38*, 259–264. [CrossRef]

53. O'Connor, W.; Dahlin, D.; Nilsen, D. Research status on the sequestration of carbon dioxide by direct aqueous mineral carbonation. In Proceedings of the 18th Annual International Pittsburgh Coal Conference, Newcastle, Australia, 3–7 December 2001; p. 12.

54. Huijgen, W.J.J.; Ruijg, G.J.; Comans, R.N.J.; Witkamp, G.J. Energy consumption and net CO_2 sequestration of aqueous mineral carbonation. *Ind. Eng. Chem. Res.* **2006**, *45*, 9184–9194. [CrossRef]

55. Fagerlund, J.; Nduagu, E.; Zevenhoven, R. Recent developments in the carbonation of serpentinite derived $Mg(OH)_2$ using a pressurized fluidized bed. *Energy Procedia* **2011**, *4*, 4993–5000. [CrossRef]

56. Yan, H.; Zhang, J.; Zhao, Y.; Liu, R.; Zheng, C. CO_2 sequestration by direct aqueous mineral carbonation under low-medium pressure conditions. *J. Chem. Eng. Jpn.* **2015**, *48*, 937–946. [CrossRef]

57. Bhardwaj, R.; van Ommen, J.R.; Nugteren, H.W.; Geerlings, H. Accelerating natural CO_2 mineralization in a fluidized bed. *Ind. Eng. Chem. Res.* **2016**, *55*, 2946–2951. [CrossRef]

58. Power, I.M.; Dipple, G.M.; Southam, G. Bioleaching of ultramafic tailings by *Acidithiobacillus* spp. for CO_2 sequestration. *Environ. Sci. Technol.* **2010**, *44*, 456–462. [CrossRef] [PubMed]

59. Lackner, K.S.; Wendt, C.H.; Butt, D.P.; Joyce, E.L.; Sharp, D.H. Carbon dioxide disposal in carbonate minerals. *Energy* **1995**, *20*, 1153–1170. [CrossRef]

60. Kwon, S.; Fan, M.; Dacosta, H.F.M.; Russell, A.G. Factors affecting the direct mineralization of CO_2 with olivine. *J. Environ. Sci.* **2011**, *23*, 1233–1239. [CrossRef]

61. Pan, S.-Y.; Ling, T.-C.; Park, A.-H.A.; Chiang, P.-C. An overview: Reaction mechanisms and modelling of CO_2 utilization via mineralization. *Aerosol Air Qual. Res.* **2018**, *18*, 829–848. [CrossRef]

62. Huijgen, W.J.J.; Witkamp, G.J.; Comans, R.N.J. Mechanisms of aqueous wollastonite carbonation as a possible CO_2 sequestration process. *Chem. Eng. Sci.* **2006**, *61*, 4242–4251. [CrossRef]

63. Yadav, S.; Mehra, A. Dissolution of steel slags in aqueous media. *Environ. Sci. Pollut. Res.* **2017**, *24*, 16305–16315. [CrossRef] [PubMed]

64. De Windt, L.; Chaurand, P.; Rose, J. Kinetics of steel slag leaching: Batch tests and modeling. *Waste Manag.* **2011**, *31*, 225–235. [CrossRef] [PubMed]

65. Han, D.R.; Namkung, H.; Lee, H.M.; Huh, D.G.; Kim, H.T. CO_2 sequestration by aqueous mineral carbonation of limestone in a supercritical reactor. *J. Ind. Eng. Chem.* **2015**, *21*, 792–796. [CrossRef]

66. Tu, M.; Zhao, H.; Lei, Z.; Wang, L.; Chen, D.; Yu, H.; Qi, T. Aqueous carbonation of steel slag: A kinetics study. *ISIJ Int.* **2015**, *55*, 2509–2514. [CrossRef]

67. Revathy, T.D.R.; Palanivelu, K.; Ramachandran, A. Direct mineral carbonation of steelmaking slag for CO_2 sequestration at room temperature. *Environ. Sci. Pollut. Res.* **2016**, *23*, 7349–7359. [CrossRef] [PubMed]

68. Ghacham, A.B.; Pasquier, L.C.; Cecchi, E.; Blais, J.F.; Mercier, G. CO_2 sequestration by mineral carbonation of steel slags under ambient temperature: Parameters influence, and optimization. *Environ. Sci. Pollut. Res.* **2016**, *23*, 17635–17646. [CrossRef] [PubMed]

69. Ukwattage, N.L.; Ranjith, P.G.; Li, X. Steel-making slag for mineral sequestration of carbon dioxide by accelerated carbonation. *Measurement* **2017**, *97*, 15–22. [CrossRef]

70. Lekakh, S.N.; Rawlins, C.H.; Robertson, D.G.C.; Richards, V.L.; Peaslee, K.D. Kinetics of aqueous leaching and carbonization of steelmaking slag. *Metall. Mater. Trans. B* **2008**, *39*, 125–134. [CrossRef]

71. Bonenfant, D.; Kharoune, L.; Sauve, S.; Hausler, R.; Niquette, P.; Mimeault, M.; Kharoune, M. CO_2 sequestration potential of steel slags at ambient pressure and temperature. *Ind. Eng. Chem. Res.* **2008**, *47*, 7610–7616. [CrossRef]

72. Kodama, S.; Nishimoto, T.; Yamamoto, N.; Yogo, K.; Yamada, K. Development of a new pH-swing CO_2 mineralization process with a recyclable reaction solution. *Energy* **2008**, *33*, 776–784. [CrossRef]

73. Doucet, F.J. Effective CO_2-specific sequestration capacity of steel slags and variability in their leaching behaviour in view of industrial mineral carbonation. *Miner. Eng.* **2010**, *23*, 262–269. [CrossRef]

74. Morales-Flórez, V.; Santos, A.; Lemus, A.; Esquivias, L. Artificial weathering pools of calcium-rich industrial waste for CO_2 sequestration. *Chem. Eng. J.* **2011**, *166*, 132–137. [CrossRef]

75. Chang, E.E.; Chen, C.H.; Chen, Y.H.; Pan, S.Y.; Chiang, P.C. Performance evaluation for carbonation of steel-making slags in a slurry reactor. *J. Hazard. Mater.* **2011**, *186*, 558–564. [CrossRef] [PubMed]

76. Chang, E.E.; Pan, S.Y.; Chen, Y.H.; Tan, C.S.; Chiang, P.C. Accelerated carbonation of steelmaking slags in a high-gravity rotating packed bed. *J. Hazard. Mater.* **2012**, *227–228*, 97–106. [CrossRef] [PubMed]

77. Pan, S.Y.; Chiang, P.C.; Chen, Y.H.; Tan, C.S.; Chang, E.E. Ex Situ CO_2 capture by carbonation of steelmaking slag coupled with metalworking wastewater in a rotating packed bed. *Environ. Sci. Technol.* **2013**, *47*, 3308–3315. [CrossRef] [PubMed]

78. Chang, E.E.; Chiu, A.C.; Pan, S.Y.; Chen, Y.H.; Tan, C.S.; Chiang, P.C. Carbonation of basic oxygen furnace slag with metalworking wastewater in a slurry reactor. *Int. J. Greenh. Gas Control* **2013**, *12*, 382–389. [CrossRef]

79. Tian, S.; Jiang, J.; Li, K.; Yan, F.; Chen, X. Performance of steel slag in carbonation–calcination looping for CO_2 capture from industrial flue gas. *RSC Adv.* **2014**, *4*, 6858–6862. [CrossRef]

80. Pan, S.Y.; Chiang, P.C.; Chen, Y.H.; Chang, E.E.; da Chen, C.; Shen, A.L. Process intensification of steel slag carbonation via a rotating packed Bed: Reaction kinetics and mass transfer. *Energy Procedia* **2014**, *63*, 2255–2260. [CrossRef]

81. Baciocchi, R.; Costa, G.; di Gianfilippo, M.; Polettini, A.; Pomi, R.; Stramazzo, A. Thin-film versus slurry-phase carbonation of steel slag: CO_2 uptake and effects on mineralogy. *J. Hazard. Mater.* **2015**, *283*, 302–313. [CrossRef] [PubMed]

82. El-Naas, M.H.; el Gamal, M.; Hameedi, S.; Mohamed, A.M.O. CO_2 sequestration using accelerated gas-solid carbonation of pre-treated EAF steel-making bag house dust. *J. Environ. Manag.* **2015**, *156*, 218–224. [CrossRef] [PubMed]

83. Pan, S.Y.; Liu, H.L.; Chang, E.E.; Kim, H.; Chen, Y.H.; Chiang, P.C. Multiple model approach to evaluation of accelerated carbonation for steelmaking slag in a slurry reactor. *Chemosphere* **2016**, *154*, 63–71. [CrossRef] [PubMed]

84. Polettini, A.; Pomi, R.; Stramazzo, A. CO_2 sequestration through aqueous accelerated carbonation of BOF slag: A factorial study of parameters effects. *J. Environ. Manag.* **2016**, *167*, 185–195. [CrossRef] [PubMed]

MDPI

St. Alban-Anlage 66

4052 Basel

Switzerland

Tel. +41 61 683 77 34

Fax +41 61 302 89 18

www.mdpi.com

Processes Editorial Office

E-mail: processes@mdpi.com

www.mdpi.com/journal/processes